DETERMINATION OF ORGANIC COMPOUNDS IN SOILS, SEDIMENTS AND SLUDGES

DETERMINATION OF ORGANIC COMPOUNDS IN SOILS, SEDIMENTS AND SLUDGES

T.R. Crompton

E & FN SPON
Taylor & Francis Group
London and New York

First published 2000 by E & FN Spon
11 New Fetter Lane, London EC4P 4EE

Simultaneously published in the USA and Canada
by E & FN Spon
29 West 35th Street, New York, NY 10001

E & FN Spon is an imprint of the Taylor & Francis Group

Printed and bound in Great Britain by St Edmundsbury Press, Bury St Edmunds, Suffolk

Publisher's Note

This book has been prepared from camera-ready copy supplied by the author.

British Library Cataloguing in Publication Data
A catalogue record for this book is available from the British Library

Library of Congress Cataloging in Publication Data
A catalogue record for this book has been requested

ISBN 0-419-25270-3

Contents

Preface

This book is concerned with a discussion of methods currently available in the world literature up to 1998 for the determination of organic substances in soils, river and marine sediments and industrial sludges.

Unlike the determination of organic substances in natural waters, the subject of the author's recently published book (*Determination of Organic Compounds in Natural and Treated Waters*, E&FN Spon, 1999), no books have been published on the determination of organic in these solids. Yet the occurrence of organic compounds, many of which are toxic, can have profound effects on the ecosystem.

In the case of soils the presence of deliberately added or adventitious organic compounds can cause contamination of the tissues of crops grown on the land or animals feeding on the land and, consequently, can cause adverse toxic effects on man, animals, birds and insects. Also drainage of these substances from the soil can cause pollution of adjacent streams, rivers and eventually the oceans. Some of the substances included in this category are pesticides, herbicides, growth regulators, organic fertilizers, crop sprays, sheep dips, etc.

The presence of organic compounds in river and oceanic sediments is due, in part, to manmade pollution and monitoring the levels of these substances in the sediment and sediment cores provides an indication of the time dependence of their concentration over large time spans. Contamination of sediments is found not only in rivers but also in estuarine and oceanic sediments and thus sediment analysis provides a means of tracking organic from their source through the ecosystem.

Another consideration is that fish, particularly bottom feeders and crustacea pick up contaminants when sediments enter their gills and the contamination of these creatures has definite toxicological implications both for the creatures themselves, for man who eats them and, in the case of fish meal, for animals.

Sediments have the property of absorbing organic contaminants from water within their bulk (accumulation) and, indeed, it has been shown that the concentration, for example, of some types of insecticide in river sediments is some 10 000 times greater than occurs in the surrounding water.

To date, insufficient attention has been given to the analysis of sediments and one of the objects of this book is to draw the attention of analysts and others concerned to the methods available and their sensitivity and limitations.

Substances that are found in sediments include all the soil pollutants mentioned above and also various organic compounds of an industrial origin such as phthalates, chorophenols, hydrocarbons and organometallic compounds to name but a few.

Organometallic compounds are included as it is becoming increasingly apparent that these compounds occur extensively throughout the ecosystem either as direct pollutants eg alkylead compounds from automobile exhausts or by biomethylation of inorganic metals occurring in sediments.

The purpose of this book is to draw together worldwide literature, up to early 1998, on the occurrence and determination of all types of organic compounds in solid samples. In this way reference to a very scattered literature can be avoided.

This is not a recipe book, ie methods are not presented in detail. Space considerations alone would not permit this. Instead, the chemist is presented with details of methods available for the determinations of all types of organic in soils, sediments and sludges. Methods are described in broad outline giving enough information for the chemist to decide whether he or she wishes to refer to the original paper. To this end, information is supplied on applicability of methods, advantages and disadvantages of one method against another, interferences, sensitivity and detection limits.

Chapter 1 discusses the principles of the various techniques now being employed in the analysis of soils, sludges and sediments and the types of determinations to which these methods can be applied. This chapter also contains a useful key system which enables the reader to quickly locate in the book sections in which are discussed the determination by various techniques of particular organic compounds in particular types of sample.

The contents are presented in as logical a manner as possible starting in chapter 2 with a discussion of hydrocarbons and polyaromatic hydrocarbons. Chapter 3 deals with various types of surface active agents while chapters 5-8 and 10, 11 and 13 discuss compounds containing oxygen, halogens, nitrogen, phosphorus, sulphur, miscellaneous organic compounds, mixtures of organic compounds and organometallic compounds. Insecticides, herbicides, growth regulators and fungicides are discussed in chapter 9. The determination of non-metals and metalloids is often a necessary preliminary to the examination of a solid sample and this is discussed in chapter 12. The book concludes in chapters 14 to 17 with discussions of various specialist aspects of the organic pollution of soils and sediments, dealing respectively with sampling procedures for solid samples, the effects of applying sewage sludge to land, the relationship between contaminant levels in soil and in crops and finally with bioaccumulation of organic in sediments.

Examination for organic substances combines all the exciting features of analytical chemistry. First, the analysis must be successful and in many cases, must be completed quickly. Often the nature of the substances to be analysed for is unknown, might occur at exceedingly low concentrations and might, indeed, be a complex mixture. To be successful in such an area requires analytical skills of a high order and the availability of sophisticated instrumentation.

The work has been written with the interests of the following groups of people in mind: management and scientists in all aspects of the water industry, river management, fishery industries, sewage effluent treatment and disposal, land drainage and water supply; also management and scientists in all branches of industry. It will also be of interest to agricultural chemists, agriculturalists concerned with the ways in which organic chemicals used in crop or soil treatment permeate through the ecosystem, the biologists and scientists involved in fish, plant, insect and plant life, and also to the medical profession, toxicologists and public health workers and public analysts. Other groups or workers to whom the work will be of interest include oceanographers, environmentalists and, not least, members of the public who are concerned with the protection of our environment.

Finally, it is hoped that the work will act as a spur to students of all subjects mentioned and assist them in the challenge that awaits them in ensuring that the pollution of the environment is controlled so as to ensure that by the dawn of the new millennium we are left with a worthwhile environment to protect.

Chapter 1

Introduction

1.1 Brief summary of methodologies

1.1.1 *Spectroscopic methods*

1.1.1.1 Visible spectrophotometry

This technique is only of value when the identity of the compound to be determined is known. There are also limitations on the sensitivity that can be achieved, usually mg L^{-1} or, occasionally, μg L^{-1}.

Some commercially available instruments, in addition to visible spectrophotometers, can also perform measurements in the UV and near IR regions of the spectrum.

Suppliers of visible spectrophotometers are reviewed in Table 1.1.

Spectroscopic methods are applicable to the determination of phenols, chlorophenols, amines, mixtures of organics, boron, halogens, total nitrogen and total phosphorus in soils, cationic surfactants, carbohydrates, total nitrogen, phosphorus and sulphur in non-saline sediments, boron, total organic carbon, total sulphur and arsenic in saline sediments, cationic surfactants, adenosine triphosphate and total organic carbon in sludges.

1.1.1.2 Spectrofluorimetric methods and luminescence spectroscopy

Spectrofluorimetric methods are applicable to the determination of aliphatic hydrocarbons, and humic and fulvic acids in soil, aliphatic hydrocarbons polyaromatic hydrocarbons, optical whiteners, and selenium in non-saline sediments, aliphatic aromatic and polyaromatic hydrocarbons and humic and fulvic acids in saline sediments. The only application found in luminescence spectroscopy is the determination of polychlorobiphenyl in soil. Generally speaking, concentrations down to the picogram (μg L^{-1}), level can be determined by this technique with recovery efficiencies near 100%.

Potentially, fluorimetry is valuable in every laboratory performing chemical analyses where the prime requirements are selectivity and

Table 1.1 Visible-ultraviolet-near infrared spectrophotometers

Spectral region	*Range (nm)*	*Manufacturer*	*Model*	*Single or double beam*	*Cost range*
UV/visible	–	Philips	PU 8620 (optional) PU 8620 scanner)	Single	Low
Visible	325–900	Cecil Instruments	CE 2343 Optical Flowcell	Single	Low
Visible	280–900	Cecil Instruments	CE 2393 (grating, digital)	Single	High
Visible	280–900	Cecil Instruments	CE 2303 (grating, non-digital)	Single	Low
Visible	280–900	Cecil Instruments	CE 2373 (grating, linear)	Single	High
UV/visible	190–900	Cecil Instruments	CE 2292 (digital)	Single	High
UV/visible	190–900	Cecil Instruments	CE 2202 (non–digital)	Single	Low
UV/visible	190–900	Cecil Instruments	CE 2272 (linear)	Single	High
UV/visible	200–750	Cecil Instruments	CE 594 (microcomputer controlled)	Double	High
UV/visible	190–800	Cecil Instruments	CE 6000 (with CE 6606 graphic plotter option)	Double	High
UV/visible	190–800	Cecil Instruments	5000 series (computerized and data station)	Double	High
UV/visible	–	Philips	PU 8800		High
UV/visible	–	Kontron	Unikon 860 (computerized with screen)	Double	High
UV/visible	–	Kontron	Unikon 930 (computerized with screen)	Double	High
UV/visible	190–1100	Perkin–Elmer	Lambda 2 (microcomputer electronics screen)	Double	High
UV/visible	190–750 or 190–900	Perkin–Elmer	Lambda 3 (microcomputer electronics)	Double	Low to High
UV/visible	190–900	Perkin–Elmer	Lambda 5 and Lambda 7 (computerized with screen)	Double Double	High
UV/visible	185–900 & 400–3200	Perkin–Elmer	Lambda 9 (computerized with screen)	UV/vis NIR	High
UV/visible	190–900	Perkin–Elmer	Lambda Array 3840 (computerized with screen)	Photo-diode	High

Source: Own files

sensitivity. While only 5–10% of all molecules possess a native fluorescence, many can be induced to fluoresce by chemical modification or tagged with a fluorescent module.

Luminescence is the generic name used to cover all forms of light emission other than that arising from elevated temperature (thermoluminescence). The emission of light through the absorption of UV or visible energy is called photoluminescence, and that caused by chemical reactions is called chemiluminescence. Light emission through the use of enzymes in living systems is called bioluminescence. Photoluminescence may be further subdivided into fluorescence, which is the immediate release (10^{-8}s) of absorbed light energy as opposed to phosphorescence which is the delayed release (10^{-6}–10^{2}s) of absorbed light energy.

The excitation spectrum of a molecule is similar to its absorption spectrum, while the fluorescence and phosphorescence emission occur at longer wavelengths than the absorbed light. The intensity of the emitted light allows quantitative measurement since, for dilute solutions, the emitted intensity is proportional to concentration. The excitation and emission spectra are characteristic of the molecule and allow qualitative measurements to be made. The inherent advantages of the techniques, particularly fluorescence, are:

1 sensitivity, picogram quantities of luminescent materials are frequently studied;
2 selectivity, derived from the two characteristic wavelengths; and
3 the variety of sampling methods that are available, i.e. dilute and concentrated samples, suspensions, solids, surfaces and combination with chromatographic methods, such as, for example is used in the HPLC separation of *o*-phthalyl dialdehyde derivatized amino acids in natural and sea water samples.

Fluorescence spectroscopy forms the majority of luminescence analyses. However, the recent developments in instrumentation and room-temperature phosphorescence techniques have given rise to practical and fundamental advances which should increase the use of phosphorescence spectroscopy. The sensitivity of phosphorescence is comparable to that of fluorescence and complements the latter by offering a wider range of molecules for study.

The pulsed xenon lamp forms the basis for both fluorescence and phosphorescence measurement. The lamp has a pulse duration at half peak height of 10μs. Fluorescence is measured at the instant of the flash. Phosphorescence is measured by delaying the time of measurement until the pulse has decayed to zero.

Several methods are employed to allow the observation of phosphorescence. One of the most common techniques is to supercool solutions to a rigid glass state, usually at the temperature of liquid nitrogen (77K). At these temperatures molecular collisions are greatly reduced and strong phosphorescence signals are observed.

Under certain conditions phosphorescence can be observed at room temperature from organic molecules adsorbed on solid supports such as filter paper, silica and other chromatographic supports.

Phosphorescence can also be detected when the phosphor is incorporated into an ionic micelle. Deoxygenation is still required either by degassing with nitrogen or by the addition of sodium sulphite. Micelle-stabilized room-temperature phosphorescence (MS RTP) promises to be a useful analytical tool for determining a wide variety of compounds such as pesticides and polyaromatic hydrocarbons.

Perkin–Elmer and Hamilton both supply luminescence instruments (see Appendix 1).

Perkin–Elmer LS–3B and LS–5B luminescence spectrometers

The LS–3B is a fluorescence spectrometer with separate scanning monochromators for excitation and emission, and digital displays of both monochromator wavelengths and signal intensity. The LS–5B is a ratioing luminescence spectrometer with the capability of measuring fluorescence, phosphorescence and bio- and chemiluminescence. Delay time (t_d) and gate width (t_g) are variable via the keypad in 10μs intervals. It corrects excitation and emission spectra.

Both instruments are equipped with a xenon discharge lamp source and have an excitation wavelength range of 230–720nm and an emission wavelength range of 250–800nm.

These instruments feature keyboard entry of instrument parameters which combined with digital displays, simplifying instrument operation. A high-output pulsed xenon lamp, having low power consumption and minimal ozone production, is incorporated within the optical module.

Through the use of an RS 232C interface, both instruments may be connected to Perkin–Elmer computers for instrument control and external data manipulation.

With the LS–5B instrument, the printing of the sample photomultiplier can be delayed so that it no longer coincides with the flash. When used in this mode, the instrument measures phosphorescence signals. Both the delay of the start of the gate (t_d) and the duration of the gate (t_g) can be selected in multiples of 10μs from the keyboard. Delay times may be accurately measured, by varying the delay time and noting the intensity at each value.

Specificity in luminescence spectroscopy is achieved because each compound is characterized by an excitation and emission wavelength. The identification of individual compounds is made difficult in complex mixtures because of the lack of structure from conventional excitation or emission spectra. However, by collecting emission on excitation spectra for each increment of the other, a fingerprint of the mixture can be obtained. This is visualized in the form of a time-dimensional contour plot on a three-dimensional isometric plot.

Fluorescence spectrometers are equivalent in their performance to single-beam UV-visible spectrometers in that the spectra they produce are affected by solvent background and the optical characteristics of the instrument. These effects can be overcome by using software built into the Perkin–Elmer LS–5B instrument or by using application software for use with the Perkin–Elmer models 3700 and 7700 computers.

Perkin–Elmer LS–2B microfilter fluorometer

The model LS–2B is a low-cost, easy-to-operate, filter fluorometer that scans emission spectra over the wavelength range 390–700nm (scanning) or 220–650nm (individual interference filters).

The essentials of a filter fluorometer are as follows:

- a source of UV/visible energy (pulsed Xenon);
- a method of isolating the excitation wavelength;
- a means of discriminating between fluorescence emission and excitation energy;
- a sensitive detector and a display of the fluorescence intensity.

The model LS–2B has all of these features arranged to optimize sensitivity for microsamples. It can also be connected to a highly sensitive 7μl liquid chromatographic detector for detecting the constituents in the column effluent. It has the capability of measuring fluorescence, time-resolved fluorescence, and bio- and chemiluminescent signals. A 40-portion auto-sampler is provided. An excitation filter kit containing six filters – 310, 340, 375, 400, 450 and 480nm – is available to enable the following assays to be performed: fluorescamine, *o*-phthaldialdehyde, 4-methyl-umbelliferone, porphyrins, dansyl derivatives, fluorescein europium and terbium organochelates.

1.1.1.3 Ultraviolet spectroscopy

This technique has found limited applications in sediment analysis and has been applied to the determination of aromatic hydrocarbons in saline sediments and mixtures of organics in non-saline sediments.

1.1.1.4 Infrared spectroscopy

This technique has found limited applications in the analysis of sediments and sludges.

Infrared spectroscopy has been applied to the determination of particulate organic carbon in non-saline sediments, aliphatic hydrocarbons and total organic carbon in saline sediments and mixtures of organics in sludges.

Fourier transform infrared spectroscopy

Fourier transform infrared spectroscopy, a versatile and widely used analytical technique, relies on the creation of interference in a beam of light. A source light beam is split into two parts and a continually varying phase difference is introduced into one of the two resultant beams. The two beams are recombined and the interference signal is measured and recorded, as an interferogram. A Fourier transform of the interferogram provides the spectrum of the detected light. Fourier transform infrared spectroscopy, a seemingly indirect method of spectroscopy, has many practical advantages, as discussed below.

A Fourier transform infrared spectroscopy spectrometer consists of an infrared source, an interference modulator (usually a scanning Michelson interferometer), a sample chamber and an infrared detector. Interference signals measured at the detector are usually amplified and then digitized. A digital computer initially records and then processes the interferogram and also allows the spectral data that results to be manipulated. Permanent records of spectral data are created using a plotter or other peripheral device.

The principal reasons for choosing Fourier transform infrared spectroscopy are: first, that these instruments record all wavelengths simultaneously and thus operate with maximum efficiency; and, second, that Fourier transform infrared spectroscopy spectrometers have a more convenient optical geometry than do dispersive infrared instruments. These two facts lead to the following advantages.

- Fourier transform infrared spectroscopy spectrometers achieve much higher signal-to-noise ratios in comparable scanning times.
- Fourier transform infrared spectroscopy spectrometers can cover wide spectral ranges with a single scan in a relatively short scan time, thereby permitting the possibility of kinetic time-resolved measurements.
- Fourier transform infrared spectroscopy provides higher-resolution capabilities without undue sacrifices in energy throughput or signal-to-noise ratios.
- Fourier transform infrared spectrometers encounter none of the stray light problems usually associated with dispersive spectrometers.
- Fourier transform infrared instruments provide a more convenient beam geometry – circular rather than slit shaped – at the sample focus.

Fourier transform Raman spectroscopy

Conventional Raman spectroscopy cannot be applied directly to aqueous extracts of sediments and soils, although it is occasionally used to provide information on organic solvent extracts of such samples. Fourier transform Raman spectroscopy, on the other hand, can be directly applied to water samples. The technique complements infrared spectroscopy in that some functional groups, e.g. unsaturation, give a much stronger response in the Raman region while others, e.g. carbonyl, give a stronger response in the infrared. Several manufacturers (Perkin–Elmer, Digilab, Bruker) now supply Fourier transform infrared spectrometers.

1.1.2 Flow-injection analysis

This technique has found very limited applications in soil and sediment analysis and is particularly useful when routine automated analyses at the mg L^{-1} level of large numbers of samples is required. The technique has been applied to the determination of total phosphorus, total organic carbon and total nitrogen in soils, total organic carbon in non-saline sediments and total sulphur in saline sediments.

Flow-injection analysis (FIA) is a rapidly growing analytical technique. Since the introduction of the original concept in 1975, about 1000 papers have been published.

Flow-injection analysis is based on the introduction of a defined volume of sample into a carrier (or reagent) stream. This results in a sample plug bracketed by carrier (Fig. 1(a)).

The carrier stream is merged with a reagent stream to obtain a chemical reaction between the sample and the reagent. The total stream then flows through a detector (Fig. 1.1(b)). Although spectrophotometry is the commonly used detector system in this application, other types of detectors have been used, namely fluorometric, atomic absorption emission spectroscopy and electrochemical, e.g. ion selective electrodes.

The pump provides constant flow and no compressible air segments are present in the system. As a result the residence time of the sample in the system is absolutely constant. As it moves towards the detector the sample is mixed with both carrier and reagent. The degree of dispersion (or dilution) of the sample can be controlled by varying a number of factors, such as sample volume, length and diameter of mixing coils and flow rates.

When the dispersed sample zone reaches the detector, neither the chemical reaction nor the dispersion process has reached a steady state. However, experimental conditions are held identical for both samples and standards in terms of constant residence time, constant temperature and constant dispersion. The sample concentration can thus be evaluated against appropriate standards injected in the same manner as samples (Fig. 1.1(c)).

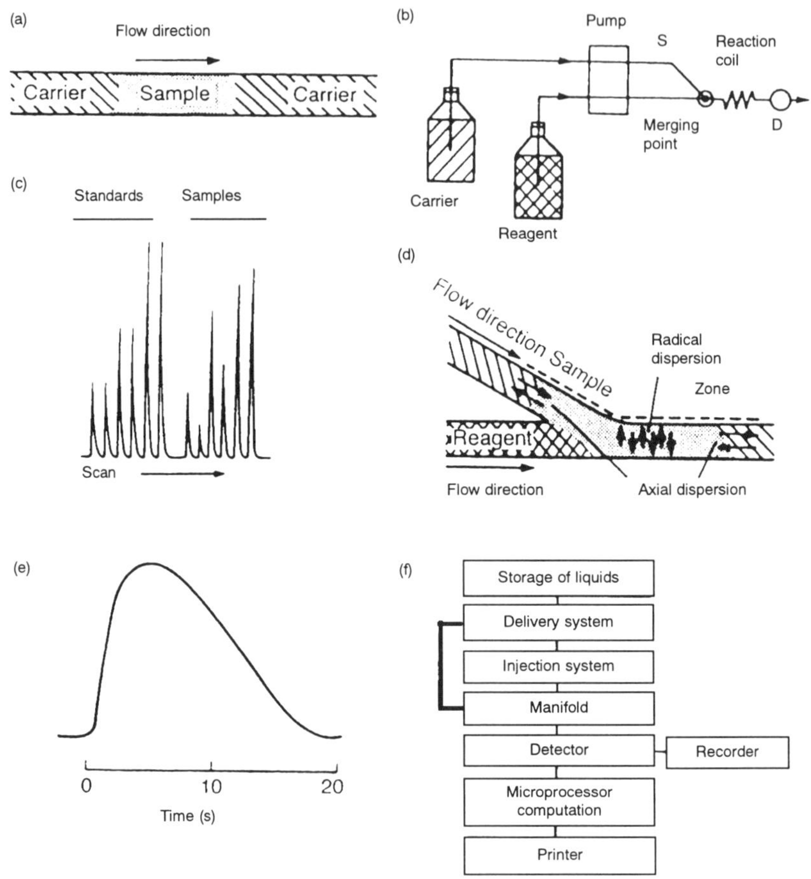

Fig 1.1 Analysis systems, flow-injection analysis
Source: Own files

The short distance between the injection site and the merging point ensures negligible dispersion of the sample in this part of the system. This means that sample and reagent are mixed in equal proportions at the merging point.

The mixing technique can be best understood by having a closer look at the hydrodynamic conditions in and around the merging point (Fig. 1.1(d)).

In Fig. 1.1(d) the hydrodynamic behaviour is simplified in order to explain the mixing process. Let us assume that there is no axial dispersion and that radial dispersion is complete when the sampler reaches the detector. The volume of the sample zone is thus 200μl after the merging point (100μl sample + 100μl-reagent as flow rates are equal). The total flow rate is 2.0ml min $^{-1}$. Simple mathematics then gives a residence time of 6s for the sample in the detector flow cell. In reality, response curves reflect some axial dispersion. A rapid scan curve is shown in Fig. 1.1(e). The baseline is reached within 20s. This makes it possible to run three samples per minute and obtain baseline readings between each sample (no carry-over), i.e. 180 samples per hour.

The configuration of an FIA system is shown schematically in Fig. 1.1(f). The (degassed) carrier and reagent solution(s) must be transported in a pulse-free transport system and at constant rate through narrow Teflon (Du Pont) tubing.

In a practical FIA system, peristaltic pumps are usually used since they have several channels, and different flow rates may be achieved by selection of a pump tube with a suitable inner diameter.

A manifold provides the means of bringing together the fluid lines and allowing rinsing and chemical reaction to take place in a controlled way. Manifolds with several lines can be assembled as required. These manifolds are mounted on plastic trays and allow the use of different reaction coils.

Flow-injection analysers available range from relatively low-cost unsophisticated instruments such as those supplied by Advanced Medical Supplies, Skalar and ChemLab to the very sophisticated instruments such as the FIA star 5010 and 5020 supplied by Tecator (Table 1.2).

1.1.3 Spectrometric methods

1.1.3.1 Atomic absorption spectrometry

This technique has been applied to the determination of arsenic, selenium, organocompounds of arsenic, mercury and tin in soils, carbohydrates, total sulphur, arsenic, antimony, bismuth, selenium and organocompounds of mercury, tin and silicon in non-saline sediments, arsenic, bismuth, selenium or organotin compounds in saline sediments and arsenic and selenium in sludges.

Basically, the atomic absorption method was designed for the determination of cations. However, it has been applied to the indirect determination of some organic substances. If an excess of a metal ion is added to a solid sample extract containing an organic substance which complexes with that metal and then the complex is extracted from the sample with a suitable organic solvent, then determination of the uncompleted excess metal in the water phase, enables one to estimate the amount of metal that has been complexed, hence the concentration of the organic substance. Alternatively the complexed metal content of the organic extract can be determined. Thus, anionic surface active agents form a chloroform soluble complex with bis(ethylene-diamine) copper II ion. Determination of copper in the chloroform extract enables one to estimate the concentration of anionic in the original water sample.

Table 1.2 Equipment for flow-injector analysis

Supplier	*Model*	*Features*	*Detectors available*
Advanced Medical Supplies	LCG 1	Relatively low-cost instrument, recorder output. No computerization on data processing (8 channels)	Colorimeter (other detectors can be used but are not linked in e.g. atomic absorption, fluorometer ion-selective electrodes
Chemlab	–	Relatively low-cost, recorder output or data analysis by microprocessor (3 channels)	Colorimeter
Skalar	–	Relatively low-cost, recorder output on data analysis by microprocessor also carries out segmented flow analysis	Colorimeter, flow cells for fluorometer and ion-selective electrodes available
Fiatron	Finlite 600	Laboratory process control and pilot plant instrument computerized	Colorimeter
	Fiatrode 400 Fiatrode 410 Fiatrode 430	Flow through analyser/controller, process control analyser	pH and ion-selective electrode
Tecator	FIA star 5025	Relatively low cost manual instrument specifically designed for fluoride, cyanide, potassium, iodide, etc.	Specially designed for use with ion-selective electrodes
	FIA star 5032	Relatively low-cost manual instrument (400–700 nm)	Spectrophotometer and/or photometer detectors
	Aquatec	Modular, semi-or fully automatic operation. Microprocessor controlled. A dedicated instrument designed for water analysis, i.e. dedicated method cassettes for phosphate and chloride, 600–100 samples h^{-1}	Flow through spectrophotometer (400–700nm)
	FIA star 5010	Modular, semi- or fully automatic operation. May be operated with process controller micro-processor. Can be set up in various combinations with 5017 sampler and superflow software which is designed to run on IBM PC/XT computer; 60–180 samples h^{-1}. Dialysis for in-line sample preparation and in-line solvent extraction. Thermostat to speed up reactions.	Spectrophotometer (400–700nm) or photometer can be connected to any flow through detector, e.g. UV/visible, inductively coupled plasma, atomic absorption spectrometer and ion-selective electrodes

Source: Own files

Since shortly after its inception in 1955, atomic absorption spectrometry has been the standard tool employed by analysts for the determination of trace levels of metals in water samples. In this technique a fine spray of the analyte is passed into a suitable flame, frequently oxygen acetylene or nitrous oxide acetylene, which converts the elements to an atomic vapour. Through this vapour radiation is passed at the right wavelength to excite the ground state atoms to the first excited electronic level. The amount of radiation absorbed can then be measured and directly related to the atom concentration: a hollow cathode lamp is used to emit light with the characteristic narrow line spectrum of the analyte element. The detection system consists of a monochromator (to reject other lines produced by the lamp and background flame radiation) and a photomultiplier. Another key feature of the technique involves modulation of the source radiation so that it can be detected against the strong flame and sample emission radiation.

A limitation of this technique is its lack of sensitivity compared to that available by other techniques (for example inductively coupled plasma atomic emission spectrometry (section 1.1.3.2)).

An additional disadvantage is that unlike ICPAES, only one element at a time can be determined. Suitable instrumentation is listed in Table 1.3.

1.1.3.2 Inductively coupled plasma atomic emission spectrometry

This technique has been applied to the determination of boron, total phosphorus and arsenic in soil, antimony and organosilicon compounds in non-saline sediments, arsenic in saline sediments and silicon and arsenic in sludges.

An inductively coupled plasma is formed by coupling the energy from a radiofrequency (1–3kW or 27–50MHz) magnetic field to free electrons in a suitable gas. The magnetic field is produced by a two- or three-turn water-cooled coil and the electrons are accelerated in circular paths around the magnetic field lines that run axially through the coil. The initial electron 'seeding' is produced by a spark discharge but, once the electrons reach the ionization potential of the support gas, further ionization occurs and a stable plasma is formed.

The neutral particles are heated indirectly by collisions with the charged particles upon which the field acts. Macroscopically the process is equivalent to heating a conductor by a radio-frequency field, the resistance to eddy-current flow producing joule heating. The field does not penetrate the conductor uniformly and therefore the largest current flow is at the periphery of the plasma. This is the so-called 'skin' effect and, coupled with a suitable gas-flow geometry, it produces an annular or doughnut-shaped plasma. Electrically, the coil and plasma form a transformer with the plasma acting as a one-turn coil of finite resistance.

Table 1.3 Available flame and graphite furnace atomic absorption spectrometers

Type instrument	*Supplier*	*Model no and type*
Flame (direct injection)	Thermo-electron	IL 157 single channel IL 357 single beam IL 457 single channel double beam Video 11 single channel single beam Video 12 single channel double beam Video 22 two double channels
Graphite furnace	Thermo-electron	IL 655 CTF
Direct injection	Perkin–Elmer	2280 single beam 2380 double beam
Graphite furnace	Perkin–Elmer	100 single beam 2100 single path double beam
Graphite furnace	Varian Associates	SpectrA A30 + 40 multi-element analysis Method storage SpectrA A10 (low cost, single beam) SpectrA A20 (medium cost, double beam)
Flame graphite furnace		SpectrA A300/400 multi-element analysis, centralized instrument control STA 95 and GTA 96 graphite tube atomizer units – compatible with all SpectrA A instruments
Graphite furnace	GBC Scientific Pty Ltd	903 single beam 902 double beam (both with impact head option)
Flame (direct injection) graphite furnace	Shimadzu	AA670 double beam AA670 G Double beam

Source: Own file

The properties of an inductively coupled plasma closely approach those of an ideal source for the following reasons:

- the source must be able to accept a reasonable input flux of the sample and it should be able to accommodate samples in the gas, liquid or solid phases;
- the introduction of the sample should not radically alter the internal energy generation process or affect the coupling of energy to the source from external supplies;
- the source should be operable on commonly available gases and should be available at a price that will give cost-effective analysis;
- the temperature and residence time of the sample within the source

Microprocessor	*Hydride and mercury attachment*	*Autosampler*	*Wavelength range*
} –	Yes	Yes	
with graphics			
with graphics			
with graphics			
computer interference			
–	–	Yes	
Yes			
Yes (with automatic background correction)	–	–	190–870
Yes	–	–	190–870
Yes	Yes	–	190–870
Yes	Yes	Yes	190–900
Yes (built in VDU)	Yes	Yes	190–900
Yes (built in VDU)	Yes	Yes	190–900
Yes (with colour graphics and 90 elements on disk)	Yes	Yes	190–900
Furnace and programmable sample dispenser operated from SpectrA A keyboard. Rapid interchange between flame and furnace operation			
Yes	Yes	Yes	176–900
Yes	Yes	Yes	170–900
Yes	Yes	Yes	190–900
Yes	Yes	Yes	190–900

should be such that all the sample material is converted to free atoms irrespective of its initial phase or chemical composition; such a source should be suitable for atomic absorption or atomic fluorescence spectrometry;

- if the source is to be used for emission spectrometry, then the temperature should be sufficient to provide efficient excitation of a majority of elements in the periodic table;
- the continuum emission from the source should be of a low intensity to enable the detection and measurement of weak spectral lines superimposed upon it;
- the sample should experience a uniform temperature field and the optical density of the source should be low so that a linear

relationship between the spectral line intensity and the analyte concentration can be obtained over a wide concentration range.

Greenfield *et al.* [1] were the first to recognize the analytical potential of the annular inductively coupled plasma.

Wendt and Fassel [2] reported early experiments with a 'tear-drop' shaped inductively coupled plasma but later described the medium power 1–3kW 18mm annular plasma now favoured in modern analytical instruments [3].

The current generation of inductively coupled plasma emission spectrometers provide limits of detection in the range of 0.1–500μg L^{-1} in solution, a substantial degree of freedom from interferences and a capability for simultaneous multi-element determination facilitated by a directly proportional response between the signal and the concentration of the analyte over a range of about five orders of magnitude.

The most common method of introducing liquid samples into the inductively coupled plasma is by using pneumatic nebulization (Thompson and Walsh [4]) in which the liquid is dispensed into a fine aerosol by the action of a high-velocity gas stream. To allow the correct penetration of the central channel of the inductively coupled plasma by the sample aerosol, an injection velocity of about 7ms $^{-1}$ is required. This is achieved using a gas injection with a flow rate of about 0.5–1lmin $^{-1}$ through an injector tube of 1.5 –2.0mm internal diameter. Given that the normal sample uptake is 1–2ml min $^{-1}$ this is an insufficient quantity of gas to produce efficient nebulization and aerosol transport. Indeed, only about 2% of the sample reaches the plasma. The fine gas jets and liquid capillaries used in inductively coupled plasma nebulizers may cause inconsistent operation and even blockage when solutions containing high levels of dissolved solids, such as sea water or particulate matter, are used. Such problems have led to the development of a new type of nebulizer, the most successful being based on a principle originally described by Babington (US Patents). In these, the liquid is pumped from a wide bore tube and thence conducted to the nebulizing orifice by a V-shaped groove (Suddendorf and Boyer [5]) or by the divergent wall of an over-expanded nozzle [6]. Such devices handle most liquids and even slurries without difficulty.

Nebulization is inefficient and therefore not appropriate for very small liquid samples. Introducing samples into the plasma in liquid form reduces the potential sensitivity because the analyte flux is limited by the amount of solvent that the plasma will tolerate. To circumvent these problems a variety of thermal and electrothermal vaporization devices have been investigated. Two basic approaches are in use. The first involves indirect vaporization of the sample in an electrothermal vaporizer, e.g. a carbon rod or tube furnace or heated metal filament as commonly used in atomic absorption spectrometry [7–9]. The second involves inserting the sample into the base of the inductively coupled plasma on a carbon rod or metal filament support [10,11]. Available instrumentation is reviewed in Table 1.4.

1.1.4 Titration procedures

As titration procedures, compared to some other procedures are relatively insensitive, it is likely that they would only be applied to those types of solid samples where the concentration to be determined is relatively high. Titration methods have been applied to the determination of carbonate, insecticides, total organic carbon and nitrogen in soil, total organic carbon in non-saline sediments and total sulphur in saline sediments.

The titration process has been automated so that batches of samples can be titrated non-manually and the data processed and reported via printouts and screens. One such instrument is the Metrohm 670 titroprocessor. This incorporates a built-in control unit and sample changer so that up to nine samples can be automatically titrated. The 670 titroprocessor offers incremental titrations with variable or constant-volume steps (dynamic or monotonic titration). The measured value transfer in these titrations is either drift controlled (equilibrium titration) or effected after a fixed waiting time; pK determinations and fixed end points (e.g. for specified standard procedures) are naturally included. End-point titrations can also be carried out.

Sixteen freely programmable computational formulae with assignment of the calculation parameters and units, mean-value calculations and arithmetic of one titration to another (via common variables) are available. Results can be calculated without any limitations.

The 670 titroprocessor can also be used to solve complex analytical tasks. In addition to various auxiliary functions which can be freely programmed, up to four different titrations can be performed on a single sample.

In addition to the fully automated 670 system, Metrohm also supply simpler units with more limited facilities which nevertheless are suitable for more simple titrations. Thus the model 682 titroprocessor is recommended for routine titrations with automatic equivalence pointer cognition or to preset end points. The 686 titroprocessor is a lower-cost version of the above instrument, again with automatic equivalence point recognition and titration to preset end points.

Mettler produce two automatic titrimeters; the DL 40 GP memotitrator and the lower-cost DL 20 compact titrator. Features available on the DL 40GP include absolute and relative end-point titrations, equivalence point titrations, back-titration techniques, multi-method applications, dual titration, pH stating, automatic learn titrations, automatic determination of standard deviation and means, series titrations, correction to printer, acid balance analogue output for recorder and correction to the laboratory

Table 1.4 Inductively coupled plasma optical emission spectrometers available on the market

Supplier	*Model*	*System*	*Number of elements claimed*
Perkin–Elmer	Plasma II	Optimized sequential system	70
Perkin–Elmer	ICP 5500	Sequential	
Perkin–Elmer	ICP 5500B	Sequential	
Perkin–Elmer	ICP 6500	Sequential	
Perkin–Elmer	ICP 5000	Can be used for flame and graphite furnace ASS and inductively coupled plasma atomic emission spectrometry (sequential)	
Perkin–Elmer	Plasma 40	Lower-cost sequential	
Labtam	Plasma Scan 8440	Simultaneous (polychromator or with optional monochromator for sequential)	60–70
Labtam	Plasma 8410	Sequential	More than 70
Thermoelectron	Plasma 300 (replacing the Plasma 200) (single (air) or double (air/vacuum) available	Sequential	Up to 63
Philips	PV 8050 series PV 8055 PV 8060 PV 8065	Simultaneous	56
Philips	PU 7450	Sequential	70
Baird	Spectrovac PS3/4 plasma hydride device option	Simultaneous and sequential	Up to 60
Baird	Plasmatest system 75	Simultaneous and sequential	Up to 64
Spectro Analytical Ltd	Spectroflame plasmahydride device option	Simultaneous and sequential	Up to 64

Source: Own file

information system. Up to 40 freely definable methods can be handled and up to 20 reagents held on store. Six control principles can be invoked. The DL 20 can carry out absolute (not relative) end-point titrations and equivalence point titrations, back-titration, series titrations, and correction to printer and balance and the laboratory information system.

Maximum analysis rate (elements min L $^{-1}$)	*Microprocessor*	*Autosampler*	*Range (nm)*
Up to 50	Yes	Yes	160–800
15	Yes	Yes	170–900
20	Yes	Yes	170–900
20	Yes	–	170–900
–	Yes	Yes	175–900
	Personal computer control	Yes	160–800
Up to 64	Yes	Yes	170–820
–	Yes	Yes	170–820
Up to 18 (single channel air); up to 24 (double channel air/vacuum)	Yes	Yes	160–900
–	Yes	Yes	165–485 and 530–860
–	Yes	Yes	190–800
Up to 80 samples h $^{-1}$ each up to 60 elements	Yes	Yes	162–766 and 162–800
–	Yes	Yes	175–768 and 168–800
–	Yes	Yes	165–800

Only one freely definable method is available. Four control principles can be invoked.

The DL 40GP can handle potentiometric, voltammetric or photometric titrations.

1.1.5 Chromatographic methods

1.1.5.1 High-performance liquid chromatography including high-performance liquid chromatography–mass spectrometry

Aliphatic hydrocarbons, triazine, substituted urea type and phenoxyacetic acid types of herbicides, Fluazifop and Fluazifop-butyl herbicides, ethylene diamine tetracetic acid salts in soil, aliphatic and polyaromatic hydrocarbons, phthalate esters, various organosulphur compounds, triazine herbicides, optical whiteners, mixtures of organic compounds and organotin compounds in non-saline sediments, aromatic hydrocarbons, humic and fulvic acids and mixtures of organic compounds in saline sediments and non-ionic surfactants and cobalamin in sludges.

This technique has also been coupled with mass spectrometry for the partial identification of organic compounds eluted from the column, e.g. the determination of phenoxy acetic acid herbicides in soil and non-saline sediments.

One of the limitations of gas chromatography and consequently of gas chromatography–mass spectrometry is that of all the organic material present in natural water samples, only a small proportion, say as low as 20%, is sufficiently volatile to be separated on gas chromatographic columns operating at even the maximum of their temperature range. Of the 275 compounds for the Appendix III list of the US Environmental Protection Agency, 150 are not amenable to gas chromatographic separation.

As a consequence of this there has, in recent years, been a growing interest in applying high-performance liquid chromatography which is not subject to this temperature limitation, to the determination of the non-volatile fractions of water.

Modern high-performance liquid chromatography has been developed to a very high level of performance by the introduction of selective stationary phases of small particle sizes, resulting in efficient columns with large plate numbers per litre.

There are several types of chromatographic columns used in high-performance liquid chromatography.

Reversed-phase chromatography

The most commonly used chromatographic mode in HPLC is reversed-phase chromatography. Reversed-phase chromatography is used for the analysis of a wide range of neutral compounds such as carbohydrates and polar organic compounds. Most common reversed-phase chromatography is performed using bonded silica-based columns, thus inherently limiting the operating pH range to 2.0–7.5. The wide pH range (0–14) of some columns (e.g. Dionex Ion Pac NSI and NS 1–5 μcolumns) removes this limitation, and consequently they are ideally suited for ion-pairing and ion-suppression reversed-phase chromatography, the two techniques which have helped extend reverse-phase chromatography to ionizable compounds.

High-sensitivity detection of non-chromophoric organic ions can be achieved by combining the power of suppressed conductivity detection with these columns. Suppressed conductivity is usually a superior approach to using refractive index or low UV wavelength detection.

Reversed-phase ion-pairing chromatography

Typically, reversed-phase ion-pairing chromatography is carried out using the same stationary phase as reversed-phase chromatography. A hydrophobic ion of opposite charge to the solute of interest is added to the mobile phase. Samples which are determined by reversed-phase ion-packing chromatography are ionic and thus capable of forming an ion pair with the added counter ion. This form of reversed-phase chromatography can be used for anion and cation separations and for the separation of surfactants and other ionic types of organic molecules. An unfortunate drawback to using silica-based columns is that ion-pairing reagents increase the solubility of silica in water, leading to loss of bead integrity and drastically reducing column life. Some manufacturers (e.g. Dionex) employ neutral macroporous resins, instead of silica, in an attempt to widen the usable pH range and eliminate the effect of ion-pairing reagents.

Ion-suppression chromatography

Ion suppression is a technique used to suppress the ionization of compounds (such as carboxylic acids) so they will be retained exclusively by the reversed-phase retention mechanism and chromatographed as the neutral species.

Column packings with an extended pH range are needed for this application as strong acids or alkalis are used to suppress ionization. In addition to carboxylic acids, the ionization of amines can be suppressed by the addition of a base to the mobile phase, thus allowing chromatography of the neutral amine.

Ion-exclusion chromatography

Unlike the pellicular packings used for ion exchange, the packings used in ion exclusion are derived from totally sulphonated polymeric materials. Separation is dependent upon three different mechanisms: Donnan exclusion, steric exclusion and adsorption/partitioning.

Donnan exclusion causes strong acids to elute in the void volumes of the column. Weak acids which are partially ionized in the eluent are not subject to Donnan exclusion and can penetrate into the pores of the packing. Separation is accomplished by differences in acid strength, size and hydrophobicity. The major advantage of ion exclusion lies in the ability to handle samples that contain both weak and strong acids. A good example of the power of ion exclusion is the routine determination of organic acids in sea water. Without ion exclusion, the high chloride ion concentration would present a serious interference.

Four basic types of elution system are used in HPLC. This is illustrated below by the systems offered by LKB, Sweden.

Isocratic system This consists of a solvent delivery for isocratic reversed-phase and gel filtration chromatography.

The isocratic system (Fig. 1.2(a)) provides an economic first step into high-performance liquid chromatography techniques. The system is built around a high-performance, dual-piston, pulse-free pump providing precision flow from 0.01 to 5ml min $^{-1}$.

Any of the following detectors can be used with this system:

- fixed wavelength ultraviolet detector (LKB Unicord 2510);
- variable UV visible (190–600nm);
- wavelength monitor (LKB 2151);
- rapid diode array spectral detector (LKB 2140);
- refractive index detector (LKB 2142);
- electrochemical detector (LKB 2143);
- wavescan EG software (LKB 2146).

Basic gradient system This is a simple upgrade of the isocratic system with the facility for gradient elution techniques and greater functionality (Fig. 1.2(b)). The basic system provides for manual operating gradient techniques such as reversed-phase, ion-exchange and hydrophobic interaction chromatography. Any of the detectors listed above under the isocratic system can be used.

Advanced gradient system For optimum functionality in automated systems designed primarily for reversed-phase chromatography and other gradient techniques, the LKB advanced-gradient system is recommended (Fig 1.2(c)). Key features include:

- a configuration that provides the highest possible reproducibility of results;
- a two-pump system for highly precise and accurate gradient formation for separation of complex samples;
- full system control and advanced method development provided from a liquid chromatography controller;
- precise and accurate flows ranging from 0.01 to 5ml min $^{-1}$.

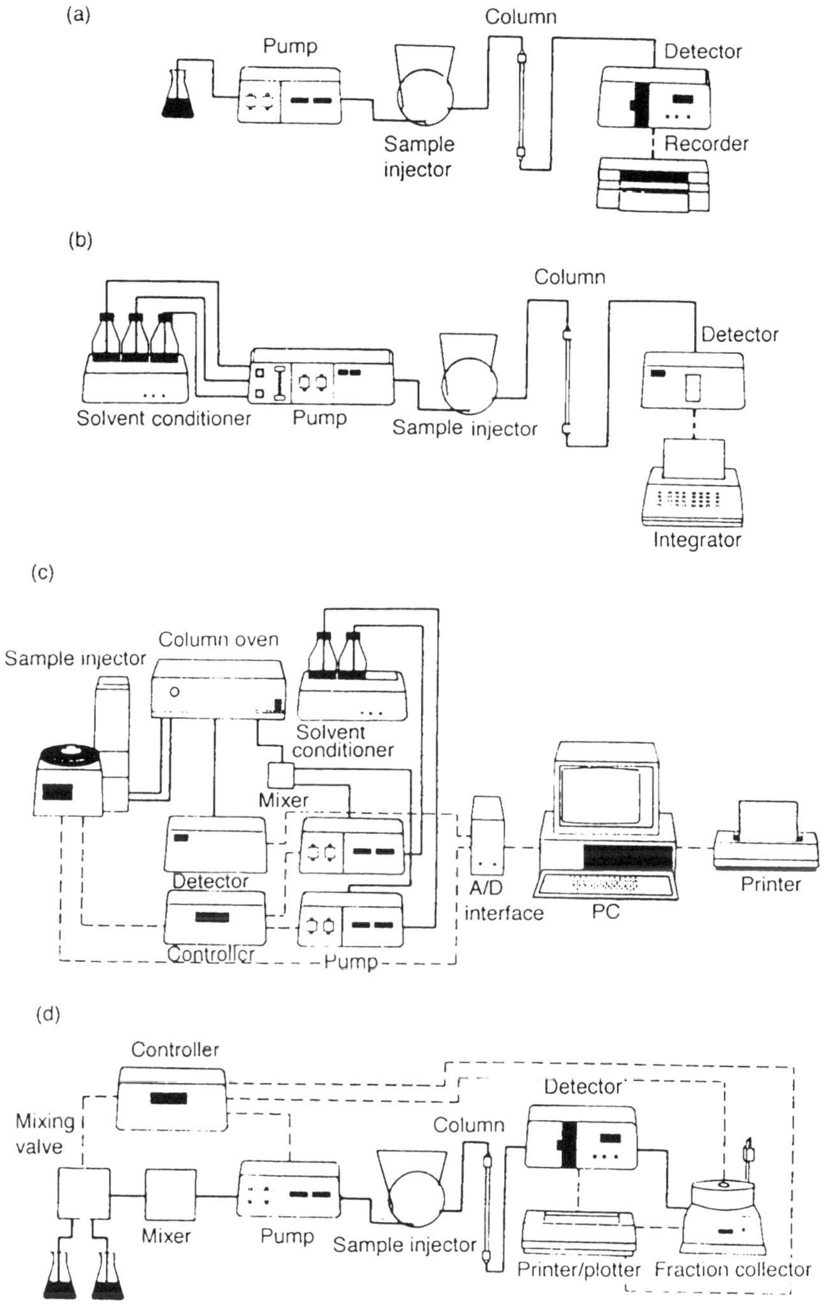

Fig. 1.2 Analysis systems, high-performance liquid chromatography
Source: Own files

Table 1.5 Detectors used in HPLC

Type of detector		*Supplier*	*Detection part no*	*HPLC instrument part no*
Spectrophotometric (variable wavelength)	190–390nm	Perkin–Elmer	LC–90	–
	193–350nm	Kontron	735 LC	Series 400
	195–350nm	Shimadzu	SPD–7A	LC–7A
	195–350nm	Shimadzu	SPD–6A	LC–8A
	195–350nm	Shimadzu	SPD–6A	LC–6A
	206–405nm (fixed wavelength choice of seven wavelengths between 206 and 405nm)	LKB	2510 Uvicord SD	–
	190–370nm	Cecil	Model 1937	Chrom-A-Scope
	190–400nm	Instruments	CE 1220	Series 100
Variable wavelength UV-visible	190–600nm	Varian	2550	2500
	190–600nm	LKB	2151	Uvicord SD
	190–700nm	Kontron	432	Series 400
	190–800nm	Kontron	430	Series 400
	185–900nm	Kontron	720 LC	Series 400
	200–570nm	Kontron	740 LC	Series 400
	190–800nm	Dionex	VDM 11	Series 400
	190–750nm	Isco	V4	
	214–660nm (18 preset wavelengths)	Isco	USA and 228	Microbo system
	195–700nm	Shimadzu	SPD 7A	LC–7A
	195–700nm	Shimadzu	SPD–6A	V LC–8A
	195–700nm	Shimadzu	SPD–6AV	LC–6A
	195–700nm	Hewlett Packard	Programmably variable wavelength detector	9050 series
	380–600nm	Cecil Instruments	CE 1200	Series 1000
	190–800nm	Applied Chromatography systems	750/16 and 5750/11	–
Conductivity	–	Dionex	CDM 11	4500i
		Roth Scientific	–	Chrom-A-Scope
Electrochemical detector	–	Dionex	PAD–11	4500i
		LKB	2143	Wave-scan EG
		Roth Scientific	–	Chrom-A-Scope
		Cecil Instruments	CE 1500	–
		PSA Inc	5100A	–
		Applied Chromatography systems	650*350*06	–

Table 1.5 continued

Type of detector	*Supplier*	*Detection part no*	*HPLC instrument part no*
Refractive index detector	LKB	2142	Wavescan EG
	Roth Scientific	–	Chrom-A-Scope
	Cecil Instruments	CE 1400	Series 1000
Differential viscosity	Roth Scientific	–	Chrom-A-Scope
mass detection (evaporative)	Applied Chromatography system	750/14	–
Diode array	Varian	9060	200L and 5001 5500 series
	Perkin–Elmer	LC 135, LC 235 and LC 480	–
	LKB	2140	
	Hewlett Packard	Multiple wavelength detector	1050 series

Source: Own files

This system is ideal for automatic method for development and gradient optimization.

Inert system By a combination of the use of inert materials (glass, titanium, and inert polymers) this system offers totally inert fluidics. Primary features of the system include (Fig. 1.2(d)):

- the ability to perform isocratic or gradient elution by manual means;
- full system control from a liquid chromatography controller;
- precise and accurate flows from 0.01–5 ml min^{-1}.

This is the method of choice when corrosive buffers, e.g. those containing chloride or aggressive solvents, are used.

Chromatographic detectors

Details concerning the types of detectors used in high-performance liquid chromatography are given in Table 1.5. The most commonly used detectors are those based on spectrophotometry in the region 185–400nm, visible ultraviolet spectroscopy in the region 185–900nm, post-column derivativization with fluorescence detection (see below), conductivity and those based on the relatively new technique of multiple wavelength ultraviolet detectors using a diode array system detector (see below).

Other types of detectors available are those based on electrochemical principles, refractive index, differential viscosity and mass detection.

Electrochemical detectors

These are available from several suppliers (Table 1.5). ESA supply the model PS 100A coulochem multi-electrode electrochemical detector. Organics, anions and cations can be detected by electrochemical means.

The Gilson Aspec automatic sample preparation system is a fully automated system for solid-phase extraction on disposable columns and on-line HPLC analysis. The Aspec system offers total automation and total control of the entire sample preparation process including clean-up and concentration. In addition, Aspec can automatically inject prepared samples into on-line HPLC systems.

Aspec is designed to receive up to 108 samples. The system is compatible with most standard disposable extraction columns. Analytichem Bond-Elut Baker SPE, Supelco Supelclean, Alltech Extract Clean, etc. There is a choice of more than 20 different stationary phases.

Spectrofluorimetric detectors

A spectrofluorimeter has been used as a detector in the high-performance liquid chromatographic separation of polyaromatic hydrocarbons in water samples [12–17]. A great improvement in sensitivity and specificity can be obtained by the correct wavelengths.

Amino acid analysers

This is an example of a dedicated application of high-performance liquid chromatography.

The most popular current techniques for amino acid analysis rely on liquid chromatography and there are two basic analytical methods. The first is based on ion-exchange chromatography with post-column derivatization. The second uses pre-column derivatization followed by reversed-phase HPLC. Derivatization is necessary because amino acids, with very few exceptions, do not absorb in the UV-visible region, nor do they possess natural fluorescence.

Each of the major methods has its own particular advantages and disadvantages. Since the variety of available chemistries can be confusing, the method itself should govern the choice that meets requirements, rather than the equipment or systems offered. The optimal method is best selected by a comprehensive and objective review of all commonly used techniques (Table 1.6).

Over-riding criteria which will influence the selection are resolution, sensitivity and speed. While the very best chromatogram for any given method will inevitably be a compromise, only the fullest evaluation of all the alternatives offered will guarantee a correct selection. The various manufacturers of this equipment are listed in Appendix 1.

Certainly, a vast amount of experience has been gained by the widespread use of conventional amino acid analysers. They offer high reliability, accuracy, reproducibility and can separate complex samples. Because conventional analysers can be fully automated, they are widely used in routine analysis. However, the method is limited by the sensitivity which can be achieved using ninhydrin as the derivatizing agent. Sensitivity can be increased by using ortho-phthaldialdehyde (OPA) instead, but where extremely high sensitivity is required, HPLC is the method of choice.

Two other reagents used in HPLC are 9-fluorenyl methoxycarbonyl chloride (FMOC) and phenylisothiocyanate (PITC). 9-fluorenyl methoxycarbonyl chloride is becoming increasingly popular in protein chemistry research because it reacts with secondary amines and also offers rapid analysis of protein hydrolysates.

One aspect governing the choice of method is the sensitivity required. If only a small amount of sample is available, then for the LKB Alpha Plus and the LKB HPLC instruments the greatest sensitivity is obtained using the Alpha Plus instrument with ion-exchange separation and post-column derivatization with *o*-phthaldialdehyde (OPA) reagent (108μg) or the HPLC instrument using reversed-phase chromatography and pre-column derivatization with OPA, or 9-fluorenyl methoxycarbonyl chloride reagents (30–33ng).

One leading supplier, LKB, is discussed below. Others are reviewed in Appendix 1.

LKB supply two instruments, the LKP 4150 Alpha HPLC, and for analysis requiring higher sensitivity and faster run times the LKB 4151 Alpha Plus.

LKB 4150 Alpha This system is a reversed-column chromatography equipped for pre-column derivatization. The column is made of glass and has solid-state heating. The detection system compromises a dual channel photometer with a high-temperature reaction coil. A single low-volume long-pathlength flow cell is employed. A fluorescence detector is available to provide an approximately tenfold increase in sensitivity over ninhydrin detection. Refrigerated sample capsule loading is supplied. Powerful programming capability permits the storage of up to 20 methods. Storage facilities for six buffers is supplied.

LKB 4151 Alpha Plus Alpha Plus is a fully automated and dedicated analyser, this turnkey system has been carefully designed to give a truly robust chromatography. Stepwise elution with up to five buffers plus flexible temperature control guarantees optimal separations from even

Table 1.6 Procedure for the determination of amino acids

Form of chromatography	*Derivatization*	*Reagent*		
Ion exchange	Post-column	Ninhydrin	Amino acids are separated in their native form on a sulphonated polystyrene resin using a system of sodium or lithium based buffers. Separation is effected by stepwise, rather than gradient elution, and the chromatography can be further optimized by carefully controlling the temperature of the analytical column. The number and nature of the buffers used depends on the complexity of the sample to be analysed. Hydrolysates of protein and foodstuff samples can be separated on a 3 or 4 buffer sodium system, whereas physiological fluids (which are inherently more complex) require the use of 5 lithium buffers. The post-column ninhydrin reaction proceeds in a reaction coil at elevated temperatures (ca. 130–135°C). Ninhydrin reacts with primary amino acids to give a chromophore with a wavelength of maximum absorption of 570nm. With secondary amino acids, ninhydrin reacts to form a yellow complex with different absorbance characteristics. For this reason, detection occurs at both 570 and 440nm.	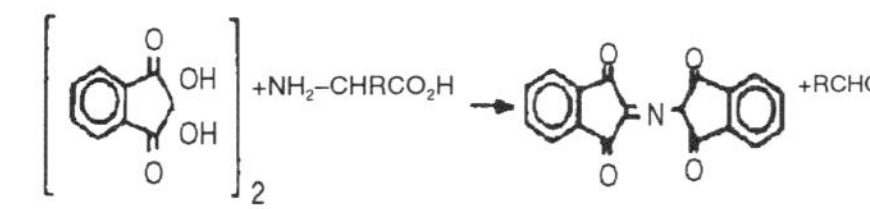 Ninhydrin Amino acid Ruhemann's purple *Simplified reaction between ninhydrin and primary amino acids* 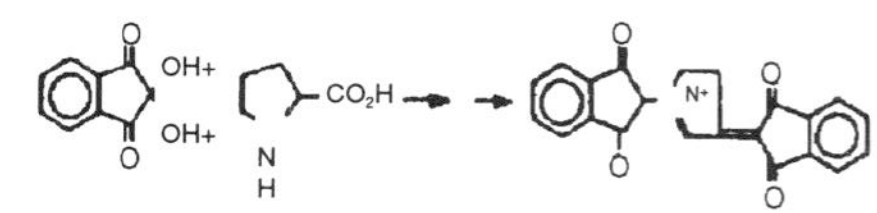 Ninhydrin Proline Yellow chromophore *Simplified reaction between ninhydrin and secondary amino acids*
Ion exchange	Post-column	*o*-phthaldialdehyde (OPA)	The reaction between OPA and primary amino acids takes place in the presence of a thiol such a 2- mercapto-ethanol, to form highly fluorescent derivatives. By using OPA as the derivatizing reagent, the sensitivity of a conventional analyser can be increased by a factor of 10.	

			On the other hand, OPA does not react with secondary amines and is therefore unsuitable if detection of proline or hydroxyproline is essential. OPA also has a low response with cystine and other disulphides. The OPA reaction proceeds rapidly at room temperature and the post-column fluidics are simplified by replacing the existing reaction module and photometer with a fluorescence detector. Excitation occurs at 330nm, and emission is measured at a wavelength of 430nm or above.	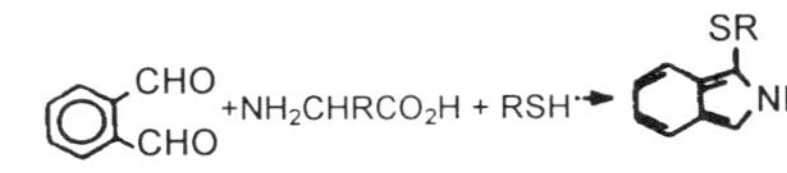 OPA Amino acid Thiol Fluorescent adduct *Simplified reaction between OPA and primary amino acids*
Reversed phase	Pre-column	*o*-phthaldialdehyde (OPA)	OPA unlike ninhydrin (and in common with other precolumn derivatizing agents), forms a different adduct with each amino acid. The resulting derivatives can be separated using reversed-phase chromatography. OPA-derivatized amino acids are usually separated on an ODS–II solid phase using a mobile phase of sodium phosphate buffer and an acetonitrile gradient. In contrast to the post-column derivatization, the inherent sensitivity of this method is higher since the column effluent is not diluted. Adequate sensitivity can therefore be achieved by UV monitoring at 330nm. For higher sensitivity, fluorescence detection is often chosen, and emission is measured at wavelengths above 430nm. moreover, baseline quality is superior. There is no necessity to remove excess OPA prior to sample injection since OPA itself will not interfere with separation or detection. However, since OPA-amino acid derivatives are unstable, complete automation of the precolumn reaction with accurate control of reaction time is essential for reproducible results.	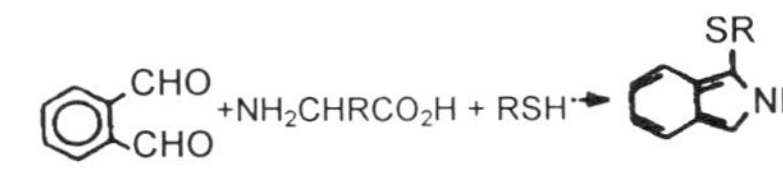 OPA Amino acid Thiol Fluorescent adduct

Table 1.6 continued

Form of chromatography	*Derivatization*	*Reagent*		
Reversed phase	Pre-column	Phenylisothiocyanate (PITC)	PITC has been used extensively in the sequencing of peptides and proteins and reactions under alkaline conditions with both primary and secondary amino acids. The methods of sample preparation and derivatization follow a stringent procedure which involves many labour-intensive stages. However, the resulting phenylthiocarbamyl-amino acids (PTC–AA's) are very stable, and the timing of the derivatization step is not as critical as when using OPA. Hardware requirements for the separation of PTC-amino acids are similar to those for OPA amino acid determination. Since the derivatives do not fluoresce the technique is limited to UV detection, which normally takes place at 245nm. Despite the problems encountered in sample preparation, the chromatography of protein hydrolysate samples is straightforward and can give good resolution in a short time.	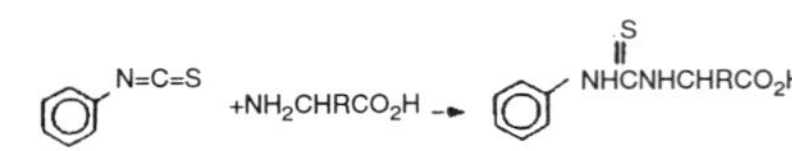 PITC Amino acid PTC–amino acid *Simplified reaction between PITC and amino acids*
Reverse phase	Pre-column	9-fluorenyl methoxycarbonyl chloride (FMOC)	FMOC is a relatively new reagent which has not only made a very significant impact in peptide synthesis, but has also become used as an effective pre-column derivatizing reagent in amino acid analysis.	

FMOC offers sensitivity comparable to OPA when used in conjunction with fluorescence detection, whilst also reacting with both primary and secondary amines. Moreover, the derivatives are stable for more than 30h in an acidic environment.

Both FMOC and its hydrolysis products have similar absorption and fluorescence spectra to FMOC-amino acids. Excess FMOC remaining after derivatization reacts with water to form 9-fluorenylmethyl alcohol (FMOC–OH), and if this is not removed prior to sample injection, it elutes as a large, broad peak in the vicinity of proline.

Removal of FMOC–OH can be achieved by liquid–liquid extraction (normally into pentane). A more elegant method of preventing interference of FMOC–OH is to react excess FMOC with a very hydrophobic amine (e.g. 1-aminoadamantane) to form a derivative which elutes after the peaks of interest.

FMOC-amino acids can be chromatographed using a C8 column and acetonitrile in sodium acetate buffer as the mobile phase. Fluorescence detection with excitation at 260nm and emission at 310nm gives the best results.

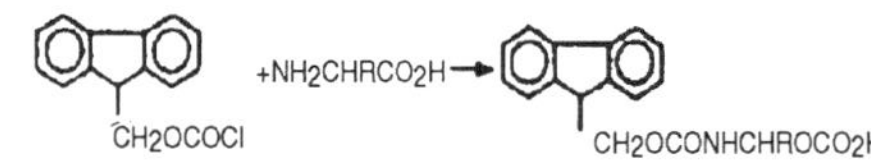

Simplified reaction between FMOC and amino acids

Source: Own files

the most complex samples. The versatile programmer monitors and controls all instrument functions and a complete fault-detection system assures absolutely safe operation while preserving the integrity of samples.

The analysis time for protein hydrolysates is 85 min using standard columns. For extra high resolution a high-resolution lithium cation exchange column is recommended which achieves baseline separation of virtually all 40 amino acids (Fig. 1.3).

High-performance liquid chromatography–mass spectrometry

Instrumentation Hewlett Packard supply the HP 5988A and HP 5987A mass-selective detectors for use with liquid chromatographs.

To date this equipment has been used extensively for identifying and determining non-volatile compounds such as diuretics and some stimulants in urine samples taken at the Olympic Games and the technique is now being introduced into the water laboratory. Particle beam technology has produced further improvements in liquid chromatography–mass spectrometry. The particle beam liquid chromatography–mass spectrometer uses the same switchable electron impact chemical ionization source and the same software and data systems that are used for a gas chromatography–mass spectrometry system. Adding a gas chromatograph creates a versatile particle-beam liquid chromatography/gas chromatography–mass spectrometry system that can be switched from liquid chromatography–mass spectrometry to gas chromatography–mass spectrometry in an instant.

Based on a new technology, particle beam enhanced liquid chromatography–mass spectrometry expands a chemist's ability to analyse a vast variety of substances. Electron impact spectra from the system are reproducible and can be searched against standard or custom libraries for positive compound identification. Chemical ionization spectra can also be produced. Simplicity is a key feature. A simple adjustment to the particle beam interface is all it takes.

The particle beam system is a simple transport device, very similar to a two-stage jet separator. The solvent vapour is pumped away, while the analyte particles are concentrated in a beam and allowed to enter the mass spectrometric source. Here they are vapourized and ionized by electron impact.

The different ways a particle beam liquid chromatography mass spectrometer can be configured reflect the versatility of the system in accommodating both the application and the availability of existing instrumentation. The system consists of these elements:

- particle beam interface mounted on the Hewlett Pack 5988A or 5987A mass spectrometer

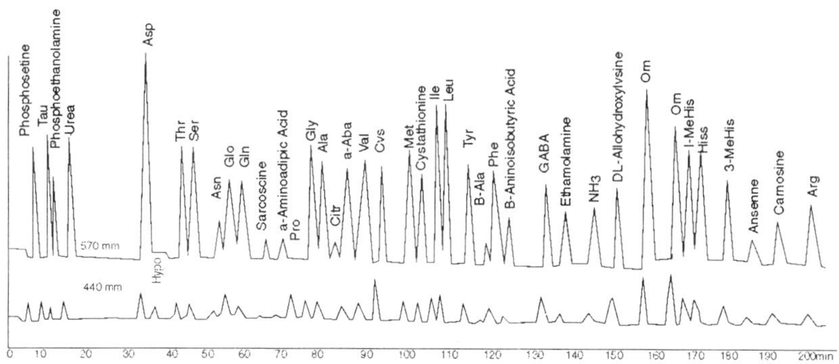

Fig. 1.3 High-performance liquid chromatography of amino acids.
Source: Own files

- liquid chromatography (either the integrated Hewlett Packard 1090 or modular Hewlett Packard 1050)
- data system (either HP 59976C chem station for single instrument operation or the Hewlett Packard 1000 RTE A-series for multi-instrument, multi-tasking, multi-user operation).

This technique is complementary to the thermospray technique. Relative advances of the particles beam technique over thermospray include library searchable electron impact spectra, improved reproducibility, easier use and increased predictability over a broad range of compounds. But since a particle beam requires same sample volatility, very large and polar compounds such as proteins may not provide satisfactory results using particle beam liquid chromatography–mass spectrometry. Additionally, certain classes of compounds such as preformed ions, azo dyes and complex sugars may not yield satisfactory electron impact spectra, but can be run on thermospray. In other words, both liquid chromatography–mass spectrometry techniques complement each other's limitations and the analyst may want to add both to address a broader range of samples.

Applications

Liquid chromatography–mass spectrometry has been used to determine triazine pesticides in land-fill soils [18] and dioctadecylmethyl ammonium in sewage sludges [19,20].

1.1.5.2 Column coupling capillary isotachoelectrophoresis

The technique offers many similar advantages to ion chromatography, namely multiple ion analysis, little or no sample pretreatment, speed, sensitivity and automation.

Separation capillary columns are made in fluorinated ethylene-propylene copolymer. Detection is achieved by conductivity cells and an a.c. conductivity mode of detection is used for making the separations visible. The driving current is supplied by a unit enabling independent currents to be preselected for the preseparation and final analytical stages. The run of the analyser is controlled by a programmable timing and control unit. The zone lengths from the conductivity detector, evaluated electronically, can be printed on a line printer.

The use of column coupling configuration of the separation unit provides the possibility of applying a sequence of two leading electrolytes to the analytical run. Therefore, the choice of optimum separation conditions can be advantageously divided into two steps, first, the choice of a leading electrolyte suitable for the separation and quantitation of the micro-constituents in the first stage (preseparation column) simultaneously having a retarding effect on the effective mobilities of micro-constituents (nitrite, fluoride, phosphate), and, second, the choice of the leading electrolyte for the second stage in which only micro-constituents are separated and quantified (macro constituents are removed from the analytical system after their evaluation in the first stage).

To satisfy the requirements for the properties of the leading electrolyte applied in the first stage and, consequently, to decide its composition, two facts had to be taken into account, i.e. the pH value of the leading electrolyte needs to be around 4 or less and at the same time the separations of the macro constituents need to be optimised by means other than adjusting the pH of the leading electrolyte (anions of strong acids).

The choice of the leading electrolyte for the second stage, in which the micro-constituents were finally separated and quantitatively evaluated, was straightforward, involving a low concentration of the leading constituent (low detection limit) and a low pH of the leading electrolyte (separation according to pK values).

1.1.5.3 Thin-layer chromatography

This technique has been applied to the determination of chlorinated insecticides, carbamate insecticides and substituted urea type herbicides in soil and chloroaliphatic hydrocarbons in non-saline sediments. Separation is usually achieved on thin layers of silica gel or alumina.

In the case of volatile compounds such as aliphatic hydrocarbons care is needed as volatiles may be lost during the separation process.

In general, the technique is limited to cases where the identity of the substance to be determined is known, although, in some cases identification of the separated compounds has been achieved by infrared spectroscopy or mass spectrometry of eluates of the individual separated spots isolated from the thin layer plate.

1.1.5.4 Supercritical fluid chromatography

This technique has been used for the determination of polychlorobiphenyls, polychlorodibenzo-*p*-dioxins, polychlorodibenzofurans, alkyl phosphates, chlorinated insecticides, organophosphorus insecticides, triazine herbicides. Dacthal insecticide, insecticide/herbicide mixtures, mixtures of organic compounds and organotin compounds in soils, and polyaromatic compounds, polychlorobiphenyls, chlorinated insecticides and organotin compounds in non-saline sediments and anionic surfactants in sludges.

Until recently the chromatographer has had to rely on either gas chromatographic or HPLC for separations, enduring the limitations of both. Lee Scientific has created a new dimension in chromatography, one which utilizes the unusual properties of supercritical fluids. With the new technology of capillary supercritical fluid chromatography (SFC) the chromatographer benefits from the best of both worlds – the solubility behaviour of liquids and the diffusion and viscosity properties of gases. Consequently, capillary SFC offers unprecedented versatility in obtaining high-resolution separations of difficult compounds.

Beyond its critical point, a substance can no longer be condensed to a liquid, no matter how great the pressure. As pressure increases, however, the fluid density approaches that of a liquid. Because solubility is closely related to density, the solvating strength of the fluid assumes liquid-like characteristics. Its diffusivity and viscosity, however, remain. SFC can use the widest range of detectors available to any chromatographic technique. As a result, capillary SFC has already demonstrated a great potential in application to water, environmental and other areas of analysis.

Suppliers of SFC instruments are reviewed in Appendix 1.

SFC is now one of the fastest growing analytical techniques. The first paper on the technique was by Klesper *et al.* [21], but supercritical fluid chromatography did not catch the analyst's attention until Novotny *et al.* [22] published the first paper on capillary SFC.

SFC finds its applications in compounds that are either difficult or impossible to analyse by liquid chromatography or gas chromatography. SFC is ideal for analysing either thermally labile or non-volatile non-chromatophoric compounds. The technique will be of interest to water chemists as a means of identifying and determining the non-volatile components of water.

Most supercritical fluid chromatographs use carbon dioxide as the supercritical eluent, as it has a convenient critical point of 31.3°C and 72.5 atmospheres. Nitrous oxide, ammonia and *n*-pentane have also been used. This allows easy control of density between 0.2g ml $^{-1}$ and 0.8g ml $^{-1}$ and the utilization of almost any detector from liquid chromatography or gas chromatography.

Wall [23] has discussed recent developments including timed-split injection, extraction and detection systems in SFC.

Time-split injection

Capillary supercritical fluid chromatography utilizes narrow 50μ m or 100μ m i.d. columns of between 3 and 20m in length. The internal volume of a 3m × 50μm i.d. column is only 5.8μl. Supercritical fluid chromatography operates at pressures from 1000kg cm ^{2} to beyond 4000kg cm ^{2}, this means that GC injection systems cannot be used. HPLC injection systems are suitable for those pressure ranges, but even using small internal loop injectors the volume introduced to the column is very large compared to the column's internal volume. To allow injections of about 10–50μl to be introduced to a capillary column, an internal loop LC injector (Valco Inst. Switzerland) has been used with a splitter (Fig. 1.4(a)), which was placed after the valve to ensure that a smaller volume was introduced onto the column. This method works well for compounds which are easily soluble in carbon dioxide at low pressures.

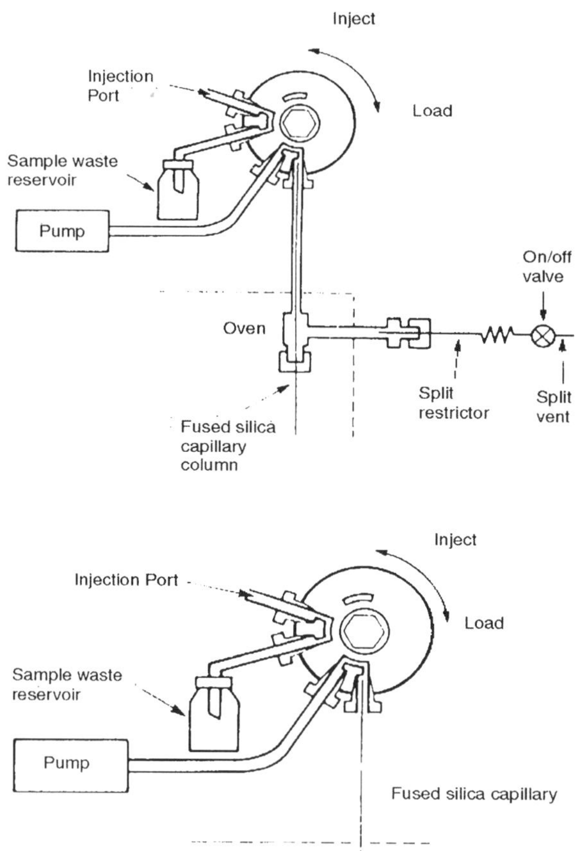

Fig. 1.4 Sample infectors: (a) split valve infector; (b) timed split and direct valve infector.
Source: Own files

However, when compounds with high molecular weights are introduced into the system they are often insufficiently soluble to remain in solution in the depressurization area of the split restrictor. The compounds then reprecipitate in the restrictor and cause a decrease in the internal diameter of the restrictor. Hence this reduces the split ratio and causes more compound to be introduced into the column on the next injection, which means that replicate injections show poor reproducibility.

Good reproducibility has been reported for capillary supercritical fluid chromatography using a direct injection method without a split restrictor. This method (Fig. 1.4(b)) utilized a rapidly rotating internal loop injector (Valco Inst. Switzerland) which remains in-line with the column for only a short period of time. This then gives a reproducible method of injecting a small fraction of the loop onto the column. For this method to be reproducible the valve must be able to switch very rapidly to put a small slug of sample into the column. To attain this a method called timed-split injection was developed (Lee Scientific). For timed split to operate it is essential that helium is used to switch the valve, air or nitrogen cannot provide sharp enough switching pulses. The injection valve itself must have its internal dead volumes minimized. Dead volumes prior to the valve allow some of the sample to collect prior to the loop, effectively allowing a double slug of sample to be injected which appears at the detector as a very wide solvent peak.

Detection systems

Supercritical fluid chromatography uses detectors from both liquid chromatography and gas chromatography. A summary of detection systems used in supercritical fluid chromatography has been documented [24].

One of the most commonly used detection systems in a gas chromatography laboratory is the electron capture detector. The first paper [25] to be published demonstrating the use of an electron capture detector with supercritical fluid chromatography showed that with supercritical fluid chromatography sensitivity to about 50pg minimum detection limit on column was obtainable.

A paper has been published showing the use of the photoionization detector [26]. Polyaromatic hydrocarbons are very sensitive using the photoionization detector and the levels detected did not break any new ground in terms of sensitivity. It did inspire HNS Systems (Newtown MA, USA), who market a photoionization detector, to try the detector with a capillary system, interfaced to a Lee Scientific 602 supercritical fluid chromatography (Lee Scientific, Salt Lake City, Utah, USA).

The photoionization detector is to a certain extent specific in that only compounds that can be ionized by a UV lamp will give a response. The solvents used were dichloromethane and acetonitrile, both of which should have little response in the photoionization detector. However, a clear sharp solvent peak was observed.

The amount detected by this system (0.3pg on column) was below the level which could have been determined using a flame ionization detector. Initial indications show that the photoionization detector may be a very useful detector for people who wish to get to lower levels on the supercritical fluid chromatography and cannot concentrate their sample.

Sulphur chemiluminescence detector The flame photometric detector commonly used in gas chromatography for sulphur specific detection has found little application in supercritical fluid chromatography. Flame photometric detection is not used in SFC because the initial results obtained with SFC using a flame photometric detector showed the response for sulphur-containing species to be very poor. Carbon dioxide has a coincident emission line which cannot be resolved from the main sulphur line, making flame photometric detection almost useless with SFC and carbon dioxide. Other sulphur detectors do exist, such as the sulphur chemiluminescence detector (CD) (Sievers Research Inc., Colorado, USA). The link to supercritical fluid chromatography has been investigated. Good sensitivities and chromatograms have been shown for standards and real samples. This detector shows no response to carbon dioxide and gives low picogram sensitivities for a wide range of sulphur compounds.

The newest developments in supercritical fluid chromatography instrumentation are the Lee Scientific 602 SFC and 622 SFC/GC. These incorporate the latest advanced technology, the latter being a dual-purpose SFC gas chromatographic instrument. They feature a pulseless high-capacity pump, a high-temperature oven for SFC and gas chromatography, compatibility with packed and capillary columns, high-sensitivity detectors (flame ionization, UV, FTIR and MS) and newly developed software capable of creating an infinite variety of simultaneous temperature and density or pressure programmes.

Some SFC chromatograms obtained using the Lee Scientific 501 SFC instrument are shown in Fig. 1.5.

1.1.5.5 Gas chromatography including gas chromatography–mass spectrometry

The identification and determination of traces of organic substances in soil, sediment and sludge samples is a subject that has made tremendous advances in recent years. The demands made on chemists in terms of specificity and sensitivity in carrying out these analyses have become greater and greater with the increasing realization that organic substances from industrial sources are permeating the ecosystem and identification and measurements of minute traces of these are required.

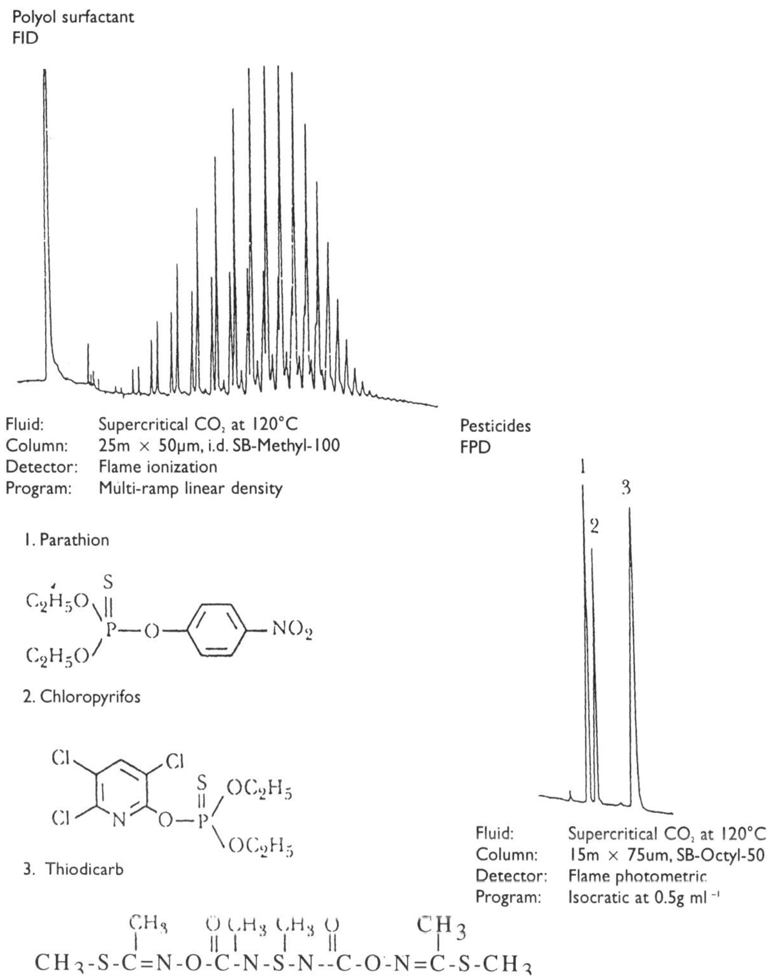

Fig. 1.5 Supercritical fluid chromatography of pesticides.
Source: Own files

For the more volatile components of water samples, i.e. those with boiling points up to about 250°C, gas chromatography has been a favoured technique for several decades. However, with the realization that retention time measurements alone are insufficient to identify organics there has been an increasing move in recent years to connect a gas chromatograph to a mass spectrometer in order to provide unequivocal identifications. Element-specific detectors are another recent development.

A limitation of gas chromatography is that it cannot handle less volatile compounds and these comprise a high proportion of the total organics content of the sample. For this reason increasing attention is being paid to the application of high-performance liquid chromatography. Again, when positive identifications are required, a mass spectrometer is connected to the outlet of the chromatograph.

This technique has been applied to the determination of:

In soils: aliphatic and polyaromatic hydrocarbons*; phenols; chloroaliphatic hydrocarbons*; chlorophenols*; methyl bromide fumigant, polychlorobiphenyls*; polychlorodibenzo-o-dioxins*; polychlorobenzofurans*; alklylphosphates*; tetrahydrothiophene*; chlorinated insecticides*; carbamate* and organophosphorus insecticides; triazine*; substituted urea and phenoxy acetic acid types of herbicides; Picloram; acaral; 2,6 dichlorobenzo-nitrile; paraquat; diquat; cyperquat*; dicamba*; Bromacil; diclofopmethyl; Diclofop; sencor, quinozoline* and Trifluralin insecticides, mixtures of insecticides and herbicides, fungicides mestranol, methoxy groups, mixtures of organic compounds, halogens, total phosphorus and sulphur, organic compounds of arsenic, lead, mercury and tin.

In non-saline sediments: aliphatic and polyaromatic hydrocarbons, phthalate esters; carboxylic acids, uronic acid; aldoses chloroaliphatics*; haloaromatics*; chlorophenols; chloroanisoles; polychlorobiphenyls; polychlorodibenzo-p-dioxins*; polychloro-dibenzofurans*; various organosulphur compounds, chlorinated insecticides, organophosphorus insecticides; mixtures of organic compounds*; triazine herbicides*; arsenic and organic compounds of mercury and tin.

In saline sediments: aliphatic and polyaromatic hydrocarbons*; carbohydrates; haloaromatic compounds; chlorophenols; basic nitrogen compounds*; various organosulphur compounds; mixtures of organic compounds*; total sulphur; arsenic and organic compounds of lead, mercury and tin*.

In sludge: anionic and non-ionic surfactants*; carboxylic acids; ßhydroxybutyrate; hydroxy valerate; chloroaliphatic compounds; chlorophenols; polychlorobiphenyls*; 4-nitrophenol*; mixtures of organic compounds*; chlorinated insecticides, phenoxy acetic acid type herbicides and organotin compounds.

The asterisked analyses were carried out by a combination of gas chromatography and mass spectrometry.

The basic requirements of a high-performance gas chromatography are:

- sample is introduced to the column in an ideal state, i.e. uncontaminated by septum bleed or previous sample components, without modification due to distillation effects in the needle and quantitatively, i.e. without hold-up or adsorption prior to the column;
- the instrument parameters that influence the chromatographic separation are precisely controlled;
- sample components do not escape detection; i.e. highly sensitive, reproducible detection and subsequent data processing are essential.

There are two types of separation column used in gas chromatography – capillary columns and packed columns.

Packed columns are still used extensively, especially in routine analysis. They are essential when sample components have high partition coefficients and/or high concentrations. Capillary columns provide a high number of theoretical plates, hence a very high resolution, but they cannot be used in all applications because there are not many types of chemically bonded capillary columns. Combined use of packed columns of different polarities often provides better separation than with a capillary column. It sometimes happens that a capillary column is used as a supplement in the packed-column gas chromatography. It is best, therefore, to house the capillary and packed columns in the same column oven and use them selectively. In the screening of some types of samples, the packed column is used routinely and the capillary column is used when more detailed information is required.

Conventionally, it is necessary to use a dual column flow line in packed-column gas chromatography to provide sample and reference gas flows. The recently developed electronic base-line drift compensation system allows a simple column flow line to be used reliably.

Recent advances in capillary column technology presume stringent performance levels for the other components of a gas chromatograph as column performance is only as good as that of the rest of the system. One of the most important factors in capillary column gas chromatography is that a high repeatability of retention times be ensured even under adverse ambient conditions.

These features combine to provide ±0.01min repeatability for peaks having retention times as long as 2h (other factors being equal).

Another important factor for reliable capillary column gas chromatography is the sample injection method. Various types of sample injection ports are available. The split/splitless sample injection port unit series is designed so that the glass insert is easily replaced and the septum is continuously purged during operation. This type of sample injection unit is quite effective for the analysis of samples having high boiling point compounds as the major components.

In capillary column gas chromatography, it is often required to raise and lower the column temperature very rapidly and to raise the sample injection port temperature. In one design of gas chromatography, the Shimadzu GC 14–A, the computer-controlled flap operates to bring in the external air to cool the column oven rapidly – only 6min from 500°C to 100°C. This computer-controlled flap also ensures highly stable column temperature when it is set to a near-ambient point. The lowest controllable column temperature is about 26°C when the ambient temperature is 20°C.

Some suppliers of gas chromatography are listed in Table 1.7.

Shimadzu gas chromatographs

This is a typical high-performance gas chromatography version (see Table 1.7 for further details).

The inner chamber of the oven has curved walls for smooth circulation of air; the radiant heat from the sample injection port units and the detector oven is completely isolated. These factors combine to provide demonstrably uniform temperature distribution. (The temperature variance in a column coiled in a diameter of 20cm is less than ±0.75°K at a column temperature of 250°C).

When the column temperature is set to a near ambient temperature, external air is brought into the oven via a computer-controlled flap, providing rigid temperature control stability. (The lowest controllable column temperature is 24°C when the ambient temperature is 18°C and the injection port temperature is 250°C. The temperature fluctuation is less than ±0.1°K even when the column temperature is set at 50°C.

This instrument features five detectors (Table 1.7). In the flame ionization detector, the high-speed electrometer, which ensures a very low noise level, is best suited to trace analysis and fast analysis using a capillary column.

Samples are never decomposed in the jet, which is made of quartz.

Carrier gas, hydrogen, air and make-up gas are separately flow-controlled. Flow rates are read from the pressure flow-rate curves.

In the satellite system, one or more satellite gas chromatographs (GC–14 series) are controlled by a core gas chromatography (e.g. GC 16A series). Since the control is made externally, the satellite gas chromatographs are not required to have control functions (the keyboard unit is not necessary).

When a GC 16A series gas chromatograph is used as the core, various laboratory-automation-oriented attachments such as bar-code reader and a magnetic-card reader become compatible: a labour-saving system can be built, in which the best operational parameters are automatically set. Each satellite gas chromatograph (GC 14A series) operates as an independent instrument when a keyboard unit is connected.

The IC card operated gas chromatography system consists of a GC–14A series gas chromatograph and a C–R5A Chromatopac data processor. All of the chromatographic and data processing parameters are automatically set simply by inserting the particular IC card. This system is very convenient when one GC system is used for the routine analysis of several different types of samples.

One of the popular trends in laboratory automation is to arrange for a personal computer to control the gas chromatography and to receive data from the GC to be processed as desired. Bilateral communication is made via the RS–232C interface built in a GC 14A series gas chromatograph. A system can be built to meet requirements.

A multidimensional gas chromatography system (multi-stage column system) is effective for analysis of difficult samples and can be built up by connecting several column ovens, i.e. tandem GC systems, each of which has independent control functions such as for temperature programming.

The Shimadzu GC 15A and GC 16A systems are designed not only as independent high-performance gas chromatographs but also as core instruments (see above) for multi-gas-chromatography systems (i.e. several gas chromatographs in the laboratory linked to a central management system) or computerized laboratory automation systems. The GC 16A has a keyboard, the GC 15A does not. Other details of these instruments are given in Table 1.7. The Shimadzu GC 8A range of instruments do not have a range of built-in detectors but are ordered either as temperature programmed instruments with TCD, FID or FPD detectors or as isothermal instruments with TCD, FID or ECD detectors (Table 1.7).

Perkin–Elmer supply a range of instruments including the basic models 8410 for packed and capillary work and the 8420 for dedicated capillary work, both supplied on purchase with one of the six different types of detection (Table 1.7). The models 8400 and 8500 are more sophisticated capillary column instruments capable of dual detection operation with the additional features of keyboard operation. Screen graphics method storage, host computer links, data handling and compatibility with laboratory automation systems. Perkin–Elmer supply a range of accessories for these instruments including an autosampler (AS–8300), an infrared spectrometer interface, an automatic headspace accessory (HS101 and H5–6), an autoinjector device (AI–I), also a catalytic reactor and a pyroprobe (CDS 190) and automatic thermal desorption system (ATD–50) (both useful for examination of sediments).

The Perkin–Elmer 8700, in addition to the features of the models 8400 and 8500, has the ability to perform multi-dimensional gas chromatography.

The optimum conditions for capillary chromatography of material heart cut from a packed column demand a highly sophisticated programming system. The software provided with the model 8700 provides this, allowing methods to be linked so that pre-column and analytical column separations are performed under optimum conditions.

Table 1.7 Commercial gas chromatographs

Manufacturer	*Model*	*Packed column*	*Capillary column*	*Detectors*	*Sample injection point system*	*Keyboard control*	*Link to computer*	*Visual display*	*Printer*	*Core instrument amenable to tap automation*	*Temperature programming/*	*Cryogenic unit (subambient chromatography)*
Shimadzu	GC–14A	Yes	Yes	FID ECD FTD FPD TCD (all supplied)	1 Split-splitters 2 Glass insert for single column 3 Glass insert tol dual column 4 Cool on column system unit 5 Moving needle system 6 Rapidly ascending temperature vaporizer	Yes	Yes	No	No	No	Yes/Yes	No
Shimadzu	GC 15A	Yes	Yes	FID ECD FTD FPD TCD	1 Split-splitters 2 Direct sample injection (capillary column) 3 Standard sample injector (packed column) 4 Moving pre-column system (capillary columns) 5 On column (capillary columns)	No	Yes	Yes	Yes	No	Yes/Yes	No
Shimadzu	GC 16A		Yes	FID	Split-splitters FCD	Yes	Yes	Yes	Yes	No	Yes/Yes	No

				dual detector instrument. (Detectors chosen from above list)								
Nordion	Micromat HRCG 412	No	Yes	Dual simultaneous detector combinations from the following: FID ECD FTD Photoionization Hall E.C.	1 Split-splitless 2 On-column injector	Yes	Yes	Yes	Yes	–	Yes/Yes	No
Siemens	SiChromat 1–4 (single oven) SiChromat 2–8 (dual oven) for multi-dimensional GC)	Yes	Yes	FID ECD FTD FPD TCD Helium detector	1 Liquid–liquid packed columns 2 Split-splitless 3 Temperature programmable 4 On-column 5 Liquid injector valve on-line 6 Gas injection valve 7 Rotary as injection valve	No	No	Yes	Yes	–	Yes/Yes	No

Source: Own files

Table 1.7 continued

Manu-facturer	*Model*	*Packed column*	*Capillary column*	*Detectors*	*Sample injection point system*	*Keyboard control*	*Link to computer*	*Visual display*	*Printer*	*Core instrument amenable to tap automation*	*Temperature programming/*	*Cryogenic unit (sub-ambient chroma-tography)*
				FID ECD FTD FPD TCD	2 Split-splitless injector 3 Direct on column injector							
Perkin–Elmer	8400 and 8500	Yes	Yes	Dual detector instrument (detectors chosen from following) FID ECD FTD FPD TCD	Can be fitted with any combination of above injection systems	Yes	Yes	Yes	Yes (CP100 printer plotter	Yes	Yes/Yes	Yes, down to –80°C
Perkin–Elmer	8700	Yes	Yes	FID ECD FPD TCD Hall E.C. photo-ionization	1 Flash vaporization 2 Split-splitless 3 Programmable temperature vaporizer 4 Gas sampling valve 5 Liquid sampling valve	Yes	Yes	Yes	Yes	–	Yes/Yes	Yes

				FTD FPD TCD (all supplied)								
Shimadzu	GC 8A	Yes	Yes	FID FCD FPD TCD Single detector instruments (detector chosen on purchase	1 Point for packed columns 2 Point for capillary columns 3 Split-splitters	Yes Not built in	Optional	No	Optional	No	Yes temp programming GC 8APT (TCD detector GC 8APF (FID detector) (FID detector) isothermal; GC 8AIT (TCD detector) GC 8AIF (FID detector) GC 8AIE (ECD detector)	No
Perkin–Elmer	8410	Yes	Yes	Single detector instrument (detector chosen on purchase) FID ECD FTD FPD TCD	1 Flash vaporization 2 Split-Splitless injector 3 Manual or automatic gas sampling valves 4 Manual or automatic liquid sampling valves	No	No	No	No	No	Yes/Yes	Yes, down to –80°C
Perkin–Elmer	8420	No	Yes	Single detector instrument (chosen on purchase)	1 Programmable temperature vaporizer	No	No	No	No	No	Yes/Yes	Yes, down to –80°C

Following the first run, in which components are transferred from the pre-column to the on-line cold trap, the system will reset to a second method and, on becoming ready, the cold trap is desorbed and the analytical run automatically started.

Other applications of the model 8700 system include fore-flushing and back-flushing of the pre-column, either separately or in combination with heart cutting, all carried out with complete automation by the standard instrument software.

There are many other suppliers of gas chromatography equipment, some of which are discussed further in Table 1.7.

Gas chromatography–mass spectrometry

The time has long since passed when one could rely on gas chromatographic or liquid chromatographic data alone to identify unknown compounds in environmental samples. The sheer number of compounds present in such materials would invalidate the use of these techniques, and even in the case of simple mixtures the time required for identification would be too great to provide essential information in the case, for example, of accidental spillage of an organic substance into a water course or inlet to a water treatment plant where information is required very rapidly.

The practice nowadays is to link a mass spectrometer or ion trap to the outlet of the gas chromatograph or high-performance liquid chromatograph so that a mass spectrum is obtained for each chromatographic peak as it emerges from the separation column. If the peak contains a single substance then computerized library-searching facilities attached to the mass spectrometer will rapidly identify the substance. If the emerging peak contains several substances, then the mass spectrum will indicate this and in many cases will provide information on the substances present.

The use of gas chromatography–mass spectrometry grew rapidly during the early 1970s as discussed by Shackleford and McGuire [27].

The first large-scale application of gas chromatography–mass spectrometers to analysis of environmental pollutants occurred in 1977 when the effluent guidelines division of EPA, under court order, began collecting data and writing regulations to limit the discharge of pollutants into surface waters. Tellaid [28] and others (Shackleford and McGuire [27]; Federal Register [29–31]; Colby [32]; Fisk *et al.* [33]; Friedman [34]) give a history of the selection of the EPA priority pollutants, the selection of gas chromatography–mass spectrometry as the technique of choice for their analysis and the problems faced in moving a research technique into production.

In addition to the processing work of the EPA many individual laboratories throughout the world concerned with the analysis of sediments, soils and water have set up equipment for gas chromatography–mass spectrometry, and what was once considered to be an expensive instrument purchased perhaps to solve a single vitally important problem has become a general workhorse instrument. No water laboratory which aims to be able to solve the kinds of problems thrust upon it can afford to be without a bench-top instrument. Instead, the problem is one of choosing the most appropriate instrumentation for their needs.

Flanigan MAT are the main suppliers of this equipment. Available instrumentation is discussed below.

SSQ 70 series single-stage quadrupole-mass spectrometer

This offers premium single-stage performance, with the option of being upgraded to a triple-stage quadrupole system (i.e. the TSQ 70). The SSQ 70 features a network of distributed microprocessors with more than 1.5 megabytes of memory linked to a powerful DEC 11/73 processor with 2.0 megabytes of memory for data-processing operations. Instrument control links can be displayed in up to eight windows on a colour display terminal. The hyperbolic quadrupole analyser gives the SSQ 70 a mass range of up to 1000μm; system performance is specified to 200m/z. The cradle vacuum system with three large inlet points at the ion source accommodates a variety of sample inlets such as capillary gas chromatography, thermospray, liquid chromatography, mass spectrometry, supercritical fluid chromatography and solids probe. Standard features of the instrument also include high-performance EI/Cl (electron impact/ chemical ionization) ion-source with exchangeable ion volumes, a PPI NICI with high-voltage conversion dynode multiplier for positive and negative ion detection and fast ion bombardment. The Varian 3400 GC gas chromatography incorporates a high-performance capillary column with multilinear temperature programming in up to eight sequences, a data-control and recording system for temperatures in the gas chromatography oven and for interface temperatures, and also for controlling and recording value timing, a data system control of optional gas chromatography accessories and a split/splitless capillary injector. The Micro VIP computer data system comprises a DEC 11/73 processor with video colour display, dot matrix printer and a data system for control of instrument control parameters and user-initiated diagnostics.

Mass spectrometry–mass spectrometry

In high-performance mass spectrometry–mass spectrometry (as opposed to gas chromatography–mass spectrometry) the separator as well as the analysis is performed by the mass spectrometer. One advantage of this technique over combined chromatography–mass spectrometry is that separation is a spatial process rather than being dependent on time. This can lead to improved analysis times and/or greater specificity. Mass spectrometry–mass spectrometry also opens up other areas such as the study of complete structures. This technique has been discussed in detail by Warburton and Millard [35].

H–SQ 30 hybrid mass spectrometer–mass spectrometer

This instrument combines a reverse-geometry (BE) magnetic section instrument with a quadrupole (QQ) analyser. This hybrid combination provides mass spectrometry–mass spectrometry operation with a high-resolution first stage (BE) and a unit resolution second stage (QQ). The four available collision regions allow experiments of low (2–100eV) and high (3keV) collision energy, as well as consecutive CID experiments using two separate collision regions. The H–SQ 30 is an ideal instrument for structural elucidation studies and ion physics.

MAT–90 high-mass–high-resolution mass spectrometer

This is a very high-performance instrument in which instrument control resides in a multiprocessor system manager leaving only the analytically important parameters to be defined by the operator. It utilizes a completely new concept of ion optics for double focusing and this gives the instrument unmatched performance.

The performance of a magnetic sector mass spectrometer depends totally on the ability to focus ions from source to detector. To produce ideal focusing a very wide range of factors must be taken into account. Modern computer simulation techniques have now been extensively applied in this instrument and have resulted in an ion optical design closer to the ideal than ever before. This configuration provides for complete image error correction in all planes.

System resolution in excess of 50 000 is achieved and excellent performance is obtained at high masses. The instrument features a novel ion source which can be exchanged in a few seconds via vacuum lock. Optimized EI and CI systems are supplied.

Optional ionization volumes are available for fast ion bombardment and alternating CC/EI. The Finnigan MAT–90 analyser has reverse Nier–Johnson geometry allowing metastable studies to be carried out using both first-order and second-order field free regions in the standard system. To extend the application of metastable techniques, the optional collision cell in the first field-free region can be used. A full range of accessories is available, including direct-probe, fast direct-probe, thermospray on-line mass spectrometry, automatic direct evaporation, fast ion bombardment, direct chemical ionization and continuous-flow fast ion bombardment.

FD/FI device

The standard MAT 90 ion source is used for optimized FD/Fl mode by means of the newly designed FD/Fl probe. Conversion from electron impact (EI), chemical ionization (CI) or fast ion bombardment (FAB) to FD/FI operation does not require the exchange of the ion source. The FD/Fl probe accommodates both the field emitter and the extraction electrodes, mounted at the probe tip. Both are introduced as a unit into the ion source through the ionization volume exchange lock without breaking vacuum. The fast and simple changeover illustrates the versatility of the Finnigan MAT 90 with no compromise on the performance.

All aspects of system control, data acquisition and processing are carried out in the integrated multiprocessor system. The primary processor is a DEC POP 11/73 with 2 megabytes (optional up to 4 megabytes) of main memory with cache and disk cache memory. The user interface is via a high-resolution colour terminal. Standard features also include computer-controlled variable entrance and exit slits, electron multiplier with ±20kV diode, direct coupled capillary gas chromatography–mass spectrometry interface with precise temperature control up to 400°C data system software including library search, quantification and data handling.

Series 700 ion-trap detector

The ion-trap detector detects any compound that can be chromatographed; it is a universal detector that can replace several conventional gas chromatography detectors such as the type used in the Varian model 3400 gas chromatogram included in the Finnigan MAT SSA–70 and TSQ–70 instruments. Electron capture, flame ionization, element specific (etc.) detectors used in the latter instruments are not universal in this sense and will not respond to all types of organic compounds, i.e. some compounds will be missed. The ion-trap detector obviates this difficulty by responding to all types of organic compounds. In the ion-trap technique one does not have to rely on retention data for identification. The mass spectrum tells you the identity with certainty. Unidentified gas chromatographic peaks are a thing of the past. Complete analysis and identification is done in one run with one detector. This makes the ion-trap detector a very attractive proposition to the water chemist involved in the analysis of environmental soiled samples such as soils, sediments and sludges. Various aspects of ion-trap detectors have been discussed by workers at Finnigan MAT and elsewhere [36–68].

During development of the ion-trap detector, it was found that the low voltage previously used for storage encouraged the production and storage of the H_2O^+ and H_3O^+ ions which occasionally led to an increase of the $M + 1$ molecular ions. This problem had been eliminated by adjusting the storage voltage such that the H_2O^+ and H_3O^+ ions are no longer stored.

This scanning method produces standard electron impact spectra which can be rapidly searched through the standard NBS library (42 222 spectra). In each case the number of ions stored in the trap would be the optimum required to produce a conventional electron impact spectrum. In order that the procedure will not affect quantitative results, ion intensities are stored after application of an adjustment factor which is always related to the true size of the peak as measured by the original fast scan. Scaling is controlled by the computer and the net result is a system with a dynamic range between 10^4 and 10^5. The efficiency of the procedure has been evaluated by measuring the signals obtained from injections of a difficult compound in quantities ranging from 10pg to 10ng. Each result has been measured three times and a log/log plot of signals against concentration shown to be a straight line over the entire range with a correlation coefficient >99%.

The ion-trap detector may be operated both as a universal detector (when full scans are stored) or, with the application of multiple ion monitoring, as a specific detector. Because approximately 50% of the ions formed in the trap are analysed, the sensitivity of the instrument in full-scan mode can be much higher than conventional mass spectrometers, in which only 0.1–0.2% of the ions formed may be detected. Thus the instrument can be used to detect 2–5pg (in full scan) of compounds eluting from the column; a performance which compares extremely favourably with those of the most sensitive specific detectors (e.g. the electron capture detector) and easily outstrips that of the flame ionization detector. As already indicated, this sensitivity is not achieved at the expense of dynamic range; as the instrument can produce linear calibration graphs for quantities within the range 5–10pg to 1000ng on column. This again compares favourably with the performance of the flame ionization detector.

When operated as a specific detector the ion-trap detector is more sensitive still but not to the extent that would be expected from the performance of other mass spectrometers operated in this mode; in view of the large number of ions monitored in full scan mode there is little more sensitivity to be gained by spending a little extra time scanning a narrow mass range, and the detection limit in this mode is in the region of 1–2pg.

The power of the system to overcome the problems associated with co-eluting compounds is demonstrated in conjunction with the use of deuterated (or ^{13}C–labelled compounds) as internal standards. Such techniques could not be used in conventional gas chromatography as the deuterated compounds often co-elute, making quantification difficult if not impossible. With the ion-trap detector, however, it is easily possible to differentiate between the ions arising from the different compounds and the intensities of these ions could then be used for quantification of the compounds involved. The application of such techniques can be shown by the quantitation of anthracene. Chromatograms of ions characteristic of anthracene and its deuterated analogue (*m/e* 178 and 188 respectively) indicated that the compounds did not in fact elute simultaneously. The retention time of the labelled compound was fractionally less than that of the unlabelled material.

In Fig. 1.6(a) is shown a partial chromatogram of a complex mixture of chlorinated biphenyls extracted from a sediment sample.

The signals from masses 292 and 326 characteristic of tetra- and pentachlorobiphenyl are shown in Fig. 1.6(b,c). The specific detection mode of the ion-trap detector can be used to improve detection limits. This detector can monitor specific masses that are characteristic of compounds of interest. The detector records the signal for only those masses and ignores all others. Interference from other compounds is virtually eliminated with the Finnigan MAT 700 detector – up to 16 different groups of masses can be monitored or a mass range of up to 40 masses can be handled. With this flexibility it is possible to monitor only the masses of interest and to improve detection limits.

Incos–50 quadrupole mass spectrometer

The Incos–50 is a relatively low-cost benchtop instrument as opposed to the research grade instruments discussed earlier. The gas chromatography–mass spectrometer transfer lines allow it to be used with either the Hewlett Packard 5890 or the Varian 3400 gas chromatographs. The Incos 50 provides data system control of the gas chromatography and accessories such as autosampler or liquid sample concentration. It can be used with capillary, wide-bore or packed columns. It performs electron ionization or chemical ionization with positive or negative detection. It also accepts desorption or other solids controls.

Finnigan MAT Chem Master Workstation

The Chem Master Workstation is a gas chromatography and gas chromatography–mass spectrometry data-processing system that speeds the flow of data through the laboratory and provides essential quality-assurance and quality-control review. It is a PC-based integrated hardware/software system that converts gas chromatographic and gas chromatography–mass spectrometric data into reliable analytical reports.

Model 1020 routine gas chromatography–mass spectrometer

This is a cost-effective completely automated system optimized for the routine analysis of complex organic samples. It is specifically designed to

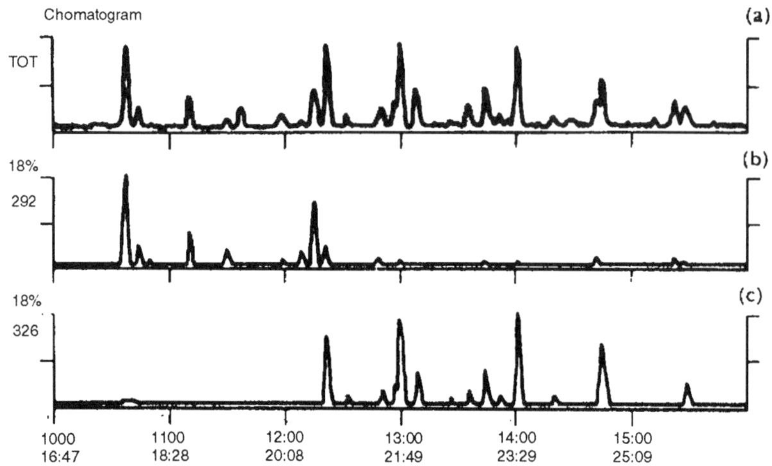

Fig. 1.6 Specific detection. Trace (a) shows a partial chromatogram for Arochlor 1254. Trace (b) shows the mass chromatogram for 292 characteristic of tetrachlorobiphenyls in the mixture. Trace (c) shows the mass chromatogram for 326 characteristic of pentachlorobiphenyls
Source: Own files

meet the needs of an analytical laboratory requiring a gas chromatography mass spectrometer with the following characteristics:

- high sample throughput
- low initial investment
- low operating costs
- ease of automation
- complete software package
- serviceability
- field-proven hardware

The model 1020 software package includes interactive programmes specifically designed for complex mixture analysis and advanced automated programmes for routine analysis. All system functions are computer controlled with minimal knowledge of mass spectrometry.

All gas chromatography parameters, including temperature programme rates and hold times, are controlled by the microprocessor and set through the CRT keyboard. Up to five sets of parameters can be stored on the computer disk for instant recall.

The mass spectrometer, when combined with a computer data system, precisely identifies and quantifies each sample component as it elutes from the gas chromatography. The model 1020 uses an electron ionization source to produce mass fragments, and a quadrupole mass filter, one of the most accurate and cost-effective devices for separating mass fragments.

For most applications, excellent spectra are produced with only a few nanograms of sample. When performing single-ion monitoring picogram and femtogram levels of many compounds can be precisely quantified. In addition to single-ion monitoring, the system's powerful software permits multiple ion monitoring of up to 25 mass ranges. This improves sensitivity and reliability of compound identifications by allowing a combination of multiple-ion mass ranges as well as single-ion monitoring to be performed in a single analysis. Mass stability of better than ±0.1μg per day ensures accurate mass assignment.

OWA–20/308 gas chromatograph–mass spectrometer

This system combines hardware and software features not found in any other low-cost gas chromatography–mass spectrometry system. The highly reliable 3000 series electron ionization source and quadrupole analyser are used to provide superior mass spectrometer performance. The software is designed with the necessary automation to perform complete quantitative analysis of any target compounds. All routine system operating parameters are adjustable through the computer's graphics display terminal. The priority interrupt foreground/background operation system allows all data-processing functions to be performed at any time with no limiting effects on data acquisition. Sophisticated data-processing programmes are readily accessible through a simple commercial structure. The simplicity of the entire system allows complete analysis with minimal operator training. Standard features of this instrument include fully automated gas chromatography–mass spectrometry, automated compound analysis and quantification, software, 4–800μ electron impact quadrupole mass spectrometer, high-capacity turbidmolecular pump vacuum system, liquid sample concentrators for volatile organics in water analysis, a sigma series programmable gas chromatography, grob-type split-splitless capillary column injector system, packed column injector with glass jet separator, Nova 4C/53K word, 16-bit minicomputer, graphics display terminal, 10-megabyte disk drive, a printer/plotter, an NBS 31000 spectra library, a full scan or multiple ion detector and a 9-track tape drive. Options include chemical ionization ion source, direct inlet vacuum lock, programmable solids probe, direct exposure probe, various GC detectors, autosampler, subambient GC operation and a 32-megabyte disk drive.

Applications

Some of the many applications of the technique are summarized in Table 1.8.

Table 1.8 Applications of gas chromatography–mass spectrometry to water analysis

Title	*Types of compounds*	*Reference*
Analysis of volatile organic compounds in industrial wastes by the Finnigan OWA GC/MS	27 halohydrocarbons and aromatics (purge and trap)	69
GC/MS method for analysis of pollutants in hazardous wastes	Volatile (purgeable) semi-volatile (solvent extractable insecticide/PCB (residue) fractions in soil, water and sediments (126 compounds mentioned)	33
GC/MS methods for the analysis of solid waste	Wide variety of organics in sludge or waste	34
Use of pollutant and biogenic markers as source discriminants of organic inputs to estuarine sediments	–	70

Source: Own file

As an example of the application of gas chromatography–mass spectrometry, Fig. 1.7 shows a reconstructed chromatograph obtained for an industrial sludge. The Finnigan MAT 1020 instrument was used in this work. Of the 27 compounds searched for, 15 were found. These data were automatically quantified. This portion of the report contains the date and time at which the run was made, the sample description, who submitted the sample and the analyst, followed by the names of the compounds. If no match for a library entry was found, the component was listed as 'not found'. Also shown is the method of quantification and the area of the peak (height could also have been chosen).

The large peak at scan #502 (Fig. 1.7) does not interfere with the ability of the software to quantify the sample. Although the compound eluting at scan #502 was not one of the target compounds in the library being reverse-searched, it was possible to identify it by forward-searching the NBS library present on the system. The greatest similarity was in the comparison of the unknown with the spectrum of benzaldehyde.

1.1.5.6 Purge and trap gas chromatography

This technique has been applied to the determination of aromatic hydrocarbons, alcohols, aldehydes, ketones, chloroaliphatic compounds, haloaromatic compounds, acrylonitrile, acetonitrile, mixtures of organic compounds and tetrahydrothiophene in soils, chloroaliphatic and haloaromatic compounds and organotin compounds in non-saline sediments, and organotin compounds in saline sediments.

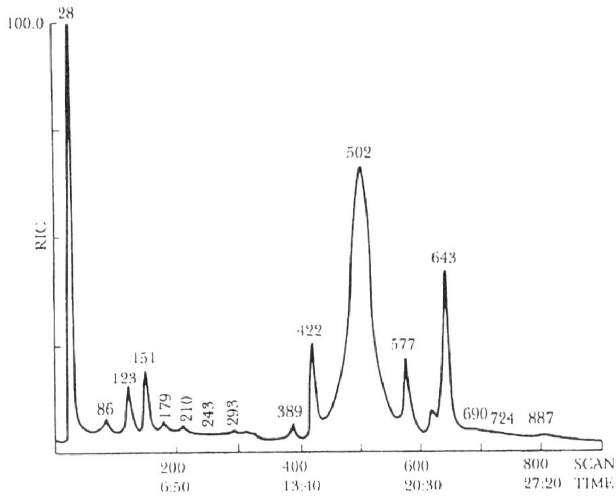

Fig. 1.7 Reconstructed ion chromatography of industrial waste sample
Source: Own files

This is an alternative technique to headspace analysis for the identification and determination of volatile organic compounds in water. The sample is purged with an inert gas for a fixed period of time. Volatile compounds are sparged from the sample and collected on a solid sorbent trap – usually activated carbon. The trap is then rapidly heated and the compounds collected and transferred as a plug under a reversed flow of inert gas to an external gas chromatograph. Chromatographic techniques are then used to quantify and identify sample components.

Instrumentation

OIC Analytical Instrument supply the 4460A purge and trap concentrator. This is a microprocessor-based instrument with capillary column capability. It is supplied with an autosampler capable of handling 76 sample vials. Two automatic rinses of sample lines and vessel purge are carried out between sample analyses to minimize carry-over.

Tekmar are another supplier of purge and trap analysis equipment. Their LSC 2000 purge and trap concentrator features glass-lined stainless steel tubing, a menu-driven programming with four-method storage and a cyrofocusing accessory.

Cyrofocusing is a technique in which only a short section of the column or a pre-column is cooled. In its simplest form a section of the column near the inlet is immersed in a flask of coolant during desorb. After desorb the coolant is removed and the column allowed to return to the oven temperature.

Performance aspects of volatiles organics analysis by purge and trap capillary column gas chromatography with flame ionization detectors has been discussed by Westendorf [71].

1.1.5.7 Pyrolysis gas chromatography including mass spectrometry

Non volatile organic compounds are not amenable for gas chromatography. However, some types of non volatile compounds, upon pyrolysis, yield volatile products which are characteristic of the original substance and can be used as a basis of a method for estimating these substances.

This technique has been applied to the determination of heteroaromatic compounds, anthropogenic hydrocarbons, polymers, haloaromatic compounds in soils, polyaromatic hydrocarbons, cationic surfactants and polychlorobiphenyls and mixtures of organic compounds in non-saline sediments and bacteria identification in sludges.

In the case of anthropogenic hydrocarbons in non-saline sediments and mixtures of organic compounds in soil the technique has been further refined by combining it with mass spectrometry (for further discussion see section 1.1.5.5 under gas chromatography–mass spectrometry).

1.1.5.8 Conventional column chromatography

Despite the advances made in high-performance liquid chromatography in recent years, there are still occasionally applications in which conventional column chromatography is employed. These methods lack the sensitivity, resolution and automation of HPLC. They include the determination of urea herbicides in soil, polyaromatic hydrocarbons, carbohydrates, chloroaliphatic compounds and humic and fulvic acids in non-saline sediments. The technique has also been applied in sludge analysis, e.g. aliphatic hydrocarbons and carboxylic acids.

1.1.6 Combustion methods

It is desirable to be able, in the case of particular elements, to be able to determine the total element content of the sample. Thus, in addition to nitrate, nitrite and ammonium it is frequently required to determine total nitrogen in the sample. In the case of halogens, for example, in addition to determining individual halogen-containing compounds, e.g. haloforms, it may be required to determine total halide or total organo-halogen. In addition to water samples measurements of total element might be required on solid samples such as river or oceanic sediments.

Available commercial instrumentation for the determination of the following total element is discussed below:

- halogen
- sulphur
- halogens and sulphur
- nitrogen
- carbon, hydrogen and nitrogen
- nitrogen, carbon and sulphur

Combustion methods have been used to determine total sulphur and total organic carbon and total halides in soil, total and particulate organic carbon, total halide, phosphorus and nitrogen, total and particulate organic carbon in saline sediments and total nitrogen in sludges.

Total halide

The Dohrmann DX 20B system is based on combustion of the sample to produce the hydrogen halide, which is then swept into a micro-coulometric cell and estimated. It is applicable at total halide concentrations up to $1000 \mu l^{-1}$ with a precision of ±2% at the $10 \mu g\ L^{-1}$ level. The detection limit is about $0.5 \mu g\ L^{-1}$. Analysis can be performed in 5 min. A sample boat is available for carrying out analysis of solid samples. The instrument has been applied to waste waters, soils and sediments.

The Dohrmann DX–20A system is the DX 20B system with an additional module which makes it possible to measure total organic halides including chlorine, bromine and iodine. It features mini-column extraction of the sample with granulated activated carbon to preconcentrate organic halides at ultra-trace levels prior to combustion of the concentrate. An optional gas sparger attachment is available for determining purgeable organic chlorine compounds. Inorganic halides are removed from sample extracts by a nitrate wash so that only organic halides are reported.

This instrument, therefore, has full capability for measurements in liquid or solid samples of total organic halogen (TOX), purgeable (volatile) organic halogen (POX), extractable organic halogen (EOX) and total halogen (TX).

Mitsubishi also supply a microprocessor-controlled automatic total halogen analyser (model TOX–10) (Fig. 1.8(b)) which is very similar in operating principles to the Dohrmann instruments discussed above, i.e. combustion at 800–900°C followed by coulometric estimation of hydrogen halide produced.

Recoveries of halogenated organics range from 92% (1,2 dibromo-ethane) to 105% (*m*-chlorobenzoic acid).

Sulphur

The Mitsubishi trace sulphur analyser models TS–02 and TN–02(S) is again a microcombustion procedure in which sulphur is oxidized to

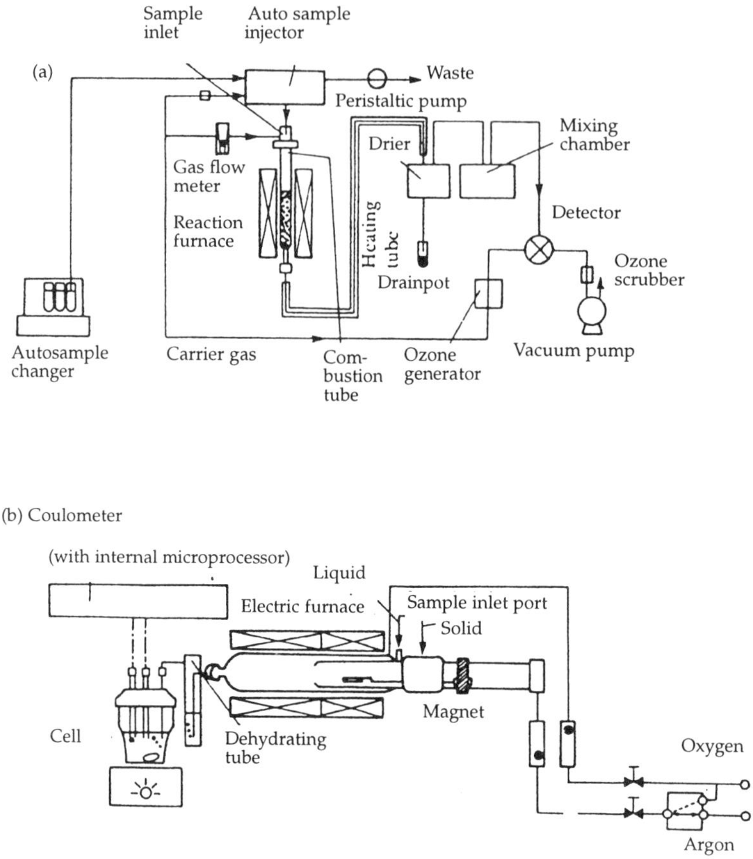

Fig. 1.8 Elemental analysis supplied by Mitsubishi (a) TN–05 nitrogen analyser; (b) TOX–10 total halogen analyser
Source: Own files

sulphur dioxide, which is then titrated coulometrically with triiodide ions generated from iodide ions:

$$SO_2 + I_3^- + H_20 \rightarrow SO_3 + 3I^1 + 2H^+$$

$$3I^1 \rightarrow I_3^- + 2e^-$$

Total sulphur/total halogen

The Mitsubishi TSX–10 halogen-sulphur analyser expands the technology of the TOX–10 to include total chlorine and total sulphur measurement. The model TSX–10, which consists of the TOX–10 analyser module and a sulphur detection cell, measures total sulphur and total chlorine in liquid and solid samples over a sensitivity range mg L^{-1} to percent.

Dohrmann also produce an automated sulphur and chlorine analyser (models MCTS 130/120). This instrument is based on combustion micro-coulometric technology.

Total bound nitrogen

Mitsubishi supply two total nitrogen analysers: the model TN–10 and the model TN–05 microprocessor control chemiluminescence total nitrogen analysers (Fig. 1.8(a)).

This instrument measures down to $\mu g\ L^{-1}$ amounts of nitrogen in solid and liquid samples.

The sample is introduced into the combustion tube packing containing oxidative catalyst under oxygen carrier gas. High-temperature oxidation (800–900°C) occurs and all chemically bound nitrogen is converted to nitric oxide (NO), R–N $\rightarrow$ CO_2 + NO. Nitric oxide then passes through a drier to remove water formed during combustion and moves to the chemiluminescence detector, where it is mixed with ozone to form excited nitrogen dioxide ($NO_2^{\cdot}$)

$$NO + 0_3 \rightarrow NO_2^{\cdot} + 0_2 \rightarrow NO_2 + O_2 + h_\upsilon$$

Rapid decay of the $NO_2^{\cdot}$ produces light in the 590–2900nm range. It is detected and amplified by a photomultiplier tube. The result is calculated from the signal produced and printed out in $mg\ L^{-1}$ or as a percentage.

A wide 0.01 to 500mg L^{-1} detection range is possible. Coefficients of variation ranged from 0.88% at the 2.54mg L^{-1} level to 3.1% at the 51mg L^{-1} level.

Dohrmann also supply an automated nitrogen analyser with video display and data processing (model DN–1000) based on similar principles which is applicable to the determination of down to 0.1mg L^{-1} nitrogen in solid and liquid samples.

Equipment for automated Kjeldahl determinations of organic nitrogen in water and solid samples is supplied by Tecator Ltd. Their Kjeltec system 1 streamlines the Kjeldahl procedure resulting in higher speed and accuracy compared to classical Kjeldahl measurements.

Perkin–Elmer supply an analyser (model 2400 CHN) suitable for determining these elements in river and oceanic sediment samples and sewage sludges.

In this instrument the sample is first oxidized in a pure oxygen environment. The resulting combustion gases are then controlled to exact conditions of pressure, temperature and volume. Finally the product gases are separated under steady-state conditions and swept by helium or argon into a gas chromatography for analysis of the components. The equipment is supplied with a 60 position autosampler and microprocessor controller covering all system functions, calculation of results and on-board diagnostics. Analysis time is 5min.

Nitrogen, carbon and sulphur

The NA 1500 analyser supplied by Carlo Erba is capable of determining these elements in 3–9min in amounts down to 10mg L^{-1} with a reproducibility of ±0.1%. A 196 position autosampler is available.

'Flash combustion' of the sample in the combustion reactor is a key feature of the NA 1500. It results when the sample is dropped into the combustion reactor which has been enriched with pure oxygen. The normal temperature in the combustion tube is 1020°C and reaches 1700–1800°C during the flash combustion.

In the chromatographic column the combustion gases are separated so that they can be detected in sequence by the thermal conductivity detector (TCD). The TCD output signal is proportional to the concentration of the elements.

A data processor plots the chromatogram, automatically integrates the peak areas and prints retention times, percent areas, baseline drift and attenuation for each run. It also computes blank values, constant factors and relative average elemental contents.

Total organic carbon

Dohrmann supply a wide range of total organic carbon analysers characteristics of which are enumerated in Table 1.9.

The operating principle of these analysers involves a process whereby a persulphate reagent is continuously pumped at a low flow rate through the injection port (and the valve of the autosampler) and then into the UV reactor. A sample is acidified, sparged and injected directly into the reagent stream. The mixture flows through the reactor where organics are oxidized by the photon-activated reagent. The light-source envelope is in direct contact with the flowing liquid. Oxidation proceeds rapidly, the resultant carbon dioxide is stripped from the reactor liquid and carried to the carbon dioxide specific non-dispersive IR detector (NDIR).

As mentioned above, many variants of the Dohrmann total organic carbon analyser are available, ranging from low-cost non-automated analysers based on sample combustion in a platinum boat (DC8JA) or using persulphate oxidation/ultraviolet irradiation (DC 88) to top-of-range fully automated and computerized systems based on combustion in a ceramic tube (DC 90) or combined simultaneous persulphate-ultraviolet oxidation (DC 180). Only one of these systems, the DC 180, is discussed below in any detail.

As shown in Fig. 1.9 sample transfer in the DC 180 is facilitated by gas pressure. Once the pick-up loop is filled a gas chase delivers the sample

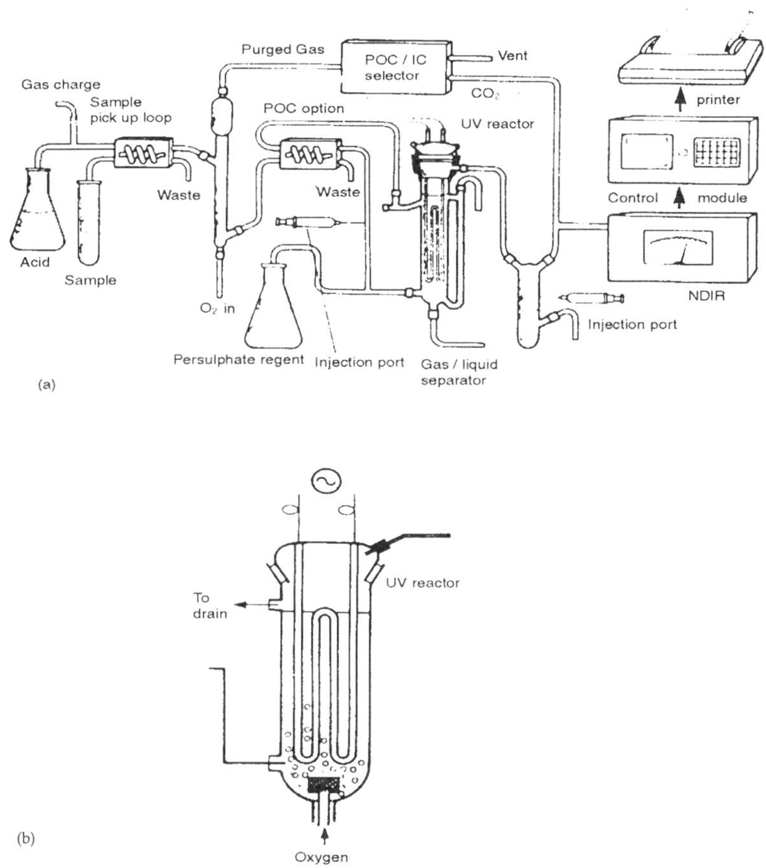

Fig. 1.9 Dohrmann DC–180 total organic carbon analyser (a) layout; (b) detail of UV reactor
Source: Own files

to the sparger. The DC 180 adds a preset amount of acid to the sample. Inorganic carbon is released in the form of carbon dioxide. Together with the purgeable organic carbon (POC) it is removed by sparging. The sample is now ready for non-purgeable organic carbon (NPOC) analysis.

Measuring non-purgeable organic carbon

A separate and independent injection loop dispenses the sample for non-purgeable organic carbon measurements.

In the reactor combined UV persulphate oxidation ensures quantitative total organic carbon recovery. The resulting carbon dioxide with entrained water goes through a gas/liquid separator, a water trap and

Table 1.9 Total organic carbon analysis as supplied by Dohrmann

Model	*Description*	*Analysis time (min)*	*Principle**	*Precision*	*Features*	*Detection limits*
DC 85 A	Low-cost analyser	5	Combustion at 800°C in platinum boat, non-dispersive IR detector	±2% at 2mg L^{-1}	Instrument expandable to include UV persulphate and vial persulphate oxidation, purgeables measurement	100μg L^{-1}
DC–88	Low-cost analyser	5	Persulphate oxidation at 100°C. Non-dispersive IR detector	100mg L^{-1} or 2%, whichever greater	Instrument expandable to include UV persulphate oxidation and fast oxidation methods	100μg L^{-1}
DC–80	Automated analyser	3–4	Low-temperature promoted persulphate oxidation	±2% wt 100–4000mg L^{-1}	120-tube autosampler date integrity review. Purgeable organic carbon measurements can be made: Operating modes; organic carbon external spare total carbon inorganic carbon purgeable carbon estimate by difference purgeable carbon determined	

DC–90	Second-generation high-temperature analyser	High-temperature combustion in ceramic tube	From RDS of 0.6% at 6mg L^{-1} to RSD of 0.5% 107mg L^{-1}	Unattended operation possible, loop-sampler option autosampler printer, option inorganic, carbon, purgeable carbon, total carbon and total organic carbon measurements possible	100µg L^{-1}
LC–180	Automated analyser	Combined UV persulphate oxidation (in combined cell)	±2% or 5mg L^{-1} whichever greater	Highly automated inorganic carbon, non-purgeable organic carbon, purgeable organic carbon measurements possible. Automatic sample size selection according to type of sample, video display/ microprocessor report on means and S.D. 45 sample autosampler	10µg L^{-1}

Source: Own files

*Inorganic carbon first removed by acidification and inert gas sparging drier before it enters the non-dispersive infrared analyser detector where the evolved carbon dioxide is measured.

Measuring inorganic carbon

If it is required to quantify inorganic carbon the sparged gas may be directed to the non-dispersive infrared analyser for quantification.

Measuring purgeable organic carbon

The volatile fraction from the sparger contains both carbon dioxide and purgeable organic carbon. If purgeable organic carbon measurements are required, carbon dioxide is removed by the lithium hydroxide scrubber. There are two options for oxidizing purgeable organic carbon. The UV persulphate reactor will convert most purgeable organic carbon except for fully halogenated organics such as freons and carbon tetrachloride. In the case of such organics, the high-temperature reactor will be required.

Measuring purgeable organic carbon inorganic carbon and non-purgeable organic carbon

In addition to giving enhanced sensitivity and greater recovery for the full range of purgeable organic carbons, the purgeable organic carbon accessory permits the analysis of purgeable organic carbon, inorganic carbon and non-purgeable organic carbon on one sample. The DC 180 reports all three of these parameters plus total organic carbon as a sum of non-purgeable organic carbon and purgeable organic carbon and total carbon as a sum of all three parameters.

Measuring total organic carbon

The DC 180 will calculate total organic carbon based on purgeable organic carbon and non-purgeable organic carbon results and include it in the report. Alternatively, total organic carbon may be determined as the difference of total carbon less inorganic carbon.

Shimadzu TOC–500 total organic carbon analyser

This is a fully automated system capable of determining between 1μg L^{-1} and 300μg L^{-1} total organic carbon. It is equipped with a 36-place auto-sampler, microprocessor and printer. Total organic carbon measurements down to 40μg L^{-1} have been achieved at a coefficient of variation of 16.3%.

OIC Analytical instruments produce the fully computerized model 700 total organic carbon analyser. This is applicable to soils and sediments. Persulphate oxidation at 90–100°C non-dispersive infrared spectroscopy is the principle of this instrument. It has the ability to measure total organic carbon, total inorganic carbon and purgeable organic carbon in the same sample. The precision is $\pm 2.0\mu g\ L^{-1}$ carbon.

1.1.7 Neutron activation analysis

This is a very sensitive and specific technique whose applications have been limited to the determination total halogen in soils and non-saline sediments.

Due to the complexity and cost of the technique no water laboratory in the UK has its own facilities for carrying out neutron activation analysis. Instead, samples are sent to one of the organizations that possess the facilities, e.g. the Atomic Energy Research Establishment at Harwell or the Joint Manchester–Liverpool University Reactor located at Risley.

As mentioned above, the technique is extremely sensitive and tends to be used when a referee analysis is required on a material which then becomes a standard for checking out other methods. Another advantage of the technique is that a foreknowledge of the elements present is not essential. It can be used to indicate the presence and concentration of entirely unexpected elements, even when present at very low concentrations.

In neutron activation analysis, the sample in a suitable container, often a pure polyethylene tube, is bombarded with slow neutrons for a fixed time together with standards. Transmutations convert analyte elements into radioactive elements, which are either different elements or isotopes of the original analyte.

After removal from the reactor the product is subject to various counting techniques and various forms of spectrometry to identify the elements present and their concentration.

1.1.8 Nuclear magnetic resonance spectroscopy and electron spin resonance spectroscopy

NMR has been applied to the determination of organomercury compounds in non-saline sediments and humic and fulvic acids in soil and saline sediments.

ESR has been used to determine anionic surfactant agents in sewage sludge.

Instrument suppliers are listed below.

- Gemini Superconducting Fourier transform NMR systems, VXR series 5, Varian Instruments, Sugar Lane, Texas, USA
- NMR imaging spectrometer systems, Vis, 1120 Auburn Road, Fremont, California 94538, USA

1.1.9 Enzymic immunoassay methods

This technique has been used to determine the following types of organic compounds in soil; polychlorobiphenyls, chlorinated insecticides, triazine herbicides, paraquat and diquat.

Biology and biochemistry laboratories perform three general types of assays:

1 Binding assays including the following: immunoassays such as radio immunoassay (RIA), fluorescence immunoassay (FIA), enzyme immunoassay (EIO), enzyme-linked immunoassay (ELISA and EMIT)
2 Enzyme assays – both kinetic and end-point radiocoordination of proteins, lipid assays, receptor binding assays and tissue-culture techniques
3 Chemical assays such as total protein assays and analytical chemistry including spectroscopy and chromatography.

Assays 1 and 2 are described below. Assay type 3 is discussed earlier under specific chemical techniques.

Immunoassays

Immunoassays (type 2) are based on the following reaction:

$$\underset{\text{antigen}}{[Ag]} + \underset{\text{antibody}}{[Ab]} \rightleftharpoons \underset{\text{complex}}{[Ab - Ag]}$$

Each of the types of immunoassay listed above (RIA, FIA, EIA, ELISA and EMIT) has its own advantages. In general immunoassays involve large numbers of samples and are a source of routine, repetitive work. Whatever the type, immunoassays require the following equipment:

- liquid handling – pipetting, dispensing, etc.
- sample conditioning – mixing, incubating, etc.
- separation – centrifugation, filtration, etc.
- measurements – spectroscopy, gamma counter, etc.

Immunoassays are nowadays performed by one of two approaches either partially automated or fully automated (robotics).

Partially automated immunoassay systems

The separate items of equipment necessary for the preliminary (i.e. sample preparation) stages of partially automated (i.e. prior to the final measurement instrument) immunoassay available from Denby Instruments Ltd are listed in Appendix 1.

Having completed the sample preparation stages the final measurement can be made by a variety of techniques.

Luminescence and fluorescence measurements in luminescent immunoassays

Perkin–Elmer LS–2B microfilter fluorometer Fluorescence is used in immunochemistry. Essentially the radioactive tag on the antigen is replaced by a fluorophore. The most commonly used tags are fluorescein and umbelliferone.

Organic chelates of certain lanthanides such as Tb^{3+} and Eu^{3+} are used as a means of removing unwanted background fluorescence in fluoro-immunoassays. The lanthanide chelates exhibit long-lived emission 50µs to 3ms. Using a gated detection system such as that of the model LS–2B the non-specific background fluorescence with a decay time of 100ns is discriminated against.

Hamilton Umicon Lumicon chemi- and biolumium assay luminometer This equipment is used in test-tube scale luminescent immunoassays. With its sample compartment (thermostatted by means of Peltier elements, which allow the temperature to be set from 15°C to 40°C with a precision of 0.1°K) this instrument is suitable for the measurement of temperature-sensitive bioluminescence resulting from enzymic reactions and also in phagocyte-mediated luminescence measurements.

This instrument can be used in two modes: the peak mode for fast-decaying pulses of light (<10s) and the repeat mode for slow kinetics (up to 100min).

Spectrophotometric plate readers Perkin–Elmer's lambda reader, an automated microprocessor-controlled, microplate reader, offers the flexibility of configuring a reliable, user-friendly, versatile system, capable of accommodating a wide variety of assays requiring calorimetric measurement on microscale (<300µl) samples. These assays include ELISA, protein determination, cytotoxicity, cytoproliferation and antibody sensitivity testing.

Radio-immunoassay (RIA) analysers Kontron instruments supply the MDA 312 multi-detector RIA analyser. This is the first multi-detector gamma counter incorporating a multi-channel instead of a conventional pulse-height analyser. During the counting period the pulse-height spectra of each individual sample (detector) is recorded and stored. Starting from the photopeak the MDA 312 now sets the energy-window levels on both sides of the peak. This way the counting window is automatically optimized for each sample.

The direct benefits of this innovative technique are twofold: the counting efficiency of an individual detector is optimal and constant.

The possibility of centring the photopeak into the counting window for each individual sample allows the selection of a relatively narrow window, whereas in the conventional technique the windows have to be set much broader in order to compensate for drifts. The result is a substantial reduction in background count rates.

Packard supply the Cobra – one Auto-gamma 5012 and 5013 instruments. This combines RiaSmart data reduction, expert quality control management and high-energy counting capabilities. It has been used in ^{125}I radio-immunoassay measurement and DNA probe analysis using ^{32}P. It features three simultaneous counting regions, half-life correction, spill-over compensation, spectral display and plotting and an on-board computer. This company also supplies the Crystal + plus benchtop manual RIA system. This is available in 24-, 12- or economical 6-detector configurations: the 24-detector model can count over 1400 samples per hour. A built-in multi-tasking microcomputer saves time by simultaneously counting samples and reporting results and the Crystal + plus can connect to a labmicro or mainframe for more extensive data reduction.

Up to 50 stored assay protocols include routines for RIA/IRMA, dual label assays, T3–uptake and FTI calculations and hepatitis, RAST and hCG screening. Quality-control charts can be printed automatically.

Biosensors Biosensors are used in the final measurement stage of immunoassays such as an electrochemistry-based enzyme-linked immunoassay (ELISA) or the measurement of catalyse-labelled antigens at an antibody-coated oxygen electrode.

Fully automated immunoassay systems

The following benefits accrue from full automation:

- improved precision by reducing human errors
- freeing sensor personnel from repetitive tasks
- isolating personnel from hazardous environments and protecting experiments from human contamination
- faster sample turn-around

The Zymark robotic laboratory automation system Although detail procedures differ in each laboratory, the basic elements of binding and enzyme assays are similar. The generalized procedure shown in Table 1.10 highlights the common steps and indicates which Zymate laboratory systems are required. These procedures are performed using common laboratory glassware such as test tubes or in multiple tube devices such as microtitre plates.

Table 1.10 Typical immunoassay procedure using Zymate robotic laboratory automation system

Step	*Procedure and comment*	*Zymate laboratory stations*
1	*Dispense samples, standards and controls* Precision dispensing or pipetting of samples, standards and controls	Z510 Master Laboratory Station, or Z190 Precision Microlitre syringe head
2	*Serial dilute – as required* Precision dispensing of reagents or buffers then transferring an aliquot from the previous dilution syringe head	Z510 Master Laboratory Station, or Z910 Precision Microlitre
3	*Mix* Vortex if possible. Vortexing is gentle, simple to perform and eliminates cross contamination	Z620 Vortex Station
4	*Transfer samples if required* Transfer sample aliquots to an syringe head with pipette appropriate tube or container tips	Z910 Precision Microlitre
5	*Incubate* Hold samples at a programmed temperature and time	Zymate controller and Z830 Power and event controller
6	*Quench or terminate reaction* Dispense reagent or buffer to terminate reaction	Z510 Master Laboratory Station
7	*Separate bound and free reactants* The following alternatives are available:	
	A Filtration – retaining either the filtrate or retentate	Several options available
	B Centrifugation	Centrifuge station 1Q84
	C Adsorption using solid phase extraction minicolumns	Z510 Master Laboratory Station and Z911 Liquid distribution hand
	D Immobilized substrate where mineral to be removed is bound to a large substrate permitting simple mechanical separation	Robotic manipulation
8	*Wash and Dry* Dispense and remove wash reagents and, if needed, control gas purge	Z510 Master Laboratory Station
9	Add scintillation cocktail or other detection amplifying reagent	Z510 Master Laboratory Station
10	Introduce sample into measurement instrument and monitor results	Z830 Power and event control station. 7820 printer

Source: Own files

Beckman Robotic Biomek 1000 automated laboratory The Biomek 1000 integrates the work formerly done by four instruments: sample preparation system, diluter/dispenser, plate washer and a spectrometer finish. In can handle assays such as radio-immunoassays (RIA), fluorescence immunoassays (FIA), enzyme immunoassays EIA and enzyme-linked immunoassays (ELISA).

1.2 Rationale, analysis of solid samples

1.2.1 Soils

Tables 1.11(a) to (c) review methods used for the determination of various types of organic compounds (1.11(a)), elements (1.11(b)) and organometallic compounds (1.11(c)) in soils.

1.2.1.1 Organic compounds

It is seen by examination of Table 1.11(a) that the most extensively used techniques are gas chromatography (28 determinants or groups of determinants including aliphatic hydrocarbons, polyaromatic hydrocarbons, phenols, chloroaliphatic compounds, chlorophenols, methyl bromide, polychlorobiphenyls, dioxins, chloro, carbamate, organophosphorus, triazine, substituted urea and phenoxyacetic acid; types of insecticides and herbicides; also picloram, Acarol, Paraquat, Diquat, Bromacil, Diclofop, Sencor, Trifluralin and mixtures of Herbicides and pesticides, fungicides, mestranol, methoxy groups and mixtures of organic compounds).

In many applications nowadays it is essential to link a mass spectrometer to the gas chromatography in order to achieve positive identification and sensitivity of analysis. Some 12 types of compounds are listed in Table 1.11(a) which are based on the application of this technique, *viz.* polyaromatic hydrocarbons, polychlorobenphenyls, dioxins, chloro, carbamate and triazine types of herbicides and pesticides, Diazinon, Dicamba, Imidazoline and Cyperquat herbicides and herbicide pesticide mixtures.

For more volatile compounds in soils, such as aromatic hydrocarbons, alcohols, aldehydes, ketones, chloroaliphatic hydrocarbons, haloaromatic hydrocarbons, acetonitrile, acrylonitrile and mixtures of organic compounds a combination of gas chromatography with purge and trap analysis is extremely useful. Pyrolysis gas chromatography has also found several applications, heteroaromatic hydrocarbons, polyaromatic hydrocarbons, polymers and haloaromatic compounds and this technique has been coupled with mass spectrometry, (aliphatic and aromatic hydrocarbons and mixtures of organic compounds).

Gas chromatography is closely followed by high-performance liquid chromatography in its popularity (eight determinants including triazine, substituted urea, phenoxy acetic acid, Fluazifop, Fluazifop methyl, herbicides and pesticides, estranol and ethylene diamine tetracetic acid and mixtures of organic compounds. Combination of HPLC with a mass spectrometric detector is gaining in popularity.

Another growing technique is super-critical fluid chromatography. Recent references to soil analysis include the following applications: aliphatic hydrocarbons, polyaromatic hydrocarbons, polychlorobiphenyls, dioxins, alkyl and aryl phosphates, chloro, organophosphorus, triazine, substituted urea, phenoxy acetic acid, Dacthal herbicides and insecticides and mixtures of herbicides and pesticides and mixtures of organic compounds.

There is no doubt that these applications will grow in the future and that the range of supercritical fluids used (carbon dioxide and methanol modified carbon dioxide, nitrogen dioxide, ammonia, fluorohydrocarbons) will increase as will the combination of this technique with mass spectrometric identification of separated compounds.

Electrophoretic and isotachoelectrophoretic techniques are gaining in popularity in soil analysis with applications to polyaromatic hydrocarbons, polychlorobiphenyls, tetrahydrothiophene and triazine herbicides, Paraquat and Diquat and growth regulators. Other lesser-used techniques include spectrophotometric methods (five determinants), spectrofluorimetric methods (two determinants), luminescence methods (one determinant), titration methods (one determinant), thin-layer chromatography (five applications), NHR spectroscopy (two applications) and enzymic immunoassays (one determinant).

1.2.1.2 Elements

It is seen by examination of Table 1.11(b) that a wide variety of techniques have been employed including spectrophotometry (four determinants), combustion and wet digestion methods and inductively coupled plasma atomic emission spectrometry (three determinants each), atomic absorption spectrometry, potentiometric methods, molecular absorption spectrometry and gas chromatography (two determinants each), and flow-injection analysis and neutron activation analysis (one determinant each). Between them these techniques are capable of determining boron, halogens, total and particulate carbon, nitrogen, phosphorus, sulphur, silicon, selenium, arsenic antimony and bismuth in soils.

1.2.1.3 Organometallic compounds

Atomic absorption spectrometry (organomercury and tin compounds) and gas chromatography (organoarsenic, lead, mercury and tin compounds) are the two most popular techniques (Table 1.11(c)) while supercritical fluid chromatography is making some inroads (organotin compounds).

Table 1.11(a) Determination of organic compounds in soils

	Spectro-photo-metry	*Spectro-fluori-metry*	*Lumin-escence*	*Titration*	*HPLC*	*HPLC-m/s*	*Isotacho electro-phoresis*	*Thin layer chroma-tography*	*Super-critical fluid chromatography*
aliphatic hydrocarbons									2.1.1.2 2.1.1.3
aromatic hydrocarbons									
heteroaromatic hydrocarbons								2.4.1.4	
polyaromatic hydrocarbons		2.4.1.8							2.4.1.7
polymers									
phenols	4.2.1.1								
alcohols, aldehydes and ketones									
chloroaliphatics									
haloaromatics									
chlorophenols	5.3.1.1								
methyl bromide									
polychlorobiphenyls			5.6.1.3						5.6.1.4
chlorodibenzo-p-dioxin and chlorodibenzofuran									5.7.1.2
amines	6.1.1.1								
acrylonitrile and acetonitrile									
alkyl phosphates									7.1.1.1
tetrahydrothiophene									
chlorinated insecticides								9.1.1.3	9.1.1.5
organophosphorus insecticides									9.3.1.2

Table 1.11(a) continued

	Spectro-photo-metry	*Spectro-fluori-metry*	*Lumin-escence*	*Titration*	*HPLC*	*HPLC-m/s*	*Isotacho electro-phoresis*	*Thin layer chroma-tography*	*Super-critical fluid chromatography*
carbamate insecticides				9.2.1.1					
triazine herbicides					9.4.1.3		9.4.1.6		9.4.1.4
substituted urea herbicides					9.5.1.2 9.5.1.3 9.5.1.5			9.5.1.1	9.5.1.6
phenoxyacetic and herbicides					9.6.1.2	9.6.1.3			9.6.1.4
misc. herbicides pesticides					9.7.1.13				9.7.1.19
diazinon									
picloram									
acarol									
dicamba									
paraquat							9.7.1.9		
diquat							9.7.1.9		
bromacil									
frenock									
diclofop and diclofopmethyl									
imidazolines									
fluazifop and fluazifopbutyl					9.7.1.13				
imigen									
propanil									
trifluralin									
cyperquat									
dacthal									9.7.1.19

Table 1.11(a) continued

	GLC	*GLC–ms*	*Purge and trap GLC*	*Pyrolysis–GLC*	*Pyrolysis–GLC–ms*	*Mass spectroscopy*	*Electrophoresis*	*Enzyme immunoassay*	*NMR*
imigen									
propanil									
trifluralin	9.7.1.17								
cyperquat		9.7.1.18							
dacthal									
multi herbicide pesticide mixtures	9.8.1.1								
dichlorobenzonitrile	9.7.1.6								
sencor	9.7.1.16								
growth regulators									
fungicides	9.10.1.2								
humic and fulvic acids									10.1.1.2
ethylene diamine tetracetic acid									
mestranol	10.15.1.1								
methoxy groups	10.6.1.1								
mixtures of organics	11.1.2		11.1.3		11.1.4				

Source: Own files

Table 1.11(a) continued

	GLC	*GLC–ms*	*Purge and trap GLC*	*Pyrolysis–GLC*	*Pyrolysis–GLC–ms*	*Mass spectro-scopy*	*Electro-phoresis*	*Enzyme immuno-assay*	*NMR*
carbamate insecticides	9.2.1.2	9.2.1.3							
triazine herbicides	9.4.1.1	9.4.1.2						9.4.1.5	
substituted urea herbicides	9.5.1.3					9.5.1.6			
	9.5.1.4								
	9.5.1.6								
phenoxyacetic and herbicides	9.6.1.1								
misc. herbicides pesticides	9.7.1.2	9.7.1.1							
	9.7.1.3	9.7.1.4							
	9.7.1.6	9.7.1.5							
	9.7.1.7								
	9.7.1.11								
	9.7.1.12								
	9.7.1.16								
diazinon		9.7.1.1							
picloram	9.7.1.2								
acarol	9.7.1.3								
dicamba		9.7.1.5							
paraquat	9.7.1.7							9.7.1.8	
diquat	9.7.1.7							9.7.1.8	
bromacil	9.7.1.1								
frenock						9.7.1.10			
diclofop and diclofopmethyl	9.7.1.12								
imidazolines		9.7.1.4							
fluazifop and fluazifopbutyl									

Table 1.11(a) continued

	GLC	*GLC–ms*	*Purge and trap GLC*	*Pyrolysis–GLC*	*Pyrolysis–GLC–ms*	*Mass spectro-scopy*	*Electro-phoresis*	*Enzyme immuno-assay*	*NMR*
aliphatic hydrocarbons	2.1.1.1 2.1.1.3				2.1.1.3				
aromatic hydrocarbons			2.2.2.1		2.2.1.2				
heteroaromatic hydrocarbons				2.3.1.1					
polyaromatic hydrocarbons	2.4.1.1	2.4.1.2		2.4.1.3		2.4.1.5	2.4.1.6	2.4.1.8	
polymers				2.5.1.1					
phenols	4.2.1.2								
alcohols, aldehydes and ketones			4.8.1.1						
chloroaliphatics	5.1.1.1	5.1.1.2	5.1.1.3						
haloaromatics			5.2.1.1	5.2.1.2					
chlorophenols	5.3.1.2	5.3.1.3							
methyl bromide	5.4.1.1								
polychlorobiphenyls	5.6.1.1	5.6.1.2				5.6.1.6		5.6.1.5	
chlorodibenzo-p-dioxin and chlorodibenzofuran	5.7.1.1	5.7.1.2							
amines									
acrylonitrile and acetonitrile			6.2.1.1						
alkyl phosphates						7.1.1.2			
tetrahydrothiophene		8.1.1.1							
chlorinated insecticides	9.1.1.1	9.1.1.2						9.1.1.4	9.1.1.6
organophosphorus insecticides									

Table 1.11(a) continued

	Spectro-photo-metry	*Spectro-fluori-metry*	*Lumin-escence*	*Titration*	*HPLC*	*HPLC-m/s*	*Isotacho electro-phoresis*	*Thin layer chroma-tography*	*Super-critical fluid chromatography*
									9.7.1.19
multi herbicide pesticide mixtures								9.8.1.2	9.8.1.3
dichlorobenzonitrile									
sencor									
growth regulators							9.9.1.1		
fungicides	9.10.1.1								
humic and fulvic acids									
ethylene diamine tetracetic acid					10.4.1.1				
mestranol									
methoxy groups									
mixtures of organics	11.1.1				11.1.5	11.1.6			11.1.7

Table 1.11(b) Determination of elements in soils

	Spectro-photo-	*Flow injection*	*Atomic absorption spectrometry*	*Inductively coupled plasma atomic emission spectrometry*	*Titration*	*GLC*	*Combustion methods*	*Wet digestion*	*Potentiometry*	*Molecular absorption spectrometry*	*NAA*
Halogens	12.2.1.1					12.2.1.2					12.2.1.3
Total organic carbon					12.3.1.1		12.3.1.2		12.3.1.3		
Nitrogen	12.5.1.2				12.5.1.1						
Total phosphorus	12.6.1.1	12.6.1.2		12.6.1.4		12.6.1.3	12.6.1.5	12.6.1			
Sulphur							12.8.4.1	12.8.1.1 12.8.4.1			
Silicon				12.9.1.1							
Arsenic			12.10.1.1								
Selenium			12.13.1.1					12.13.1.2			12.13.1.2
Boron	12.1.1.1			12.1.1.2						11.1.1.3	

Source: Own files

Table 1.11(c) Methods used in the determination of organometallic compounds in soils

	Spectro-photometry	*Atomic absorption spectrometry*	*Supercritical fluid chroma-tography*	*GLC*
Organoarsenic				13.1.1.2
Organolead				13.2.1.1
Organomercury	13.3.1.2	13.3.1.4		13.3.1.3
Organotin		13.4.1.3	13.4.1.2	13.4.1.1

Source: Own files

1.2.2 Sediments

1.2.2.1 Organic compounds

Again gas chromatographic (20 determinants) and high-performance liquid chromatographic (12 determinants) are by far the most extensively used techniques as is shown in Table 1.12(a). A combination of these techniques with mass spectrometry is growing in importance, as is evidenced by the fact that 11 applications to gas chromatography are listed. Again supercritical fluid chromatography is finding applications, as is evidenced by this application to the determination of polyaromatic hydrocarbons, chlorophenols, polychlorobiphenyls, chloro, carbamate and organophosphorus insecticides and herbicides. The next two most important techniques are thin-layer chromatography and spectrofluorimetric methods (five determinants each) while most other techniques mentioned have very limited applications.

1.2.2.2 Elements

The situation regarding the determination of elements in sediments is very similar to that occurring in the case of soils (Table 1.12(b)).

1.2.2.3 Organometallic compounds

Again, as in the case of soil analysis, atomic absorption spectrometry and gas chromatography are the methods of choice (Table 1.12(c)).

1.2.3 Sludges

1.2.3.1 Organic compounds

Gas chromatography has been applied to the determination of 17 types of organic compounds in sludges (Table 1.13(a)), including carboxylic acids,

Table 1.12(a) Methods used for the determination of organic compounds in sediments

	Spectro-photometry	*Spectro-fluorimetry*	*UV spectro-scopy*	*Infra red spectro-scopy*	*Atomic absorption spectro-scopy*	*HPLC*	*HPLC–m/s*	*GPC*	*Thin layer chroma-tography*	*Super critical fluid chroma-tography*
Aromatic hydrocarbons		2.2.2.1(s)	2.2.2.3(s)			2.2.2.2(s)				
Polyaromatic hydrocarbons		2.4.3.4(s) 2.4.2.6(n/s) 2.4.3.2(s)	2.4.2.5(n/s)			2.4.2.4(n/s) 2.4.3.3(s)			2.4.3.4(s) 2.1.3.4(s)	2.4.2.7 (n/s)
Anthropogenic compounds										
Cationic surface active agents	3.1.1.1(n/s)					3.1.1.3(n/s)				
Phthalate esters						4.1.1.2(n/s)				
Carboxylic acids										
Carbohydrates						4.4.1.2(n/s)				
Uronic acids aldoses										
Chloroaliphatics						5.1.2.4(n/s)			5.1.2.5(n/s)	
Haloaromatics										
Chlorophenols										
Aliphatic hydrocarbons		2.1.2.2(n/s) 2.1.3.2(s)		2.1.3.3(s)						
Chloroanisole										
Polychlorobiphenyls										
Chlorodibenzo-p-dioxins and chlorodibenzofurans										
Basic nitrogen compounds										
Alkyl and aryl phosphates										

	Spectro-photometry	*Spectro-fluorimetry*	*UV spectro-scopy*	*Infra red spectro-scopy*	*Atomic absorption spectro-scopy*	*HPLC*	*HPLC–m/s*	*GPC*	*Thin layer chroma-tography*	*Super critical fluid chroma-tography*
Tetrahydrothiophene and sulphur compounds										
Chlorinated insecticides									9.2.2.2(n/s)	9.1.2.2(n/s) 9.2.2.1(s)
Carbamate insecticides									9.2.2.2(n/s)	
Organophosphorus insecticides										9.3.2.2(n/s)
Ditallow dimethyl ammonium										6.7.1.1(s)
Triazine herbicides						9.4.2.1(n/s)				
Phenoxyacetic acid herbicides							9.6.2.1(n/s)			
Multiherbicide-insecticide mixtures										9.8.2.1(n/s)
Humic and fulvic acids		10.1.3.1(s)			10.1.2.2(n/s) 10.1.3.2(s)	10.1.2.1(n/s)				
Optical whiteners		10.3.1.1(n/s)				10.3.1.2(n/s)				
Mixtures of organic compounds			11.2.5(n/s)			11.3.2(n/s)		11.2.4(n/s)		

n/s = non-saline sediments s = saline sediments

Table 1.12(a) continued

	GLC	*GLC–ms*	*Purge and trap GLC*	*Pyrolysis GLC*	*Pyrolysis GLC–m/s*	*Mass spectro-metry*	*NMR*	*Coulometry*	*Misc*
Aromatic hydrocarbons									
Polyaromatic hydrocarbons	2.4.3.4(s) 2.4.2.1(n/s) 2.4.3.1(s) 2.4.2.8(n/s)	2.4.2.2(n/s)		2.4.2.3(n/s)					
Anthropogenic compounds					10.2.1.1(n/s)				
Cationic surface active agents		3.1.1.2(n/s)							
Phthalate esters	4.1.1.1(n/s)								
Carboxylic acids	4.3.1.1(n/s)								
Carbohydrates	4.4.2.1(s)								
Uronic acids aldoses	4.6.1.1(n/s)								
Chloroaliphatics	5.1.2.1(n/s)	5.1.2.2(n/s)	5.1.2.3(n/s)						
Haloaromatics	5.2.2.1(n/s) 5.2.3.1(s)	5.2.2.2(n/s)	5.2.2.3(n/s)						
Chlorophenols	5.3.2.1(n/s)	5.3.2.2(n/s)						5.3.2.2(n/s)	
Aliphatic hydrocarbons	2.1.2.1(n/s) 2.1.3.1(s)								
Chloroanisole	5.5.1.1(n/s)								
Polychlorobiphenyls	5.6.2.1(n/s)	5.6.2.2(n/s)		5.6.2.3(n/s)					5.6.2.5(n/s)
Chlorodibenzo-p-dioxins and chlorodibenzofurans	5.6.3.1(s) 5.6.2.5(n/s)	5.7.2.1(n/s)				5.7.2.1(n/s)			
Basic nitrogen compounds		6.6.6.1(s)							
Alkyl and aryl phosphates		7.1.2.1(n/s)							

	GLC	*GLC–ms*	*Purge and trap GLC*	*Pyrolysis GLC*	*Pyrolysis GLC–m/s*	*Mass spectro-metry*	*NMR*	*Coulometry*	*Misc*
Tetrahydrothiophene and sulphur compounds	8.2.1.1(n/s) 8.2.2.1(s)	8.2.1.2(n/s)							
Chlorinated insecticides	9.1.2.1(n/s) 9.2.2.1(s)								
Carbamate insecticides	9.2.2.1(n/s)								
Organophosphorus insecticides	9.3.2.1(n/s) 9.3.3.1(s)								
Ditallow dimethyl ammonium									
Triazine herbicides									
Phenoxyacetic acid herbicides									
Multiherbicide-insecticide mixtures									
Humic and fulvic acids							10.1.3.3(s)		
Optical whiteners									
Mixtures of organic compounds	11.2.1(n/s) 11.3.1(s)	11.2.2(n/s) 11.3.2(s)		11.2.6(n/s)					

Source: Own files

n/s = non-saline sediments s = saline sediments

Table 1.12(b) Methods used for determination of elements in sediments

	Spectro-photo-metry	*Spectro-fluori-metry*	*Infra-red spectro-scopy*	*Atomic absorption spectro-metry*	*Inductively coupled plasma atomic emission spectro-metry*	*GLC*
Boron	12.1.2.1(s)					
Halogens						
Total organic carbon	12.3.3.1(s) 12.3.2.1(n/s)		12.3.3.2(s)			
Particulate organic carbon			12.4.1.1(n/s)			
Nitrogen	12.5.2.1(n/s)					
Total phosphorus	12.6.2.1(n/s)					
Particulate phosphorus						
Sulphur	12.8.2.1(n/s) 12.8.3.2(s)					12.8.3.3(s)
Arsenic	12.10.3.1(s)			12.10.2.2 (n/s) 12.10.3.3(s)	12.10.2.3 (n/s) 12.10.3.4(s)	12.10.2.1 (n/s) 12.10.3.2 (s)
Antimony	12.11.1.1(n/s)				12.11.1.2 (n/s) 12.11.2.1(s)	2.11.1.3(s)
Bismuth				12.12.1.1(n/s) 12.12.2.1(s)		
Selenium	12.13.2.1 (n/s) 12.13.3.1 (s)	12.13.2.2(s)		12.13.2.3 (n/s) 12.13.3.3(s)	12.13.3.4(s) 12.13.2.4(s) (n/s)	12.13.3.2

Source: Own files

n/s = non-saline sediment s = saline sediment

ß hydroxy butyrate, ß hydroxy valerate, chloroaliphatics, chlorophenols, phenyl sulphone sarcosine, chlorinated, organophosphorus, phenoxy-acetic acid herbicides and insecticides, cationic* and anionic* surface active agents, polychlorobiphenyls*, 4-nitrophenol*, azarenes* and nitro-azarenes*, mixtures of organic compounds* and repirometric measurements. Those asterisked above have been determined by the gas chromatography–mass spectrometric technique.

High-performance liquid chromatography has, to date, only found limited applications in the analysis of sludges (aliphatic hydrocarbons, cationic and non-ionic surface active agents, carboxylic acids and cobalamin).

Mass spectrometry	*Titration*	*X-ray methods*	*Scanning electron microscopy*	*NAA*	*Combustion methods*	*Wet oxidation methods*
				12.2.2.1(n/s)		
		12.3.3.5(s)			12.3.2.2(n/s)	12.3.3.4(s)
					12.3.3.3(s)	
					12.4.1.2(n/s)	12.4.1.3(n/s)
						12.4.2.1(s)
					12.5.2.2(n/s)	12.5.2.3(n/s)
					12.6.2.3(n/s)	12.6.2.4(n/s)
						12.7.1.1(n/s)
	12.8.3.1(s)		12.8.3.4(s)			
12.10.2.4 (n/s)						

The situation regarding elements and organometallic compounds in sludges is reviewed respectively in Tables 1.13(b) and 1.13(c).

1.2.4 Resume

In Tables 1.11–1.13 analytical techniques are cross-referenced with organic compound element or organometallic compound determined in soil, sediment or sludge and the section number in the book. If the reader finds that a method is not listed for determining a particular compound in the particular type of sample, then by examination of the table he may find a

Table 1.12(c) Methods used in the determination of organometallic compounds in sediments

	Spectro-photo-metry	*Atomic absorption spectro-metry*	*Inductively coupled plasma emission spectro-metry*	*HPLC*	*Super critical fluid chroma-tography*	*GLC*	*GLC–m/s*	*Purge and trap GLC*	*Mass spectro-metry*	*NMR*	*Wet digestion methods*
Organoarsenic		12.1.2.2(s)									
Organolead						13.2.2.2(n/s) 13.2.2.1(n/s) 13.2.3.1(s)					
Organomercury	12.3.2.2 (n/s)	12.3.2.4 (n/s)				13.3.2.3(n/s) 13.3.3.1(s)				13.3.2.5 (n/s)	13.4.2.7 (n/s)
Organotin		12.4.2.5 (n/s) 12.4.3.3(s)		12.4.2.4 (n/s)	12.4.2.6 (n/s)	13.4.2.1 (n/s) 13.4.3.1 (s)	13.4.2.2 (n/s)	13.4.2.7 (n/s) 13,4,2,3 (n/s) 13.4.3.2(s)	13.4.2.7 (n/s)		
Organosilicon		12.5.1.1 (n/s)	12.5.1.2 (n/s)						13.5.1.3 (n/s)		

Source: Own files

n/s = non-saline sediment s = saline sediment

Table 1.13(a) Methods used in the determination of organic compounds in sludges

	Spectro-photometry	*Lumines-cence*	*Infrared spectro-scopy*	*HPLC*	*Super critical fluid chroma-tography*	*GLC*	*GLC–m/s*	*Pyrolysis GLC*	*ESR*	*Thin layer chromatography*
Aliphatic hydrocarbons				2.1.4.1						
Polyaromatic hydrocarbons										2.4.4.1
Cationic surface active agents	3.1.2.1			3.1.2.3			3.1.2.2			
Anionic surface active agents					3.2.1.2		3.2.1.1		3.2.1.3	
Non-ionic surface active agents				3.3.1.1						
Carboxylic acids				4.3.2.2		4.3.2.1				
ß hydroxy butyrate and ß hydroxy valerate						4.7.1.1				
Chloroaliphatic hydrocarbons						5.1.4.1				
Chlorophenols						5.3.4.1				
Polychlorobiphenyls							5.6.4.1			
4-nitrophenol							6.3.1.1			
Phenyl sulphone sarcosine						6.8.1.1				
Azarenes and nitroazarenes							6.9.1.1			
Adenosine triphosphate	7.2.1.1	7.2.1.2								
Chlorinated insecticides						9.1.4.1	9.1.4.2			
Organophosphorus insecticides						9.3.4.1				
Phenoxyacetic acid herbicides						9.6.3.1				
Cobalamin				10.9.1.1						
Mixtures of organic compounds			11.4.2				11.4.1			

Source: Own files

Table 1.13(b) Methods used in the determination of elements in sludges

	Spectro-photometry	*Atomic absorption spectro-metry*	*Inductively coupled plasma atomic emission spectrometry*	*X-ray methods*	*Ion selective electrodes*	*Wet digestion methods*	*Miscellaneous*
Halogens					12.2.3.1		
Total organic carbon	12.3.4.1						
Nitrogen						12.5.3.1	
Total phosphorus				12.6.3.1			
Sulphur							12.8.4.1
Silicon							12.9.2.1
Arsenic		12.104.1	12.10.4.2				
Selenium		12.13.4.1					

Source: Own files

Table 1.13(c) Methods used in the determination of organometallic compounds in sludges

	GLC
Organomercury	13.4.4.1

Source: Own files

method that is listed under another type of sample that is applicable to the type of sample in which he is interested. Thus, if a method is not available for determining a particular compound in say soil, he may find one is listed under sediments. Obviously this approach will not always be applicable. Thus a method listed under sewage is not likely to be applicable to the ultra low level analysis of marine sediments.

By far the most extensively used techniques employed in the analysis of solids are gas chromatography, high-performance liquid chromatography and, increasingly, supercritical fluid chromatography.

In a well-equipped laboratory it is mandatory that these three techniques be coupled with a mass spectrometric detector in order to achieve a combination of resolution of mixtures, positive identification of separated organics and the high sensitivity that is essential when dealing with environmental samples. The penetration of mass spectrometers in recent years is indicated by the fact that of the 50 types of organic compound that have been determined by gas chromatography in 21 cases mass spectrometric detection is discussed. This trend will, no doubt, continue.

In the case of high-performance liquid chromatography and mass spectrometry applications to date have been more limited. In only 10% of the types of organics discussed was mass spectrometry invoked.

The technique involving pyrolysis of the organic compound followed by gas chromatography or gas chromatography–mass spectrometry of the pyrolysis products has to date found limited but important applications (heteroaromatic compounds, polyaromatics, polychlorobiphenyls, polymers, haloaromatics, bacteria identification, anthropogenic compounds, aliphatic hydrocarbons, aromatic hydrocarbons and mixtures of organic compounds). A combination of purge and trap analysis with gas chromatography is extremely useful to the determination of volatile organic compounds such as aromatic hydrocarbons, alcohols, aldehydes, ketones, chloroaliphatics and haloaromatics.

Isotachoelectrophoresis or capillary column isotachoelectrophoresis has been employed in a number of analyses (triazine, herbicides, paraquat, diquat, growth regulators, polyaromatic hydrocarbons, polychlorobiphenyls and tetrahydrothiophene).

Spectrophotometric methods have found fairly extensive applications but are limited in sensitivity and specificity (cationic surface active agents, adenosine triphosphate, phenols, chlorophenols, amines, fungicides, carbohydrates, mixtures of organics, boron, halogens, nitrogen and phosphorus). The same comments apply to spectro-fluorimetric methods (polyaromatic hydrocarbons, humic and fulvic acids, aliphatic and aromatic hydrocarbons and optical whiteners) and thin-layer chromatographic methods (polyaromatic hydrocarbons, chlorinated, carbamate and substituted urea herbicides and pesticides and mixtures of the latter).

Atomic absorption spectrometric methods and, more recently, the inductively coupled plasma atomic emission method, are, of course, mandatory if determination of elements is required (arsenic, selenium, boron, phosphorus and silicon).

Other types of analysis with very limited applications include luminescence analysis (adenosine triphosphate, polychlorobiphenyls), neutron activation analysis (halogens, selenium), infrared spectroscopy (respirometry and mixtures of organics), gel permeation chromatography (mixtures of organics), ultraviolet spectroscopy (aromatic hydrocarbons, polyaromatic hydrocarbons and mixtures of organics), nuclear magnetic resonance spectroscopy (humic and fulvic acids, chlorinated insecticides), electron spin resonance spectroscopy (anionic surface active agents), coulometry (chlorophenols), titration methods (carbamate insecticides/ herbicides), enzyme immunoassays (polyaromatic hydrocarbons), flow-injection analysis (phosphorus) and combustion techniques (total organic carbon, phosphorus and sulphur).

References

1 Greenfield, S., Jones, I.L., Berry, C.T. (1964) *Analyst* (London), **89**, 713.
2 Wendt, R.H. and Fassell, V.A. (1965) *Analytical Chemistry*, **37**, 920.
3 Scott, R.H. (1974) *Analytical Chemistry*, **46**, 75.
4 Thompson, M. and Walsh, J.N. (1983) In *A Handbook of Inductively Coupled Plasma Spectrometry*, Blackie, London & Glasgow, p.55.
5 Suddendorf, R.F. and Boyer, K.V. (1978) *Analytical Chemistry*, **50**, 1769.
6 Sharp, B.L. *The Conespray Nebulizer*. British Technology Group, Patent Assignment No 8,432,338.
7 Gunn, A.M., Milland, D.L. and Kirkbright, G.F. (1978) *Analyst* (London), **103**, 1066.
8 Matusiewicz, H. and Barnes, R.M. (1984) *Applied Spectroscopy*, **38**, 745.
9 Tikkonen, M.W. and Niemczyk, T.M. (1984) *Analytical Chemistry*, **56**, 1997.
10 Salin, E.D. and Horlick, G. (1979) *Analytical Chemistry*, **51**, 2284.
11 Salin, E.D. and Szung, R.L.A. (1984) *Analytical Chemistry*, **56**, 2596.
12 Vaughan, C.G., Wheets, B.B. and Whitehouse, M.J.J. (1973) *Journal of Chromatography*, **28**, 203.
13 Lewis, N.M. (1975) *Water Treatment and Examination*, **14**, 243.
14 Sorrell, R.K. and Reding, R. (1979) *Journal of Chromatography*, **185**, 655.
15 Sorrell, R.K., Dressman, R.L. and McFarrer, E.F. (Proceedings of the Water Quality Technology Conference, Kansas City, Mo. December 5–7 1977, American Water Works Association, Denver, Colorado, pp.3A–3 (1978).
16 Das, B.S. and Thomas, G.H. (1978) *Analytical Chemistry*, **50**, 967.
17 Schonmann, M. and Kern, H. (1981) *Varian Instrument Applications*, **15**, 6.

18 Hewlett Packard (1988) *Peak*, Autumn, 10.

19 Rivera, J., Carxach, S., Ventura, F. *et al.* FAB–CAD–MIKES. *Analysis of non-ionic surfactant in raw sewage sludges*, (ed J.F.F. Todd), Proceedings of the 10th International Mass Spectrometry Conference, Swansea September 9–13 1985, J. Wiley, Basingstoke, UK (1985).

20 Righton, M.J.G. and Watts, C.D. (1986) *Identification of surfactants using sublation extraction and fast atom bombardment mass spectrometry.* Water Research Center Report. ER. 1194–M. December.

21 Klesper, E., Corwin, A. and Turner, D. (1962) *Journal of Organic Chemistry*, **27**, 700.

22 Novotny, M., Springstron, P.J. and Lee, M. (1981) *Analytical Chemistry*, **53**, 407A.

23 Wall, R.J. (1988) In *Chromatography and Analysis.* J. Wiley & Sons, London.

24 Later, D., Bornhof, D., Lee, E. *et al.* (1987) *Liquid Chromatography – Gas Chromatography*, **444**, 804.

25 Kennedy, S. and Wall, R. (1988) *Liquid Chromatography – Gas Chromatography*, **445**, 10.

26 Sim, P., Elson, C. and Quilliam, M. (1988) *Journal of Chromatography*, **445**, 239.

27 Shackleford, W.W. and McGuire, J.M. (1986) *Spectra*, **10**, 17.

28 Tellaid, W.A. (1986) *Spectra*, **10**, 4.

29 Federal Register (1979), 44, 94666.

30 Federal Register (1984) 49, 38801.

31 Federal Register (1984) Method 624, 49, 43234.

32 Colby, B.N. (1986) *Spectra*, **10**, 49.

33 Fisk, J.F., Haeberer, A.M. and Kovell, S.P. (1986) *Spectra*, **10**, 22.

34 Friedman, D. (1986) *Spectra*, **10**, 40.

35 Warburton, G. and Millard B. (1984) *International Laboratory*, issue 7, xii.

36 Kelly, P.E. New Advances in the operation of the ion trap mass spectrometry. *Finnigan MAT IDT* 10.

38 Kelly, P.E. Ion trap detector literature reference list. *Finnigan MAT IDT* 21.

39 Campbell, C. The ion trap detector for gas chromatography: technology and application. *Finnigan MAT IDT* 15.

39 Stafford, G.C. Recent improvements in and analytical applications of advanced ion trap technology. *Finnigan MAT IDT* 16.

40 Stafford, G.C. Advanced ion trap technology in an economical detector from GC. *Finnigan MAT IDT* 20.

41 Stafford, G.C. The Finnigan MAT ion trap mass spectrometry – new developments with ion trap technology. *Finnigan MAT IDT* 24.

42 Rordorf, B.F. An automated flow tub kinetics instrument with integrated GC–IDT analysis. *Finnigan MAT IDT* 13.

43 Syka, J.E.P. Positive ion chemical ionization with an ion trap mass spectrometer. *Finnigan MAT IDT* 19.

44 Yost, R.A., McClennan, W. and Menzzelaar, H.L.C. Enhanced full scan sensitivity and dynamic range in Finnigan MAT ion trap detector with automatic gain control of software. *Finnigan MAT IDT* 22.

45 Camp, C. Ion trap advancements. Higher sensitivity and greater dynamic range with automatic gain control software. *Finnigan MAT IDT* 23.

46 Richards, J.M. and Bradford, D.C. Development of a Curie Point pyrolyser inlet from the Finnigan MAT ion-trap detector. *Finnigan MAT IDT* 25.

47 Bishop, P. The ion trap detector, universal and specific detection in one detector. *Finnigan MAT IDT* 28.

48 Bishop, P. The Use of an ADT 50 GLC ion trap detector combination. *Finnigan MAT IDT* 36.

49 Bishop, P. Low cost mass spectrometer for GC. *Finnigan MAT IDT* 42.

50 Campbell, C. and Evans, S. The ion-trap detector – techniques and its application. *Finnigan MAT IDT* 29.

51 Olsen, E. Serially interfaced gas chromatography/Fourier Transform infrared spectrometer/ion trap detector. *Finnigan MAT IDT* 35.
52 Allison, J. The hows and whys of ion trapping. *Finnigan MAT IDT* 41.
53 Todd, J., Mylchreest, I., Berry, T. and Games, D. Supercritical chromatography mass spectrometry with an ion-trap detector. *Finnigan MAT IDT* 46.
54 Eichelberger, J.W. and Budd, W.L. Studies in mass spectrometry with an ion trap detector. *Finnigan MAT* 47.
55 Eichelberger, J.W. and Slivon, L.E. Existence of self chemical ionization in the ion-trap detector. *Finnigan MAT IDT* 48.
56 Genin, E. Le détecteur à plegeage d'ions de chromatogeraphie en phas gaseux. Technologie et applications. *Finnigan MAT IDT* 53.
57 Lehair, M. The use of the IDT low cost GC/MS system for the identification of trace compounds. *Finnigan MAT IDT* 51.
58 Richards, J.M, McClennan, W.H., Burger, J.A. and Menza, H.H.C. Pyrolysis short column GC/MS as by the IDT and ITMS. *Finnigan MAT IDT* 56.
59 Complex flavour analysis with the ion trap detector. *Finnigan MAT Application Data Sheet* ADS 7.
60 Determination of Δ9 carboxyl THC by GC/MS with the ion-trap detector. *Finnigan MAT Application Data Sheet* ADS 8.
61 Analysis of commercially abused drugs by GC/MS with the ion trap detector. *Finnigan MAT Application Data Sheet* ADS 9.
62 Analysis of base neutrals by GC/MS with the ion-trap detector. *Finnigan MAT Application Data Sheet* ADS 10.
63 Analysis of coal extract with the ion-trap detector. *Finnigan MAT Application Data Sheet* ADS 12.
64 Gas chromatographic analyses of China White (Fentanyl) with the ion-trap detector. *Finnigan MAT Application Data Sheet* ADS 13.
65 High sensitivity full scan analysis of D3–11 NOR-9-carboxyl Δ-9 tetrahydrocannabinol by the ion-trap detector. *Finnigan MAT Application Data Sheet* ADS 14.
66 High sample levels with the ion-trap detector. *Finnigan MAT Application Data Sheet* ADS 24.
67 Calibration of the ion-trap detector for quantification of tetrafluor–1.4 dicyanobenzene and 2,6 dichloro-4-nitroaniline. *Finnigan MAT Application Data Sheet* ADS 27.
68 Dynamic range of the ion-trap detector. *Finnigan MAT Application Data Sheet* ADS 29.
69 Schnute, W.C. Analysis of volatile organic compounds, industrial wastes by the Finnigan MAT OWA GC/MS. *Finnigan Corporation Applications Report* No. AR 8018.
70 Steadman, J. and Matnura, R.F.C. The use of pollutant and biogenic markers as source discrimination of organic inputs to estuarine sediments. *Finnigan MAT IDT* 8.
71 Westendorf, R.G. (1986) Presented at 1986 Water Quality Technology Conference of the American Water Works Association. Portland, Oregon, November. Tekmar Company, PO Box 371865, Cincinatti, OH 45223–11856, USA.

Chapter 2

Hydrocarbons

2.1 Aliphatic hydrocarbons

2.1.1 Soils

2.1.1.1 Gas chromatography

Karasek *et al.* [1] determined hydrocarbons in benzene water extracts (pH7) of soil and in incinerator or fly ash by a variety of techniques including gas chromatography with flame ionization, electron capture and mass spectrometric detectors. Benzene water extractants were adjusted to pH4, 7 and 10 before the extraction in order to selectively extract various types of acidic and basic organic compounds in addition to hydrocarbons.

2.1.1.2 Supercritical fluid extraction chromatography

Yang *et al.* [2] have compared sorbent trapping with solvent trapping after the supercritical fluid extraction of volatile petroleum hydrocarbons in soil. Sorbent trapping yielded quantitative collections of *n*-alkanes as volatile as *n*-hexane, while solvent trapping effectively collected *n*-alkanes as volatile as *n*-octane.

Burford *et al.* [3] reported a coupled supercritical extraction-gas chromatographic method that can quantitatively extract and determine both gasoline and diesel range hydrocarbons from contaminated soils. The direct transfer of the extract to a gas chromatograph reduced analysis times to about 80min, compared to the 18h required for conventional sonication analysis.

Liang and Tilotta [4] have described the use of supercritical argon for the extraction of petroleum hydrocarbons from soil samples. Argon is an attractive solvent because it is inexpensive and inert. Additionally, it has a clear spectral window in the infrared region which makes it useful for on-line (i.e. directly coupled) experiments. Spiking studies conducted with gasoline, no. 1 fuel oil, and no. 5 fuel oil on sand, loam and clay show that component recovery rates for argon supercritical fluid extraction generally increase with increasing pressure and/or temperature. The highest recovery rates (and recoveries) were obtained for argon supercritical fluid extraction at 500atm and 150°C. Under these conditions, the components of the gasoline and no. 1 fuel oil spikes could be recovered in as little as 12min. However, the no. 5 fuel oil components could not be quantitatively removed from the loam and clay matrixes even for extraction times as long as 100min. It is shown in this work that argon supercritical fluid extraction performs similarly to carbon dioxide supercritical fluid extraction for petroleum hydrocarbon contamination in real-world soil samples under moderate pressure and temperature conditions. Specifically, argon supercritical fluid extraction and carbon dioxide supercritical fluid extraction have similar recoveries and reproducibilities, but argon supercritical fluid extraction requires a slightly longer extraction time.

2.1.1.3 Miscellaneous

Commonly used methods for the determination of petroleum hydrocarbon contamination in soil are modifications of Environmental Protection Agency method 418.1, which use sonication or a Soxhlet apparatus for analyte extraction and either infrared spectrometry [5] or gas chromatography with flame ionization detection [6–7] for extract analysis. Regardless of the analytical method following the extraction, both modifications use Freon–113, which has been implicated as a cause of ozone depletion. Therefore, alternative methods are being sought for the determination of hydrocarbon contamination in environmental samples that reduce the need for this halogenated solvent.

Concawe [8] have described a method for the determination of aliphatic hydrocarbons in soil based on carbon tetrachloride extraction followed by infrared spectroscopy or gas chromatography.

Thermal desorption mass spectrometry is a rapid technique for the determination of oil in soils and sediments [9]. This method exhibited lower analytical variance compared to Soxhlet extraction, i.e. followed by conventional analysis. The analysis time for wet soil samples was about 20min.

Thermogravimetric methods such as pyrolysis gas chromatography–mass spectrometry have been used to characterize hydrocarbon sludges from polluted soils [10]. In combination with conventional extraction and supercritical fluid extraction followed by gas chromatography–mass spectrometry, over 100 constituents were identified in soil samples. Thermogravimetric analysis–mass spectrometric results distinguished between the release of a component by thermosorption and by pyrolysis.

2.1.2 *Non-saline deposited and suspended sediments*

2.1.2.1 Gas chromatography

Gas chromatography has been used to distinguish between fossil fuels added to sediments through oil pollution and those hydrocarbons present in low concentrations as natural biogenic products (Blumer and Sass [11], Farrington and Quinn [12]).

Guiney *et al.* [13] showed that fused silica capillary gas chromatography offered high resolution and precision and provided reasonable detection limits for analysing kerosine range hydrocarbons in non-saline sediments isolated from rivers and streams.

2.1.2.2 Fluorescence spectrometry

Wakeham [14] has discussed the application of synchronous fluorescence spectroscopy to the characterization of indigenous and petroleum derived hydrocarbons in lacustrine sediments. The author reports a comparison, using standard oils, of conventional fluorescence emission spectra and spectra produced by synchronously scanning both excitation and emission monochromators.

2.1.2.3 Miscellaneous

Broman *et al.* [15] have discussed methods of fingerprinting petroleum hydrocarbons in bottom sediments.

Petroleum pollution monitoring laboratories in the Mediterranean region participated (1984–1986) in two intercalibration exercises (MEDCAL I and II) to evaluate the International Oceanographic Commission (IOC) Manual for petroleum hydrocarbon determination in sediment (IOC, Manuals and Guides, No. 11). The main source of error in the analysis was the extraction/partition step. When the results were corrected for recoveries, relative standard deviations for *n*-alkanes, UCM (unresolved complex mixture) and total aromatics, which had previously been 60, 56 and 49%, respectively, were reduced to 17, 30 and 6%, respectively.

2.1.3 *Saline deposited and suspended sediments*

2.1.3.1 Gas chromatography

Gas chromatography has also been used to distinguish between fossil fuels added to sediments through oil pollution and those hydrocarbons present in low concentrations as natural biogenic products (Blumer and Sass [16]; Farrington and Quinn [12]).

Walker *et al.* [17] studied profiles of hydrocarbons in sediment according to depth in sediment cores collected at Baltimore Harbour in

Chesapeake Bay, Massachusetts. Gas liquid chromatography was used to detect hydrocarbons present at different depths in the sediment, while low resolution mass spectrometry was employed to measure concentrations of paraffins, cycloparaffins, aromatics and polynuclear aromatics. Their data show that the concentrations of total and saturated hydrocarbons decreased with increased depth, and it was noted that identification and quantification of hydrocarbons in oil-contaminated sediments is required if the fate of these compounds in dredge spills is to be determined.

May *et al.* [18] have described a gas chromatographic method for analysing hydrocarbons in marine sediments and sea water which is sensitive at the submicrogram per kilogram level. Dynamic head space sampling for volatile hydrocarbon components, followed by coupled-column liquid chromatography for analysing the non-volatile components, requires minimal sample handling, thus reducing the risk of sample component loss and/or sample contamination. The volatile components are concentrated on a Tenax gas chromatographic precolumn and determined by gas chromatography or gas chromatography–mass spectrometry.

Brown *et al.* [19] have described a gas chromatography–mass spectrometry technique for fingerprinting petrogenic hydrocarbons. The technique identified and quantified *n*-alkanes, the isoprenoids pristane and phytane, pentacyclic triterpanes, the unresolved complex mixture, and total hydrocarbon content. Results obtained using sediments preserved with chloroform during sediment trap collection were compared with those for unpreserved anoxic sediments and anoxic bottom surface sediment. Petrogenic hydrocarbons were detected at all stations, concentrations decreasing with increasing distance from an urban area. Carbon preference index values increased along the transect, indicating a greater dominance of biogenic hydrocarbons further out in the archipelago. The compositions of preserved and unpreserved anoxic samples were very similar. These results indicated that the sediment trap technique was a useful method of collecting and preserving material for fingerprinting petrogenic hydrocarbons.

Takada and Ishimatari [20] extracted alkylbenzenes with normal C_{10}–C_{14} and branched C_{11}–C_{13} alkyl chains from marine and coastal sediment and suspended matter in benzene methanol. The extract in benzene was then applied to a Florisil column for removal of copper sulphide and polar materials, and then subjected to silica gel column chromatography. Alkyl benzenes were quantified and identified using gas chromatography with flame ionization detection. The recoveries of alkyl benzenes were 81–94%.

Page *et al.* [21] used capillary gas chromatography and capillary gas chromatography–mass spectrometry to determine aliphatic hydrocarbons in interstitial sediments collected on the French coastline following the Amoco Cadiz disaster.

2.1.3.2 Spectrofluorimetry

Brown *et al.* [22] have described a rapid field method for detecting down to 2µg of oil in sediments associated with marine oil spills. The method was employed in connection with the Argo Merchant oil spill off Nantuckett in December 1976.

In this method the sediment is mixed with sodium sulphate and extracted with *n*-hexane. A portion of the extract is applied to a paper strip which is then eluted with petroleum ether:benzene (35:65) for 60 seconds. Viewing of the strip under ultraviolet light reveals a blue fluorescent spot indicating the presence of oil in the sediment.

Fluorescence spectroscopy has been adapted as an alternative analytical method for estimating oil in sediments [23,24].

Interlocutory, comparisons have been performed on the determination of selected trace aliphatic and aromatic hydrocarbons in marine sediments [25–27].

2.1.3.3 Infrared spectroscopy

Mark [28] has described an infrared method for the determination of the oil content of marine sediments. He showed that the magnitude of the CH_2 stretching band at 2925cm $^{-1}$, normally used to determine oil in a sediment, is enhanced when biological matter is also present. The concentration of this material can generally be estimated from the magnitude of the protein-NH band at 1650cm $^{-1}$ with the use of a calculated correction to the total absorption at 2925cm $^{-1}$, but the oil must contribute less than 10% to the total absorption at 2925cm $^{-1}$. It is desirable, however, that the nature of the organic matter be determined by means of a study of the complete infrared spectrum.

2.1.3.4 Miscellaneous

Whittle [29] has described a thin-layer chromatographic method for the identification of hydrocarbon marker dyes in oil polluted waters.

McLeod *et al.* [25] conducted interlaboratory comparisons of methods for determining traces of aliphatic and aromatic hydrocarbons in marine sediments. Agreement within a factor of 2 to 3 was obtained between the 12 participating laboratories.

2.1.4 Sludges

2.1.4.1 Column chromatography

In a method described by Ryzhova *et al.* [30] for the determination of aliphatic hydrocarbons in wastewater sludges the hydrocarbons were extracted from sludge using a mixture of pentane and hexane. Metal salts and compounds containing oxygen were removed by successive treatment with 30% sodium hydroxide and 85% phosphoric acid. After concentration and drying, extracts were evaporated under nitrogen. Hydrocarbons were separated by eluent-adsorption chromatography on ASK silica. Hydrocarbons formed 95% of the total organic content of the extracted material. Quantitative gas chromatographic results for 41 compounds are presented. Saturated compounds present in the locally treated sludge and water were principally alkanes. Cycloalkanes did not exceed 1%. Over 25 alkyl benzenes were also quantified.

2.2 Aromatic hydrocarbons

2.2.1 Soil

2.2.1.1 Purge and trap gas chromatography

Kester [31] has discussed the application of purge and trap gas chromatography to the determination of aromatic hydrocarbons such as benzene, ethyl benzene, toluenes and xylene in soil. In this method, a 4g portion of the soil is dispersed in 9ml methanol and 1ml of a methanoic surrogate spike containing deuterated compounds for mass spectrometric recovery analysis in a 15ml screw capped vial. Volatile compounds are dissolved in the solvent by shaking for 1min or by sonicating for 30min. The slurry is allowed to settle centrifuging if necessary and 10 to 100µl aliquot of the extract is added to organic-free water and dry pumped and ambient temperature into a Tenax trap. The purge gases are analysed by gas chromatography using a photoionization detector or can be analysed by mass spectrometry.

2.2.1.2 Pyrolysis gas chromatography–mass spectrometry

de Leeuw *et al.* [32] have described a method based on Curie Point flash evaporation-pyrolysis gas chromatography–mass spectrometry for the fast screening of anthropogenic aliphatic hydrocarbons in soils. The detection limit is in the low $\mu g\ kg^{-1}$ range. Polycyclic aromatic hydrocarbons, heteroaromatic hydrocarbons and haloorganic compounds and pyrolysis products of polymers can also be screened by this method.

These workers observed that during Curie Point flash pyrolysis compounds which are reasonably volatile at elevated temperatures do not fragment on the pyrolyser wire but simply evaporate from it. Thus it appeared possible for them that the organic matter present in soils can be characterized and identified very rapidly without any sample pretreatment by direct evaporation/pyrolysis of whole samples.

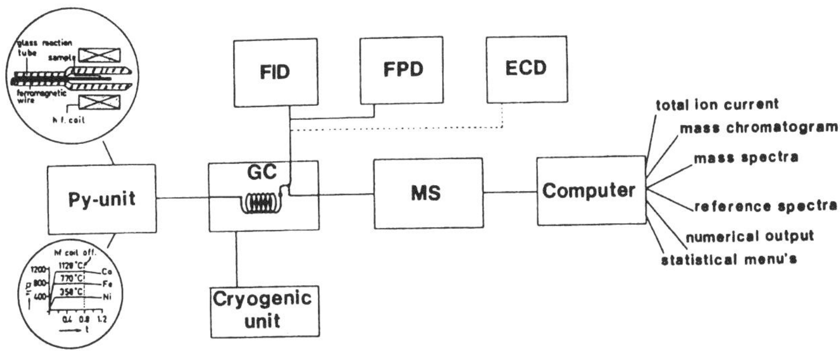

Fig. 2.1 Instrumental setup for screening analysis by evaporation/pyrolysis gas chromatography
Source: Reproduced with permission from American Chemical Society [32]

Method

The soil sample was homogenized by grinding before analysis.

The instrumental setup for the screening analysis is shown in a very schematic way in Fig. 2.1. A ferromagnetic wire with a selected Curie temperature (see insert) is coated with a methanol suspension of the soil or sediment sample to be analyzed and allowed to dry in the air. The coated wire is inserted into a Pyrex tube and positioned in the pyrolysis unit between the high-frequency coils. Powering of this high-frequency coil causes a rapid temperature rise of the wire containing the sample to the Curie temperature within 0.1–0.2s (see Fig. 2.1). The compounds generated, either by pyrolysis or evaporation, are fed to a capillary column [33] that is cooled by means of a cryogenic device where they are trapped on the liquid phase of the capillary column. These compounds are then separated on the capillary column, and the individual compounds can be monitored by common flame ionization detectors, flame photometric detectors, or electron capture detectors, or they can be monitored and identified by a mass spectrometer (GC–MS mode).

Figure 2.2 shows the total ion current trace and a number of appropriate mass chromatograms obtained from the pyrolysis gas chromatography–mass spectrometry analysis of the polluted soil sample. The upper trace represents a part of the total ion current magnified eight times. The peak numbers correspond with the numbers mentioned in Table 2.1 and refer to the identified compounds. The identification was based on manual comparison of mass spectra and relative gas chromatographic retention times with literature data [34, 35] and with data of standards available. In some cases unknown compounds were tentatively identified on the basis of *a priori* interpretation of their mass spectra (labelled 'tentative' in Table 2.1).

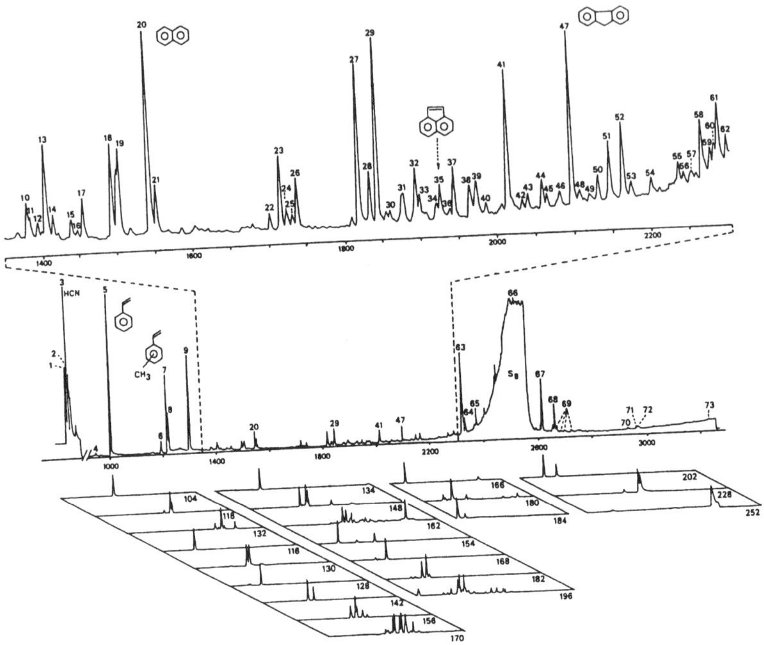

Fig. 2.2 TIC of Ev/Py GC–MS analysis of the polluted soil sample. The upper trace represents a part of the TIC magnified eight times. The number in the mass chromatograms represent the *m*/*z* values indicative of the following classes of compounds: *m*/*z* 104, 118, 132 (C_0–C_2 styrenes); *m*/*z* 116, 130 (C_0–C_1 indenes); *m*/*z* 128, 142, 156, 170 (C_0–C_2 naphthalenes); *m*/*z* 134, 148, 162 (C_0–C_2 benzo[*b*]thiophenes), *m*/*z* 154 (biphenyl and acenaphthene); *m*/*z* 168, 182, 196 (C_1–C_2 biphenyls and C_0–C_2 dibenzofurans); *m*/*z* 166, 180 (C_0–C_1 fluorenes and 9-fluorenone); *m*/*z* 184 (dibenzothiophene); *m*/*z* 202 (fluoranthene and pyrene); *m*/*a* 228 (benzo[*c*]phenanthrene, benz(a)anthracene, chrysene, and triphenylene); *m*/*z* 252 (benzo[e]pyrene and benzo[*a*]pyrene). The *x* axes of the mass chromatograms correspond exactly with the appropriate parts of the TIC *x* axis directly above them (e.g. the major peak in the mass chromatogram of *m*/*z* 168 corresponds with peak 41 in the total ion current trace).
Source: Reproduced with permission from American Chemical Society [32]

Because no pretreatment of the samples was carried out, the peaks present in the total ion current trace reflect components generated by pyrolysis of primary compounds ('real pyrolysis products') and components that are present as such in the sample and simply evaporate ('free products'). If desired these two types of products may be differentiated using wires with a Curie temperature of 358°C [36]. It was demonstrated in separate analyses (not shown here) that most compounds were not generated by pyrolysis but were present as such in the sample and 'thermally extracted'. Compounds 1–8 and 10–17, 27, 37, 38, 54 and 65 were only present in pyrolysis gas chromatographic analyses carried out

Table 2.1 Identified evaporation and pyrolysis products of the soil sample

peak no.	*compd*	*peak no.*	*compd*	*peak no.*	*compd*
1	H_2S, CO_2, CO	25	methylbenzo[*b*]thiophene	49	dimethylbiphenyl
2	dicyanogen	26	1-methylnaphthalene	50	unknown organic sulphur compound (*m/z* 67, 100, 196)
3	hydrogen cyanide	27	1-phenylethyl *tert*-butyl ether (tentative)	51	methylbenzofuran
4	ethylbenzene	28	bipheny	52	methylbenzofuran
5	styrene	29	unknown organic sulphur compound (*m/z* 67, 100, 164)	53	methylbenzofuran
6	α-methylstyrene	30	1-ethylnaphthalene + dimethyl-benzo[*b*]thiophene	54	α-l-phenyl ethyl ether (tentative)
7	3-methylstyrene	31	2,6- and/or 2,7-dimethylnaphthalene	55	2-methylfluorene
8	4-methylstyrene	32	1,3-dimethylnaphthalene	56	1-methylfluorene
9	indene	33	1,7- and/or 1,6-dimethylnaphthalene	57	C_2-benzofuran
10	α,3-dimethylstyrene	34	2,3- and/or 1,4-dimethylnaphthalene	58	9-fluorenone
11	3-ethylstyrene	35	acenaphthalene	59	C_2-benzofuran
12	α,4-dimethylstyrene	36	1,2-dimethylnaphthalene	60	C_2-benzofuran
13	3,5-dimethylstyrene	37	(C_1-phenyl)ethyl *tert*-butyl ether (tentative)	61	dibenzothiophene
14	α,2- or 2,5- or 2,4-dimethyl-styrene	38	(C_1-phenyl)ethyl *tert*-butyl ether (tentative)	62	C_2-benzofuran
15	phenyl ethyl ether	39	acenaphthene + 4-methylbiphenyl	63	phenanthrene
16	2,3-dimethylstyrene	40	3-methylbiphenyl	64	anthracene
17	3,4-dimethylstyrene	41	dibenzofuran	65	bis(1-phenylethyl) thioether (tentative)
18	methylidene	42	C_3-naphthalene	66	elemental sulphur
19	isomeric methylidenes	43	C_3-naphthalene	67	fluoranthene
20	naphthalene	44	C_3-naphthalene	68	pyrene
21	benzo[*b*]thiophene	45	C_3-naphthalene	69	isomeric naphthobenzofurans
22	methylbenzo[*b*]thiophene	46	C_3-naphthalene	70	benzo[*c*]phenanthrene
23	2-methylnaphthalene	47	fluorene	71	benzo[*a*]anthracene
24	methylbenzo[*b*]thiophene	48	C_3-naphthalene + dimethylbiphenyl	72	chrysene + triphenylene
				73	benzo[*a*]pyrene + benzo[e]pyrene

Source: Reproduced with permission from American Chemical Society [32]

with wires having a Curie point of 510°C and therefore are considered to be real pyrolysis products.

2.2.2 *Saline deposited and suspended sediments*

2.2.2.1 Spectrofluorimetry

Hargrave and Phillips [37] have used fluorescence spectroscopy to evaluate concentrations of aromatic constituents in aquatic sediments. The oil concerned, a Venezuelan crude, contained about 35% by weight of aromatic constituents. Aromatic substances were extracted with n-hexane and fluorescence spectroscopy was used to produce a series of contour diagrams of fluorescence intensity at various excitation and emission wavelengths, in order to compare fluorescence spectral patterns of sample extracts and standard oils. Petroleum residues were determined and it was found that total oil concentrations ranged from 10 to 3000μg g^{-1} wet sediment, with the highest concentrations occurring in sediment particles.

2.2.2.2 High-performance liquid chromatography

Vowles and Mantoura [38] determined sediment-water partition coefficients and the high-performance liquid chromatography capacity factors for 14 alkylbenzene and polyaromatic hydrocarbons. The partition coefficients correlated well with the alkyl-cyano capacity factors, and it was concluded that this phase gave a better indication of sorption on sediment than either the octanol or octadecylsilane phases.

Krahn *et al.* [39] have described a high-performance liquid chromatographic method for the determination of 127 aromatic hydrocarbons and 21 chlorinated hydrocarbons in solvent extracts of marine sediments.

2.2.2.3 Ultraviolet spectroscopy

Hennig [40] has applied ultraviolet spectroscopy to the determination of aromatic constituents of residual fuel oil in hexane extracts of marine sediment samples. Examination of the ultraviolet spectra of samples of an oil pollutant from a beach and crude oil, at various concentrations, revealed strong absorption maxima at approximately 228nm and 256nm. The ratio of the peak heights at these wavelengths is constant for a particular oil, and is independent of concentration. These permit quantitative analysis of sediment samples many months after an oil spill.

A recovery of 97% was obtained in this method using 10g of clean sediment samples spiked with 777μg oil. The standard deviation was 0.014, and the coefficient of variance = 3.47% with $n = 10$.

The ultraviolet absorption spectra of extracts of oil-contaminated sediment samples and unpolluted beach are shown in Fig. 2.3. Levy [41]

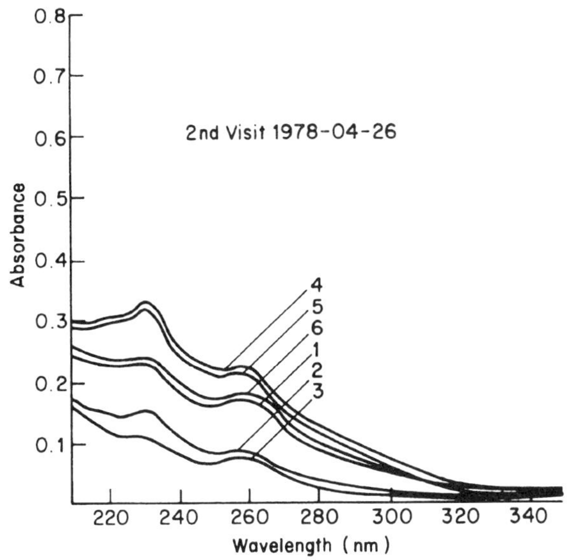

Fig. 2.3 Ultraviolet absorption spectra of field sediment: (1) Vic 2/2/3; (2) Vic 2/2/10; (4) Vic 2/3/5; (6) Vic 2/3/10
Source: Reproduced with permission from Elsevier Science Ltd [40]

has shown the *R* value for a given oil to be constant for more than one year. This fact makes the method particularly useful in long-term pollution studies.

2.3 Heteroaromatic hydrocarbons

2.3.1 Soil

2.3.1.1 Pyrolysis gas chromatography

The Curie Point flash evaporation-pyrolysis gas chromatography–mass spectrometric method [32] described in section 2.2.1.2 for the analysis of aromatic hydrocarbons in soils has also been applied to the determination of heteroaromatic compounds (Table 2.2) such as methyledene, isomeric methylidenes, biphenyl and methylbenzofurans.

2.4 Polyaromatic hydrocarbons

2.4.1 Soil

2.4.1.1 Gas chromatography

The extraction procedure described by Karasek *et al.* [1] in section 2.1.1.1 has been applied to the determination of polyaromatic hydrocarbons in

Table 2.2 Concentration[a] of several contaminants in the soil sample

compd	*concn, ppm*
soil	
phenanthrene	506
anthracene	144
fluoranthene	340
pyrene	208
benz[*a*]anthracene	79
chrysene/triphenylene	103
benzo[e]pyrene	35
benzo[*a*]pyrene	38
perylene	12
sediment	
pentadecane	300

Source: Reproduced with permission from American Chemical Society [32]

[a]Based on wet weight

soils. In this procedure the soil is Soxhlet extracted for 48h with a mixture of benzene and water at pH7. Gas chromatographic techniques using a variety of detectors were used to identify and determine polyaromatic hydrocarbons, also phenols, hydrocarbons and polychlorinated dibenzofurans.

2.4.1.2 Gas chromatography–mass spectrometry

Robbat *et al.* [42] carried out on-site detection of polyaromatic hydrocarbons in soils using thermal desorption gas chromatography–mass spectrometry on hexane extracts of soils.

2.4.1.3 Pyrolysis gas chromatography

The Curie Point flash evaporation-pyrolysis gas chromatography–mass spectrometric method [32] described in section 2.2.1.2 for the analysis of aliphatic hydrocarbons in soils has also been applied to the determination of polyaromatic hydrocarbons (see Table 2.1). Table 2.2 lists the polyaromatic hydrocarbon contents found by this method in a soil sample.

Analysis of variance was used to assess the effects on polyaromatic hydrocarbons extraction at the 99% confidence level for the four factors varied. The percentage of ^{14}C in the extract and soil residue does not total 100% because of degradation and volatilization during incubation and due to losses during analysis. The data are presented in Table 2.3 and represent the average of the three replicates for the extract or soil residue.

Table 2.3 Percent recovery of 14C in extract and residue[a]

extraction	*PAH*[b]	*concn, µg/g*	*$HgCl_2$,%*	*extract*	*residue*
Soxhlet	BaP	50	0.5	79.3	3.1
			0	77.1	2.5
		5	0.5	74.9	9.3
			0	77.8	6.6
	Anthracene	50	0.5	76.4	3.3
			0	75.5	3.0
		5	0.5	75.8	5.3
			6	61.1	4.5
Polytron	BaP	50	0.5	68.2	4.6
			0	71.9	4.6
		5	0.5	63.0	12.7
			0	64.9	9.6
	Anthracene	50	0.5	67.5	7.4
			0	68.3	3.8
		5	0.5	52.2	10.5
			0	50.5	8.9

Source: Reproduced with permission from the American Chemical Society [32]

[a]Each value is the mean of 3 replicates. [b]BaP, benzo[*o*]pyrene

Significant differences at the 99% confidence level were observed for the extraction technique and for the polyaromatic hydrocarbons concentration in the soil. The average recovery by the Soxhlet technique was 74.5% whereas 62.8% was the average Polytron recovery. A much higher proportion was extracted with polyaromatic hydrocarbons at the 50µg/g level (72.6%) than at the 5µg/g level (64.6%) suggesting that the extraction efficiency is not constant with concentration.

2.4.1.4 Thin layer chromatography

Fowlie and Bulman [43] have carried out a detailed study of the extraction of anthracene and benzo[*a*]pyrene from soil. They carried out a replicated [24] factorial experiment using Soxhlet extraction and Polytron techniques. Soxhlet extraction followed by thin layer chromatography gave higher recoveries of the two polyaromatic hydrocarbons.

In this study the soil samples were spiked with labelled and unlabelled benzo[*a*]pyrene, or anthracene at 5 and 50µg/g soil. The samples were incubated in biometer flasks at 20°C for three and five months for anthracene and benzo[*a*]pyrene, respectively, allowing degradation to be monitored and the polyaromatic hydrocarbon to interact with the soil matrix.

Subsamples were taken from each flask and extracted by one of two methods: by overnight Soxhlet extraction with 1:1 hexane:acetone [44] or by extraction in a Polytron homogenizer (Brinkman Instruments Ltd) with three successive 24mL portions of acetone [45].

Thin layer chromatography of the solvent extracts was carried out on Whatman LKSDF silica gel 250µm plates. The plates were developed in ascending fashion with 1:1:1 hexane:acetone:toluene or hexane. After drying, the lanes were divided into 1cm sections and the silica gel recovered for counting of the labelled benzo[*a*]pyrene and anthracene.

2.4.1.5 Mass spectrometry

Hankin *et al.* [46] have used spacially residued time of flight mass spectrometry for quantification studies on polyaromatic hydrocarbons. Deuterated polyaromatic hydrocarbons were used as internal standards, chrysene-d_{12} being adopted in the final method. Theoretical values were obtained by this procedure on standard reference soils.

2.4.1.6 Electrophoresis

Brown *et al.* [47] have described a cyclodextran modified capillary electrophoretic method for the determination of polyaromatic hydrocarbons in contaminated soils.

The soil was extracted by one of two methods:

1 direct extraction with dichloromethane followed by dilution with methanol; or
2 extraction of soil with carbon dioxide supercritical fluid and dilution of extract with dichloroethane or methanol or 1:1 mixtures thereof.

In the electrophoretic method negatively charged sulfobutyl ether ßcyclodextrin ($S_ß$BCD) and neutral methyl ßcyclodextrin (MßCD) were added to the running buffer, and separation of polyaromatic hydrocarbons was effected on the basis of differential distributions (partitioning) of the polyaromatic hydrocarbons components between the two cyclodextrin (CD) types. Satisfactory separation of 16 polyaromatic hydrocarbons was achieved in under 20m with 35mm SBßCD and 1mM MßCD with efficiencies of 10^5 theoretical plates. Laser induced fluorescence detection provided sensitive detection of 11 of the 16 compounds. Detection limits were in the low µg L^{-1} range.

2.4.1.7 Supercritical fluid extraction

Lagenfeld *et al.* [48] studied the effect of temperature and pressure on the supercritical fluid extraction efficiencies of polyaromatic hydrocarbons and polychlorobiphenyls in soils. At 50°C raising the pressure from 350 to 650atm had no effect on recoveries.

Burford *et al.* [49] studied extraction rates of spiked versus native polyaromatic hydrocarbons from heterogeneous environmental samples using supercritical fluid extraction and sonication in methylene chloride. Relative extraction rates of native polyaromatic hydrocarbons ranging from naphthalene M = 128 to benzo[*b*]fluoranthrene M = 252, and those of spiked deuterated polyaromatic hydrocarbons from heterogeneous railroad bed soil compared well with results obtained using sequential extraction with pure supercritical carbon dioxide or modified (10% methanol) supercritical carbon dioxide and using sonication with methylene chloride. Extraction rates of spiked deuterated polyaromatic hydrocarbons were up to ten-fold higher than with the same native polyaromatic hydrocarbons. 30min extraction with pure carbon dioxide recovered >90% of polyaromatic hydrocarbon in most cases, but only recovered 25–80% of native polyaromatic hydrocarbon.

Reindt and Hoffler [50] optimized parameters in the supercritical fluid extraction of polyaromatic hydrocarbons from soil. These workers used carbon dioxide –8% methanol for extraction and obtained 88–101% recovery of polyaromatic hydrocarbons in the final high-performance liquid chromatography.

Barnabas *et al.* [51] have discussed an experimental design approach for the extraction of polyaromatic hydrocarbons from soil using supercritical carbon dioxide. They studied 16 different polyaromatic hydrocarbons using pure carbon dioxide and methanol modified carbon dioxide. The technique is capable of determining down to 100mg kg^{-1} polyaromatic hydrocarbons in soils.

Tena *et al.* [52] carried out a screening of polyaromatic hydrocarbon types in soil by on-line fibre optic interfaced supercritical fluid extraction spectrofluorimetry.

The apparatus incorporates a fibre optic interface for the spectrofluorimetric measurement on the supercritical carbon dioxide emerging from the extraction cell of a supercritical fluid extractor, prior to depressurization from up to 350 bar. Recoveries of polyaromatic hydrocarbons are between 89 and 107%, and measurements can be carried on with a relative standard deviation of less than 5%.

Accelerated solvent extraction is a new technique for the extraction of a range of organic pollutants from soils and related material. The technique is based on the use of a solvent or combination of solvents to extract organic pollutants at elevated pressure and temperature from a solid matrix. The range of organic pollutants for which the technique is proposed includes semivolatile compounds, organochlorine pesticides, organophosphorus pesticides, chlorinated herbicides, polychlorinated biphenyls and polycyclic aromatic hydrocarbons [53–56].

Huettenhain and Windrich [57] described a novel extraction method for polyaromatic hydrocarbons in soil in which the samples were first modified by grinding with silica gel to destroy the interaction between the analytes and soil humic matter. The homogeneous phase produced is then used as a stationary phase of a medium-pressure liquid chromatographic system. Accelerated solvent extraction, Soxhlet extraction, microwave-assisted extraction and supercritical fluid extraction were compared for recovery of polyaromatic hydrocarbons in contaminated soils. Accelerated solvent extraction exhaustively extracted all available polyaromatic hydrocarbons using a 5min static extraction.

Saim *et al.* [58] also studied the application of accelerated solvent extraction for the determination of polyaromatic hydrocarbons from soil. Their experimental design approach, based on central composite design, was used to investigate the dependence of accelerated solvent extraction operating variables (pressure, temperature, extraction time) on the recovery of 16 polycyclic aromatic hydrocarbons from native, contaminated soil. At the 95% confidence interval, no significance in terms of the three operating parameters was found when considering the total polyaromatic hydrocarbon recovery. However, when individual polyaromatic hydrocarbons were considered, some compounds were found to be dependent on operating variables. The most significant operating variable was extraction temperature. Low extraction temperature (40°C) was found to be significant for naphthalene, chyrsene and benzo[*b*]fluoranthene. Using constant operating conditions (100°C, 14 MPa and an extraction time of 5min plus 5min equilibration time), the influence of extraction solvent was evaluated. No dependence on recovery was found when polar organic solvents, i.e. a dipole moment of >1.89, were used.

Supercritical fluid extraction with carbon dioxide has been applied to the determination of polyaromatic hydrocarbons in soil.

2.4.1.8 Miscellaneous

Lopez-Avila *et al.* [59] used microwave assisted extraction to assist the extraction of polyaromatic hydrocarbons from soils. Another extraction method was described by Hartmann [60] for the recovery of polyaromatic hydrocarbons in forest soils. The method included saponification of samples in an ultrasonic bath, partitioning of polyaromatic hydrocarbons into hexane, extract cleanup by using solid-phase extraction, and gas chromatography–mass spectrometric analysis using deuterated internal standards. Polyaromatic hydrocarbons were thermally desorbed from soils and sediments without pretreatment in another investigation [61].

Bublitz [62] used time resolved laser induced fluorescence spectroscopy and fibre optics to determine polyaromatic hydrocarbons in oil polluted soils. The detection limit was 5mg kg^{-1} oil in soil.

Immunochemical methods have been employed to determine polyaromatic hydrocarbons in soils [63, 64]. On-site analysis is possible by this technique.

Micellar electrokinetic capillary chromatography with photodiode array detection was used for the determination of polyaromatic hydrocarbons in soil [65]. A detection limit of 10pg and linear calibration over five orders were observed. Compared to a standard gas chromatographic analysis method, the miscellar electrokinetic chromatographic method is faster, has a higher mass sensitivity and requires smaller sample sizes.

2.4.2 Non-saline deposited and suspended sediments

Following an aviation kerosin spill, hydrocarbons were detected in trout stream sediments and fish up to 14 months after the spill [13]. After a fire at a weed treatment plant in 1970 a large area of mixed forested ecosystem became contaminated with polycyclic aromatic hydrocarbons and creosote [66]. High polyaromatic concentrations in stream sediments adversely affected micro- and meiobenthic communities at all trophic levels. Stein *et al.* [67] have studied the uptake by bethnic fish (English sole, *Parophrys vetulus*) of benzopyrene and polychlorinated biphenyls from sediments. Accumulation of contaminants from sediments was a significant route of uptake by English sole.

Following a fire at a wood products treatment plant in 1970, a large area of Bayou Bonfouca, La., a mixed-forested ecosystem in the Lake Pontchartrain basin, and of its surrounding drainage areas, became heavily contaminated with polycyclic aromatic hydrocarbons in a coal-tar mixture (creosote). The effects of this contamination at four sites in the Bayou were investigated by Catallo and Gambrell [66]. Tabulated data are included on the concentrations of a number of polyaromatic hydrocarbon compounds in sediment and water at the study sites, on selected physicochemical properties of the sediment, and on microbenthic and meiobenthic communities at these sites. It was demonstrated that the high polyaromatic hydrocarbon concentrations in the sediments had adversely affected micro- and meiobenthic communities at all trophic levels, as well as affecting sediment properties. With increasing concentrations of creosote, detrital accumulation increased and redox potentials became more oxidizing; these phenomena were related to reductions in specific biomass of fungi and bacteria.

2.4.2.1 Gas chromatography

Bjorseth *et al.* [68] described a capillary gas chromatographic method for determining polyaromatic hydrocarbons in sediments. Up to 34 polyaromatic hydrocarbons were identified, some carcinogenic.

Giger and Schnaffer [69] described a glass capillary gas chromatographic method for the determination of polyaromatic hydrocarbons in lake and river sediments. Polyaromatic hydrocarbons are isolated by a sequence of solvent extraction, gel filtration, and adsorption chromatography, and individual concentrations determined by gas chromatography.

Readman *et al.* [70] used flame ionization capillary gas chromatography to determine polyaromatic hydrocarbons in extracts of rivers Mersey, Dee and Tamar estuary sediments.

2.4.2.2 Gas chromatography–mass spectrometry

Tan [71] devised a rapid simple sample preparation technique for analysing polyaromatic hydrocarbons in sediments. Polyaromatic hydrocarbons are removed from the sediment by ultrasonic extraction and isolated by solvent partition and silica gel column chromatography. The sulphur removal step is combined into the ultrasonic extraction procedure. Identification of polyaromatic hydrocarbon is carried by gas chromatography alone and in conjunction with mass spectrometry. Quantitative determination is achieved by addition of known amounts of standard compounds using flame ionization and multiple ion detectors.

Robbat *et al.* [42] carried out on-site detection of polycyclic aromatic hydrocarbons in hexane extract of sediments using thermal desorption gas chromatography–mass spectrometry.

The thermal desorption gas chromatography–mass spectrometry [42] described in section 2.4.1.2 has been applied to the determination of polyaromatic hydrocarbons in sediments.

2.4.2.3 Pyrolysis gas chromatography

de Leeuw *et al.* [32] screened anthropogenic compounds, including polyaromatic hydrocarbons, in polluted sediments by flash evaporation/pyrolysis gas chromatograph–mass spectrometry. Sediments were homogenized by sonication. Aliquots of samples were then suspended in methanol and drops applied to a pyrolysis wire, the Curie point of which was 510°C. The pyrolysis unit was mounted on the detector block of a gas chromatograph at a temperature of 300°C. Separation was achieved on a fused silica column coated with CP–SIL5. Flame ionization, flame photometric or electron capture detectors were used to monitor individual compounds which could be identified by mass spectrometry using 80*e*V EI ionization. Polyaromatic hydrocarbons, haloorganics, aliphatic hydrocarbons, heteroaromatics, elemental sulphur and cyanides were identified.

Thomas *et al.* [72] used pyrolysis gas chromatography–mass spectrometry as a fast economic screening technique for polyaromatic hydrocarbons. Thomas used reverse-phase liquid chromatography with atmospheric pressure chemical ionization mass spectrometry/mass spectrometry for the determination of polycyclic aromatic sulphur heterocycles in sediments.

2.4.2.4 High-performance liquid chromatography

Marcomini *et al.* [73] applied gradient elution reversed-phase high-performance liquid chromatography coupled with a variable wavelength adsorption detector to 23 selected polycyclic hydrocarbons (including most of those on the Environmental Protection Agency priority pollutant list) in radiodated sediment core. Compounds which were not separated by chromatography were adequately resolved and quantified by performing three runs per analysis using characteristic UV-visible absorption maxima. Detection limits ranged from 0.1 to 1μg kg $^{-1}$ dry weight, and the average recovery in spiking experiments was approximately 87%, with the lowest yield for naphthalene (56%).

2.4.2.5 Ultraviolet spectroscopy

Lee *et al.* [74] used UV spectroscopy to identify polyaromatic hydrocarbons in river sediments. The procedure involved the collection of sediments, air drying in the dark, sieving, and extraction for organic content. This was followed by column chromatography (silica gel with cyclohexane as eluent), followed by a second chromatographic step with Sephadex LH–20 and propan-2-ol as eluent. The eluate was then concentrated under vacuum and prepared for ultraviolet analysis.

2.4.2.6 Spectrofluorimetry

Saber *et al.* [75] reported on the quantitative determination of polyaromatic hydrocarbons in extracts of lacustrine sediments using high resolution Shpol'skii spectrofluorimetry at 10°K.

Garrigues and Emald [76] give details of a procedure for the determination of polycyclic aromatic hydrocarbons in sediment samples by high resolution spectrofluorimetry in *n*-alkane matrices.

2.4.2.7 Supercritical fluid chromatography

Langenfeld *et al.* [48] studied the effect of temperature and pressure on supercritical fluid extraction efficiencies of polyaromatic hydrocarbons and polychlorobiphenyls in river sediments. At 50°C, raising the pressure from 350 to 650atm was without effect on recovery from sediments.

Langenfeld *et al.* [48] also compared supercritical monochlorofluoromethane, nitrogen dioxide and carbon dioxide for the extraction of polyaromatic hydrocarbons from sediments. Monochlorodifluoromethane provided the highest recoveries.

Hawthorne *et al.* [77] compared supercritical chlorodifluoromethane, nitrous oxide and carbon dioxide for the extraction of polychlorobiphenyls from sediments. Chlorodifluoromethane provided the highest recoveries, while methanol modified carbon dioxide gave 90% recovery of polychlorobiphenyls from sediments.

2.4.2.8 Miscellaneous

Lopez-Avila *et al.* [59] have described a microwave assisted extraction procedure for the separation of polyaromatic hydrocarbons from sediments. Tan [71] described a rapid sample preparation technique for analysing polyaromatic hydrocarbons in sediments. Polyaromatic hydrocarbons are removed from the sediment by ultrasonic extraction and isolated by solvent partition and silica gel column chromatography. The sulphur removal step is combined into the ultrasonic extraction procedure. Identification of polyaromatic hydrocarbon is carried out by gas chromatography alone and in conjunction with mass spectrometry. Quantitative determination is achieved by addition of known amounts of standard compounds using flame ionization and multiple ion detectors.

2.4.3 Saline deposited and suspended

2.4.3.1 Gas chromatography

Walker *et al.* [17] studied profiles of hydrocarbons in sediment according to depth in sediment cores collected at Baltimore Harbour in Chesapeake Bay, Massachusetts. Gas liquid chromatography was used to detect hydrocarbons present at different depths in the sediment, while low resolution mass spectrometry was employed to measure concentrations of paraffins, cycloparaffins, aromatics and polynuclear aromatics. Their data show that the concentrations of total and saturated hydrocarbons decreased with increased depth, and it is commented that identification and quantitation of hydrocarbons in oil-contaminated sediments is required if the fate of these compounds in dredge spoils is to be determined.

Readman *et al.* [70] selected capillary gas chromatography using a flame ionization detector as the method for quantifying sterols, in particular coprostanol, as a marker of faecal pollution. The hydrocarbon fraction produced as a by-product of the sterol analysis was used for quantifying 'oil derived' and polycyclic aromatic hydrocarbons. Analyses of sediments from estuaries of the Mersey, Dee and Tamar rivers are given as examples of how to interpret results of the method. Petrogenic and biogenic inputs of saturated hydrocarbons could be distinguished.

2.4.3.2 Spectrofluorimetry

Saber *et al.* [75] used high resolution Shpol'skii spectrofluorimetry at 10°K to quantitatively determine polyaromatic hydrocarbons in lacustral sediments. Polyaromatic hydrocarbons incorporated into *n*-alkane matrix at low temperature yielded high resolution fluorescence spectra of quasi lines with a multiplet structure related to several insertion sites. Samples required extraction and purification, and the choice of sample treatment, which depended on the total organic pollution levels, is discussed.

2.4.3.3 High-performance liquid chromatography

The high-performance liquid chromatographic procedure [38] described in section 2.2.2.2 has been applied to the determination of polyaromatic hydrocarbons in saline sediments.

2.4.3.4 Miscellaneous

Dunn and Stich [78] and Dunn [79] have described a monitoring procedure for polyaromatic hydrocarbons, particularly benzo[*a*]pyrene in marine sediments. The procedures involve extraction and purification of hydrocarbon fractions from the sediments and determination of compounds by thin layer chromatography and fluorometry, or gas chromatography. In this procedure, the sediment was refluxed with ethanolic potassium hydroxide, then filtered and the filtrate extracted with isooctane. The isooctane extract was cleaned up on a florisil column, then the polyaromatic hydrocarbons were extracted from the isoactive extract with pure dimethyl sulphoxide. The latter phase was contacted with water, then extracted with isooctane to recover polyaromatic hydrocarbons. The overall recovery of polyaromatic hydrocarbons in this extract by fluorescence spectroscopy was 50–70%.

2.4.4 *Sludge*

2.4.4.1 Thin layer chromatography

McIntyre *et al.* [80] have described a method for the determination of polynuclear aromatic hydrocarbons in sewage sludges, using a thin layer chromatography procedure. The method involves solvent extraction with cyclohexane, clean-up of the extract by silica gel thin layer chromatography and final separation of the purified polyaromatic hydrocarbons by two-dimensional thin layer chromatography. Sample chromatograms are evaluated by comparison with standard plates under ultra-violet light. Good, reproducible separation of the six polyaromatic hydrocarbons specified by the World Health Organization is achieved, using relatively simple and inexpensive apparatus. The results were comparable with those obtained by other methods, at the µg per litre level. The novel feature of the extraction procedure is the mixing of the homogenized sludge with cyclohexane in a laboratory disperser in glass centrifuge tubes, giving an extraction efficiency of 80–100%.

2.4.4.2 Miscellaneous

Aichberger and Reifenauer [81] have reviewed methods for the determination of polyaromatic hydrocarbons in sewage sludge.

2.5 Polymers

2.5.1 Soil

2.5.1.1 Pyrolysis gas chromatography–mass spectrometry

The Curie Point flash evaporation-pyrolysis gas chromatography–mass spectrometric method [32] described in section 2.2.1.2 for the analysis of aliphatic hydrocarbons in soil has also been applied to the determination of polystyrenes in soil via identification and determination of their unzipping pyrolysis products, such as styrene monomer, α-methyl styrene, 3-methyl styrene, 4-methyl styrene, α-3 dimethyl styrene, 3-ethylstyrene, α-4 dimethyl styrene, 3.5 dimethyl-styrene, α-2 or 2,5 or 2.4 dimethyl styrene also various phenyl ethers.

References

1 Karasek, F.W., Charbonneau, G.H., Revel, G.J. and Tong, H.Y. (1987) *Analytical Chemistry*, **59**, 1027.
2 Yang, Y., Hawthorne, S.B. and Miller, D.J. (1995) *Journal of Chromatography*, **A699**, 265.
3 Burford, M.D., Hawthorne, S.B. and Miller, D.J. (1996) *Journal of the American Environmental Laboratory*, **8**, 1.
4 Liang, S. and Tilotta, D.C. (1998) *Analytical Chemistry*, **70**, 616.
5 US Environmental Protection Agency (1979) *Methods for Chemical Analysis of Water and Wastes*, EPA 600/14–79/020, Washington, DC.
6 State Water Resources Control Board (1988) *Leaking Underground Fuel Tank (LUFT) Field Manual*; State of California, Sacramento.
7 American Society for Testing and Materials (1984) *Annual Book of ASTM Standards*, Philadelphia, Vol 11.02
8 Concawe Report 9/72 (1972) Hydrocarbons in Soil. *Method* v/72 v/1–6, November.
9 Geerdink, M.J., Erkelens, C., Van Dam, J.C. *et al.* (1995) *Analytica Chimica Acta*, **315**, 159.
10 Remmler, M., Kopinke F.D. and Stottmeister, G. (1995) *Thermochim Acta* **263**, 101.
11 Blumer, M. and Sass, J. (1972) *Marine Pollution Bulletin*, **3**, 92.
12 Farrington, J.W. and Quinn, J.G. (1973) *Estuary and Coast Marine Science*, **1**, 71.
13 Guiney, P.D., Sykora, J.L. and Keleti, G. (1987) *Environmental Toxicology and Chemistry*, **6**, 105.
14 Wakeham, S.G. (1977) *Environmental Science and Technology*, **11**, 272.
15 Broman, D., Colmsjo, A., Ganning, B. *et al.* (1987) *Marine Pollution Bulletin*, **18**, 380.
16 Blumer, M. and Sass, J. (1972) *Marine Pollution Bulletin*, **3**, 92.
17 Walker, J.D., Caldwell, R.R., Hamming, M.C. and Ford, H.T. (1975) *Environmental Pollution*, **9**, 231.

18 May, W.E., Chesler, S.N., Cram, S.P. *et al.* (1975) *Journal of Chromatographic Science*, **13**, 535.
19 Brown, D., Colsmio, A., Ganning, B. *et al.* (1987) *Marine Pollution Bulletin*, **18**, 380.
20 Takada, H. and Ishimatari, R. (1985) *Journal of Chromatography*, **346**, 281.
21 Page, D.S., Foster, C., Fickett, P.M. and Gilfillan, E.S. (1988) *Marine Pollution Bulletin*, **19**, 107.
22 Brown, L.R., Pabst, G.S. and Light, M. (1978) *Marine Pollution Bulletin*, **9**, 81.
23 Zitko, V. and Carson, W.V. (1970) Technical Report Fisheries Research Board, No. 217, Ottawa, Canada.
24 Scarrett, D.J. and Zitko, V. (1972) *Journal of Fisheries Research Board, Canada*, **29**, 1347.
25 McLeod, W.D., Prohaska, P.G., Gennero, D.D. and Brown, D.W. (1982) *Analytical Chemistry*, **54**, 386.
26 Hilpert, L.R., May, W.E., Wise, S.A. *et al.* (1982) *Analytical Chemistry*, **54**, 458.
27 Albaiges, J. and Grimalt, J. (1987) *International Journal of Environmental Analytical Chemistry*, **31**, 281.
28 Mark, H.B. (1972) *Environmental Science and Technology*, **6**, 833.
29 Whittle, P.J. (1977) *Analyst (London)*, **107**, 976.
30 Ryzhova, G.L., Slizhov, Y.G., Borodina, O.I. *et al.* (1986) *Soviet Journal of Water Chemistry and Technology*, **8**, 106.
31 Kester, P.E. (1987) *Analysis of Volatile Organic Compounds in Soils by Purge and Trap Gas Chromatography*. Tekmar Company, PO Box 371856, Cincinnati, Ohio, 45222–1856.
32 de Leeuw, J.W., de Leer, E.W.B., Sinninghe Damsté, J.S. and Schuyl, P.J.W. (1986) *Analytical Chemistry*, **58**, 1852.
33 Van de Meent, D., Brown, S.C., Philip, R.P. and Simoneet, B.R.T. (1980) *Geochimica Cosmochimica Acta*, **44**, 999.
34 Stenhagen, E., Abrahamson, S., McLafferty, F.W. (eds) (1974) *Registry of Mass Spectra Data*, John Wiley, New York, Vol 1–4.
35 Lee, M.L. Vasillaros, D.L., White, C.M. and Novotny, M. (1979) *Analytical Chemistry*, **51**, 768.
36 Crisp, P.T., Ellis, J., de Leeuw, J.W. and Schenck, P.A. (1986) *Analytical Chemistry*, **58**, 258.
37 Hargrave, R.T. and Phillips, G.A. (1975) *Environmental Pollution*, **8**, 193.
38 Vowles, P.D. and Mantoura, R.F. (1987) *Chemosphere*, **16**, 109.
39 Krahn, M.M., Moore, L.K., Bogar, R.G. *et al.* (1988) *Journal of Chromatography*, **437**, 161.
40 Hennig, H.F.O. (1979) *Marine Pollution Bulletin*, **10**, 234.
41 Levy, E. (1971) *Water Research*, **5**, 723.
42 Robbat, A., Liu Tyng-Liu and Abraham, B.M. (1992) *Analytical Chemistry*, **64**, 1477.
43 Fowlie, P.J.A. and Bulman, T.L. (1986) *Analytical Chemistry*, **58**, 721.
44 Maybury, R. (1984) *Laboratory Manual for Pesticide Residue Analysis in Agricultural Products*, Food Production and Inspection Branch, Agriculture, Canada, revised.
45 Afghan, B.K. and Wilkinson, R.J. (1981) Method for determination of Polynuclear Aromatic Hydrocarbons in Environmental Samples HPLG-multi-detection system. Environment Canada, Manuscript 20–AMD–3–81–BKA.
46 Hankin, S.M., John, P., Simpson, A.W. and Smith, G.P. (1996) *Analytical Chemistry*, **68**, 3235.
47 Brown, R.S., Luang, J.H.T., Szolar, O.H.J. *et al.* (1996) *Analytical Chemistry*, **68**, 287.
48 Langenfeld, J.J., Hawthorne, S.B., Miller, D.J. and Rawliszyn, J. (1993) *Analytical Chemistry*, **65**, 338.

49 Burford, M.D., Hawthorne, S.B. and Miller, D.J. (1993) *Analytical Chemistry*, **65**, 1497.
50 Reindt, S. and Hoffler, F. (1994) *Analytical Chemistry*, **66**, 1808.
51 Barnabas, J.J., Dean, J.R., Tomlinson, W.R. and Owen, S.P. (1996) *Analytical Chemistry*, **68**, 2064.
52 Tena, M.T., Luque de Castro, M.D. and Valcarcel, M. (1996) *Analytical Chemistry*, **68**, 2386.
53 Environmental Protection Agency SW–846. (1995) *Test Methods for Evaluating Solid Wastes*, Method, 3545, US, 3rd edn, Update 111, US GPO, Washington, DC.
54 Ezzell, J.L., Richter, B.E., Felix, W.D. *et al.* (1995) *Liquid Chromatography–Gas Chromatography*, **13**, 390.
55 Richter, B.E., Jones, B.A., Ezzell, J.L. *et al.* (1996) *Analytical Chemistry*, **68**, 1033.
56 Dean, J.R. (1996) *Analytical Communications*, **33**, 191.
57 Huettenhain, S.H. and Windrich, J. (1996) *International Journal of Environmental Analytical Chemistry*, **63**, 245.
58 Saim, N., Dean, J.R., Abdullah, M.P. and Zakaria, Z. (1998) *Analytical Chemistry*, **70**, 420.
59 Lopez-Avila, V., Young, R. and Beckert, W.F. (1994) *Analytical Chemistry*, **66**, 1097.
60 Hartmann, R. (1996) *International Journal of Environmental Analytical Chemistry*, **62**, 161.
61 Medina-Vera, M. (1996) *Journal of Applied Pyrolysis*, **36**, 27.
62 Bublitz, J., Christopherson, A. and Schade, W. (1996) *Fresenius Journal of Analytical Chemistry*, **355**, 684.
63 Sparrevik, M. and Jonassen, H. (1995) *Soil Environment*, **5**, 537.
64 Hudak, R.T., Melby, J.M. and Stave, J.W. (1994) 87th, 14B. Paper 94–R, P143.06. Proceedings Annual Meeting – Air Waste Management Association.
65 Brueggemann, O. and Freitag, R. (1995) *Journal of Chromatography*, **717**, 309.
66 Catello, W.J. and Gambrell, R.P. (1987) *Chemosphere*, **16**, 1053.
67 Stein, J.E., Ham, E., Casillas, E. *et al.* (1987) *Marine Environmental Research*, **22**, 123.
68 Bjorseth, A., Knutsen, J. and Skei, J. (1979) *Science of the Total Environment*, **13**, 71.
69 Giger, W. and Schnaffer, C. (1978) *Analytical Chemistry*, **50**, 243.
70 Readman, J.W., Preston, M.R. and Mantoura, R.F.C. (1986) *Marine Pollution Bulletin*, **17**, 298.
71 Tan, Y. (1979) *Journal of Chromatography*, **176**, 319.
72 Thomas, D., Crain, S.M., Sim, P. and Benoit, F.M. (1995) *Journal of Mass Spectrometry*, **30**, 1034.
73 Marcomini, A., Sfriso, A. and Pavoni, B. (1987) *Marine Chemistry*, **21**, 15.
74 Lee, H.K., Weight, G.J. and Swallow, W.H. (1988) *Environmental Pollution*, **49**, 167.
75 Saber, A., Jarocz, J., Marin-Bouer, M. *et al.* (1987) *Journal of Environmental Analytical Chemistry*, **28**, 171.
76 Garrigues, P. and Emald, M. (1987) *Chemosphere*, **16**, 485.
77 Hawthorne, S.B., Lagenfeld, J.T., Miller, D.J. and Burford, M.D. (1992) *Analytical Chemistry*, **64**, 1614.
78 Dunn, B.P. and Stich, H.F.J. (1976) *Journal of Fisheries Research Board, Canada*, **33**, 2040.
79 Dunn, B.P. (1976) *Environmental Science and Technology*, **10**, 1018.

80 McIntyre, A.E., Perry, R. and Lester, J.N. (1981) *Analytical Letters (London)*, **14**, 29.
81 Aichberger, K. and Reifenauer, D. (1984) Landlwirtschaftlich-Chemische. Bundesanstalt, Linz Processing and Use of Sewage Sludge: Proceedings Third International Symposium, Brighton, 1983. (P.L. Hermitte and H. Ott eds.). D. Reidel Publishing Co., Dordrecht, 161–163 (08BHER).

Chapter 3

Surface active agents

3.1 Cationic surfactants

3.1.1 Non-saline deposited and suspended sediments

3.1.1.1 Spectrophotometry

Ambe and Hanya [1] have combined the Longwell and Maniece [2] methods using methylene blue with the infrared spectroscopic method of Sallee [3] to devise a method for the determination of alkylbenzene sulphonates. Methylene blue alkylbenzene sulphonate complexes give absorption peaks at 890 and 1010cm $^{-1}$, the ratio of the heights being proportional to the ratio of the amount of sulphonate to the total amount of methylene blue sensitive substances in the complex. The filtered sample is shaken (50ml) with 0.1N sulphuric acid (1ml), 0.025% methylene blue solution (1ml) and 1,2-dichloroethane (20ml) for 1min. After washing the separated organic layer twice with 20ml of 0.0013% solution of methylene blue in 0.004N sulphuric acid also containing 0.022% of silver sulphate, its extinction is measured at 655nm to give the total amount of substances active towards methylene blue. The organic layer is evaporated to dryness prior to pelleting with potassium bromide and examination by infrared spectroscopy.

This method has been applied to bottom sediments and muds [4]. The mud sample is centrifuged to separate the water, dried at room temperature, ground and sieved. This residue is extracted for 1h at 80°C with methanol-benzene (1:1), the extraction is repeated twice, and the combined extracts are evaporated and the residue dissolved in water. Alkylbenzenesulphonates are then determined by infrared spectroscopy as described above.

3.1.1.2 Gas chromatography–mass spectrometry

Trehy *et al.* [5] determined linear alkyl benzene sulphonates in sediments in amounts down to 0.5mg kg $^{-1}$ using this technique.

3.1.2 Sludge

3.1.2.1 Spectrophotometry

Hallmann [6] showed that cationic surfactants of the dialkyldimethylammonium chloride type may be extracted from sewage sludges combination of organic solvents and concentrated hydrochloric acid. The determination of the surfactant in the resultant extract by formation of a complex with disulphide blue requires the separation of anionic and non-ionic surfactants as well as other interfering substances such as humic acids, and dissolved iron and manganese. Methods of purifying the extract are described, together with results of analyses performed on various sludges. Typical values were between 500 and 1000ppm with the majority closer to 1000ppm.

3.1.2.2 Gas chromatography–mass spectrometry

McEvoy and Giger [7] give details of techniques developed for determination of linear alkylbenzenesulphonates in stabilized sewage sludges, including a rapid screening procedure to detect whether these compounds were present, and a confirmatory method involving formation of the sulphonyl chlorides and subsequent high resolution gas chromatography with flame ionization detection and directly coupled mass spectrometry employing both electron-impact and chemical-ionization modes. The linear alkylbenzenesulphonates concentrations found in several digested sludges ranged from 0.3–1.2% of dry sludge. The implications of these high concentrations for the use of sludge in agriculture are considered.

Simms *et al.* [8] discuss the quantitative determination of cationic surface active agents at the sub-ppb level in sewage sludges using fast atom bombardment mass spectrometry.

3.1.2.3 High-performance liquid chromatography

This technique has been used to determine linear alkyl benzene sulphonates in sludges [9, 11].

3.2 Anionic surfactants

3.2.1 Sludge

3.2.1.1 Gas chromatography–mass spectrometry

The technique [7] discussed above in section 3.1.2.2 has been applied to anionic surfactants.

3.2.1.2 Supercritical fluid chromatography

Fernandez *et al.* [9] used supercritical fluid extraction combined with ion pair liquid chromatography to determine quaternary ammonium in digested sludges and marine sediments. Carbon dioxide modified with 30% methanol was used as the extractant at an operating pressure of 380atm. Between 0.2 and 3.7g kg $^{-1}$ surfactant was found in Swiss works effluent sludges, determined with a relative standard deviation of 7%.

3.2.1.3 Electron spin resonance spectroscopy

Senesi and Sposito [10] studied copper II-anionic surfactant complexes using electron spin resonance (ESR) spectrometry. The surfactants used were linear alkyl aryl sulphonate, sodium dodecyl benzene sulphonate, and sodium lauryl sulphate. Copper-ligand molar ratios ranged from 0.1 to 1.0 ERS spectra of frozen (77K) aqueous solutions showed that all three surfactants formed inner sphere complexes with copper(II) ions held in four oxygen-ligand, square planar binding sites. The sulphonate type surfactants had a higher affinity for copper(II) than did the ester sulphate type. The copper(II)-anionic surfactant complexes were different from, and less stable than, copper(II)-fulvic acid complexes. Solution spectra yielded more structural information on the complexes than did solid state spectra. It was concluded that undegraded anionic surfactants in sewage sludge did not participate as isolated, independent ligands, but may participate as co-ligands with other oxygen-containing functional groups, or as moieties incorporated into the fulvic acid structure.

3.3 Non-ionic surfactants

3.3.1 Sludge

3.3.1.1 High-performance liquid chromatography

Various workers have studied the application of this technique [11–13]. Applying this technique to the determination of alkylphenol mono and diethoxylates and alkyl phenols in sewage sludge Abel and Giger [12] obtained recoveries exceeding 80% with relative standard deviation better than 8% and a detection limit of 0.5μg L $^{-1}$. The same workers studied the determination of alkylphenol polyethoxylates in sewage and sewage sludge. In this procedure alkylphenol polyethoxylates in wastewater samples were stripped into ethyl acetate. Normal high-performance liquid chromatography phase using bonded phase aluminium silicate columns separated alkylphenol polyethoxylates and allowed their quantification. Alkylphenol ethoxylates were selectively determined by absorption at 277nm. Relative deviations were 2–10% for major oligomers. Limits of detection for individuals were estimated at 1μg per litre. Total recovery of alkylphenol polyethoxylates was 87%. Reverse-phase high-performance liquid chromatography was used to determine alkyl substituents and as a rapid screening method. Combined normal and reverse phase high-performance liquid chromatography gave information on structure and concentration of alkylphenol polyethoxylate surfactants in sewage and sewage sludges. Nonlyphenol ethoxylates with 1–18 ethoxy units were found in untreated wastes at 0.8–2 3mg per litre.

References

1 Ambe, Y. and Hanya, T. (1972) *Japan Analyst*, **21**, 252.
2 Longwell, N. and Maniece, O. (1955) *Analytical Abstracts*, **2**, 2244.
3 Sallee, O. (1956) *Analytical Chemistry*, **28**, 1822.
4 Ambe, Y. (1973) *Environmental Science and Technology*, **7**, 542.
5 Trehy, M.L., Gledhill, W.E. and Orth, R.G. (1990) *Analytical Chemistry*, **62**, 2581.
6 Hellmann, H. (1983) *Fresenius Zeitschrift für Analytische Chemie*, **315**, 425.
7 McEvoy, J. and Giger, W. (1986) *Environmental Science and Technology*, **20**, 376.
8 Simms, J.R., Kevugh, T., Ward, S.R. and Bandaurraga, M.M. (1988) *Analytical Chemistry*, **60**, 2613.
9 Fernandez, P., Alder, C.A., Suter, M.J.F. and Giger, W. (1996) *Analytical Chemistry*, **68**, 921.
10 Sevesi, N. and Sposito, G. (1987) *Water, Air and Soil Pollution*, **35**, 147.
11 HMSO (1993) Linear alkylbenzene sulphonates and alkylphenolethoxylates in waters, waste waters and sludges by high-performance liquid chromatography, HMSO, London.
12 Abel, M. and Giger, W. (1985) *Analytical Chemistry*, **57**, 1577.
13 Abel, M. and Giger, W. (1985) *Analytical Chemistry*, **57**, 2584.

Chapter 4

Oxygen containing compounds

4.1 Phthalate esters

4.1.1 Non-saline deposited and suspended sediments

4.1.1.1 Gas chromatography

Thuren [1] determined phthalates in sediment using solvent extraction (acetonitrile, petroleum ether), clean-up with deactivated Florisil, and quantitative analysis by gas chromatography. The detector response was linear between 0.5 and 100ng. The detection limit (signal:noise ratio 2:1) was 0.1ng for dimethylphthalate, dibutylphthalate and di(2-ethylhexyl) phthalate, and 0.05ng for benzoylbutylphthalate. Recovery was between 30% and 130% depending on the ester. Low recovery for dimethylphthalate (30%) was probably due to pyrolysis in the detector (detector temperature was 320°C).

4.1.1.2 High-performance liquid chromatography

Schwartz *et al.* [2] have described a high-performance liquid chromatographic method for determining di-2-ethylhexyl and di-2-butyl phthalate in river sediments. This method requires no sample clean-up and consists of a single extraction step using *n*-hexane:acetone:methanol (8:1:1 *v/v*) followed by quantitative analysis using high-performance liquid chromatography.

Following the procedure described above, it is possible to detect down to 10ng of both esters, i.e. equivalent to 0.5mg kg^{-1}. Fig. 4.1 shows a typical chromatogram of a hexane extract of sediment taken from the River Rhine.

Schwartz *et al.* [2] investigated the biodegradation of phthalic acid esters adsorbed in river sediments by repeated analysis of the sediment over a two-week period. The di-2-ethylhexyl phthalate content turned out to remain essentially constant (s.d., 5%; n = 10),irrespective of the absence or presence of a microbial inhibitor (500ppm of sodium azide or mercuric chloride added immediately after sampling); i.e. no marked biodegradation occurred.

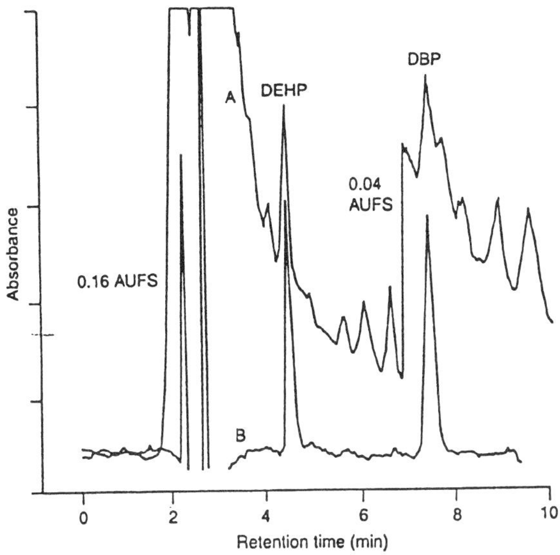

Fig. 4.1 HPLC chromatography of (a) hexane extract of a sediment; (b) a standard solution of DEHP and DBP. Retention times: DEHP, 4.5min; DPB,7.5 min
Source: Reproduced with permission from Gordon and Breach [2]

The results of analyses of sediment samples taken from the River Rhine are listed in Table 4.1. From Table 4.1 it is seen that for the sediment of the River Rhine di-2-ethylhexyl phthalate and di-*n*-butyl phthalate concentrations generally are between 2 and 50ppm.

The phthalate levels of the sediment samples should actually be regarded as minimum values, since only the amount of extractable phthalates has been determined. Eglinton *et al.* [3] report that some organic pollutants in sediments may be converted into insoluble complexes, such as humates. On the other hand, data by Cifrulak [4] suggest that the use of a methanol-containing solvent mixture, rather similar to the one employed by Schwartz *et al.* [2], effectively removes all phthalates from sediment and soil samples.

4.2 Phenols

4.2.1 Soil

4.2.1.1 Spectrophotometric methods

Talsky [5] has described a higher order derivative spectrometric method for the determination of phenols in soils.

Table 4.1 DEHP and DBP content (ppm) and composition of sediments taken from the Fiver Rhine in 1977

Sample station	*Date*	*DEPH*	*DBP*	*% dry weight*	*Fraction with d_p <16μm*
01	10–05	15.0	5.0	–	–
	22–08	70.5	15.5	34	–
	03–10	9.5	ND	57	0.12
02	10–05	20.5	7.5	–	–
	22–08	25.0	4.5	43	–
	03–10	19.0	ND	34	0.40
03	22–08	26.0		34	–
	03–10	27.0	ND	33	0.40
04	22–08	10.0	ND	58	–
	03–10	17.0	ND	48	0.36
05	10–05	13.0	4.0	–	–
	22–08	21.0	3.0	37	–
	03–10	27.0	ND	24	0.51
07	22–08	10.5	ND	46	–
	03–10	6.5	ND	59	0.11
08	10–05	0.1*	ND	77	–
09	22–08	14.5	ND	45	–
10	10–05	9.0	2.0	–	–
	22–08	28.5	0.5	32	–
	03–10	25.0	ND	25	–
11	22–08	45.0	0.5	34	–

Source: Reproduced with permission from Gordon and Breach [2]

ND not detectable; *Procedure: 25 instead of 10g dry sample; 5 instead of 30ml *n*-hexane

4.2.1.2 Gas chromatography

Karasek *et al.* [6] determined phenols in soils by extraction with a mixture of benzene and water modified to pH10 by the addition of 2-methoxy-ethylamine. The phenol in the extract was identified and determined by gas chromatography using a variety of detectors including flame ionization, electron capture and mass spectrometry.

4.2.1.3 Miscellaneous

Lopez-Avila *et al.* [7] used microwave assisted extraction to assist the extraction of phenols from soils.

4.2.2 Non-saline deposited and suspended sediments

4.2.2.1 Miscellaneous

Goldberg and Weiner [9] have described methods for the extraction and concentration of phenolic compounds from sediment. Lopez-Avila *et al.* [8]

have described a microwave assisted extraction for the separation of phenols from sediments.

4.3 Carboxylic acids

4.3.1 Non-saline deposited and suspended sediments

4.3.1.1 Gas chromatography

Procedures for determining fatty acids in sediments involved liquid–liquid extraction, liquid–solid adsorption chromatography followed by gas liquid chromatographic analysis [10–12]. Liquid extractions have been performed with methanol-chloroform [13], methylene chloride [14] and benzene-methanol [15,16]. Typical liquid–solid adsorbents are silicic acid. Standard gas chromatographic separations for complex mixtures employ non-polar columns packed with OV–1, OV–17, OV–101, SE–30, or glass capillary columns containing similar phases.

4.3.1.2 Miscellaneous

Farrington and Quinn [17] gave details of procedures involving saponification and extraction. Between 32 and 65% of the fatty acids were not released from sediments by organic solvent extraction.

Mendoza *et al.* [18] determined carboxylic acid compounds in a 5mol L^{-1} lacustrine sediment core taken in Leman Lake. Unbound and tightly bound compounds were not converted from one form to another. The abundance profiles below 30cm were not only similar but showed no decreasing trend, suggesting a common origin in three forms. The presence of unsubstituted monounsaturated acids in the C20–C32 range suggested a possible origin for long chain fatty acids other than from higher plants. Nothing was known of the origins of (omega-1)-hydroxy acids longer than C20, or those of 2-methylnonacosanoic acid.

4.3.2 Sludge

4.3.2.1 Gas chromatography

Analysing volatile acids in aqueous systems, resulting mainly from the presence of water, have been reported [19]. The volatile acids' high polarity as well as their tendency to associate and to be adsorbed firmly on the column require esterification prior to gas chromatographic determination. The presence of water interferes in esterification so that complex drying techniques and isolation of the acids by extraction, liquid solid chromatography, distillation, and even ion exchangers had to be used [20–23].

The introduction of the more sensitive hydrogen flame ionization detector has made possible the analysis of dilute aqueous solutions of organic acids by gas liquid chromatography. Problems, such as 'ghosting' at high acid concentrations and an excessive tailing effect of the water in dilute solutions, masking the components, have been reported for aqueous solutions [24]. Subsequently phosphoric [25] or metaphosphoric acids [26] were added to the liquid phase, resulting in more reproducible column performance and reduced 'ghosting'. Addition of formic acid to the carrier gas was recommended by Cochrane [27] to overcome all the problems normally associated with analysing free fatty acids by gas chromatography.

Baker [28] used FFAP column for direct injection of dilute aqueous solutions of acids (FFAP, a reaction product of polyethylene glycol 20 000 and 2-nitrophthalic acid developed by Varian Aerograph). The acetic acid peak was not clear and the ability of this column to separate normal and iso fatty acids was not reported. Van Huyssteen [19] successfully used a Chromosorb 101 column coated with 3% FFAP for separation of volatile acids by direct injection of synthetic aqueous solutions and anaerobic digesters samples, which were first centrifuged and acidified with hydrochloric acid. He injected 1μl at acid concentration greater than 50mg L^{-1} and 2μl below 50mg L^{-1}. Ghosting was observed upon injecting 2μl 25mg L^{-1} C_2–C_6 acid solutions. Van Huyssteen did not try to inject volumes greater than 2μL. His column affected complete separation of the C_2–C_6 straight and branched short chain fatty acids from synthetic aqueous solutions, but less sharpened peaks were obtained from anaerobic digester samples. The response with acetic acid approximated that of the other acids.

An official gas chromatographic method [29] is available from the determination of volatile fatty acids in sewage sludge. This method is based on gas liquid chromatographic estimation with a flame ionization detector, and is applicable up to 2000mg total volatile fatty acids per litre, while the concentrations of individual fatty acids can also be determined. Where this method is not practicable an empirical method based on the spectrophotometric determination of ferric hydroxamates can be used, giving a value for total fatty acids expressed as acetic acid. For control purposes a rapid test is described in which the volatile fatty acids are determined by electrometric titrimetry on the neutralized sludge obtained from the determination of alkalinity.

4.3.2.2 Column chromatography

Methods have been described for the determination of total fatty acids in raw sewage sludge. These methods [30–32] require a concentration steps such as simple distillation, steam distillation, evaporation, or extraction [33–35] which resulted in great losses of the volatile matter [36,37].

Straight distillation or steam distillation of volatile acids and the chromatographic separation have been proposed in *Standard Methods* [38] for the organic acids in sludge. In this method an acidified aqueous sample, containing relatively high concentrations of organic acids, is adsorbed on a column of silicic acid and the acids are eluted with n-butanol in chloroform. The eluate is collected and titrated with standard base. All short chain (1–6 carbon) organic acids are eluted, but so are crotonic, adipic, pyruvic, phthalic, fumaric, lactic, succinic, malonic, aconitic and oxalic acids, as well as alkyl sulphates and alkyl-aryl sulphonates. No information on the individual volatile acids is obtained by this method and the results are reported collectively as total organic acids. Various chromatographic methods, such as paper [39, 40] and gel chromatography [32], have been used for the analysis of sludge digester liquor. Mueller *et al.* [37] have modified the indirect chromatographic method for samples of raw sewage and river water involving tedious concentration steps, leading to losses. In paper chromatography individual volatile acid concentrations should be higher than 600mg L^{-1}, while in other methods the minimum detectable level is 1000mg L^{-1} [39,40].

4.4 Carbohydrates

4.4.1 Non-saline deposited and suspended sediments

4.4.4.1 Spectrophotometric method

McQuaker and Fung [41] determined carbohydrates in sediments spectrophotometrically at 485nm by reacting with phenol and concentrated sulphuric acid.

4.4.4.2 Column chromatography

Mopper and Regeus [42] determined monosaccharides in lake sediments with a sensitivity of 0.1nmol, using an automated chromatographic sugar analyser. The test solution is forced by nitrogen pressure into a nylon column (110cm × 2.8mm) packed with Echnicon type S resin (sulphate form) and maintained at 76°C. The sugars are separated by pumping 89% ethanol through the column and the eluate is mixed with an alkaline solution of tetrazolium blue which is considerably more sensitive and less corrosive than other dyes. The extinction is monitored at 520nm.

4.4.2 Saline deposited and suspended sediments

4.4.2.1 Gas chromatography

Crowie and Hedges [43] have described a flame ionization gas chromatographic method for the determination of equilibrated isomeric mixtures of six monosaccharides (galactose, glucose, xylose, mannose, rhamnose, fucose, arabinase and lyxose) in saline sediments. Acid hydrolysis yields monomeric sugars which may exist in up to five isomeric forms when in solution. Lithium perchlorate is used to catalytically equilibrate sugar isomer mixtures in pyridine prior to conversion to their trimethylsilyl ether derivatives. Analysis is carried out by use of gas liquid chromatography on fused-silica capillary columns. Quantification on the basis of a single clearly resolved peak for each sugar is made possible by the equilibration step. Sugar losses and optimal conditions for maximum reproducible sugar recovery are determined for each extraction stage.

Carbohydrate recovered through the elusive analytical procedures were in the range 101% (lyxose, xylose, galactose) to 108% (glucose). Reproducibility data is quoted in Table 4.2. Down to 0.1μg of each monosaccharide can be determined in a sample hydrolysate.

4.5 Sterols

4.5.1 Non-saline deposited and suspended sediments

4.5.1.1 Miscellaneous

Dreier *et al.* [44] determined sterols in lacustrine sediments. Samples of wet lacustrine sediments were heated under anoxic conditions at 150, 175, 200 and 250°C for five days; at 175°C for five days with influx of potassium hydroxide and methanol to remove sterols; and at 175°C for 12, 18, 24 and 48h, after which extraction was performed. Heating the sediment increased the amounts of extractable sterols provided that the temperature did not exceed 200°C, because degradation became rapid above that temperature. The behaviour of sterol ketones was similar, but the temperature limit was slightly higher. The various levels of the sterols extracted are tabulated; 4-methylsterols had a high stability towards thermal degradation under the conditions used.

4.6 Uronic acids and aldoses

4.6.1 Non-saline deposited and suspended sediments

4.6.1.1 Gas chromatography

These substances can be determined by a procedure [45] involving preliminary hydrolysis with hydrofluoric acid at 135°C producing N-alkyl aldonamide and alditolacetates. These substances are then determined by capillary column gas chromatography.

Table 4.2 Reproducibility[a]

	mg of sugar/100mg of organic carbon									
	LYX	*ARA*	*RHA*	*RIB*	*XYL*	*FUC*	*MAN*	*GAL*	*GLU*	*TCH$_2$O*
					Sediment					
SMD	0.16± 0.02	1.16± 0.10	1.33± 0.05	0.37± 0.07	1.18± 0.06	0.99± 0.08	2.31± 0.18	2.38± 0.15	4.28± 0.07	1.42± 0.40
%SMD	12.5	6.62	3.76	18.9	5.08	8.08	7.79	6.30	1.64	2.82

Source: Reproduced with permission from American Chemical Society [43]

[a]Sediment, 10–12cm interval from Dabob Bay box-core, triplicate, 600mg, 2.54% organic carbon.

4.7 βhydroxy butyrate and βhydroxy valerate

4.7.1 Sludge

4.7.1.1 Gas chromatography

Comeau *et al.* [46] have described a simple assay for separating and quantifying poly-beta-hydroxybutyrate and poly-beta-hydroxyvalerate in activated sludge samples involved sludge lyophilization, purification of the chloroform extract by re-extraction with water, and capillary gas liquid chromatography. The detection limit, estimated by using hydroxybutyric acid standards, was approximately 10ug per litre.

4.8 Alcohols, ketones and aldehydes

4.8.1 Soil

4.8.1.1 Purge and trap gas chromatography

This technique has been applied [47] to the determination of ethanol, methylethyl ketone, paraldehyde and acrolein in soils. Following extraction of the soil with methanol and gas purging the purge gas is trapped on a Tenax column. The purgate obtained by heating the Tenax column is analysed by gas chromatography and/or mass spectrometry.

References

1 Thuren A. (1986) *Bulletin of Environmental Contamination and Toxicology*, **36**, 33.
2 Schwartz, H.W., Anzion, G.J.M., Van Vleit, H.P.M. *et al.* (1979) *International Journal of Environmental Analytical Chemistry*, **6**, 133.
3 Eglinton, G., Simoneit, B.R.T. and Zoro, J.A. (1975) *Proceedings of Royal Society (London)*, **B189**, 145.
4 Cifrulak, S.D. (1969) *Soil Science*, **107**, 63.
5 Talsky, G. (1983) *International Journal of Environmental Analytical Chemistry*, **14**, 81.

6 Karasek, F.W., Charbonneau, S.H., Renel, G.J. and Tong, H.Y. (1987) *Analytical Chemistry*, **59**, 1027.
7 Lopez-Avila, V., Young, R. and Beckert, W.F. (1994) *Analytical Chemistry*, **66**, 1097.
8 Lopez-Avila, V., Northcutt, R., Oustat, J. and Wickham, M. (1983) *Analytical Chemistry*, **55**, 881.
9 Goldberg, M.C. and Weiner, E.R. (1980) *Analytica Chemica Acta* **112**, 373.
10 Ishmatari, R. and Hanya, T. (1974) Gas chromatography–mass spectrometric detection of organic compounds in a river water. Proceedings International Meeting Ed. Technig. 6th, 1051.
11 Baedeckce, M.J., Nissenbaum, A. and Kaplan, I.R. (1972) *Geochimica Cosmochimica Acta*, **38**, 1185.
12 Carrol, K.K. (1976) In *Lipid Chromatographic Analysis* (ed G.V. Marinetti), Academic Press, New York, vol. 1, pp.174–212.
13 Johnson, R.W. and Calder, J.A. (1973) *Geochomica Cosmochimica Acta*, **37**, 264.
14 Thompson, S. and Eglinton, G. (1978) *Geochimica Cosmochimica Acta*, **42**, 199.
15 Van Hoevan, W., Maxwell, J.R. and Calvin, M. (1969) *Geochimica Cosmochimica Acta*, **33**, 877.
16 Nishimura, M. (1977) *Geochimica Cosmochimica Acta*, **41**, 1817.
17 Farrington, J.W. and Quinn, J.G. (1971) *Geochimica Cosmochimica Acta*, **35**, 735.
18 Mendoza, Ya., Gulacar, F.O., Hu, Z.L. and Bucks, A. (1987) *International Journal of Environmental Analytical Chemistry*, **31**, 107.
19 Van Huyssteen, J.J. (1970) *Water Research*, **4**, 645.
20 Hunter, I.R., Orgeren, V.H. and Pence, J.W. (1960) *Analytical Chemistry*, **32**, 682.
21 Murtaugh, J.J. and Bunch, R.L. (1965) *Journal Water Pollution Control Fed.*, **37**, 410.
22 Gehrke, G.W. and Larkin, W.M. (1961) *Journal of Agricultural Food Chemistry*, **9**, 85.
23 Harivank, J. (1964) *Vodni Hsparastvi Chem. (Hungary)*, Abstr., 14.
24 Smith, S. and Dila, R. (1965) *Journal of Pharmacy, Belgium*, **20**, 225.
25 Emery, E.M. and Koenrner, W.E. (1961) *Analytical Chemistry*, **33**, 146.
26 Erwin, E.S., Marco, G.C. and Emery, E.M. (1961) *Journal of Dairy Science*, **44**, 1768.
27 Cochrane, G.C. (1975) *Journal of Chromatographic Science*, **13**, 440.
28 Baker R.A. (1960) *Journal of Gas Chromatography*, 418.
29 Department of the Environment (1980) National Water Council Standing Committee of Analysts (1979) Determination of volatile fatty acids in sewage sludge. Examination of waters and associated materials. Her Majesty's Stationery Office, London.
30 Andrews, J.F. and Pearson, E.A. (1965) *International Journal of Air and Water Pollution*, **9**, 439.
31 McCarty, P.L., Jens, J.S. and Murdoch, W. (1962) *The Significance of Individual Volatile Acids in Anaerobic Treatment*. Proceedings of the 17th Purdue Industrial Waste Conference.
32 Mueller, H.F., Buswell, A.M. and Larsen, T.E. (1956) *Sewage IndustrialWastes*, **28**, 255.
33 Hunter, J.V. (1962) The Organic Composition of Various Domestic Sewage Fractions. PhD Thesis, Rutgers University.
34 Hunter, J.W. and Heukelekian, H. (1965) *Journal Water Pollution Control Federation*, **37**, 1142.
35 Painter, H.A. and Viney, M. (1959) *Journal Biochemistry Microbiolgy Technology Engineering*, **1**, 143.
36 Hindin, E. (1964) *Water Sewage Works*, **92**, 94.
37 Mueller, H.F., Larson, T.E. and Lennarz, W.J. (1958) *Analytical Chemistry*, **30**, 41.

38 American Public Health Association, WPCF and AWWA (1976) *Standard Methods for the Determination of Water and Wastewater*, 14th edn, New York.
39 Buswell, A.M., Gilcreas, F.M. and Morgan, G.B. (1962) *Journal of Water Pollution Control Federation*, **34**, 307.
40 Manganelli, R.M. and Brofazi, F.R. (1957) *Analytical Chemistry*, **29**, 1441.
41 McQuaker, N.R. and Fung, T. (1975) *Analytical Chemistry*, **47**, 1435.
42 Mopper, K. and Regeus, E.T. (1972) *Analytical Biochemistry*, **45**, 147.
43 Cowie, G.L. and Hedges, J.I. (1984) *Analytical Chemistry*, **56**, 497.
44 Dreier, F., Bucks, A. and Gulacar, F.O. (1988) *Geochimica Cosmochimica Acta*, **52**, 1663.
45 Walter, J.S. and Hedges, J.I. (1988) *Analytical Chemistry*, **60**, 988.
46 Comeau, Y., Hall, K.J. and Oldham, W.K. (1988) *Applied and Environmental Microbiology*, **54**, 2325.
47 Kester, P.E. (1987) *Analysis of Volatile Organic Compounds in Soils by Purge and Trap Chromatography*. Tekmar Company, PO Box 371856, Cincinnati, Ohio 45222–1856.

Chapter 5

Halogen containing compounds

5.1 Chloroaliphatic compounds

5.1.1 Soil

5.1.1.1 Gas chromatography

Deetman *et al.* [1] have devised an electron capture gas chromatographic technique, applicable to mud samples, for the determination of down to 1ng L^{-1} of 1,1,1-trichloroethane, trichloroethylene, perchloroethylene, 1,1,1,2-tetrachloroethane, 1,1,2,2-tetrachloroethane, pentachloroethane, hexachloroethane, pentachlorobutadiene, hexachlorobutadiene, chloroform and carbon tetrachloride. These workers used extraction of the samples with *n*-pentane as a means of isolating the chlorinated compounds from the sample. Recoveries of 95% were obtained in a single extraction. To dry the extract anhydrous sodium sulphate was found to be effective. Furthermore this drying agent could be freed from electron-capturing contaminants by heating [2] and did not absorb the chlorinated compounds. Under the specific conditions (i.e. using a temperature programmed Dexsil-300 column) all the compounds are separated with the exception of carbon tetrachloride and 1,1,1-trichloroethane which are resolved only on the Apiezon–L column. This column is an alternative with the proviso that it is not suitable for samples containing the less volatile compounds. If the sample contains chlorobromomethanes which can interfere with the determination of chloroform and trichloroethylene, it is advisable to augment the analyses by repeating the chromatography with a column containing oxydipropionitrile packing which will separate the bromine compounds from the chlorinated solvents. To avoid contamination use of a glove box is recommended for the preparation of samples. In general, it is wise to exclude chlorinated solvents from the laboratory and if the ambient air is suspect, to blanket the inject port of the chromatography with clean nitrogen.

Neumayr [3] carried out soil atmosphere studies using capillary gas chromatography and electron capture and flame ionization sequential detection and used this as a means of pinpointing zones of soil and groundwater contamination.

5.1.1.2 Gas chromatography–mass spectrometry

Methods have been described for determining chlorinated aliphatic hydrocarbons in soil and chemical waste disposal site samples. The latter method involves a simple hexane extraction and temperature programmed gas chromatographic analysis using electron capture detection and high resolution glass capillary columns. Combined gas chromatography–mass spectrometry was used to confirm the presence of the chlorocarbons in the samples [4].

5.1.1.3 Purge and trap gas chromatography

Kester [5] has discussed the application of the purge and trap gas chromatographic method to the determination of aliphatic chloro-compounds in soil. Following methanol extraction of the soil the extract is gas purged and the purge gases trapped on a Tenax silica gel/charcoal trap followed by thermal desorption from the trap and examination by gas chromatography and mass spectrometry. Compounds that have been determined by this method are listed in Table 5.1.

In this method a 4g portion of the soil is dispersed in 9ml methanol and 1ml of a methanoic surrogate spike in a 15ml screw-capped vial. Volatile compounds are dissolved in the solvent by shaking for 1min or by sonicating for 30min. The slurry is allowed to settle (centrifuge if necessary) and an aliquot; up to 100ul, of the extraction solvent is added to organic-free water and purged at ambient temperature.

Because an aliquot of the extraction medium is used in the analysis, the detection limits suffer. The solvent/sample ratio is defined by protocol, reducing the analyst's ability to decrease detection limits. Samples are spiked to provide recovery data. A detection limit of 1mg kg^{-1} was achieved for each analyte in the matrix.

5.1.1.4 Miscellaneous

Kerfoot [6] examined the performance of a grab sampling technique for soil-gas measurement analyses, at a site with groundwater known to be contaminated with chloroform. The study assessed the correlation between soil-gas and groundwater analyses with chloroform as a model volatile organic compound. Chloroform concentration in soil gas increased linearly with depth in the unsaturated zone.

A study of the vertical profile of chlorinated solvents in the soil, enables the source of contamination to be distinguished; for atmospheric inputs a peak occurred a short distance below ground, whereas for inputs from groundwater the concentration increased progressively as the water table was approached.

Table 5.1 Method 8010 Halogenated volatile organic compounds in groundwater, liquid or solid matrices

Compounds	
Bis (2-chloroexthoxy)methane	1,1-Dichloroethane
Bis (2-chloroisopropyl)ether	1,2-Dichloroethane
Bromodichloromethane	1,1-Dichloroethylene (Vinylidene chloride)
Bromoform	trans-1,2-Dichloroethylene
Bromomethane	Dichloromethane
Carbon tetrachloride	1,2-Dichloropropane
Chloracetaldehyde	(trans)-1,2-Dichloropropylene
Chlorobenzene	1,1,2,2-Tetrachloroethane
Chloroethane	1,1,1,2-Tetrachloroethane
Chloroform	Tetrachloroethylene
1-Chlorohexane	1,1,1-Trichloroethane
2-Chloroethyl vinyl ether	1,1,2-Trichloroethane
Chloromethane	Trichloroethylene
Chloromethyl methyl ether	Trichlorofluoromethane
Dibromochloromethane	Trichloropropane
Dibromomethane	Vinyl chloride
Dichlorodifluoromethane	

Source: Reproduced with permission from Techmar Company, USA [5]

Mehran *et al.* [7] determined the distribution coefficient of trichloroethylene in soil water systems. The distribution coefficient of trichloroethylene could be used to define the retardation factor, which expressed the velocity of trichloroethylene migration relative to an advancing water front. The two methods used to obtain the distribution coefficient were field measurements based on trichloroethylene concentrations in soil at various depths, and theoretical methods based on total organic carbon content of the soil and octanol–water partition coefficient for trichloroethylene. The average distribution coefficient was 0.18ml per g and the average retardation factor was 2.48 (19 field samples). Theoretical methods were valid for soils with greater than 1% organic carbon. Reasonable estimates for actual migration rates could be provided for soils low in organic carbon. Field methods were still preferred as the effect of various factors on partitioning of trichloroethylene were integrated.

5.1.2 Non-saline deposited and suspended sediments

5.1.2.1 Gas chromatography

Murray and Riley [8, 9] described gas chromatographic methods for the determination of trichloroethylene, tetrachloroethylene, chloroform and carbon tetrachloride in sediments. These sediments were separated and determined on a glass column (4m × 4mm) packed with 3% of SE–52 on Chromosorb W (AW DMCS) (80–100 mesh) and operated at 35°C, with argon (30ml min $^{-1}$) as carrier gas. An electron capture detector was used, with argon-methane (9:1) as quench gas. Chlorinated hydrocarbons were stripped from water samples by passage of nitrogen and removed from solid samples by heating in a stream of nitrogen. In each case, the compounds were transferred from the nitrogen to the carrier gas by trapping on a copper column (30cm × 6mm) packed with Chromosorb W (AW DMCS) (80–100 mesh) coated with 3% of SE–52 and cooled at –78°C, and subsequently sweeping on to the gas chromatographic column with the stream of argon. A limitation of this procedure was that compounds which boil considerably above 100°C could not be determined [10].

Amin and Narang [11] closed loop stripped volatile haloparaffins from sediments and adsorbed the volatiles on Poropak N. The compounds were eluted with methanol and the elute analysed for organic compounds by gas chromatography with electron capture and photoionization detection. A detection limit of 7μkg $^{-1}$ for each photoionization active and 1ng g $^{-1}$ for each electron capturing compound was achieved. Samples could be stored in methanol for up to 90 days without significant loss of the volatile compounds. Recoveries ranged from 71% (bromoform) to 111% (fluorobenzene).

Zitko [20] has described a confirmatory method in which the chloroparaffins in sediments are reduced to normal hydrocarbons which are then analysed by gas chromatography. This method lacks sufficient sensitivity for trace (sub-ppm) analysis and the confirmatory method may be difficult to apply. Friedman and Lombardo [21] have described a gas chromatographic method applicable to chloroparaffins that are slightly volatile; the method is based on microcoulometric detection and photochemical elimination of chlorinated aromatic compounds that otherwise interfere.

The application of gas chromatography to the determination of chlorinated hydrocarbons in water and sediments, with particular reference to the types of these compounds used in industry, has been reviewed by Hassler and Rippa [12]. Glaze *et al.* [13] used flame ionization, electron capture and Coulson electrolytic detectors with gas chromatography to study the formation of chlorinated aliphatics during the chlorination of waste waters. Chlorinated normal paraffins up to C_{30} carbon number range are of low volatility and are thermally unstable, producing hydrogen chloride on decomposition; hence direct gas chromatography is not attractive.

Amin and Narang [11] stripped volatile compounds from sediment samples and absorbed on cartridges filled with Porapak N. The compounds were eluted from the cartridges with methanol. The eluate was assayed for various organic compounds by gas chromatography with electron capture and photoionization detection. A detection limit of 7ng/g for each photoionization-active and 1ng/g for electron-capturing compound was achieved. Time study experiments showed that, while untreated samples should be analyzed within seven days after collection, samples stored in methanol could be held for up to 90 days without significant loss of the volatile compounds.

The following compounds were determined by this procedure: chloroform; bromoform; 1,1,1-trichloroethane; 1,1,2,2-tetrachloroethane; trichloroethylene; benzene; carbon tetrachloride; toluene; bromodichloromethane; chlorobenzene; 1,1,2-trichloroethane; *o,p*-xylene; tetrachloroethylene; *o,p*-chlorotoluene; 1,2-dibromoethane and fluorobenzene (used as an internal standard).

Onuska and Terry [14] have described a method for the determination of chlorinated benzenes in bottom sediment deposits. Sample preparation methods using Soxhlet extraction, ultrasonic extraction or steam distillation were compared. The chlorinated benzenes were characterized by open tubular column gas chromatography with electron capture detection. In recovery studies using sediments with different organic matter contents, the steam distillation method was the most efficient. Detection limits were in the range 0.4–10μg kg^{-1}.

Lee *et al.* [15] described an acetone-hexane extraction procedure followed by electron capture gas chromatography for the determination of down to 1μg kg^{-1} chlorinated phenols in sediments.

5.1.2.2 Gas chromatography–mass spectrometry

Gas chromatography–mass spectrometry has been, applied to the determination of volatiles in river sediment samples [16].

Carey and Hart [17] collected samples of 'pools' of non-aqueous material on the surface of sediments in the St. Clair river, Ontario, and analysed by gas chromatography and mass spectrometry. A large number of compounds were identified, including tetrachloroethane, tetra-, penta- and hexachloroethanes, chlorobutanes, chlorobutadienes, chlorohexadienes, heptachlorostyrene, octachlorostyrene and octachlornaphthalene. The results suggested that the source of the pools was not just a simple spill of perchloroethylene, as previously thought.

5.1.2.3 Purge and trap gas chromatography

Charles and Simmons [18] obtained overall recoveries of 38%, 48% and 54% respectively, for chloroform, trichloroethylene and chlorobenzene from sediments using purge and trap methods.

5.1.2.4 Column chromatography

Zitko [19] has devised a method based on column chromatography [20,21] followed by microcoulometric detection. The procedure is not specific.

5.1.2.5 Thin layer chromatography

Hollies *et al.* [22] have carried out an extensive study of the determination of chlorinated long chain (C_{13}–C_{20}) normal paraffins (cerechlors) in river sediments. They considered liquid chromatography, gas chromatography and thin layer chromatography. Chlorinated paraffins are separated from the sediment by Soxhlet extraction with petroleum ether. A concentrate of the extract is then cleaned up on an alumina column which adsorbs these compounds, allowing impurities to pass through. The chlorinated paraffins are then desorbed with toluene. Analysis of the extract is carried out by thin layer chromatography on silica. The plate is developed by covering with a second plate coated with alumina and heating face to face at 240°C. The alumina plate is then sprayed with silver nitrate to visualize the separated chloroparaffins as grey-black spots. Any chloroparaffins present in the extract are then identified by reference to the R_f values which are approximately 0.74 and 0.80 for C_{13}–C_{17} and C_{20}–C_{30} chloroparaffins respectively.

In this method a container of sediment is thoroughly stirred and mixed. 20g of sediment is taken in a tared 250ml beaker, placed in a vacuum oven at 70°C and dried to constant weight. The weight of dried sediment is calculated.

A 10.0 ± 0.1g sample of dried sediment is weighed into a tared Soxhlet thimble, enclosed with a pad of silica wool and transferred to the Soxhlet apparatus. The sample is extracted with about 60ml of petroleum spirit for 24h. The extract is transferred to a 100ml beaker and evaporated to about 5ml on a steam bath. At this stage, 60ml of petroleum spirit are introduced as a blank and, starting with a dummy Soxhlet extraction, using duplicate apparatus, treated in the same way as a sample extract in this and subsequent stages.

The petroleum spirit concentrate from the evaporation step is transferred into a prepared aluminium oxide column using a 10ml pipette. The excess of solvent is slowly run off until the meniscus just touches the surface of the sodium sulphate plug. The remains of the concentrate are washed from the beaker with two consecutive 5ml portions of petroleum spirit, adding them to the adsorption column as above. Petroleum spirit (100ml) is passed through the column to remove gross impurities from the chlorinated paraffin that remains adsorbed. This eluent is discarded. The chlorinated paraffin is desorbed by eluting with 50ml of toluene and collecting in a 100ml beaker. The toluene eluent is evaporated on the steam bath, assisting the evaporation with a jet of clean nitrogen and the concentrate is transferred quantitatively to a 2ml vial, using toluene for washing. The concentrate is evaporated to dryness on a steam bath, cooled and 0.2ml of petroleum spirit added from a calibrated pipette and swirled to dissolve the residue. This clean-up stage is performed on the 'blank'.

Two thin layer chromatography (t.l.c.) tanks are used; into the first tank is poured *n*-hexane to a depth of 10mm and into the second toluene similarly. The ends of the tanks are lined with filter paper wetted with the solvent in the tank. The tank covers are replaced and allowed to equilibrate for 30min.

Spots of 1µl, 10µl and 20µl of sediment extracts and 20µl of blank extract on the adsorption chromatography clean-up stage are applied to the silica gel t.l.c. plate using 1µl micropipettes and the spot drier.

For development, the plate is transferred to the *n*-hexane tank and developed to the first solvent limit line. The plate is removed and dried in the t.l.c. plate drier until no odour of solvent is apparent (*ca.* 10min). The plate is cut in half and the lower half is placed in the toluene tank, and eluted up to the second solvent limit line. The plate is removed and dried as above. The plate is reversed by putting it back in the hexane tank with the cut edge dipping in the hexane. The plate is eluted back to the origin line, removed and dried as above.

The positions of the origin and solvent limit lines of an aluminium oxide half-plate are marked in exactly the same way as for the silica gel half-plate. The two half-plates are clamped face to face with spring clips and the chloroparaffin spots 'printed' on to the aluminium oxide plate by heating the plates at 240°C for 8min. The half-plates are cooled to ambient temperature, undamped and the silica gel half-plate discarded. The aluminium oxide half-plate is sprayed evenly with the silver nitrate reagent and then placed under the UV lamp for 10min to develop the chromatogram. The half-plate is removed and inspected under ordinary light. Any spots from chlorinated compounds will have a grey-to-black colour on a nearly white background. Any chlorinated paraffin in the sample is identified by reference to the R_f values which are approximately 0.74 and 0.80 for C_{13}–C_{17} and C_{20}–C_{30} chlorinated paraffin respectively.

Within 30min (the plate turns grey in strong light) any chlorinated paraffin in the sample is estimated by visual comparison between the sample and standard spot intensities.

5.1.3 Saline deposited and suspended sediment

5.1.3.1 Miscellaneous

Gron [23] has reviewed methods for the determination of halogenated organic compounds (adsorbable, volatile and extractable), with particular reference to their applicability to wastewaters and marine samples (marine sediments and marine organisms). Typical analytical results for marine samples are tabulated. The determination of extractable halogenated compounds was of greatest importance for marine samples.

5.1.4 Sludge

5.1.4.1 Gas chromatography

The Yorkshire Water Authority has published a method for the determination of haloforms in sewage sludge [24].

Heckel [25] has discussed the determination of adsorbable organohalogen compounds in sewage.

Yong and Rawliszyn [26] used a multiplex gas chromatograph with a hollow fibre membrane interface in a solventless method for the determination of traces of aliphatic chlorocompounds such as trichloroethane in raw sewage sludge. Down to 0.4μg L^{-1} of these compounds could be determined.

5.2 Haloaromatic compounds

5.2.1 Soil

5.2.1.1 Purge and trap gas chromatography

Kester [5] has discussed the application of the purge and trap gas chromatographic method discussed in section 5.1.1.3 to the determination of chloroaromatic compounds such as chlorobenzene, 1,2-dichlorobenzene, 1,3-dichlorobenzene, bromobenzene in soils. Following methanol extraction of the soil the extract is gas purged and the purge gases trapped on a Tenax 1 silica gel-charcoal trap followed by thermal desorption from the trap and examination by gas chromatography and/or mass spectrometry.

5.2.1.2 Pyrolysis gas chromatography

The Curie Point flush evaporation-pyrolysis gas chromatography–mass spectrometric method [27] described in section 2.2.1.2 for the analysis of aliphatic hydrocarbons in soils has also been applied to the determination of haloorganic compounds such as di- and trichlorobenzenes.

5.2.2 Non-saline deposited and suspended sediments

5.2.2.1 Gas chromatography

Chlorinated hydrocarbons that have been determined in extracts of river sediments by gas chromatography include higher chlorinated aromatic hydrocarbons, alpha and gamma hexachlorocyclohexanes and dichlorobenzenes in amounts down to 0.5μg kg^{-1} in the sediment [28–30].

Onuska and Terry [14] developed and integrated analytical procedure for determining chlorinated benzene contaminants that enables quantitation of individual isomers as low as 0.4μg kg $^{-1}$ in sediment samples. Preparation of the sample can be performed by using one of three techniques, namely Soxhlet extraction with carbon tetrachloride, ultrasonic extraction or steam distillation. Chlorinated benzenes are then characterized and quantified by open tubular column gas chromatography with electron capture detection. Recoveries of individual chlorinated benzene isomers at three different levels from two different types of sediment, one low and one high in organic matter, were evaluated. Although all three methods are quantitative, the steam distillation method was found to be the most efficient for the determination as far as time and simplicity were concerned. Data presented indicated that detection limits of this method are 0.4–10μg kg $^{-1}$ of individual chlorobenzene isomers. Chlorobenzene recovery from bottom sediment samples at concentration levels between 1 and 100μg kg $^{-1}$ was 86 ± 14%.

The following chlorinated benzenes were determined in a sediment; 1,3-dihexachlorobutadiene, 1,3,5-trihexachlorobutadiene, 1,2,4-trihexachlorobutadiene, 1,2,3-trihexachlorobutadiene, 1,2,3,5- tetrahexachlorobutadiene, 1,2,4,5-tetrahexachlorobutadiene and pentahexachlorobutadiene.

5.2.2.2 Gas chromatography–mass spectrometry

Lee *et al.* [31] demonstrated that the chlorobenzene and hexachlorobutadiene contents of a lake sediment remained unchanged after four years storage in the dark at 4°C. Subsamples were analysed periodically over the four years by Soxhlet extraction, Florsail cleaning and analysis by capillary column gas chromatography with electron capture or mass spectrometric detection.

5.2.2.3 Purge and trap gas chromatography

Charles and Simmons [18] obtained overall recoveries from sorbed compounds on three different sediments of 38% for chloroform, 48% for trichloroethylene and 54% for chlorobenzene.

Hites [32] used this technique to investigate the occurrence of chlorobenzene, chlorotoluenes and chlorophenols in hazardous waste dumps in Niagara Falls.

5.2.2.4 Miscellaneous

Bierl [28] has described a procedure based on the micro-steam distillation and extraction technique for recovery and determination of low and medium-boiling chlorinated organic compounds. Recoveries of around 90% were obtained for a wide variety of chlorinated aliphatic and aromatic hydrocarbons, including alpha- and gamma-hexachlorocyclohexanes. For 10g samples the detection limits were 0.5ng per g for low-boiling derivatives, 2.5ng per g for the dichlorobenzenes and 1–2ng per g for a range of higher chlorinated aromatics. Factors affecting reproducibility of results are briefly indicated.

As discussed in section 5.2.2.1 Hellman [30] has studied the adsorption and desorption of hexachlorobenzene on sediments.

5.2.3 Saline deposited and suspended sediments

5.2.3.1 Gas chromatography

The isooctane extraction gas chromatographic procedure [14] described in section 5.2.2.1 has been applied to the determination of various (0.003–0.07mg kg $^{-1}$) chlorobenzenes in estuarine sediments.

5.3 Chlorophenols

5.3.1 Soils

5.3.1.1 Spectrophotometry

The higher order derivative spectrophotometric method described by Talsky [33] for the determination of phenols in soils has been used to determine pentachlorophenol.

5.3.1.2 Gas chromatography

Stark [34] has described a gas chromatographic method for the determination of pentachlorophenol as the trimethylsilylether in amounts down to 0.5mg kg $^{-1}$ in soil.

Renberg [35] used an ion-exchange technique for the determination of chlorophenols and phenoxy acetic acid herbicides in soil. In this method the soil extracts are mixed with Sephadex QAE A–25 anion exchanger and the adsorbed materials are then eluted with a suitable solvent. The chlorinated phenols are converted into their methyl ethers and the chlorinated phenoxy acids into their methyl or 2-chloroethyl esters for gas chromatography.

The sample is shaken with 0.2m sodium hydroxide (4ml g $^{-1}$ soil) in a test tube for 30min. After centrifugation the liquid is removed and re-extracted with a new portion of sodium hydroxide solution. The volume of the combined alkaline extracts is estimated. The extract (2ml) and 8ml of water are shaken for 10min with Sephadex QAE,A–25 ion exchanger (3ml bed volume). After centrifugation, the liquid is discarded and the ion exchanger rinsed with 5ml of distilled water. The water is discarded and the procedure continued as described below.

A suspension of the ion exchanger is transferred into a 10cm × 1cm (id) column). The ion exchanger is allowed to settle. The column is eluted with 10ml of acidified methanol (1g sulphuric acid/50ml methanol). To one part of the eluate (in a test tube) is added an equal volume of benzene and four parts of the hydrochloric acid solution (1.0M). The test tube is shaken and, after centrifugation, the benzene phase is transferred into a graduated test tube and the substances converted into suitable derivatives as described under the preparation of derivatives below.

An etherial solution of diazomethane is added to the benzene extract to form the methyl ester of the chlorophenols. After about 1h, the solution is evaporated to the original volume. The extract is injected into the gas chromatograph and the result compared with a standard treated the same way.

The results obtained in determining various chlorophenols in soil are shown in Table 5.2.

5.3.1.3 Gas chromatography–mass spectrometry

Stable-isotope dilution analysis is an analytical technique in which a known quantity of a stable-labelled isotope is added to a sample prior to extraction, in order to quantitate a particular compound. The ratio of the naturally abundant and the stable-labelled isotope is a measure of the naturally abundant compound and can be determined only by gas chromatography–mass spectrometry since the naturally abundant and the stable-labelled isotope cannot be completely separated gas chromatographically.

Lopez-Avila *et al.* [36] used a stable isotope dilution gas chromatography–mass spectrometric technique to determine down to 0.1ppb of pentachlorophenol (also Atrazine, Diazinon and lindane) in soil. Soil samples are extracted with acetone and hexane. Analysis is performed by high-resolution gas chromatography–mass spectrometry with mass spectrometer operated in the selected ion monitoring mode. Accuracy greater than 86% and a precision better than 8% were demonstrated by use of spiked samples.

A Finnigan 4021 quadruple mass spectrometer coupled to a 9610 gas chromatography and an Incos 2300 data system was used for the measurements. Calibration standards and sample extracted were injected automatically by a fused silica Varian autosampler. Compound separations were performed on a fused silica capillary column 30m × 0.25mm i.d. (DB–5, 0.25μm film thickness, J&W Scientific, Rancho, Cordova, CA) at the following conditions: splitless injection at 50°C followed by temperature programming to 300°C at 15°C/min; injector temperature of 260°C; transfer line temperature of 280°C; carrier gas (He) at 10psi pressure. The mass spectrometer operating conditions were ion source temperature at 300°C, electron energy of 70eV, and selected ion

Table 5.2 Levels of substances in fortified samples and corresponding recoveries

	Level (ppm)	*Soil 1g Humus rec (%)*	*Clay soil rec(%)*
Fungicides			
2,4,6-Trichlorophenol	0.50	>97	>97
2,3,4,6-Tetrachlorophenol	0.50	>97	>97
pentachlorophenol	1.5	>97	>97

Source: Reproduced with permission from American Chemical Society[35]

monitoring mode for ions at *m/z* 188, 200, 205, 181, 224, 304, 314, 266 and 272.

The total acquisition times, total scan times and the dwell times for the various ions mentioned are shown in Table 5.3.

Working calibration standards of pentachlorophenol in methylene chloride were prepared by serial dilution of the composite standards with methylene chloride, at five concentrations, 0.1, 1.0, 2.0, 5.0 and 10µg/ml $^{-1}$, for the unlabelled pentachlorophenol and 1µg/ml $^{-1}$ for the stable-labelled pentachlorophenol.

Standard soil samples were prepared as follows: 50g aliquots of the fresh sandy loam soil were slurried with 10mL of organics-free water and were spiked with various amounts of pentachlorophenol. The spike was added in 100µL of methanol to the wet soil and was allowed to equilibrate with the soil for 1h. Stable-labelled pentachlorophenol was added at 4ppb and was also allowed to equilibrate with the soil for 1h.

Following equilibration, the soil slurry was extracted at an acidic pH with 1:1 acetone-hexane as shown in Fig. 5.1.

Results obtained by this technique in the chromatographic separation of pentachlorophenol and pentachlorophenol-$^{13}C_6$ and phenanthrene-d_{10}

Table 5.3 SIM Descriptors used in GC/MS analysis

Descriptor*		dwell time, s	
benign mass	*end mass*	*set*	*actual*
187.5	88.5	0.050	0.052
265.5	266.5	0.120	0.105
271.5	272.5	0.120	0.105

Source: Reproduced with permission from American Chemical Society [36]

*Total acquisition time, 0.262s; total scan time, 0.300 ; centroid sampling intensity, 0.200ms

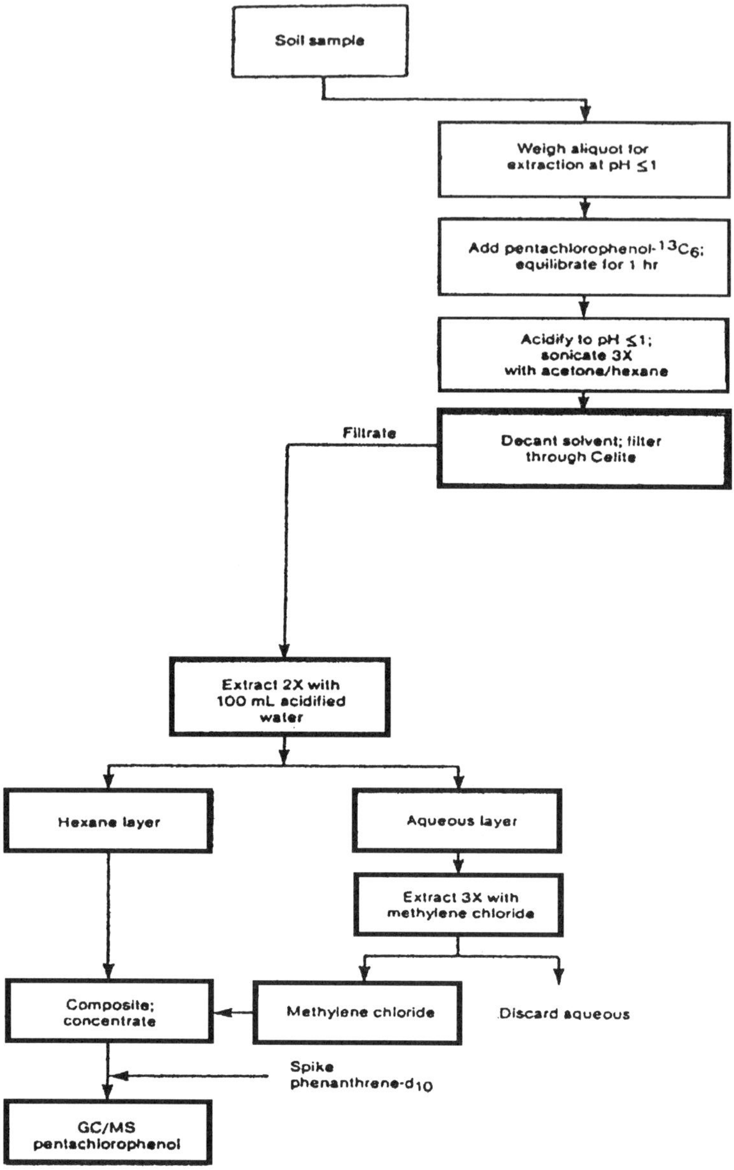

Fig. 5.1 Extraction of Atrazine, lindane, pentachlorophenol and Diazinon from soil
Source: Reproduced with permission from the American Chemical Society [36]

(internal standard) is presented in Fig. 5.2. Each unlabelled compound in Fig. 5.2 was present at a concentration of 0.1–1.0μg/mL, while the stable-labelled isotopes were present at 1μg/mL (sample size 1μL).

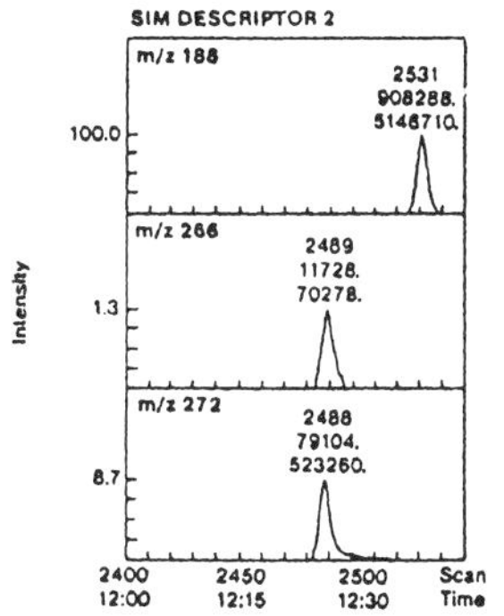

Fig. 5.2 Selected ion monitoring chromatograms for pentachlorophenol (*m/z* 266), pentachlorophenol-$^{13}C_6$ (*m/z* 272), phenanthrene-d_{10} (*m/z* 188)
Source: Reproduced with permission from the American Chemical Society [36]

5.3.2 *Non-saline deposited and suspended sediments*

5.3.2.1 Gas chromatography

Lee [42] determined pentachlorophenol and 19 other chlorinated phenols in sediments. Acidified sediment samples were Soxhlet extracted (acetone–hexane), back extracted into potassium bicarbonate, acetylated with acetic anhydride and re-extracted into petroleum ether for gas chromatographic analysis using an electron capture or a mass spectrometric detector. Procedures were validated with spiked sediment samples at 100, 10 and 1ng chlorophenols per g. Recoveries of monochlorophenols and polychlorophenols (including dichlorophenols) were 65–85% and 80–95%, respectively. However, chloromethyl phenols were less than 50% recovered and results for phenol itself were very variable. The estimated lower detection limit was about 0.2ng per g.

5.3.2.2 Miscellaneous

Wegman and Greve [37] have described a microcoulometric method for determining extractable pentachlorophenols in sediments. Hites [32] examined creek sediments by gas chromatography–mass spectrometry to identify unusual long-range marker compounds. The mass spectra of several common halogenated compounds (chlorobenzenes, chlorotoluenes and chlorophenols) were interpreted. Structures and reaction pathways were postulated for these three compounds. Their structures were sufficiently unusual for them to serve as tracers of material leaking from a dump site. All the compounds were in sediment and fish samples from Ontario lake taken at distances exceeding 300km from the source.

Schllenberg *et al.* [38] studied the adsorption of chlorophenols on sediments.

5.3.3 Saline deposited and suspended sediments

5.3.3.1 Gas chromatography

Xie [39] determined trace amounts of chlorophenols and chloroguaiacols in marine sediments collected off the Swedish coast. The compounds were desorbed from sediment surfaces by a mixture of acetic anhydride and hexane, after buffering with 0.1mol L^{-1} sodium carbonate. The optimal pH was achieved by a 1:4 ratio of buffer to acetic anhydride. The acetylated extracts were analysed by glass capillary gas chromatography with electron capture detection. The recoveries, at the $\mu g\ kg^{-1}$ level, ranged from 85–100% with standard deviations of 4–11%.

5.3.4 Sludge

5.3.4.1 Gas chromatography

Baird *et al.* [40] utilized gas chromatography in his study of the biodegradability of chlorinated phenols in sewage sludge.

5.4 Methyl bromide

5.4.1 Soil

5.4.1.1 Gas chromatography

Kerwin *et al.* [41] determined methyl bromide soil fumigant by cyrotrapping and electron capture gas chromatography. Down to 0.23pM of methyl bromide could be detected by this procedure. Kerwin *et al.* [41] found levels of methyl bromide in the stratosphere and claimed that this contributed to ozone destruction.

5.5 Chloroanisole

5.5.1 Non-saline deposited and suspended sediments

5.5.1.1 Gas chromatography

Lee [42] has described a simple, sensitive and isomer specific method for analysing 19 chloroanisoles and two chloromethylanisoles in sediments using capillary gas chromatography columns with electron capture detection or mass spectral detection. Sediments were Soxhlet extracted with a hexane/acetone mixture. OV–I or SPB–5 columns were used for separation. Detection limits for sediment matrices were 0.002ug per g for mono- and dichloroanisoles and 0.001ug per g for tri-, tetra- and pentachloroanisoles. The method was used for pathway and degradation studies of chlorophenols originating in the Fraser river wood preserving industry.

5.6 Polychlorobiphenyls

5.6.1 Soil

5.6.1.1 Gas chromatography

Jensen *et al.* [43, 44] pointed out that bottom soils in rivers contain elementary sulphur which greatly interferes in gas chromatographic methods for the determination of polychlorobiphenyls and chlorinated insecticides. They discuss methods of overcoming such interferences. Chiraenzelli [45] found that air drying soils and sediments for 24h at ambient conditions resulted in validization losses of 14–23%, with most occurring within the first 8h. Polychlorobiphenyl loss was strongly correlated with water loss. Microwave-assisted extraction with electron capture gas chromatography has been compared to ELISA for the determination in the field of polychlorobiphenyls in soils and sediments. Both techniques were found to be amenable to field screening and monitoring applications [46].

5.6.1.2 Gas chromatography–mass spectrometry

Teichman *et al.* [16] separated polychlorobiphenyls from chlorinated insecticides in soil samples using gas chromatography coupled to mass spectrometry. Polychlorobiphenyls were separated from DDT and its analogues and from the other common chlorinated insecticides by adsorption chromatography on columns of alumina and charcoal. Elution from alumina columns with increasing fractional amounts of hexane first isolated Dieldrin and Heptachlor from a mixture of chlorinated insecticides and polychlorobiphenyls. The remaining fraction, when added to a charcoal column, could be separated into two fractions, one containing the chlorinated insecticides, the other containing the polychlorobiphenyls by eluting with acetone-dimethyl ether (25:75) and benzene, respectively. The polychlorobiphenyls and insecticides were then determined by gas chromatography on the separate column eluates without cross-interference.

Teichman *et al.* [16] used a gas chromatography (Aerograph 1200) containing a glass column (180cm × 3mm) packed with 4% SE–30, 6% SP–4201 on Chromosorb W (100–120 mesh). They also used an Aerograph 204 gas chromatography containing a glass column (180cm × 3mm) with 4% SE–30, 6% QF–1 on Chromosorb W (80–100 mesh). The operating conditions were:

Model:	1200	204
Column temperature	180°C	185°C
Injector temperature	215°C	200°C
Detector temperature	200°C	200°C
Nitrogen gas flow rate	25ml min $^{-1}$	30ml min $^{-1}$

Both instruments contained an electron capture detector with a tritium foil source. For gas chromatography–mass spectrometry, a Varian 1400 gas chromatography coupled to a Finnegan 3000 mass spectrometer was used. The 1400 was equipped with a glass column (180cm × 2mm i.d.) packed with 4% SE–30, 6% SP–4201 on Supelcoport (100–120 mesh). The operating conditions were: column temperature, 210°C; transfer-line temperature, 250°C, gas jet separator temperature, 255°C, flow rate of helium gas, 12ml min $^{-1}$, sensitivity, 10^{-7} A/V, electron multiplier voltage, 2.25kV; electron ionization current, 6.95eV.

Recovery of polychlorobiphenyls from soil samples obtained in spiking experiments was 100% while that of chlorinated insecticides ranged from 81.5% (Heptachlor) to 96.3% (Dieldrin). A limit of detection of 6.5ppb was obtained from aroclors 1254 and 1260.

Lopshire [47] explored the exchange reaction of chlorine by oxygen with polychlorobiphenyl anions as a method for compound-selective polychlorobiphenyl congener detection in a gas chromatography–mass spectrometry–mass spectrometric system. Multiple reaction monitoring allowed separate chromatograms to be detected for each different polychlorobiphenyl composition from tetra- through nonachloro. Polychlorobiphenyls were recovered from sediments in another investigation by steam distillation/solvent extraction, followed by enantioselective analysis using multidimensional electron capture gas chromatography [48].

Alford Stevens *et al.* [49] carried out a multi-laboratory study of automated gas chromatography–mass spectrometric determinations of polychlorinated biphenyls in soil. The influence of various factors on the accuracy of analytical results were studied. Shaker extraction for 12.5h followed by Florisil chromatography were demonstrated to be the most reliable methods for extraction and clean-up.

Rabbat *et al.* [50] carried out an evaluation of a thermal desorption gas chromatographic–mass spectrometric method for the on-site detection of polychlorobiphenyls in hexane extracts of soils. Down to 35mg kg $^{-1}$ of polychlorobiphenyl was detected in soil samples.

5.6.1.3 Luminescence method

A photoactivated luminescence method has been described for the rapid screening of polychlorinated biphenyls in soil [51].

5.6.1.4 Supercritical fluid chromatography

Brady *et al.* [52] have discussed pressure-temperature phase diagrams for carbon dioxide polychlorobiphenyls and examined the rate process of desorption from soils. Supercritical carbon dioxide was used to extract polychlorobiphenyls and DDT and Toxaphene from contaminated soils.

Supercritical carbon dioxide at 100atm at 40°C was continuously passed through a fixed bed of soil.

Hawthorne *et al.* [53] compared supercritical extraction with chlorodifluoromethane, nitrous oxide and carbon dioxide for the extraction of polychlorobiphenyls and polyaromatic hydrocarbons from soil. Chlorodifluoromethane provided the highest recoveries while methanol modified carbon dioxide gave a 90% recovery of polychlorobiphenyls from soil.

Lagenfeld *et al.* [116] studied the effect of temperature and pressure on the supercritical fluid extraction of polychlorobiphenyls and polyaromatic hydrocarbons from soil. At 50°C raising the pressure from 356 to 650atm had no effect on recovery of polychlorobiphenyls. A temperature of 200°C was necessary for effective extraction.

Von Bavel *et al.* [55] used a solid phase carbon trap in conjunction with supercritical fluid chromatography for the simultaneous determination of polychlorobiphenyls, pesticides, polychlorodibenzo-*p*-dioxins and polychlorodibenzofurans in soils.

Yang *et al.* [56] used subcritical water to extract polychlorobiphenyls from soil and sediments. Quantitative recovery of polychlorobiphenyls was observed at water temperatures of 250 and 300°C when the pressure was reduced to 50atm at 300°C the extraction was complete within 5min.

Supercritical fluid extraction with carbon dioxide has been applied to the determination of polychlorobiphenyls in soil [113].

5.6.1.5 Enzyme based immunoassay

Johnson and Van Emon [57] have described a quantitative enzyme based immunoassay procedure for the determination of polychlorinated biphenyls in soils and sediments and compared the results with those obtained by a gas chromatographic method. The soil is extracted with methanol, or Soxhlet extracted or extracted with a supercritical fluid. In the case of the latter two extractants good agreement was obtained between immunoassay and gas chromatographic methods. Spiking recoveries from soil achieved ranged from 104% (Aroclor 1248) to 107% (Aroclor 1242). Detection limits were 9µg kg^{-1} (Aroclor 1245) and 10.5µg kg^{-1} (Aroclor 1242). Chlorinated anisoles, benzenes or phenols did not interfere.

5.6.1.6 Miscellaneous

Hellmann [30] studied the adsorption and desorption of polychlorobiphenyls and hexachlorobenzene for clays in contact with water. Appreciable adsorption of these compounds on to clays occurred.

Yu and Bayne [58] differentiated different aroclors in soil using linear discrimination and analyses by electron capture negative ion chemical ionization mass spectrometry.

5.6.2 Non-saline deposited and suspended sediments

5.6.2.1 Gas chromatography

Gas chromatography has been used extensively for the determination of polychlorobiphenyls in river sediments [30,38,59–67]. Both capillary [60–63] and packed [38,42,64–67] columns have been used.

Goerlitz and Law [59] determined chlorinated insecticides in sediment and bottom material samples, which also contained polychlorobiphenyls by extracting the sample with acetone and hexane. The combined extracts were passed down an alumina column. The first fraction (containing most of the insecticides and some polychlorinated biphenyls and polychlorinated naphthalenes) was eluted with hexane and treated with mercury to precipitate sulphur. If the polychlorinated hydrocarbons interfered with the subsequent gas chromatographic analysis, further purification on a silica gel column was necessary.

Kerkhoff *et al.* [60] used capillary gas chromatography to determine polychlorobiphenyls in various Dutch river sediments.

Kominar [61] has described a method for the determination of polychlorobiphenyls in river sediments in which samples were extracted using ultrasonics into 1:1 *n*-hexane/acetone. The extract was partitioned with water and back extracted into benzene. Combined organic extracts were dried on sodium sulphate, reduced in volume and cleaned up by gel permeation chromatography and silica gel partitioning. Analysis of polychlorobiphenyls was carried out by gas chromatography with electron capture detection. Packed columns and fused silica wall coated open tubular (WCOT) columns were compared. WCOT columns had lower loading values but identified individual congeners and characteristic groups based on the number of chlorine atoms. This allowed study of long term degradation and bio-accumulation as well as providing data for pattern recognition profiles to assist in identification of sources of pollution.

Brown *et al.* [62] showed that agents capable of attacking polychlorinated biphenyls might leave residues that exhibit characteristic signatures in their capillary gas chromatographic patterns. Chromatograms of polychlorobiphenyl residues in aquatic sediments from six polychlorobiphenyl spill sites are reviewed. Preferential reductive chlorination of the more heavily chlorinated polychlorobiphenyl homologues was observed. This group of congeners included all those that were persistent in man, inducers of P-448-type cytochromes, or thyrotoxic in rats. The polychlorobiphenyl dechlorinated exhibited several distinct congener selection patterns, indicative or mediation by several different local populations of anaerobic micro-organisms. The lower polychlorobiphenyl congeners formed by reductive dechlorination were oxidatively biodegradable by aerobic bacteria.

Maris *et al.* [63] used an on-line narrow base column liquid chromatography–capillary gas chromatography to determine polychlorobiphenyls in sediments. This system was compared to an alternative method, involving two off-line column clean-up steps and subsequent capillary gas chromatographic analysis, for the determination of polychlorinated biphenyls in sediment samples from the Meuse river. In both cases the sediment samples were extracted and desulphurized prior to analysis via a heart-cutting technique. The former method gave recoveries of 90–100% which were in good agreement with recoveries from the longer procedure. A liquid chromatographic electron capture detector method was also incorporated into the liquid chromatographic– gas chromatographic procedure as a useful polychlorobiphenyl broad spectrum screening technique.

The electron capture gas chromatographic and ELISA procedures described in section 5.6.1.1 have been applied to the field determination of polychlorobiphenyls in sediments [46].

5.6.2.2 Gas chromatography–mass spectrometry

Eichelberger *et al.* [68[applied gas chromatography–mass spectrometry, with computer controlled repetitive data acquisition from selected specific ions, to the analyses of polychlorobiphenyls in lake sediments. The polychlorinated biphenyl mixtures were separated by gas chromatography at 180°C in a coiled glass column (180cm × 2cm) packed with 1.5% OV–17 plus 1.95% QF–1 on Gas-Chrom Q (100–120 mesh), with helium (30ml min^{-1}) as carrier gas. Effluent is passed via a glass jet enrichment device into a quadruple mass spectrometer controlled by a mini-computer in such a way that only selected ions of specific m/e pass through the quadruple field. There is a substantial gain in sensitivity, without loss of qualitative information contained in the complete mass system. This technique provides a basis for a sensitive qualitative and quantitative (from ion-abundance chromatograms obtained from subset scanning) analysis for polychlorinated biphenyls.

McMurtrey *et al.* [65] investigated the feasibility of determining polychlorinated biphenyls adsorbed on sediments by a procedure involving pyrolytic desorption at 1000°C, followed by gas chromatography and mass spectrometry. The procedure was capable of detecting polychlorinated biphenyl in sediment at the 10mg kg^{-1} level.

Robbat *et al.* [50] evaluated a thermal desorption gas chromatographic–mass spectrometric technique for the detection of polychlorobiphenyl in sediments and soils.

Alford Stevens *et al.* [49] have reported on an inter-laboratory study of the determination of polychlorinated biphenyls in environmental sediments. Electron capture gas chromatography and mass spectrometry were used to identify and determine polychlorinated biphenyls. For electron capture, an overall standard deviation of 30% was achieved while mass spectrometry gave 38%.

5.6.2.3 Pyrolysis gas chromatography–mass spectrometry

McMurtrey *et al.* [65] investigated the feasibility of determining polychlorinated biphenyls adsorbed on soils and sediments, by a procedure involving pyrolytic desorption at 1000°C, followed by gas chromatography and mass spectrometry. The procedure was capable of demonstrating the presence of polychlorobiphenyls on air-dried sediment at the 10ppm level.

5.6.2.4 Supercritical fluid chromatography

Hawthorne *et al.* [53] studied the effect of temperature and pressure on supercritical fluid extraction efficiencies of polychlorinated biphenyls in river sediments. At a temperature of 50°C, raising the pressure from 350 to 650atm, had a beneficial effect on recovery of polychlorinated biphenyls from sediments. Recovery was improved however as the extraction temperature was increased from 50–200°C.

Hawthorne *et al.* [53] compared supercritical monochloride-fluoromethane, nitrogen dioxide and carbon dioxide for the extraction of polychlorobiphenyl from sediments. Monochlorodifluoro methane provided the highest recovery. Methanol modified carbon dioxide provided a 90% recovery of polychlorobiphenyls from sediments.

The procedure [56] involving subcritical water extraction of polychlorobiphenyls from soils described in section 5.6.1.4 has also been applied to sediments.

5.6.2.5 Miscellaneous

Lee and Chau [66] have discussed the development and certification of a sediment reference material for total polychlorobiphenyls. Alford Stevens *et al.* [49] in an inter-laboratory study on the determination of polychlorobiphenyls in environmentally contaminated sediments showed the mean relative standard deviation of measured polychlorobiphenyl concentrations was 34%, despite efforts to eliminate procedural variations. Eganhouse and Gosset [67] have discussed the sources and magnitude of bias associated with the determination of polychlorobiphenyls in environmental sediments. Hellman [30] studied the adsorption and desorption of polychlorobiphenyl on sediments.

Quensen *et al.* [69] showed that micro-organisms isolated from Hudson river sediment dechlorinated most polychlorinated biphenyls in Aroclor 1242 under anaerobic conditions in the laboratory. The higher the polychlorobiphenyl concentration, the more rapid the rate of dechlorination. The possible mechanisms involved are discussed. The products of dechlorination were less toxic than the original compounds and were more readily degraded by aerobic bacteria; wastewater containing polychlorobiphenyl could be treated biologically by sequential anaerobic and aerobic processes.

Bertoni *et al.* [70] used electron capture gas chromatography to determine 2,3,7,8-tetrachlorodibenzo-*p*-dioxin in Sevesco soil at the pbb level with a 15% standard deviation.

Korfmacher *et al.* [71] employed a short clean-up procedure followed by electron capture gas chromatography for the determination of octachlorodibenzo-*p*-dioxin in soils using a furzed silica capillary gas chromatographic column. The technique was suitable as a routine screening procedure for samples taken from contaminated sites.

Christman *et al.* [72] gave details of procedures for extraction, clean-up, and concentration of samples of soil prior to the determination of their content of polychlorinated dibenzo-*p*-dioxins and polychlorinated dibenzofurans by gas chromatography and by gas chromatography–mass spectrometry. Instrumental parameters are also included. Some typical results are tabulated.

The benzene–water extraction gas chromatographic procedure described in section 2.1.1.1 for the determination of aliphatic hydrocarbons in soil has also been applied to the determination of polychlorinated dibenzo-*p*-dioxins in soil [73].

The enzyme immunoassay procedure [57] discussed in section 5.6.1.5 for the determination of polychlorobiphenyls in soils has also been applied to sediments.

5.6.3 Saline deposited and suspended sediments

5.6.3.1 Gas chromatography

Three different detection methods (gas chromatography with electron capture, mass spectrometric and atomic emission detectors) have been compared for the determination of polychlorobiphenyls in highly contaminated marine sediments [74]. Only atomic emission detection in the chlorine-selective mode provided excellent polychlorobiphenyl profiles without interferences. However, the lower sensitivity of the atomic emission detector, compared to the other two detectors required a 10 to 20g sample size for most analyses.

5.6.4 Sludge

5.6.4.1 Gas chromatography–mass spectrometry

Erikson and Pellizari [75] applied gas chromatography to show that neutral extracts of sewage sludge contained appreciable amounts of polychlorobiphenyls (Table 5.4).

Gutierrez *et al.* [76] give an account of an investigation to compare packed-column gas chromatography, capillary column gas chromatography

Table 5.4 Quantitation of PCBs in hexane eluate of neutral extract of sewage sludge

Compound	*Amount found* *(µg ml⁻¹) (µg L ⁻¹)*	*Sludge concentration*
$C_{12}H_9Cl$	8.5[a]	57
$C_{12}H_8Cl_2$	220[a]	1500
$C_{12}H_7Cl_3$	760[a]	5100
$C_{12}H_6Cl_4$	470[a]	3100
$C_{12}H_5Cl_5$	57	380
$C_{12}H_4Cl_6$[b]	76	510
$C_{12}H_3Cl_7$[b]	24	160
$C_{12}H_2Cl_8$		
$C_{12}HCl_9$		
$C_{12}Cl_{10}$[b]	14	93
Total PCBs	1600	10 800

Source: Reproduced with permission from Springer Verlag [75]

[a]Average of two determinations

[b]Not identified in mass spectra. Identification confirmed by comparison of the intensities of two or more ions in the parent cluster.

and capillary column gas chromatography–mass spectrometry techniques in the analysis of complex sewage sludge samples for polychlorinated biphenyls and the organochlorine insecticides gamma-HCH, Aldrin, Dieldrin and Endrin. The results were found to differ among the three methods. It was concluded that capillary column gas chromatography was the best method for analysing these residues in complex sewage samples.

Polychlorobiphenyl cogeners 77, 126 and 169 have been determined in sewage sludge at detection limits as low as 100kg by gas chromatography negative ion mass spectrometry [115].

The electron capture negative ion chemical ionization mass spectrometric method [58] discussed in section 5.6.1.6 for the determination of polychlorobiphenyls in soils has also been applied to sludges.

5.7 Polychlorodibenzo-*p*-dioxins and polychlorodibenzo-furans

Polychlorinated dibenzo-*p*-dioxins, polychlorinated dibenzofurans and ortho-unsubstituted polychlorinated biphenyls (non-ortho polychlorobiphenyls) are three structurally and toxicologically related families of anthropogenic chemical compounds that have in recent years been shown to have the potential to cause serious environmental contamination due to their extreme toxicity [77–82]. These substances are trace-level components or byproducts in several large-volume and widely used synthetic chemicals, principally polychlorobiphenyls and chlorinated phenols [83, 84] and can also be produced during combustion processes [79, 85–87] and by photolysis [88, 89]. In general, polychlorodibenzo-*p*-dioxins and dibenzofurans and non-ortho polychlorobiphenyls are classified as highly toxic substances [90], although the toxicities are dramatically dependent on the number and positions of the chlorine substituents [91]. About ten individual members of a total of 216 polychlorodibenzo-*p*-dioxins and dibenzofurans, and non-ortho polychlorobiphenyls are among the most toxic man-made or natural substances to a variety of animal species [77–80]. The toxic hazards posed by these chemicals are exacerbated by their propensity to persist in the environment [92–94] and to readily bioaccumulate [95–97], and although the rate of metabolism and elimination is strongly species dependent [96], certain highly toxic isomers have been observed to persist in the human body for more than ten years [98].

The majority of scientific and governmental concerns for the hazards of these compounds have been directed toward analytical methodologies, toxicology, epidemiology and determination of the disposition in the environment of the single most toxic isomer, 2,3,7,8-tetrachlorodibenzo-*p*-dioxin [77–82, 84].

More recently, however, investigations into the formation and occurrence of polychlorodibenzofurans suggest that this family of toxic compounds may commonly occur at comparable or greater levels than the dibenzo dioxins and could generally pose a greater hazard than polychlorodibenzo-*p*-dioxins. Polychlorodibenzofurans are often found as cocontaminants in and are readily produced from pyrolysis of polychlorobiphenyls [83, 99–102]. Most important, the polychlorodibenzofurans produced from pyrolysis of polychlorobiphenyls are predominantly the most toxic isomers, those having a 2,3,7,8-chlorine substitution pattern [81]. A number of recent fires involving electrical transformers and capacitors have demonstrated the potential for formation of hazardous levels of polychlorodibenzofurans from pyrolysis of polychlorobiphenyls [102–105].

Pentachlorophenol, a large-volume fungicide and wood preservative, contains relatively high levels of hexa-, hepta- and octachlorodibenzo dioxins and essentially no tetrachlorodibenzo-*p*-dioxins [83 85]; and polychlorodibenzo-*p*-dioxin incineration of materials containing chlorophenols readily produces mixtures of polychlorodibenzo-*p*-dioxins, but 2,3,7,8-tetrachlorodibenzo-*p*-dioxin is a minor component. On the other hand, the highly toxic 1,2,3,7,8-pentachloro isomer is a major component of polychlorodibenzo-*p*-dioxins and polychlorodibenzofurans usually produce mixtures of distinctly different relative component abundances [83]. On the other hand, the preferential accumulation of certain isomers in animals may prevent source identification from analyses of biological samples.

5.7.1 Soil

5.7.1.1 Gas chromatography–mass spectrometry

Tong *et al.* [106] have described a high-resolution gas chromatographic mass spectrometric method for the determination of monobromopoly-chlorodibenzo-*p*-dioxin in soils and incinerator wastes.

A good example of the application of gas chromatography–mass spectrometry to the determination of polychlorodibenzo-*p*-dioxin and dibenzofurans up to the octochlorocogeners in soils and sediments is that of Smith *et al.* [110] which, it is claimed, is sufficiently sensitive to determine down to 1–5 parts per trillion of these substances.

This method permits determinations of parts-per-trillion levels and lower of tetrachloro through octachloro congeners of dibenzo-*p*-dioxins and dibenzofurans in various types of sediments. Preliminary tests also indicated the method is applicable to determinations of tetrachloro through hexachloro congeners of ortho-unsubstituted polychlorinated biphenyls. Interferences are reduced to extremely low levels. The procedure has an extremely low susceptibility to false-positive determinations which could result from the presence of a wide variety of contaminants. A modular approach to contaminant enrichment has permitted the integration of seven processes into a two-step procedure, significantly reducing time requirements and the number of sample manipulations and making the procedure amenable to automation. The reliability and accuracy of the procedure was demonstrated by the results of intra-laboratory and inter-laboratory studies and by successful analyses of over 200 samples of a wide variety of types.

Sample preparation

Sediment samples are combined with at least four times their weight of anhydrous sulphate. The mixture is then spread out to a depth of less than 3cm so that the mass, which solidifies after 3–6h, can be readily broken up after drying overnight. The mixture is then dry-blended (any kitchen model blender) in a glass jar to yield a fine powder. Samples of low water content did not require overnight equilibration with sodium sulphate before blending. A second blending of the mixture 4–6h after the first is often required to produce a more homogeneous and finely divided sample.

In this procedure the soil sample (spiked with isotopic marker compounds) is processed in a two-part enrichment procedure (Fig. 5.3). In part I, a mixture of the sample and sodium sulphate is subject to solvent extraction, and the extract is, in the same process, passed through a series of silica-based adsorbents and then through the carbon/glass fibre adsorbent. The extract passes through the adsorbents in the following order: potassium silicate, silica gel, cesium or potassium silicate, silica gel

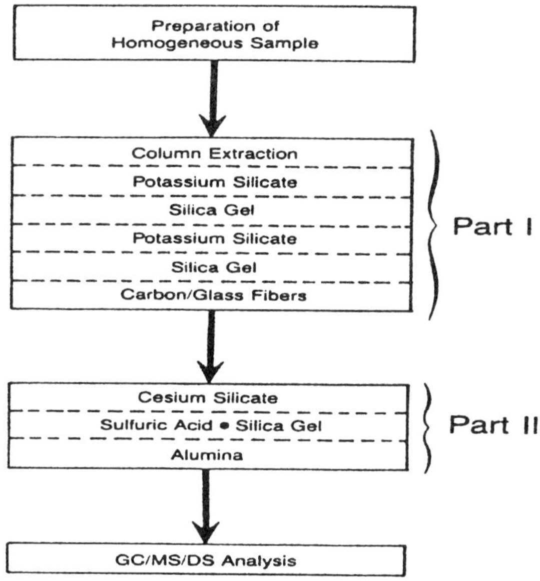

Fig. 5.3 Flow chart of total procedure
Source: Reproduced with permission from American Chemical Society [110]

and finally an activated-carbon adsorbent. The residues of interest (polychlorodibenzo-*p*-dioxins and polychlorodibenzofurans and non-ortho polychlorobiphenyls, as well as other chemical classes such as polychlorinated naphthalenes, polychlorinated biphenylenes and certain polynuclear aromatic hydrocarbons) are retained on the carbon adsorbent and subsequently recovered by reverse elution with toluene.

In part II of the procedure, following a change of solvent to hexane, the sample is applied to a second series of adsorbents contained in two tandem columns. The first column contains small amounts of cesium or potassium silicate and sulphuric acid impregnated silica gel. The effluent from this column flows directly onto the second column containing activated alumina on which the final fractionation of several classes of residues is accomplished. Following reduction of sample volume, analyses are carried out by high-resolution gas chromatography–low-resolution mass spectrometry-computer data system analysis and under some circumstances by gas chromatography–electron capture detector analysis.

Determinations of polychlorodibenzo-*p*-dioxins and polychlorodibenzofurans were carried out with a Finnigan 4023 GC–MS system equipped with an INCOS data system and with negative and positive chemical ionization options. Methane was used as the reagent gas for the negative ion chemical ionization analyses. The gas chromatograph was usually fitted with either a 30m × 0.25mm DB–5 fused-silica capillary column or a 55m × 0.27mm Silar 10c column prepared by H.R. Buser,

Swiss Federal Research Station, Wadenswil, Switzerland. The carrier gas was helium and the following temperature programme was used with *o*-xylene solvent: 150–255°C at 3°C/min and then 12°C/min to 290°C and hold for 10min. The electron impact mode (EI) and multiple ion detection were routinely used for gas chromatography–mass spectrometry, identification and quantitation of polychlorodibenzo-*p*-dioxins and polychlorodibenzofurans including isotopic marker compounds ([^{13}C]-TCDD, [^{37}Cl]-TCDF and [^{37}Cl]-OCDD). By use of DB–5 column, a series of either 8 or 12 mass-to-charge ratio (*m/z*) values were monitored within each of five or six chromatographic windows, each window being defined by the lower and upper elution limits of a particular group of congeners. The multiple ion detection analysis involved the monitoring of four or five members of a molecular ion cluster and occasionally of the fragment ion cluster resulting from the loss of COCl, M–63.

Gas chromatographic analyses employing a packed column [2m × 2mm 3% OV–17 on 100/120 Supelcoport] were carried out on a gas chromatograph equipped with an electron capture detector. Nitrogen was used as the carrier gas with the following temperature programme: 180–270°C at 8°C/min and hold for 15min.

Representative multiple ion mass chromatograms of soil samples are presented in Fig. 5.4. These gas chromatography–mass spectrometric determinations of polychlorodibenzo-*p*-dioxin and polychlorodibenzofurans, and non-ortho polychlorobiphenyls in differing types of samples serve to exemplify the versatility of the procedure for such analyses. The gas chromatography–mass spectrometric data were usually uncluttered by extraneous components, and interpretation of the data was routinely straightforward.

5.7.1.2 Supercritical fluid chromatography

Von Bavel *et al.* [55] have developed a solid phase carbon trap (PX–21 active carbon) for the simultaneous determination of polychlorodibenzo-*p*-dioxins and polychlorodibenzofurans also polychlorobiphenyls and chlorinated insecticides in soils using superfluid extraction liquid chromatography for the final determination. Supercritical fluid extraction with carbon dioxide has been applied to the determination of dioxins in soil [114].

5.7.1.3 Miscellaneous

Walters and Guiseppe-Elle [108] studied the sorption of 2,3,7,8-tetrachlorodibenzo-*p*-dioxin to soils from aqueous methanol mixtures and evaluated the applicability of the cosolvent theory to such sorption. Sorption kinetics were influenced by the fraction of methanol in the liquid phase and the soil type. Linear equilibrium sorption isotherms were

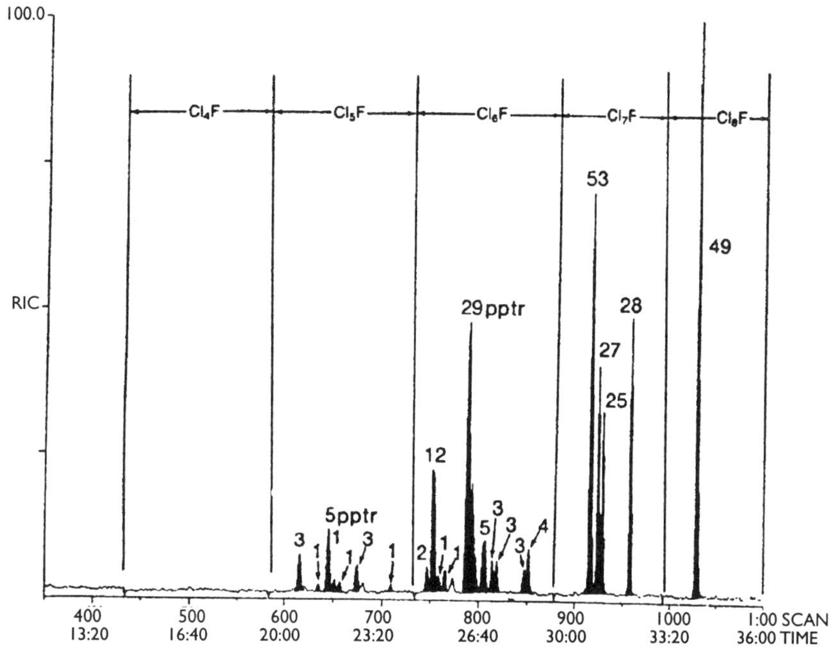

Fig. 5.4 Representative analyses of environmental samples of GC/NICI–MS–MID PCB contaminated soil from Fountain City, WI
Source: Reproduced with permission from the American Chemical Society [110]

obtained for all soils and liquid systems studied. Although it appeared that the presence of relatively high concentrations of miscible solvent in the liquid phase would increase the mobility of 2,3,7,8-tetrachlorodibenzo-*p*-dioxin, methanol concentrations of 1g per litre or less would not significantly increase dioxin mobility.

di Domenico *et al.* [107] have discussed analytical techniques used for the determination of 2,3,7,8-tetrachlorodibenzo-*p*-dioxin in environmental samples taken after the industrial accident at Sevesco, Italy. Detection thresholds of 2–50ppt were achieved for agricultural soil samples.

5.7.2 Non-saline deposited and suspended sediments

5.7.2.1 Gas chromatography–mass spectrometry

Various workers have discussed methodology for the determination of polychlorodibenzo-*p*-dioxins and dibenzofurans in sediments [110,111,115] and silts [109]. The method described by Smith *et al.* [110] for the determination of polychlorodibenzo-*p*-dioxins and dibenzofurans in soils and discussed in section 5.7.1.1 is also applicable to sediments. Taguchi [112] presented preliminary results obtained in a round robin inter-laboratory study on various techniques using gas chromatography and mass spectrometry for the determination of polychlorodibenzo-*p*-dioxins and dibenzofurans in sediments. Clean-up techniques are discussed.

References

1 Deetman, A.A, Demeulemeester, P., Garcia, M. *et al.* (1976) *Analytical Chemistry*, **82**, 1.
2 Guam, C.S. and Wong, M.K. (1972) *Journal of Chromatography*, **72**, 283.
3 Neumayr, V. (1986) *Soil and Groundwater Protection*. (eds. G. Milde and R. Leschber), Gustav Fischer Verlag, Stuggart, pp.65–84.
4 De Leon, I.R., Maberry, M.A., Overton, E.B. *et al.* (1980) *Journal of Chromatographic Science*, **18**, 85.
5 Kester, P.E. (1987) *Analysis of Volatile Organic Compounds in Soils by Purge and Trap Gas Chromatography*. Techmar Company, PO Box 371856, Cincinnati, Ohio 452222–1856.
6 Kerfoot, H.B. (1987) *Environmental Science and Technology*, **21**, 1022.
7 Mehran, M., Olsen, R. and Rector, B.M. (1987) *Groundwater*, **25**, 275.
8 Murray, A.J. and Riley, J.P. (1973) *Analytica Chimica Acta*, **65**, 261.
9 Murray, A.J. and Riley, J.P. (1973) *Nature*, **242**, 37.
10 Novak, J., Zluticky, J., Kubulka V. and Mostecky, J. (1973) *Journal of Chromatography*, **76**, 45.
11 Amin, T.A. and Narang, R.S. (1985) *Analytical Chemistry*, **57**, 648.
12 Hassler, J. and Rippa, F. (1977) *Vyskummy Ustai*, Vodneko Hospodarstra; Veda a Vyskun. Praxi. No 50.
13 Glaze, W., Henderson, J.E., Bell, J.E. and Van Wheeler, A. (1972) *Journal of Chromatographic Science*, **11**, 590.
14 Onuska, F. and Terry, K.A. (1985) *Analytical Chemistry*, **57**, 801.
15 Lee, H.B., Stokker, Y.D. and Chan, A.S.Y. (1987) *Journal of Association of Official Analytical Chemists*, **70**, 1003.
16 Teichman, J., Bevenue, A. and Hylin, J.W. (1978) *Journal of Chromatography*, **151**, 155.
17 Carey, J.H. and Hart, J.H. (1986) *Water Pollution Research Journal of Canada*, **21**, 309.
18 Charles, M.J. and Simmons, M.S. (1987) *Analytical Chemistry*, **59**, 1217.
19 Zitko, V. (1973) *Journal of Chromatography*, **81**, 152.
20 Zitko, V. (1974) *Journal of Association of Official Analytical Chemists*, **57**, 1253.
21 Friedman, D. and Lombardo, P. (1975) *Journal of Association of Official Analytical Chemists*, **58**, 703.
22 Hollies, J.I., Pinnington, P.F., Handley, A.J. *et al.* (1979) *Analytical Chemistry*, **11**, 201.
23 Gron, C. (1988) *Vatten*, Technical University of Denmark, Lyngby, **44**, 205.
24 Yorkshire Water Authority (1974) Method YWA501–01. *Determination of Haloforms in Sewage Sludge*.
25 Heckel, E. (1986) *Umwelt*, **31**, 39.
26 Yong, M.J. and Rawliszyu, J. (1992) *Analytical Chemistry*, **65**, 1758.
27 de Leeuw, J.W., de Leer, E.W.B., Sinningh, L. *et al.* (1986) *Analytical Chemistry*, **58**, 1852.
28 Bierl, F. (1988) *Fresenius Zeitschrift für Analytische Chemie*, **330**, No 415.
29 Wegman, R.C.C. and Hafster, A.W.M. (1982) *Water Research*, **16**, 1265.
30 Hellman, H. (1985) *Deutsch Gewässerkundliche Mitteilungen*, **29**, 111.
31 Lee, H.B., Hong-you, R. and Chau, A.S.Y. (1986) *Analyst (London)*, **111**, 81.
32 Hites, R.A. (1988) *Analytical Chemistry*, **60**, 647A.

33 Talsky, G. (1983) *International Journal of Environmental Analytical Chemistry*, **14**, 81.
34 Stark, A. (1969) *Journal Agriculture and Food Chemistry*, **17**, 871.
35 Renberg, L. (1974) *Analytical Chemistry*, **46**, 459.
36 Lopez-Avila, V., Hirata, P., Kroska, S. *et al.* (1985) *Analytical Chemistry*, **57**, 2797.
37 Wegman, R.C.C. and Greve, P. (1977) *Science of the Total Environment*, **7**, 235.
38 Schellenberg, K., Lenenberger, C. and Schwartzenbach, R.D. (1984) *Environmental Science and Technology*, **18**, 652.
39 Xie, T.H. (1983) *Chemosphere*, **12**, 1183.
40 Baird, R.B., Kuo, C.L., Shapiro, J.S. and Yanko, W.A. (1974) *Archives of Environmental Contamination and Toxicology*, **2**, 165.
41 Kerwin, R.A., Crill, P.M., Tabot, R.W. *et al.* (1996) *Analytical Chemistry*, **68**, 899.
42 Lee, H.B. (1988) *Journal of Association of Official Analytical Chemists*, **71**, 803.
43 Jensen, G., Renberg, L. and Reutergard, L. (1977) *Analytical Chemistry*, **49**, 316.
44 Southwest Water Laboratory (1971) Method No. SP 8/71. *Sediment Extraction Procedures*, Athens, Georgia, USA.
45 Chiarenzelli, J., Serndato, R., Arnold, G. *et al.* (1996) *Chemosphere*, **33**, 899.
46 Lopez-Avila, V., Benedicto, J., Charon, C. **et al**. (1996) *Environmental Science and Technology*, **29**, 271.
47 Lopshire, R.F., Watson, J.T. and Enke, C.G. (1996) *Toxicology Industrial Health*, **12**, 375.
48 Glausch, A., Blanch, G.P. and Schurig, V. (1996) *Journal of Chromatography A*, **723**, 399.
49 Alford Stevens, A.L., Eichelberger, J.W. and Budde, W.L. (1988) *Environmental Science and Technology*, **22**, 304.
50 Robbat, D., Tyng-Yumi, L. and Abraham, B.M. (1992) *Analytical Chemistry*, **64**, 358.
51 Vo-Dihn, T., Pal, A. and Pal, J. (1994) *Analytical Chemistry*, **66**, 1264.
52 Brady, B.O., Kao, C.C., Doolen, K.M. *et al.* (1987) *Industrial and Engineering Chemistry*, **26**, 261.
53 Hawthorne, S.B., Lagenfeld, J.J., Miller, D.J. and Burford, M.D. (1992) *Analytical Chemistry*, **64**, 1614.
54 Hawthorne, S.B., Lagenfeld, J.J., Miller, D.I. and Rawliszyn, J. (1993) *Analytical Chemistry*, **65**, 338.
55 Von Bavel, B., Jaremo, M., Karlsson, L. and Lindstrom, G. (1996) *Analytical Chemistry*, **68**, 1279.
56 Yong, Y., Bowadt, S., Hawthorne, S.B. and Miller, D. (1995) *Journal American Chemistry*, **67**, 4571.
57 Johnson, J.C. and Van Emon, J.M. (1996) *Analytical Chemistry*, **68**, 162.
58 Yu, Ma, C. and Bayne, C.K. (1993) *Analytical Chemistry*, **65**, 772.
59 Goerlitz, D.F. and Law, L.M.J. (1974) *Journal of Association of Official Analytical Chemists*, **57**, 176.
60 Kerkhoft, M.A.T., de Vries, A., Wegman, R.C.C. and Hotske, A.W.M. (1982) *Chemosphere*, **11**, 165.
61 Kominar, R.J., Onuska, F.L. and Terry, K.A. (1985) *Journal of High Resolution Chromatography and Chromatography Communications*, **8**, 585.
62 Brown, J.F., Bedard, D.L., Brannan, M.J. *et al.* (1987) *Science*, **236**, 709.
63 Maris, F.A., Noroozian, E., Otten, R.R. *et al.* (1988) *Journal of High Resolution Gas Chromatography and Chromatography Communications*, **11**, 197.
64 Alford Stevens, A.L. and Budde, T.A. (1985) *Analytical Chemistry*, **57**, 2452.
65 McMurtrey, K.D., Wildman, N.J. and Tal, H. (1983) *Bulletin of Environmental Science and Toxicology*, **31**, 734.
66 Lee, H.B. and Chau, A.S.Y. (1987) *Analyst (London)*, **112**, 37.
67 Eganhouse, R.E. and Gossett, R.W. (1991) *Analytical Chemistry*, **63**, 2130.

68 Eichelberger, J.W., Harris, L.E. and Budde, W.J. (1974) *Analytical Chemistry*, **46**, 227.
69 Quensen, J.F., Tiedje, J.M. and Boyd, S.A. (1988) *Science*, **242**, 752.
70 Bertoni, G., Brocco, D., Di Palo, V. *et al.* (1978) *Analytical Chemistry*, **50**, 732.
71 Korfmacher, W.A., Rushing, L.G., Nestorick, D.M. *et al.* (1985) *Chemosphere*, **14**, 841.
72 Christman, W., Rotard, W., Schinz, V. and Bode, H. (1986) *Chemosphere*, **15**, 2077.
73 Karasak, F.W., Charbonneau, S.H., Renel, G.J. and Tong, H.Y. (1987) *Analytical Chemistry*, **59**, 1027.
74 Pedersen-Bjergaard, S., Semb, S.I., Vedde, J. and Brevic, E.M. (1996) *Chromatographia*, **43**, 44.
75 Erikson, M.D. and Pellizzari, E.D. (1979) *Bulletin of Environmental Contamination and Toxicology*, **22**, 688.
76 Gutierrez, A.G., McIntyre, A.E., Lester, J.N. and Perry, R. (1983) *Environmental Technology Letters*, **4**, 521.
77 Tucker, R.E., Young, A.L. and Grey, A.P. (eds) (1983) *Human and Environmental Risks of Chlorinated Dioxins and Related Compounds*, Plenum Press, New York.
78 Kimbrough, R.D. (ed) (1980) *Halogenated Biphenyls, Terphenyls, Naphthalenes, Dibenzodioxins and Related Products*, Elsevier, North Holland, New York.
79 Hutzinyer, O., Frei, R.W., Merian, E. and Pocchiara, F. (eds) (1982) *Chlorinated Dioxins and Related Compounds. Impact on the Environment*. Pergamon Press, New York.
80 Nicolson, W.J., Moore, J.A. (eds) (1979) *Health Effects of Halogenated Aromatic Hydrocarbons*, New York Academy of Sciences, New York.
81 Lee, D.H.K., Falk, H.L. (eds) (1973) *Environmental Health Perspectives Experimental Issue No 5*. US Department of Health, Education and Welfare, Publication No. (NIH) 74–218, September.
82 Huff, J.R., Moore, J.A., Saracci, D.R. and Tomatis, L. (1980) *Environmental Health Perspectives*. (ed. D.P. Rial) US Department of Health and Human Services Publication No. 80–218, No 36, pp.221–240.
83 Rappe, C., Buser, H.R. and Bosshardt, H.P. (1979) *Health Effects of Halogenated Aromatic Hydrocarbons*, (eds. W.J. Nicolson and J.A. Moore), New York Academy of Sciences, pp.1–18.
84 Esposito, M.P., Tiernan, T.O. and Dryden, F.E. (1980) *Dioxins*. US Environmental Protection Agency Report No EPA–600/2–80–197. November.
85 Olie, K., Vermeullen, P.L. and Hutzinger, O. (1977) *Chemosphere*, **6**, 455.
86 Ahling, B., Lindskog, A., Jansson, B. and Sundstrom, G. (1977) *Chemosphere*, **8**, 461.
87 Buser, H.R., Bosshardt, H.P., Rappe, C. and Lindahl, R. (1978) *Chemosphere*, **7**, 419.
88 Crosby, D.G. and Wong, A.S. (1976) *Chemosphere*, **5**, 327.
89 Lamparski, L.L., Stehl, R.H. and Johnson, R.L. (1980) *Environmental Science and Technology*, **14**, 196.
90 McConnell, F.E. (1980) *Halogenated Biphenyls, Terphenyls, Naphthalenes, Dibenzodioxins and Related Products* (ed. R.D. Kimbrough), Elsevier, North Holland, New York, pp.109–150.
91 Goldstein, J.A. (1980) *Halogenated Biphenols, Terpyhenyls, Napthalenes, Dibenzodioxins and Related Products*. (ed R.D. Kimbrough). Elsevier, North Holland, New York, pp.151–190.
92 di Domenico, A., Vivcano, G. and Zapponi, G. (1982) *Chlorinated Dioxins and Related Compounds. Impact on the Environment*. (eds. O. Hutzinger, R.W. Frei, E. Merian and F. Pocchiari), Pergamon Press, New York, pp.105–114.

93 Ward, C.T. and Matsumura, F. (1978) *Archives of Environmental Contamination and Toxicology*, **7**, 349.
94 Young, A.L. (1983) *Human and Environmental Risks of Chlorinated Dioxins and Related Compounds*, (eds. R.E. Tucker, A.L. Young, and A.P. Gray), Plenum Press, New York, pp.173–190.
95 Bickel, M.H. and Muhlback, S. (1982) *Chlorinated Dioxins and Related Compounds. Impact on the Environment*, (eds. O. Hutzinger, R.W. Frei, E. Merican and F. Pocchiari), Pergamon Press, New York, pp.303–306.
96 Decad, G.M., Birnbaum, L.S. and Matthews, S. (1982) *Chlorinated Dioxins and Related Compounds. Impact on the Environment*, (eds. O. Hutzinger, R.W. Frei, E. Merian and F. Pocchiari), Pergamon Press, New York, pp.307–315.
97 Isensee, A.R. (1978) *Ecological Bulletin*, **27**, 255.
98 Masuda, Y. and Kuroki, H. (1980) *Halogenated Biphenyls, Terphenyls, Naphthalenes, Dibenzodioxins and Related Compounds*, (ed. R.D. Kimbrough), Elsevier, North Holland, New York, pp.561–569.
99 Kuratsune, M. (1980) *Halogenated Biphenyls, Terphenyls, Napthalenes, Dibenzodioxins and Related Products*, (ed. R.D. Kimbrough), Elsevier, North Holland, pp.287–302.
100 Vos, J.G., Keoman, J.H., Van der Maas, H.L. *et al.* (1970) *Food and Cosmetics Toxicology*, **8**, 625.
101 Buser, H.R., Bosshardt, H.P. and Rappe, C. (1978) *Chemosphere*, **7**, 109.
102 Smith, R.M., O'Keefe, P., Hilker, D.R. *et al.* (1982) *Chemosphere*, **11**, 715.
103 Janssen, R. and Sundstrom, G. (1982) *Chlorinated Dioxins and Related Compounds. Impact on the Environment*, (eds. O. Hutzinger, R.W. Frei, E. Merian and F. Pocchiari), Pergamon Press, New York, pp.201–208.
104 Rappe, C., Markland, S., Bergqvist, P.A. and Hansson, M. (1982) *Chem Ser.*, **20**, 56.
105 Rappe, C., Markland, S., Buser, H.R. and Bosshardt, H.P. (1978) *Chemosphere*, **3**, 269.
106 Tong, H.I., Monson, S.J., Gross, M.L. and Huang, L.Q. (1991) *Analytical Chemistry*, **63**, 2697.
107 di Domenico, A., Merli, F., Boniforti, L. *et al.* (1979) *Analytical Chemistry*, **51**, 735.
108 Walters, R.W. and Guiseppi-Elle, A. (1988) *Environmental Science and Technology*, **22**, 819.
109 Tong, H.Y., Giblin, D.E., Lapp, R.I. *et al.* (1991) *Analytical Chemistry*, **63**, 1772.
110 Smith, L.M., Stalling, D.L. and Johnson, J.L. (1984) *Analytical Chemistry*, **56**, 1830.
111 Lawrence, J., Onuska, F., Wilkinson, R. and Afghan, B.K. (1986) *Chemosphere*, **15**, 1085.
112 Taguchi, V. (1986) *Chemosphere*, **15**, 1147.
113 Alexandrou, N. and Pawliszyn, J. (1989) *Analytical Chemistry*, **61**, 2770.
114 Onuska, F.I. and Terry, K.A. (1991) *Journal of High Resolution Chromatography and Chromatography Communications*, **14**, 829.
115 Raverdino, V., Holzer, R. and Berset, J. (1996) *Fresenius Journal of Analytical Chemistry*, **354**, 477.
116 Lagenfeld, J.J., Hawthorne, S.B., Miller, D.J. and Rawliszyn, J. (1993) *Analytical Chemistry*, **65**, 338.

Chapter 6

Nitrogen containing compounds

6.1 Amines

6.1.1 Soil

6.1.1.1 Spectrophotometry

Talsky [1] has used higher order derivative spectrophotometry to determine aniline in soil.

6.2 Acrylonitrile and acetonitrile

6.2.1 Soil

6.2.1.1 Purge and trap gas chromatography

Kester [2] has discussed the application of purge and trap gas chromatography in the determination of acrylonitrile in soil. In this method the soil sample is heated for 30min to 85°C and dry purged with dry helium and the volatiles collected in a Tenax trap. Subsequent release of acrylonitrile and acetonitrile by heating the Tenax trap to 100 to 180°C is followed by collection of the volatiles and analyses by gas chromatography using Chromosorb 101 to column programmed from 80 to 150°C and a flame ionization detector.

6.3 4-Nitrophenol

6.3.1 Sludge

6.3.1.1 Gas chromatography–mass spectrometry

In a method for the determination of 4-nitrophenol in sewage sludge (Lee and Peart [3]) the sludge is subjected to supercritical carbon dioxide extraction and on-line acetylation. The extract is analysed by gas chromatography–mass spectrometry. Down to 0.1mg kg^{-1} of 4-nitrophenol could be detected in sludge.

6.4 Nitrosamines

6.4.1 Soil

6.4.1.1 Miscellaneous

Many *N*-nitrosamines are toxic and carcinogenic, and furthermore the carcinogenic action exhibits a high degree of organ specificity. Nitrosamines are formed by interaction between nitrite and an amine with varying ease, depending on the nature of the amine and the prevailing conditions. The reaction is not restricted to secondary amines, but also occurs with primary and tertiary amines and even quaternary ammonium salts. Thus, the precursors are widespread, both as naturally occurring compounds and in many commercial and industrial processes nitrosamines are generated and it is therefore conceivable that trace amounts may be present in air and water in the vicinity of industrial sites. Nitrosamines in minute amounts have been found in deionized water, generated from the resins.

Mills and Alexander [4] have discussed the factors affecting the formation of dimethylnitrosamine in samples of soil. Dimethylnitrosamine was formed as readily in sterilized samples as in non-sterile samples, indicating that, although micro organisms can carry out an enzymatic nitrosation in some soils, dimethylnitrosamine can be formed by a non-enzymatic reaction, even at near neutral conditions. The presence of organic matter appears to be important in promoting nitrosation in the presence of the requisite precursors.

6.5 Diazo compounds

6.5.1 Non-saline deposited and suspended salines

6.5.1.1 Miscellaneous

Weber and Wolfe [5] have shown that aromatic diazocompounds in sediments were readily degraded by an abiotic surface-mediated reaction. The exact nature of the reducing agent was not determined, but it appeared to be associated with the sediment. There was no apparent correlation between the rate of degradation and the measured reduction potential of the dyes. The rate of degradation appeared to be controlled by the amount of partitioning on the sediment, with increasing partitioning inhibiting the reduction process. The experimental results were used to develop a model for the reduction process.

6.6 Basic nitrogen compounds

6.6.1 *Saline deposited and suspended sediments*

6.6.1.1 Gas chromatography–mass spectrometry

Kido *et al.* [6] determined basic organic compounds such as quinoline, acridine, aza-fluorene, and their *N*-oxides in marine sediments found in an industrial area. The sediments were extracted with benzene by using a continuous extractor for 12h. Hydrochloric acid solution (1N) was added to the benzene extracts, and the mixture was shaken for 5min; the acid layer separated from the benzene layer was made alkaline by the addition of sodium hydroxide, and the alkaline aqueous solution was extracted with diethyl ether; the ether extracts were then dehydrated with anhydrous sodium sulphate and concentrated with a Kuderna-Danish evaporator. The concentrations were separated and analysed by gas chromatography–mass spectrometry and gas chromatography high-resolution mass spectrometry.

Krone *et al.* [7] used capillary column gas chromatography with nitrogen specific detection and gas chromatography–mass spectrometry to determine nitrogen-containing aromatics originating from creosote oil in solvent extracts of sediments taken in Eagle Harbour, Puget Sound and in uncontaminated areas. Organic sediment extracts and the commercial creosote oil were fractionated by silica alumina column chromatography. No nitrogen-containing aromatics were detected in sediments from a pristine reference area. Over 90 nitrogen-containing aromatics were identified in the sediments from Eagle Harbour and in the creosote oil. The total nitrogen-containing aromatics concentration in Eagle Harbour sediments ranged from 200μg to 1200mg kg^{-1} sediment (dry weight). Primarily, three ring and four ring nitrogen-containing aromatics were identified, thought to originate from a wood creosoting facility on the shores of the harbour.

6.7 Ditallow dimethyl ammonium

6.7.1 *Saline deposited and suspended sediments*

6.7.1.1 Supercritical fluid extraction

Supercritical carbon dioxide modified with methanol has been used to extract ditallowdimethylammonium from marine sediments [8].

6.7.2 *Sludges*

6.7.2.1 Supercritical fluid extraction

Supercritical carbon dioxide modified with methanol has been used to extract ditallowdimethylammonium from sludges [8].

6.8 (Phenylsulphonyl) sarcosine

6.8.1 Sludge

6.8.1.1 Gas chromatography

(Phenylsulphonyl) sarcosine has been identified in sewage sludge using derivitivization and gas chromatography–mass spectrometry [9].

6.9 Azarenes and nitroazarenes

6.9.1 Sludge

6.9.1.1 Gas chromatography–mass spectrometry

Azarenes and nitroazarenes have been determined in sewage sludge by gas chromatography–mass spectrometry [10].

References

1 Talsky, G. (1983) *International Journal of Environmental Analytical Chemistry*, **14**, 81.
2 Kester, P.E. (1987) *Analysis of Volatile Organic Compounds in Soil by Purge and Trap Gas Chromatography*. Teekmar Company, PO Box 371856 Cincinnati, Ohio 45222–1856.
3 Lee, H.B. and Peart, T.E. (1996) *Analytical Chemistry*, **68**, 1976.
4 Mills, A.I. and Alexander, M. (1976) *Journal of Environmental Quality*, **5**, 437.
5 Weber, E.J. and Wolfe, N.L. (1987) *Environmental Toxicology and Chemistry*, **6**, 911.
6 Kido, A., Shinohara, R., Eto, S. *et al.* (1979) *Japan Journal of Water Pollution Research*, **2**, 245.
7 Krone, C.A., Burrows, D.W., Brown, D.W. *et al.* (1986) *Environmental Science and Technology*, **20**, 1144.
8 Fernandez, P., Alder, A.C., Suter, M.J.F. and Giger, W. (1996) *Analytical Chemistry*, **68**, 921.
9 Heberer, T. and Stan, H.J. (1994) *Fresnius Environmental Bulletin*, **3**, 639.
10 Bodzek, D., Janoszka, B. and Warzecha, L. (1996) *Water, Air and Soil Pollution*, **89**, 417.

Chapter 7

Phosphorus containing compounds

7.1 Alkyl and aryl phosphates

7.1.1 Soil

7.1.1.1 Supercritical fluid extraction

David and Seilier [1] compared the efficiencies of various extraction techniques including supercritical fluid [2], high pressure solvent and Soxhlet extraction for the removal of organophosphorus hydraulic fluids from soil. High pressure solvent extraction was at temperatures up to 200°C and pressures up to 170 bar was the favoured technique. Extraction efficiencies were similar in all three methods, but the favoured method was more rapid and cheaper to operate.

7.1.1.2 Mass spectrometry

Ingram *et al.* [2] applied static secondary ion mass spectrometry to determine down to 70pg m^{-2} of tributyl phosphate in soil surfaces.

7.1.2 Non-saline deposited and suspended sediments

7.1.2.1 Gas chromatography–mass spectrometry

Ishikawa *et al.* [3] developed procedures to determine the trialkyl and triaryl phosphate esters in sediment, involving extraction with dichloromethane (for water) or acetone (for sediment), followed by gas chromatography using a flame photometric detector and gas chromatography–mass spectrometry after clean-up through a Florisil column.

7.1.2.2 Miscellaneous

Sediments containing 50–1600mg kg^{-1} of triphenyl phosphate altered the drift dynamics of benthic invertebrates. Invertebrates exposed to contaminated sediments drifted almost immediately when threshold toxicity was reached [4].

7.2 Adenosine phosphate ester

7.2.1 Sludge

7.2.1.1 Spectrophotometry

Smirnova *et al.* [5] have described a simple non-enzymatic method of quantitative determination of adenosine triphosphate in activated sludge from aeration tanks. Extraction of the nucleotides in boiling distilled water was followed by removal of the protein impurities by acidification. Barium salts of di- and triphosphates of the nucleotides were precipitated and the precipitate was washed and dissolved in acid to convert the barium salts to sodium salts. The quantity of adenosine triphosphate was determined quantitatively by inorganic phosphorus in the liquid over the precipitate before and after acid hydrolysis, and by ultraviolet absorption spectra. The method was tested in activated sludge from operational sewage works. There was good agreement between the adenosine triphosphate content determined spectrophotometrically and by phosphorus, despite the presence of small quantities of secondary impurities.

7.2.1.2 Luminescence methods

Patterson *et al.* [6] carried out determinations of adenosine triphosphate in activated sludge. The method involved the use of firefly lantern extract.

Kucnerowcz and Verstraete [7] carried out direct measurements of microbial adenosine-5′-triphosphate in activated sludge samples. The method uses an activated sludge apparatus designed for determining the biodegradability of anionic detergents. Mixed liquors are diluted with triethanolamine buffer, homogenized, mixed with adenosine-5′-triphosphate releasing agent and the luminescence of the mixture is measured after addition of luciferase. Results obtained were in agreement with literature data for adenosine-5′-triphosphate in activated sludge. Relationships established between adenosine-5′-triphosphate content of sludges and other sludge parameters indicate that adenosine-5′-triphosphate determination could be used as a method of monitoring activated-sludge treatment processes.

André *et al.* [8] discuss the determination of adenosine-5′-triphosphate by luciferin–luciferase assay. This method was applied to the determination of adenosine-5′-triphosphate in bacterial colonies filtered from samples of polluted water after incubation for different periods. The adenosine-5′-triphosphate was extracted from the residue in the filter and the amount compared with the biochemical oxygen demand of the filtered water. The oxygen uptake rate and the rate of formation of adenosine-5′-triphosphate were then plotted against time, the two curves being similar in up to three to four days' incubation, after which adenosine-5′-triphosphate production declined markedly, although oxygen uptake continued to increase.

Hysert *et al.* [9] state that the firefly bioluminescence adenosine-5′-triphosphate assay has several attractive features including high sensitivity, selectivity and freedom from sample interferences are a consequence of the use of purified luciferase and synthetic D-luciferin at optimum concentrations [10]. The high assay sensitivity permitted very high sample dilutions thus further reducing the possibility of interference and/or inhibition of the bioluminescent reaction by sample components [10–12]. Typically, samples were diluted ten-fold in extraction, a further 100-fold prior to assay, and 100-fold in the assay itself – total of 10 000. Furthermore, the assay is fast (approximately 1min per assay) and reproducible (relative standard deviations for standard adenosine-5′-triphosphate solutions vary from 2–4%). Its wide dynamic range (over five decades of adenosine-5′-triphosphate concentration) permits direct adenosine-5′-triphosphate determination over a broad range.

These workers [9] demonstrated that interference and inhibition of the bioluminescence assay by extract components was negligible by a standard adenosine-5′-triphosphate method [12]. This circumstance undoubtedly resulted from the aforementioned high extract dilution as well as from the use of purified luciferase. The reproducibility of the overall method for determining the adenosine triphosphate content of activated sludge, which includes sampling, dimethyl sulphoxide extraction and adenosine-5′-triphosphate assay, was considerably poorer than that observed for the adenosine-5′-triphosphate assay alone or for adenosine-5′-triphosphate assays of pure cultures. The relative standard deviations for the latter assays were 2–4%, whereas those for the activated-sludge determinations were in the 7–11% range. This greater method variance no doubt results from difficulties of reproducibly sampling the heterogenous, clumped biological flocs that comprise activated sludge.

7.2.1.3 Miscellaneous

Higgins [13] has compared four methods of extracting adenosine-5′-triphosphate using activated sludge samples from an aeration basin: dilution with cold nitric acid and extraction with cold Tris buffer; dilution with distilled water and extraction with cold nitric acid; dilution with distilled water followed by extraction with boiling Tris buffer; extraction using a nucleotide releasing agent for bacteria. This releasing agent only extracted 32% of the adenosine-5′-triphosphate extracted using boiling Tris buffer; the other two methods were slightly less efficient than the boiling Tris buffer, but not significantly so. The uncomplicated extraction with cold nitric acid was suitable for use on-site. Adenosine-5′-triphosphate levels in activated sludge dropped by 25% during the first hour after collection so it was important that adenosine-5′-triphosphate was extracted on site. Extracts were stable for at least 7h. There was no significant difference in the adenosine-5′-triphosphate content between activated sludge samples collected at four different sites in an aeration basin.

7.2.2 Non-saline deposited and suspended sediments

7.2.2.1 Luminescence method

Tobin *et al.* [14] give details of two extraction procedures for the determination of adenosine-5′-triphosphate in sediment samples by luciferin–luciferase assay.

7.3 Inositol phosphate ester

7.3.1 Non-saline deposited and suspended sediments

7.3.1.1 Miscellaneous

Inositol phosphate esters have been determined in lake sediments [15].

References

1 David, M.J. and Seiber, J.N. (1996) *Analytical Chemistry*, **68**, 3038.
2 Ingram, J.C., Groenwald, G.S., Appelhans, A.D. *et al.* (1996) *Analytical Chemistry*, **68**, 1309.
3 Ishikawa, S., Taketami, M. and Shinohara, R. (1985) *Water Research*, **19**, 119.
4 Fairchild, J.F., Boyle, T., English, W.R. and Rabeni, C. (1987) *Water, Air and Soil Pollution*, **36**, 271.
5 Smirnova, L.A., Filenko, S.N. and Shchetinni, A.I. (1985) *Soviet Journal of Water Chemistry and Technology*, **7**, 76.
6 Patterson, J.W., Brezonik, P.L. and Putnam, H.D. (1970) *Environmental Science and Technology*, **4**, 569.
7 Kucnerowcz, F. and Verstraete, W. (1979) *Journal of Chemical Technology and Biotechnology* **29**, 707.
8 André, M., Van Beneden, P. and Bassleer, J. (1978) *Tribune du Cebedeau*, **31**, 251.
9 Hysert, D.W., Knudson, F.B., Morrison, M.N. *et al.* (1979) *Biotechnology and Bioengineering*, **21**, 1301.
10 Hysert, D.W., Kovecses, F. and Morrison, N.M. (1976) *Journal of American Brewing Chemistry*, **34**, 145.
11 Lundin, A. and Thore, A. (1975) *Applied Microbiology*, **30**, 713.
12 Lundin, A. and Thore, A. (1975) *Analytical Biochemistry*, **66**, 47.
13 Higgins, J. (1987) *Water (Australia)*, **14**, 16.
14 Tobin, S.R., Ryan, J.F. and Afghan, B.K. (1978) *Water Research*, 7813.
15 Weimer, W.C. and Armstrong, D.E. (1977) *Analytica Chemica Acta*, **94**, 35.

Chapter 8

Sulphur containing compounds

8.1 Tetrahydrothiophen

8.1.1 Soil

8.1.1.1 Gas chromatography–mass spectrometry

Garlucci *et al.* [1] discuss a method for determining tetrahydrothiophen contaminant in soil using headspace high-resolution gas chromatography together with mass spectrometry. Down to 10ng of this substance could be determined.

8.2 Miscellaneous sulphur compounds

8.2.1 Non-saline deposited and suspended sediments

8.2.1.1 Gas chromatography

The analysis of organosulphur compounds has been greatly facilitated by the flame photometric detector [2]. Volatile compounds can be separated by a glass capillary chromatographic column and the effluent split to a flame ionization detector and a flame photometric detector. The flame photometric detector response is proportional to $\approx[S^2]$ [3–6]. The selectivity and enhanced sensitivity of the flame photometric detector for sulphur permits quantitation of organosulphur compounds at relatively low concentrations in complex organic mixtures. The flame ionization detector trace allows the organosulphur compounds to be referenced to the more abundant aliphatic and/or polynuclear aromatic hydrocarbons.

Reliable flame photometric detector quantification of organosulphur compounds requires careful optimization of the gas chromatograph parameters. Although the relative response of the flame photometric detector to various sulphur compounds remains somewhat controversial [7], analysis of organosulphur compounds by flame photometric detector is now relatively straightforward.

In one method dichloroethane extraction of the sediment, followed by elimination of elemental sulphur, mercaptans, disulphide and dibenzothiophene on a copper column is followed by a gas chromatographic analysis with flame photometric detection of the organosulphur compounds. The detection limit is 1ng as sulphur with a precision of ±10% [8].

8.2.1.2 Gas chromatography–mass spectrometry

Trehey *et al.* [9] determined alkyl benzene sulphonates and dialkyltetralin sulphonates in sediments by this technique with a detection limit of 0.5g kg $^{-1}$.

8.2.1.3 High-performance liquid chromatography

Shea and MacCrehan [10] determined hydrophillic thiols in sediment pore water using ion-pair chromatography coupled to an electrochemical detector. Down to 2p mole absolute of these compounds could be determined including cysteine, monothiogylcerol, glutathione, mercapto-pyruvic acid, 3-mercaptopropionic acid and 2-mercaptopropionic acid.

8.2.2 Saline deposit and suspended sediments

8.2.2.1 Gas chromatography

In a method described by Bates and Carpenter [8] for the characterization of organosulphur compounds in the lipophilic extracts of marine sediments these workers showed that the main interference is elemental sulphur (S_8). Techniques for its elimination are discussed. Saponification of the initial extract is shown to create organosulphur compounds. Activated copper removes S_8 from an extract and appears neither to create nor to alter organosulphur compounds. However, mercaptans and most disulphides are removed by the copper column. The extraction efficiency of several other classes of sulphur compounds is 80–90%. Extracts are analyzed with a glass capillary gas chromatograph equipped with a flame photometric detector. Detection limit is 1g S and precision ±10%.

In this method, sediment samples were freeze-dried (Virtis, Unitrap II), weighed (80g) and Soxhlet extracted in pre-extracted paper thimbles (43 × 123mm Whatman single thickness) with methylene chloride for 24h. Quantitation of the total extract was obtained by weighing an aliquot. After weighing, the extract was concentrated to ≈10mL by rotary evaporation (ambient temperature and 50cm of Hg vacuum) and eluted through a column of activated copper powder to eliminate S_8. The column was prepared by passing 2N hydrochloric acid, water, methyl acetate and methylene dichloride through a 10mm i.d. column containing 5cm of copper powder. The column eluate was then reconcentrated to 10mL, reweighed, and evaporated to dryness under nitrogen for gas chromatographic analysis. The recovery of 100ng of individual sulphur standards

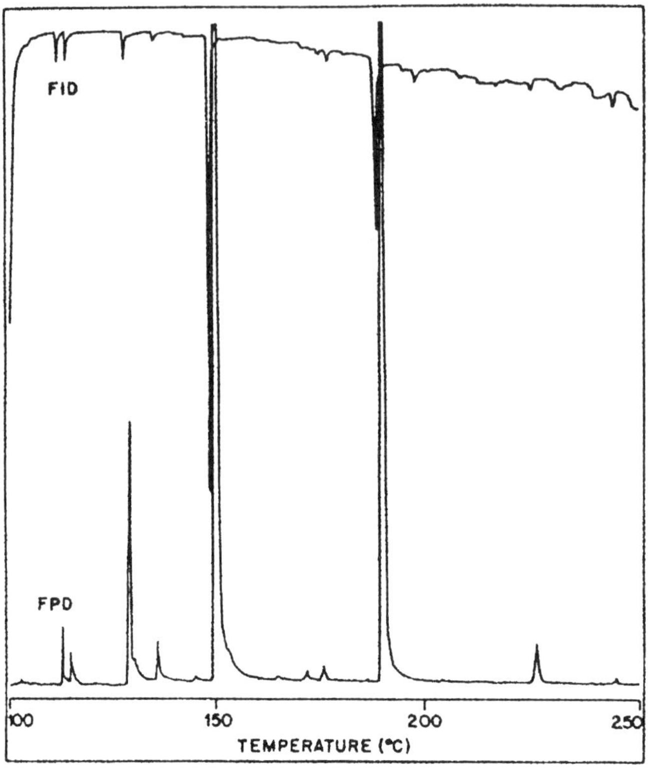

Fig. 8.1 Gas chromatogram of a saponified sediment sample extract
Source: Reproduced with permission from the American Chemical Society [8]

added to pre-extracted sediment varied from 80–90% for sulphides, sulphones and aromatic sulphur compounds. Mercaptans and disulphides were not recovered since these classes of compounds were retained in the copper column. The trace obtained by applying this method to a sediment sample is shown in Fig. 8.1. All the major components contained sulphur.

References

1 Garlucci, G., Airoldi, L. and Fanelli, R. (1984) *Journal of Chromatography*, **287**, 425.
2 Brody, S.S. and Chaney, J.E. (1966) *Journal of Gas Chromatography*, **4**, 42.
3 Greer, D.G. and Bydalek, T.J. (1973) *Environmental Science and Technology*, **7**, 153.
4 Sugiyama, T., Suzuki, Y. and Takeuchi, T. (1973) *Journal of Chromatography*, **77**, 309.
5 Bentz, A.P. (1976) *Analytical Chemistry*, **48**, 454A.
6 Garza, M.E. Jr. and Muth, J. (1974) *Environmental Science and Technology*, **8**, 249.

7 Burnett, C.H., Adams, D.F. and Farwell, S.O. (1978) *Journal of Chromatographic Science*, **16**, 68.
8 Bates, T.S. and Carpenter, R. (1979) *Analytical Chemistry*, **51**, 551.
9 Trehey, M.L., Gledhill, W.E. and Orth, R.G. (1990) *Analytical Chemistry*, **62**, 2581.
10 Shea, D. and MacCreehan, W.A. (1988) *Analytical Chemistry*, **60**, 1449.

Chapter 9

Insecticides, herbicides, growth regulators and fungicides

9.1 Chlorinated insecticides

9.1.1 Soil

9.1.1.1 Gas chromatography

Deubert [1] has discussed the sources of compounds which interfere in the analyses in water and soil extract for DDT and Dieldrin by gas electron capture chromatography. Nitration of these insecticides eliminated their peaks so that background interference peaks could be studied.

The solvent extraction of chlorinated pesticide residues from soil is often achieved by using mixtures of solvents such as hexane-isopropanol or hexane acetone, but can be unsatisfactory owing to the emulsification problems [2,3] or, with hexane-isopropanol, poor recovery [2,4]. Acetone extraction of soil is efficient [4,5] but problems can arise from large amounts of coextracted material unless an efficient clean-up technique [6] is used prior to analysis by gas chromatography.

Hesselberg and Johnson [7] used Florisil column extraction followed by gas chromatography to determine DDT, Dieldrin, Endrin and Methoxychlor in fish and mud. Samples are prepared by blending with sodium sulphate (plus solid carbon dioxide for fish) until a free-flowing dry mixture is obtained. A glass column (400mm × 20mm) is packed with sodium sulphate (2g), the sample mixture is added and tamped down, and the resulting column is washed with solvent (200ml) at 3–6ml min $^{-1}$. The solvents used are 1% methanoic phosphoric acid to elute 2,4-D, cyclohexane for DDT, Dieldrin, Endrin and Methoxychlor, ethyl ether for Simazine and 10% of ether in light petroleum for Parathion. The eluates are cleaned up by solvent partitioning and column chromatography on Florisil and the pesticides are then determined by gas chromatography on a packed OV–7 column. Recoveries of added pesticides (>5ng g $^{-1}$) were 95–100%.

Woodham *et al.* [8] converted Dieldrin and Endrin to chemical derivatives prior to gas chromatographic determination of these substances in soil and sediments. An aliquot of extract after appropriate clean-up and evaporation to dryness, is treated with 1ml of conversion reagent (10% solution of boron trichloride in 2-chloroethanol) in a centrifuge tube, which is then placed, unstoppered, in a water bath at 90°C and left for 2h or for 10min for a sample that may contain Dieldrin, or Endrin, respectively, or for 2h for a mixture. The tube is cooled, 5ml of hexane and 10ml of 7% aqueous sodium sulphate are added, and the contents are mixed and left for the phases to separate. The hexane phase is analysed on a column of OV–17–QF–1 on Gas-Chrom Q with electron capture detection (tritium source). Down to 0.01ppm of either pesticides (0.01 part per 10^9 in water) could be detected.

Gooding *et al.* [9] used DDT-dehydrochlorinase for the identification of DDT in soils. The enzyme converts DDT to DDE which is then determined gas chromatographically on Chromosorb WHMDS at 190°C using an electron capture detector.

Mahel'ova *et al.* [10] determined BHC isomers in soil by gas–liquid and thin layer chromatography after extraction with light petroleum. An air-dried, ground sample (18 mesh) (20g) was deactivated by addition of 25% of water and set aside for 24h. Siloxid (active silica) was added to form a powdery mixture, which was extracted with 250ml of light petroleum (boiling range 35–50°C) in a Soxhlet apparatus for 12h. The extract was evaporated to 5ml, and purified on a column of Celite 545 mixed with fuming sulphuric acid. The insecticides were eluted with light petroleum (250ml), which was evaporated to dryness; the residue was dissolved in hexane and an aliquot of the solution was subjected to gas chromatography on a column packed with 1.5% OV–17 and 2% QF–1 on Chromosorb (80–100 mesh). The column was operated at 190°C, nitrogen (60–80ml min^{-1}) was used as carrier gas with an electron capture detector. Recoveries of the isomers, of *p,p'*-DDT and of 1,1-dichloro-2,2-bis-(4-chlorophenyl)ethylene at the 0.1 ppm level ranged from 82–106% with a coefficient of variation between 4.3 and 11.9%. The purified extracts were also examined by thin layer chromatography.

Suzuki *et al.* [11,12] studied the determination of chlorinated insecticides in soils using high-resolution electron capture gas chromatography with glass capillary columns. They compared resolution efficiencies of organochlorine insecticides and their related compounds with wall-coated open tubular (WCOT) and support-coated open tubular (SCOT) glass capillary columns with those of conventional packed glass columns. These columns were coated with silicone OV–101 as the liquid phase. Applicabilities of the glass capillary column to environmental samples were investigated. An all-glass system was used to prevent thermal decomposition. The 'resolution index', i.e. peak height/half-width of peak of standard injected, generally increased in the following order, conventional packed glass column ≤WCOT glass capillary column ≤SCOT glass capillary column. Excellent resolution of insecticides was obtained with SCOT glass capillary columns and WCOT glass capillary columns. Log-log plots of the resolution index *vs* relative retention times compared to Aldrin were linear. These workers used a Shimadzu GC–5AIEE glass chromatograph equipped with a dual electron capture detector (^{3}H, 300mCi, foil type). Coiled SCOT glass capillary columns (23m × 0.28mm) were used. With SCOT glass capillary columns, the liquid phase was coated to the salt layer adhered to the inner wall of the glass capillary column. With WCOT glass capillary columns, the liquid phase was directly coated to the inner wall of the glass capillary column. Therefore, the surface area per unit of length was broader in the SCOT glass capillary column than in the WCOT glass capillary columns. The glass capillary column was connected to a holder. An OV–101 PCG (3% on Gas Chrom Q, 80–100 mesh, U-shaped, 2m × 3mm) was used. The gas chromatographic conditions used were as follows: temperatures of column, injector and detector, for both conventional packed glass column chromatography and glass capillary column chromatography were 190, 210, and 200°C, respectively. For the packed column a flow rate of carrier gas (highly purified nitrogen gas, 99.9999 + %) was 60ml min $^{-1}$. The flow rates through the glass capillary column were both adjusted to 2ml min $^{-1}$. The effluent from both glass capillary columns was scavenged at 60ml min $^{-1}$ and entered into the electron capture detector. The splitter ratio was 1:25 was SCOT glass capillary column and 1:40 for WCOT glass capillary column, respectively. All columns used were well conditioned before use. The chart drive was 2cm min $^{-1}$ for estimation of the 'resolution index', i.e. peak height/half-width of peak, which showed the sharpness of a peak and the degree of resolution efficiencies, and 1cm min $^{-1}$ for the analyses of environmental samples. Attenuations for the conventional packed glass columns, SCOT glass capillary column, and WCOT glass capillary columns were 8 × 10, 16 × 102, and 16 × 10^2, respectively. Under these conditions, duplicate injections of 1–5µl of each standard showing 30% in full-scale deflection were made, and the resolution index was calculated. Also, 5µl of extracts from samples were injected. Minimum detectable levels of α-BHC, ß0BHC, γBHC, ∂BHC, Heptachlor, Heptachlor epoxide, Aldrin, Dieldrin, Endrin, *p,p'*-DDE, *p,p'*-TDE, and *p,p'*-DDT in 100g samples of field soil and bottom sediment were 0.0005, 0.0032, 0.0014, 0.0040, 0.0012, 0.0020, 0,0014, 0,0020, 0.0056, 0.0032, 0.0080, and 0.0120 ppm, respectively, on SCOT glass capillary columns.

Mangani *et al.* [13] have described a method for determining extract chlorinated insecticides in soil. In this procedure a short column is packed with the soil sample. The insecticides in the soil are desorbed by a suitable solvent mixture chosen for its polarity characteristics.

In the case of soil analysis, a 2.0cm i.d. × 30cm column is used. Stones, roots and other gross impurities were removed, and the soil was reduced to a size between 30 and 60 mesh. 20g of soil was packed into the column between two glass wool plugs. Insecticides were desorbed from the soil by passing small volumes of acetone:toluene (1:1) through the column at a rate of 2ml m^{-1}. The results in Table 9.1 show that this is the most efficient extraction solvent of those studied by Mangani *et al.* [13].

Gas chromatography of the extracts was performed as follows. A glass column, 3m long, 2mm i.d. containing Supelcoport coated with 1.5% SP2250 + 1.95% SP2401 was used for the analyses. The column was maintained at 200°C and flow rate was 40mL/min. This ensured complete separation of most pesticides in the test mixture. High-purity nitrogen was the carrier gas. Measurements were obtained with a DANI (Milan, Italy) Model 3600 dual column gas chromatograph equipped with a frequency modulated ^{63}Ni electron capture detector. The electron capture detector was operated in its linear range in all cases. Peak areas were measured with an integrator.

In order to check different adsorption of pesticides on artificially and actually polluted soil, Mangani *et al.* [13] prepared elution and recovery curves for the pesticides present in an actual soil sample, using a toluene:acetone mixture (1:1). These are reported in Fig. 9.1. Although the overall recovery using 25mL of solvent for the extraction is the same in both instances, a very significant difference is observed in the elution curves. These tail to a greater degree in the naturally polluted soil than in the artificially polluted soil. This shows definitely that pesticides can be adsorbed on higher energy sites due to the porous structure of the material. Longer time is required by the organic molecules to be occluded with the pores of the soil.

Gambrell *et al.* [14] have described a procedure for the recovery of DDT (also kepone and Permethrin) added to soil suspensions incubated under controlled redox potential and pH conditions. They studied the effect of time on the levels of the insecticides and their breakdown products.

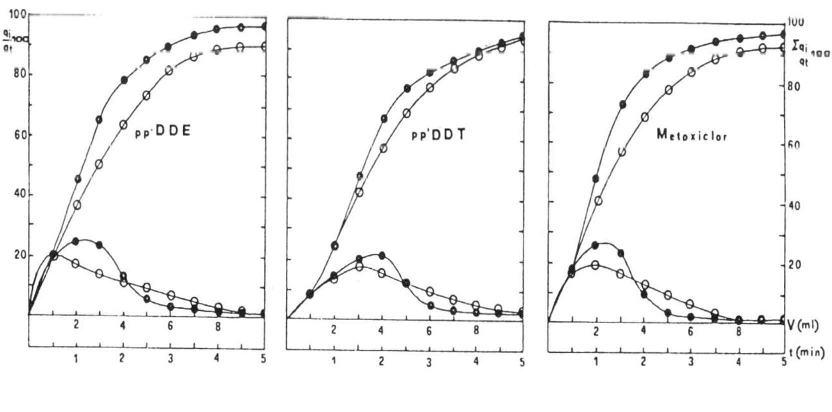

Fig. 9.1 Elution and recovery curves for an actually polluted soil (O) and an artificially polluted soil (•) using acetone–toluene (1:1)
Source: Reproduced with permission from the American Chemical Society [13]

Table 9.1 Recovery of pesticides from soil with this method using different solvents compared with the Soxhlet extraction method[a] (Column D)

	A	*B*	*C*	*D*	*E*	*F*	*G*	*H*
α-hexachlorocyclohexane (α-BHC)	84	86	84		24	87	99	85
lindane	85	85	86	85.2	23	89	99	86
ß-hexachlorocyclohexane (ß-BHC)	85	87	86		54	93	100	85
heptachlor	86	90	82		10	85	101	92
aldrin	81	76	83		32	97	100	77
endosulfan I	91	88	87		0	97	98	89
p,p'-dichlorodiphenyltrichloroethylene (*p,p*'-DDE)	82	87	88	83.1	3.5	94	02	85
dieldrin	96	81	93	88.2	0	92	98	88
o,p'-dichlorodiphenyldichloroethane (*o,p*'-DDD)	90	93	86		0	90	99	88
endrin	89	91	95		0	92	102	89
p,p'-dichlorodiphenyldichloroethane (*p,p*'-DDD)	88	84	87	91.2	0	91	99	90
p,p'-dichlorodiphenyltrichloroethane (*p,p*'-DDT)	92	89	88	94.2	10	94	102	91
metoxichlor	87	90	85			92	100	89

Source: Reproduced with permission from the American Chemical Society [13]

[a] A = 25mL of toluene; B = 25ml of petroleum ether–acetone (1:1); C = 25mL of diethyl ether–petroleum ether (40–60°C) (1:1); D = 65mL of acetone–hexane (1:1) in Soxhlet; E = 10mL of petroleum ether (40–60°C); F = 10mL of acetone; G = 25mL of toluene–acetone (1:1); H = 25mL of hexane–acetone (1:1)

Samples were analysed using gas chromatography, pH and redox potential affected the persistence of insecticides to different degrees. The recovery of DDT was effected by redox potential but not by pH.

Cooke and Western [2] have pointed out that although acetone is a good solvent for the extraction of chlorinated insecticides from soil it suffers from the disadvantage of contributing coextracted material in the extract unless an efficient clean up of the acetone is carried out before the soil extraction is carried out.

The analysis of the oily residue by gas chromatography, utilising flame-ionisation detection, yielded a single peak with the same retention time as diacetone alcohol, whereas electron capture detection gave inconclusive results. Examination by gas chromatography–mass spectrometry using electron-impact ionisation produced a spectrum with a base ion at m/e 43 (the same as that for acetone) but no signal at m/e 116 corresponding to the parent ion. Chemical ionisation produced a spectrum with a base ion at m/e 99 and a parent ion at m/e 117 (M + 1). Diacetone alcohol (4-hydroxy-4-methylpentane-2-one) gave the highest purity and fit value.

The identity of this compound was confirmed from analyses by nuclear magnetic resonance and infrared spectroscopy. Both methods gave spectra identical with those from an authentic sample.

Cooke and Western [2] postulate that alumina in the soil promotes the formation of diacetone alcohol out in the acetone extract and thus render acetone a doubtful solvent for soil extraction.

9.1.1.2 Gas chromatography–mass spectrometry

Teichman *et al.* [15] separated polychlorobiphenyls from chlorinated insecticides in soil samples using gas chromatography coupled to mass spectrometry. Polychlorobiphenyls were separated from DDT and its analogues and from the other common chlorinated insecticides by adsorption chromatography on columns of alumina and charcoal. Elution from alumina columns with increasing fractional amounts of hexane first isolated Dieldrin and Heptachlor from a mixture of chlorinated insecticides and polychlorobiphenyls. The remaining fraction, when added to a charcoal column, could be separated into two fractions, one containing the chlorinated insecticides, the other containing the polychlorobiphenyls, by eluting with acetone–diethyl ether (25:75) and benzene, respectively. The polychlorobiphenyls and the insecticides were then determined by gas chromatography on the separate column eluates without cross-interference.

Teichman *et al.* [15] used a gas chromatograph (Aerograph 1200) containing a glass column (180cm × 3.1mm) packed with 4% SE–30, 6% SP–4201 on Chromosorb W (100–120 mesh). They also used an Aerograph 204 gas chromatograph containing a glass column (180cm × 3.1mm) with 4% SE–30, 6% QF–I on Chromosorb W (80–100 mesh). The operating conditions were:

Model:	1200	204
Column temperature	180°C	185°C
Injector temperature	215°C	200°C
Detector temperature	200°C	200°C
Nitrogen gas flow rate	25ml min^{-1}	30ml min^{-1}

Both instruments contained an electron capture detector with a tritium foil source. For gas chromatography–mass spectrometry, a Varian 1400 gas chromatograph coupled to a Finnegan 3000 mass spectrometer was used. The 1400 was equipped with a glass column (180cm × 2mm i.d.) packed with 4% SE–30, 6% SP–4201 on Supelcoport (100–120 mesh). The operating conditions were: column temperature, 210°C; transfer-line temperature, 250 ¯C, gas jet separator temperature, 255°C, flow rate of helium gas, 12ml min^{-1}; sensitivity, 10^{-7}A/V; electron multiplier voltage, 2.25kV; electron ionization current, 6.95eV.

A summation of the elution of the chlorinated organic insecticides and the polychlorobiphenyl from the alumina column is given in Table 9.2. Heptachlor epoxide and Dieldrin were removed from the column by extending the elution solvent beyond the 30ml volume with an additional, but separate, elution volume of 30ml. The polychlorobiphenyls remained an integral part of the mixture containing the insecticides in the first 30ml of eluate. The elution pattern of alumina column fraction one on the charcoal column, Table 9.3, shows that the insecticides were separated from the polychlorobiphenyls by means of acetone–diethyl ether eluent. The polychlorobiphenyls were subsequently removed from the charcoal column with benzene. Known amounts of insecticides and polychlorobiphenyls (Aroclor 1254) were added to soils and oyster samples; the samples were analysed as described above to check the efficiency of the analytical procedure. Recoveries of the added chemicals to soils were consistent and acceptable (Table 9.4). The limits of delectability of the chemicals examined (Table 9.5) refer to those obtained from pure solutions and they are also applicable to samples extracts.

The isotope dilution gas chromatography–mass spectrometry method described by Lopez-Avila *et al.* [16] and fully discussed in section 5.3.1.3, has been applied to the determination of 0.1^{-1}µg kg^{-1} Lindane in soil. Accuracy was greater than 86% and precision better than 8%.

9.1.1.3 Thin layer chromatography

Mahel'ova *et al.* [10] determined BHC isomers in soil by gas–liquid and thin layer chromatography after extraction with light petroleum. An air-dried, ground sample (18mesh) (20g) was deactivated by addition of 25% of water and set aside for 24h. Siloxid (active silica) was added to form a powdery mixture, which was extracted with 250ml of light petroleum

Table 9.2 Percentage recovery of insecticides eluted from neutral alumina

Compound	*First fraction* 0–15ml	15–20ml	20–25ml	25–30ml	*Second fraction* 30–60ml
Lindane			10	90	
Heptachlor	100				
Aldrin	100				
Heptachlor epoxide					100
p,p'-DDE	100				
Dieldrin					100
p,p'-DDD			50	50	
p,p'-DDT	100				
PCBs	100				
γ-Chlordane	10	80	10		
α-Chlordane	80	20			

Source: Reprinted with permission from Elsevier Science Publishers B.V. [15]

Table 9.3 Percentage recovery of insecticides eluted from charcoal

Compound	*First fraction: 90ml acetone-diethyl ether (25:75)* 0–30ml	30–60ml	60–90ml	*Second fraction: 60ml benzene* 0–30ml	30–60 ml
Lindane	30	40	30		
Heptachlor	100				
Aldrin	100				
p,p'-DDE	50	50			
p,p'-DDT	80	20			
PCBs				80	20
Chlordane	100				

Source: Reprinted with permission from Elsevier Science Publishers B.V. [15]

Table 9.4 Recovery of insecticides and PCBs from fortified soil samples

Insecticide	*Fortified (ppm)*	*Recovered (ppm)*	*Recovered (%)*
Lindane	0.0013	0.0011	84.6
Heptachlor	0.027	0.0022	81.5
Aldrin	0.0027	0.0023	85.2
Heptachlor epoxide	0.0027	0.0025	92.6
p,p'-DDE	0.0027	0.0025	92.6
Dieldrin	0.0027	0.0026	96.3
p,p'-DDD	0.0027	0.0027	100
p,p'-DDT	0.0027	0.0027	100
γ-Chlordane	0.0033	0.0035	106
α-Chlordane	0.0033	0.0037	112
PCB (Aroclor 1254)	0.066	0.066	100

Source: Reprinted with permission from Elsevier Science Publishers B.V. [15]

Table 9.5 Limits of detectability of insecticides and PCBs using the described procedure under conditions

Insecticide	*Detectability (pp 10^9)*
Lindane	0.04
Heptachlor	0.05
Aldrin	0.06
Heptachlor epoxide	0.10
p,p'-DDE	0.14
Dieldrin	0.14
p,p'-DDD	0.25
p,p'-DDT	0.33
γ-Chlordane	0.10
α-Chlordane	0.11
Arochlors 1254, 1250 (PCBs)	6.5

Source: Reprinted with permission from Elsevier Science Publishers B.V. [15]

(boiling range 35–50°C) in a Soxhlet apparatus for 12h. The extract was evaporated to 5ml, and purified on a column of Celite 545 mixed with fuming sulphuric acid. The insecticides were eluted with light petroleum (250ml), which was evaporated to dryness; the residue was dissolved in hexane and an aliquot of the solution was subjected to gas chromatography on a column packed with 1.5% OV–17 and 2% QF–1 on Chromosorb (80–100 mesh). The column was operated at 190°C, nitrogen (60–80ml min $^{-1}$) was used as carrier gas with an electron capture detector. Recoveries of the isomers, of *p,p'*-DDT and of 1,1-dichloro-2,2-*bis*-(4-chlorophenyl)ethylene at the 0.1ppm level ranged from 82–106% with a coefficient of variation between 4.3 and 11.9%. The purified extracts were also examined by thin layer chromatography.

9.1.1.4 Enzyme-based immunoassay

Gooding *et al.* [9] used DDT-dehydrochlorinase for the identification of DDT in soils. The enzyme converts DDT to DDE which is then determined gas chromatographically on Chromosorb WHMDS at 190°C using an electron capture detector.

Gillespie *et al.* [17] have pointed out that static bioassays for DDT and others toxins give relatively imprecise data and suggest improvements in methodology to overcome this.

9.1.1.5 Supercritical fluid chromatography

Brady *et al.* [18] have discussed temperature/pressure phase diagrams for carbon dioxide-DDT and carbon dioxide-polychlorobiphenyls and examined the rate process of desorption from soils. Supercritical carbon dioxide was used to extract DDT, polychlorobiphenyls and Toxaphene from contaminated soils. Supercritical carbon dioxide at 100atm and 40°C was continually passed through a fixed bed of soil. 70% of DDT and 75% of Toxaphene was leached from a topsoil contained with 1000mg kg^{-1} DDT and 400mg kg^{-1} supercritical carbon dioxide at a rate of 0.7g s^{-1}.

As discussed in section 5.6.1.4 a solid phase carbon trap has been used in conjunction with supercritical fluid extraction liquid chromatography for the simultaneous determination of organochlorine insecticides, polychlorobiphenyls, polychlorodibenzo-*p*-dioxins and polychlorodibenzofurans in soils [19].

Snyder *et al.* [20] have compared supercritical fluid extraction with classical sonication and Soxhlet extraction for the extraction of selected pesticides from soils. Samples extracted with supercritical carbon dioxide modified with 3% methanol at 350atm and 50°C gave a ≥85% recovery of organochlorine insecticides including Dichlorvos, Endrin, Endrin aldehyde, *p,p'*-DDT mirex and decachlorobiphenyl (and organophosphorus insecticides).

Supercritical fluid extraction with carbon dioxide has been applied to the determination of chlorinated insecticides in soil [259].

9.1.1.6 Miscellaneous

Novikova [21] has reviewed the literature (209 references) covering the extraction, clean-up and analysis of organochlorine (and organophosphorus) insecticides in soil. Johnson and Starr [22] and Chiba and Morley [4] have studied factors affecting the extraction of Dieldrin and Aldrin from different soil types; ultrasonic extraction was recommended by these workers. Lopez-Avila *et al.* [23] used microwave assisted extraction to extract chlorinated insecticides from soils.

Residues of DDT and its degradation products DDE, DDD and 4,4'-dichlorobenzophenone have been found in soil samples collected in vineyards and cornfields where DDT has not been used for four years [24]. During a study of the DDT content of orchard soils sampled at different depths, involving extractions with acetone, Cooke and Western [2] observed that when the solvent was removed using a rotary evaporator a sweet-smelling oily residue remained in the flask and its amount increased with the depth of the soil. Application of a variety of techniques (gas chromatography–mass spectrometry, infrared spectroscopy, nuclear magnetic resonance spectroscopy) proved that this residue was diacetone alcohol (4-hydroxy-4 methyl pentone-2-ore), a dimer of acetone, produced by the condensation reaction of acetone by the presence of various metal oxides (iron, aluminium, magnesium, barium). Pesticide extraction from soil samples using elution with appropriate solvent mixtures has also been reported [25–29]. Such methods show some advantages both on the usual solvent extraction using a separatory funnel for water and on Soxhlet extraction for soil.

Mangani *et al.* [13] used Carbopack B columns to recover chlorinated insecticides in soil samples. These workers noted that, although the principles governing the adsorption and extraction process in the extraction in soil analysis are the same as those that govern liquid–solid chromatography, the main feature of a chromatographic column, i.e. separation efficiency, is almost completely absent.

Thus, the 'columns' used for the extraction should be regarded rather as an extraction apparatus than actual chromatographic columns. For soil analysis, a column is packed with soil. The soil behaves as an adsorbent that retains pesticides on its surface. These are eluted by a solvent mixture, that should be chosen for appropriate polarity characteristics.

The two elution processes differ in the nature of the adsorbents and also in the story of the adsorption process. In fact, pesticides might have been spread over the soil several years before the analysis. The successive action of watering modifies the original adsorption energy distribution, making available deeper and stronger adsorption sites in the structure of the material. Several solvents or solvent mixtures were tested by Mangani *et al.* [13] in order to find the one with the best extracting power for the highest number of compounds.

The best results are obtained with a mixture of petroleum ether–toluene (2:1). Using only 10mL of this mixture gives a recovery higher than 90% in all instances. In several cases values very close to a complete recovery are obtained. It is interesting to note that, in general, the recovery is higher with the mixture than when using the two solvents separately. By use of 25mL of this solvent mixture a recovery very close to 100% is obtained in all cases. One should note that a slightly polar solvent, such as ethyl ether, is inadequate, not only for non-polar or aromatic compounds but also for compounds containing double bonds or polar groups. In some cases the *n*-hexane ethyl ether mixture [27] ensures a higher recovery than with the two separate solvents; but in several other cases, such as with *p,p'*-DDE, Dieldrin, Endrin, and *p,p'*-DDD, the recovery is much lower than when using the two solvents separately. In fact, the extraction efficiency depends upon various factors, namely the structure of the compounds to be recovered, the structure of the adsorbent, and the structure of the solvent mixture. The best result is obtained with a compromise among all these factors, and this should be kept in mind when selecting a solvent mixture to extract pesticides from an adsorbent.

Since both *n*-hexane and ethyl ether have almost no similarities in structure with the adsorbent and with the molecules to be extracted, a very good recovery cannot be expected. The results obtained with the mixture of column A (Table 9.6), the one adopted, are fully comparable with those obtained by using a separatory funnel extraction (column H) but the amount of solvent required is about ten times less.

Table 9.6 Percent recovery of pesticides from water with the present method[a]

Solvent	*A*	*B*	*C*	*D*	*E*	*F*	*G*	*H*	*I*
α-hexachlorocyclohexane (α-BHC)	98	97	95	83	82	95	95	100	94
lindane	97	94	91	81	80	92	92	98	89
ß-hexachlorocyclohexane (ß-BHC)	98	94.5	88	80	81	91	93	97	
heptachlor	100	95	87	35	33	90	91	100	90
aldrin	100	99	66	62	67	70	88	100	82
endosulfan I	100	98	78	62	68	73	83	101	90
p,p'-dichlorodiphenyltrichloroethylene (*p,p*'-DDE)	97	94	95	94	87	92	97	98	80
dieldrin	99	97	96	90	85	94	94	104	95
o,p'-dichlorodiphenyldichloroethane (*o,p*'-DDD)	95	91	73	78	73	86	79		80
endrin	96	92	95	90	85	93	96	100	86
p,p'-dichlorodiphenyldichloroethane (*p,p*'-DDD)	97	95	96	93	89	95	97	99	75
p,p'-dichlorodiphenyltrichloroethane (*p,p*'-DDT)	100	99	81	90	86	96	75	100	
metoxichlor	99	97	78	86	81	93	81	93	

Source: Reproduced with permission from the American Chemical Society [13]

[a] A= 25mL of petroleum ether (40–60°C)-toluene (2:1); B = 10mL of petroleum ether (40–60°C)-toluene (2:1); C = 10mL of ethyl ether; D = 10mL of *n*-hexane; E = 10mL of petroleum ether; F = 10mL of toluene; G = 10mL of *n*-hexane–diethyl ether (1:1); H = 180mL of methylene chloride in there successive steps in separatory funnel; I = 25mL of *n*-hexane diethyl ether (1:1)

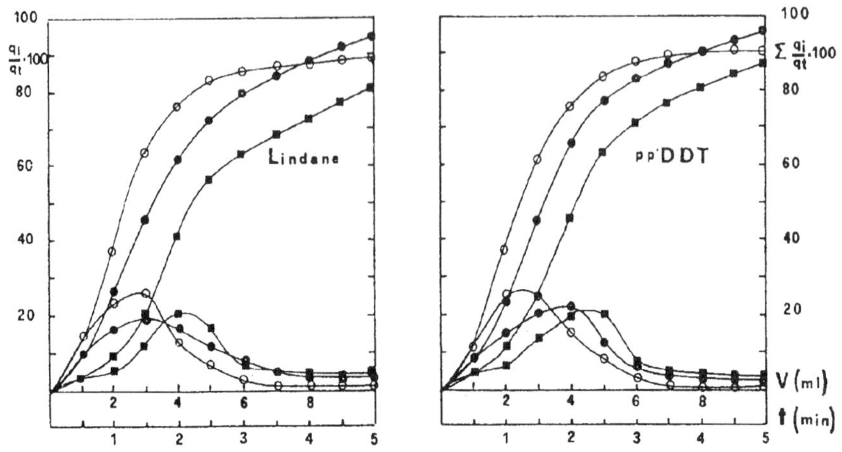

Fig. 9.2 Elution and recovery curves for two pesticides with different solvents on artificially polluted soil: (O) acetone; (•) acetone–toluene (1:1): (•) toluene
Source: Reproduced with permission from American Chemical Society [13]

The elution curves, reported in Fig. 9.2, clearly indicate the effects of the two solvents. Acetone shows a peak of shorter retention time than toluene and this means a high extraction power for most of the active sites. However, after 6mL, the recovery reaches a steady state. Toluene gives a poor recovery when used first, and the average retention is higher. However, it turns out to be very useful in recovering the last traces.

By comparing the two adsorbents, it turns out that the non-polar character of the petroleum ether is exploited in the extraction from the strong non-specific active sites of carbon black. On the other hand, the polar character of acetone makes the extraction from the highly strong specific active sites of the siliceous material of the soil possible. The retention time obtained with the mixture is intermediate between those two separate solvents, and the recovery, in turn, is higher. The proper choice of the solvent mixture plays a very important role on the size of the final volume of solution in which the pesticides are collected.

9.1.2 Non-saline deposited and suspended sediments

9.1.2.1 Gas chromatography

The gas chromatographic procedure described by Woodham *et al.* [8] and discussed in section 9.1.1.1 for the determination of Aldrin, Endrin and Dieldrin in soils has also been applied to sediments.

Sackmauerevá *et al.* [30,42] have described the method, given below, for the determination of chlorinated insecticides (BHC isomers, DDE, DDT and hexachlorobenzene) in water, fish and sediments.

The sediment sample is allowed to dry in open air and then sieved. To 20g of the sample 20% distilled water is added for deactivation purposes and the excess water is then bound to active silica (Siloxid), so that a powdery consistency is obtained. The insecticides studied are extracted with petroleum ether (b.p. 30–60°C) in a Soxhlet apparatus. The extract is concentrated using the vacuum rotary evaporator and the coextractants are separated on a Celite oleum column. The petroleum ether eluate is then concentrated to a volume of 1ml and used for gas chromatography under the following conditions [10,31–33].

Working conditions: temperature of the column 180–200°C, temperature of the injection port 210°C, temperature of the electron capture detector (63Ni) 200–225°C, nitrogen flow rate 60–80ml min $^{-1}$, EC detector voltage 20–70V. The optimum operating voltage is to be found experimentally to obtain the highest response towards the components. One µl of the concentrated sample is injected into the gas chromatograph (Carlo Erba, type 452 GI used). When necessary, the sample is diluted with hexane. Then, identical volumes of standard compound, solution mixtures are placed to the apparatus under standard conditions. Under the above conditions, the insecticide concentration is in linear proportion to the peak height over the following range:

α-BHC	0.03–0.12µg ml $^{-1}$
ß-BHC	0.15–0.60µg ml $^{-1}$
γ-BHC	0.04–0.18µg ml $^{-1}$
∂-BHC	0.03–0.12µg ml $^{-1}$
p,p'-DDE	0.15–0.60µg ml $^{-1}$
p,p'-DDT	0.30–1.2µg ml $^{-1}$
p,p'-DDD	0.30–1.2µg ml $^{-1}$
p,p'-DDT	0.30–1.2µg ml $^{-1}$

When the individual insecticides are present in the solution in such a concentration range, the electron capture responds nearly uniformly to all insecticides. A column filled with 1.5% silicone OV–17 plus silicone oil (fluoralchylsiloxane) on Chromasorb W (80–100 mesh) is used for separation of the BHC alpha, beta, gamma and delta isomers (hexachlorocyclohexane), *o,p*'-DDT, *p,p*'-DDE, *p,p*'-DDD, and *p,p*'-DDT. α-BHC and hexachlorobenzene (HCB) have a common peak. They can be separated on a column filled with 2.5% Silicone Oil XE–60 (ß-cyanoethyl-methyl-silicone) on Chromosorb W (80–100 mesh).

Sackmauerevá *et al.* [30,42] used thin layer chromatography on silica gel plates to confirm the identity of chlorinated insecticides previously identified by gas chromatography. The compounds can be separated by single or repeated one-dimensional development in *n*-heptane or in *n*-heptane containing 0.3% ethanol. The plate is dried at 65°C for 10min and detected by spraying with a solution of silver nitrate plus 2 phenoxy-ethanol [34,35]. Thereafter, the plate was dried at 65°C for 10min and illuminated with an ultraviolet light (λ = 254nm) until spots representing the smallest amounts of standards were visible (10–15min). The pesticide residues may be evaluated semi-quantitively by simple visual evaluation of the size and of the intensity of spot colouration and by comparing extracts with standard solutions.

Using the gas chromatography methods Sackmayerevá *et al.* [30] obtained from spiked samples the four BHC isomers at 93–103.5% recovery. Both DDT and DDE were yielded in 85.6–94%, 90–93.2%, 90–102.4% and 92–105.8% from sediment. Purification on a Florisil column was used in determining chlorinated insecticides unstable at low pH (Aldrin, Dieldrin). The type and activity of Florisil influence the yield and accuracy of the method. Therefore, the activity of this adsorbent had to be verified and adjusted [36,37].

The average concentration of the BHC andß isomer and $\alpha + \gamma$ isomers, and of DDE and DDT in sediment was found to be 0.010, 0.010, 0.016, 2.11 and 0.70mg kg^{-1}, respectively. These results suggest that chlorinated insecticides, due to their physical and chemical properties, can accumulate and adsorb on to solid particles.

The procedure described by Suzuki *et al.* [11,12], discussed in section 9.1.1.1 for the determination of chlorinated insecticides in soils has also been applied to hexane extracts of river sediments using high-resolution gas chromatography with glass capillary columns. Minimum detectable levels of α-BHC, ß-BHC, γ-BHC, ∂-BHC, Heptachlor, Heptachlor epoxide, Aldrin, Dieldrin, Endrin, *p,p'*-DDE, *p,p'*-TDE and *p,p'*-DDT in 100g samples of bottom sediment were 0.0005, 0.0032, 0.0014, 0.0040, 0.0012, 0.0020, 0.0014, 0.0020, 0.0056, 0.0032, 0.0080 and 0.0120mg kg^{-1} respectively.

Frank *et al.* [38] have described a method for the determination of DDT and TDE and Dieldrin in sediments. Air-dried mineral sediments (25g) were brought to within 50% of field storage capacity and left for 24h. Acetone and hexane in the ratio 1:1 (250ml) were added to the sediment and the mixture was shaken for 2h. An aliquot (100ml) was filtered off, water was added, and the organochlorine insecticides were partitioned into hexane. Air-dried organic sediments (25g) were blended with acetonitrile and water (2:1) for 5min and an aliquot (10g) was filtered off. The filtrate was partitioned into hexane. This mixture was transferred to a Soxhlet extraction apparatus and subjected to exhaustive extraction with hexane for 7h. Extracts from sediments were evaporated to dryness by rotary vacuum at 45°C. A one-step column clean-up method [39] was used for the isolation of organochlorine insecticides. Florisil (60–100 mesh) activated at 650°C was reheated at 135°C for a minimum of 24h. After cooling, the adsorbent was partly deactivated by the addition of water at the rate of 5ml 100g^{-1} and allowed to equilibrate. Eluates were concentrated just to dryness with rotary vacuum evaporation at 45°C, the residue was redissolved in 5ml hexane and used for subsequent chromatographic analysis. All solvents had been redistilled from glass. Varian Aerograph Models 204 and 1200 gas chromatographs, equipped with 250 millicurie tritium electron capture detectors, were used for qualitative and quantitative assays. Operating parameters were as follows:

Column: 150cm × 3mm – pyrex packed with 4% SE–30 + 6% QF–1 on Chromosorb W preconditioned 72h at 225°C.
Temperatures: column, 175°C; detector, 200°C; injection block, 225°C.
Carrier gas: nitrogen at 40ml min $^{-1}$.

Injection volumes of 5µl were used for both sample solutions and comparison standards. Qualitative residue confirmation was accomplished with thin layer chromatography using silica gel. Plates were developed with 1% chloroform in *n*-heptane, and visualized with alkaline silver nitrate spray as the chromogenic agent. Alternatively, *p,p*′-DDT and *p,p*′-TDE were confirmed by treatment with 5% methanoic potassium hydroxide [40]. Partial confirmation of Dieldrin was achieved by fractionating the analysis solution on a Mills column, thus isolating Dieldrin in the second fraction [35].

Goerlitz and Law [41] determined chlorinated insecticides in sediment and bottom material samples which also contained PCBs by extracting the sample with acetone and hexane. The combined extracts were passed down an alumina column. The first fraction (containing most of the insecticides and some polychlorinated biphenyls and polychlorinated naphthalenes) is eluted with hexane and treated with mercury to precipitate sulphur. If the polychlorinated hydrocarbons interfere with the subsequent gas chromatographic analysis, further purification on a silica gel column is necessary.

The method described by Teichman *et al.* [15] and discussed in section 9.1.1.2 for the determination of chlorinated insecticides and PCBs in soils has also been applied to sediments. The procedure involves adsorption chromatography on alumina and charcoal, elution with increasing fractional amounts of hexane on alumina columns, and with acetone–diethyl ether and benzene on charcoal columns. The polychlorobiphenyl and pesticides are then determined by gas chromatography on the separate elutes without interference.

Wegmann and Hofstee [43] have developed a capillary gas chromatographic method for the determination of organochlorine insecticides in river sediments. Bottom soils from rivers, collected in slow current areas may contain high concentrations of organochlorine insecticides and polychlorobiphenyls. When the current moves more rapidly or benthic animals become more active, these compounds are stirred into the water along with suspended particles and become accessible to organisms that live in the bottom layer.

Bottom soil is quite different from soil on land, particularly if it is collected from an anaerobic zone. Bottom soil specimens also have varying composition. The presence of elementary sulphur and organic compounds of sulphur greatly complicates analysis of the residual

Table 9.7 Influence of desulphurizing agents on extraction of OCP and PCB from bottom oils

Pesticide	*Extraction of OCP and PCB from bottom soils, %, after purification of extract by*		
	mercury	*copper*	*TBA*
Hexachlorobenzene	81.2	98.1	94
Gamma-HCCH	75.7	94.8	79
Heptachlor	39.8	5.4	–
Aldrin	95.5	93.3	73
Heptachlor epoxide	69.1	96.6	97
n,n'-DDE	92.1	102.9	95
n,n'-DDD	–	–	94
n,n'-DDT	79.8	85.1	94
Dieldrin	79.1	94.9	–
Endrin	90.8	89.3	–
PB	–	–	–
Arochlor 1254	97.1	104.3	–
Chlophen A–50	–	–	95–101

Source: Reproduced with permission from Elsevier Science Ltd [43]

organochlorine insecticides and polychlorobiphenyls, rapidly poisoning the packing of chromatographic columns. It is completely impossible to determine the isomers of HCCH, hexachlorobenzene, Aldrin and Heptachlor in unpurified bottom soil extracts. Raw bottom soil extracts are treated with highly purified copper powder or metallic mercury to facilitate analysis or are sulphurized with sodium sulphate in the presence of tetrabutyl ammonium sulphate [44].

Table 9.7 presents the results of analysis of organochlorine insecticides and polychlorobiphenyls using copper, mercury and tetrabutyl ammonium sulphate to desulphurize bottom soil extracts.

Testing of the desulphurizing agents indicated that tetrabutyl ammonium sulphate in combination with sodium sulphate was the most effective.

The method for analysing sediment involves extraction of organochlorine insecticides and polychlorobiphenyls with a mixture of acetone and hexane together with 1% aqueous ammonium chloride. The extracts are then concentrated for purification with concentrated sulphuric acid and aqueous sodium sulphite in the presence of tetrabutylammonium sulphate and finally gas chromatographic analysis is applied.

The minimum detectable quantities are: HCCH isomers – 0.01ng; 4,4'-DDE (*n,n'*-DDE) – 0.05ng; 4.4'-DDD (*n,n'*-DDD) – 0.01ng; 4,4'-DDT (*n,n'*-DDT) – 0.20ng; and PCB (chlophen A–50) – 1.0ng. The volume of the aliquot injected into the chromatograph is 4µl, the final volume of the extract is 3–5 ml. 80–90% of the organochlorine pesticides and polychlorobiphenyl present in marine bottom soils can be determined by this method (in water 90–95%). The measurement error is 15–20% for soils and 10% for water.

To determine the effect of storage time of the bottom soil specimen on the degree of extraction of pesticides and polychlorobiphenyl, centrifuged sediments were stored for two months at room temperature and in a freezer at –10 to –15°C. Storage of centrifuged bottom soils for a month both in the freezer and at room temperature had little influence on the degree of extraction of organochlorine pesticides. However, freezing of the bottom soils at –10 to –15°C is preferable, since preparation of specimens for analysis takes less time: a less persistent emulsion is formed during extraction, and separation of the organic and inorganic layers requires significantly less time. Two months storage of pesticide-contaminated bottom soils at room temperature resulted in a decrease in the degree of their extraction by 10–40%, and in the freezer by 5–20%.

If the sediment sample contains elemental sulphur, which hinders the determination of chlorinated organic insecticides, then this can be removed by a process involving treatment with concentrated sulphuric acid and tetrabutyl ammonium sulphate.

Jensen *et al.* [45,46] also discuss complications in analysis due to the presence of elementary sulphur and organosulphur compounds in the gas chromatographic determination of DDT and polychlorobiphenyls in sediments and sewage sludges.

The method can also be used for a search for both volatile and/or polar pollutants. The sulphur interfering in the gas chromatographic determination is removed in a non-destructive treatment of the extract with tetrabutylammonium sulphite. This lipophilic ion pair rapidly converts the sulphur to thiosulphate in an organic phase. The recovery of added organochlorines was above 80% and the detection limit in the range of 1–10ppb from a 10g sample. Elemental sulphur present in most sediment and digested sludge has caused significant problems in residue analysis [63,64]. If the sulphur level is high, the electron capture detector will be saturated for a considerable period of time, and if the level of sulphur is low, it gives three or more distinct peaks on the chromatogram which can interfere with BHC isomers and Aldrin. Treatment of the crude extract with potassium hydroxide in ethanol [65] or Raney nickel [66] will quantitatively destroy all sulphur, but will at the same time convert DDT and DDD to DDE and DDMU (1-chloro-2,2-*bis*(4-chlorophenyl)ethane), respectively, and most BHC isomers are lost. Metallic mercury has also been used for removal of sulphur [67]. Jensen *et al.* [45] described an efficient, rapid, non-destructive method to remove the sulphur according to the reaction:

$$(TBA^+)_2SO_3^{2-} + S(s) \rightarrow 2TBA^+ + S_2O_3^{2-}$$

where TBA^+ = tetrabutylammonium ion.

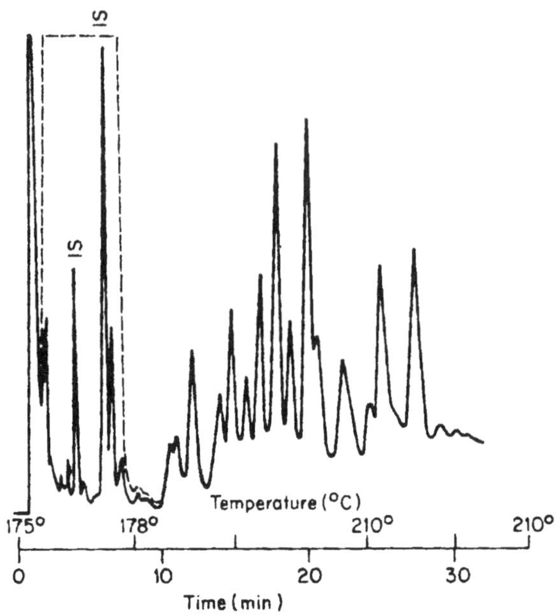

Fig. 9.3 Gas chromatogram of extract from sediment before (– – –) and after (——) TBA–sulphite treatment (the PCB level is 240ppb on a wet weight basis). · = PCB components. IS = internal standard.
Source: Reproduced with permission from American Chemical Society [45]

Figure 9.3 shows the gas chromatogram obtained for a sulphur containing sediment sample before and after treatment with tetrabutylammonium sulphate–sodium sulphate. It is seen that the effects of sulphur and sulphur containing organic compounds in the sample are completely eliminated by this treatment.

9.1.2.2 Supercritical fluid chromatography

The supercritical fluid chromatographic procedure [20] described in section 9.1.1.5 for the determination of organochlorine insecticides in soils has also been applied to river sediments. Snyder *et al.* [20] compared supercritical fluid extraction with classical sonication and Soxhlet extraction for selected organochlorine insecticides. Samples of sediments extracted with supercritical carbon dioxide modified with 3% methanol at 350atm and 50°C gave ≥85% recovery of organochlorine insecticides including Dichlorvos, Diazinon, Endrin, Endrin aldehyde, decachlorobiphenyl, *p,p′*-DDT and Mirex.

Grob *et al.* [47] compared supercritical extraction with classical sonication and Soxhlet extraction from river sediment for selected organochlorine insecticides including Endrin, Endrin aldehyde, *p,p'*-DDT Mirex and decachlorobiphenyl. The sample was extracted with carbon dioxide modified with 3% methanol at 350atm and 50°C, and 85% recovery of these compounds was achieved.

9.1.2.3 Miscellaneous

Goldberg [48] studied the relationship between pesticide concentrations in water and in sediments and its dependence on the specific surface area of the sediment.

Lopez-Avila *et al.* [23] have described a microwave assisted extraction procedure for the separation of chlorinated insecticides from sediments.

9.1.3 Saline deposited and suspended sediments

9.1.3.1 Miscellaneous

Picer *et al.* [49] described a method for measuring the radioactivity of labelled DDT contaminated sediments in relatively large volumes of water, using a liquid scintillation spectrometer. Various marine sediments, limestone and quartz in sea water were investigated. External standard ratios and counting efficiencies of the systems investigated were obtained, as was the relation of efficiency factor to external standard ratios for each system studied.

9.1.4 Sludges

9.1.4.1 Gas chromatography

Various workers [50–62] have reported methods for the determination of polychlorobiphenyls and organochlorine insecticides in sewage and sewage sludges.

Mattson and Nygren [57] have described the chromatographic procedure for the determination of polychlorobiphenyls and some chlorinated insecticides in sewage sludge. The capillary column is coated with silicone oil SF 96.

The sample is extracted with a mixture of hexane, acetone and water. After separation, the hexane phase is reduced in volume and divided into two aliquots, one of which is first shaken with 7% fuming sulphuric acid to remove lipids, and then with cyanide to eliminate interference by elemental sulphur. The other aliquot is evaporated to dryness and heated with ethanolic potassium hydroxide. The two aliquots are injected into a gas chromatograph fitted with a glass capillary column and an electron capture detector. Hexabromobenzene is used as an internal standard. Polychlorinated biphenyls are determined quantitatively by comparing the peaks of the sample with those of Clophen A 50 or A 60. The individual percentage composition of the chlorobiphenyls in the polychlorinated biphenyl oils is used.

The concentration levels of chlorinated hydrocarbons in sewage sludge allowed the use of a splitter injection technique [68]. These impurities, together in the first part of the column, cause poor separations and lower sensitivity. The effect of changing the glass tubes of the splitter is shown in Fig. 9.4. Nearly 70 peaks were detectable when a polychlorobiphenyl oil (Clophen A 50) was chromatographed. The polychlorobiphenyls are quantitated by using the percentage composition of the individual components in the polychlorobiphenyl oils (Fig. 9.5).

This method has good reproducibility and has a detection limit for the total mount of polychlorobiphenyls in the dried sample of at least 0.1mg kg^{-1} and for DDT, DDD and DDE limits of 0.01, 0.005 and 0.005mg kg^{-1} respectively.

Mattson and Nygren [57] point out that lipids and some other impurities in the crude extracts of sewage sludge can be destroyed by treatment with fuming sulphuric acid, either by shaking the acid [65] or by eluting on a fuming sulphuric acid–Celite column [69,70]. Dieldrin is decomposed by this treatment but DDT and its metabolites, DDD and DDE, are not (Table 9.8). Extracts of sewage sludges often contain large amounts of elemental sulphur, particularly after treatment with sulphuric acid. These interfere with early eluting compounds in the gas chromatographic step (Fig. 9.6(a)).

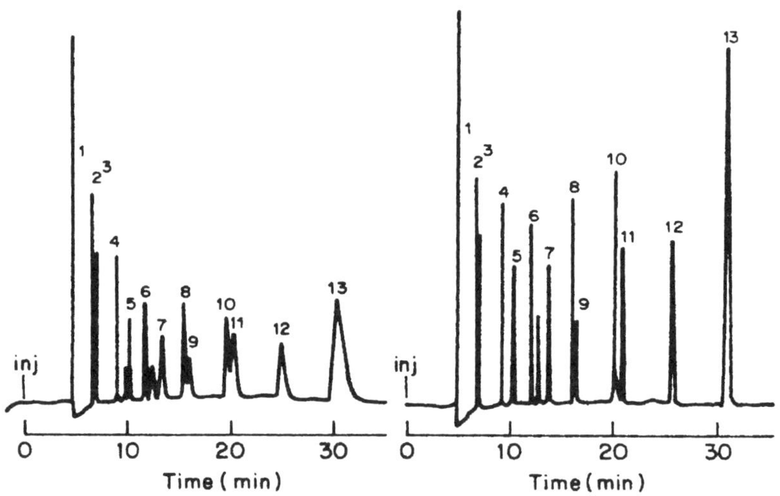

Fig. 9.4 Illustration of the effect of changing injector glass tube. Left: standard solution run after about 100 injections of sewage sludge extracts. Right: the same solution run after changing the injector glass tubes.
Source: Reproduced with permission from Elsevier Science Publishers B.V. [57]

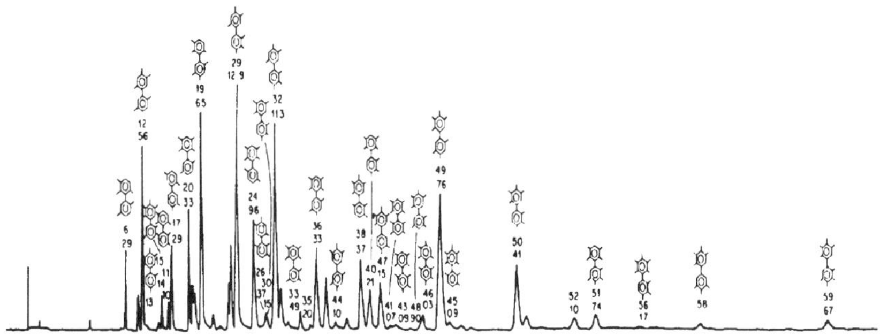

Fig. 9.5 Chromatogram of Clophen A 60 run on a 60m SF 96 glass capillary column at 185°C.
Source: Reproduced with permission fromElsevier Science Publishers B.V. [57]

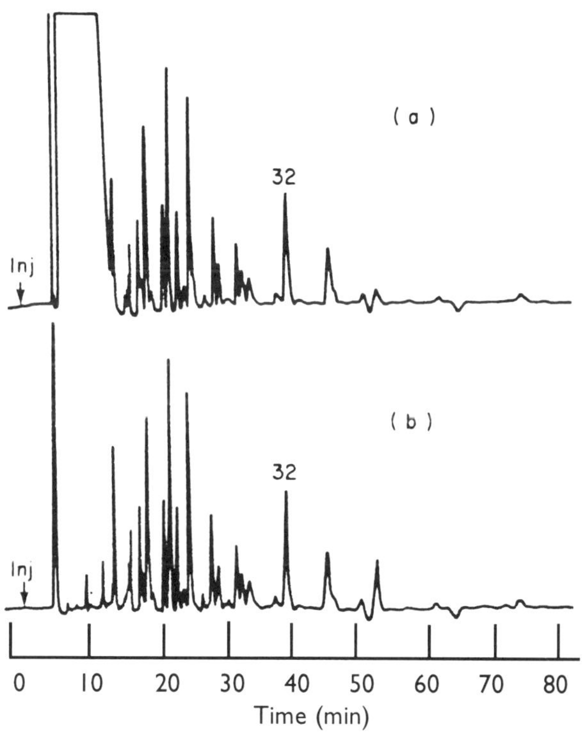

Fig. 9.6 Interference by elemental sulphur. Extract of sewage sludge sample treated with 7% fuming sulphuric acid(a) and also treated with cyanide (b).
Source: Reproduced with permission from Elsevier Science Publishers B.V. [57]

Table 9.8 Effect of treatment of a solution of chlorinated hydrocarbons and the internal standard hexabromobenzene with fuming sulphuric acid (I), fuming sulphuric acid plus potassium cyanide (II), and potassium hydroxide (III) expressed as percentages of the compounds in an untreated solution

Compound name (no)	*Concentration* (ng ml $^{-1}$)	*Method of treatment* *I*	*II*	*III*
α-BHC (1)	0.022	94	59	0
ß-BHC (2)	0.072	94	70	6
Lindane (3)	0.026	92	55	11
Heptachlor (4)	0.026	94	90	104
Aldrin (5)	0.020	92	87	104
Heptachlor epoxide (6)	0.033	83	79	100
p,p'-DDMU (7)	0.090	100	91	190
Dieldrin (8)	0.048	0	0	103
p,p'-DDE (9)	0.022	104	100	437
p,p'-DDD (10)	0.085	100	95	0
o,p'-DDT (11)	0.096	98	95	0
p,p'-DDT (12)	0.103	100	95	0
Hexabromobenzene (13)	0.57	(100)	(100)	(100)

Source: Reproduced with permission from Elsevier Science Publishers B.V. [57]

Sulphur was removed by the Bartlett and Skoog [71] method in which the sulphur is reacted with cyanide in acetone solution to produce thiocyanate (Fig. 9.6(b)). BHCs are decomposed to some extent, probably to pentachlorocyclohexane. An alternative procedure for the removal of sulphur utilizing barium hydroxide is also described. Alkali hydroxides should not be used as they cause dehydrochlorination of BHCs [65]. Lindane and its isomers are dehydrochlorinated to trichlorobenzenes [72] and are eluted together with the solvent. Mattsson and Nygren [57] have also tested a column with a packed alkaline post-column to remove the sulphur peak from the chromatogram. In the post-column DDT and DDD are dehydrochlorinated but this does not effect their retention times. Cochrane and Maybury [73] have used the reaction with sodium hydroxide in methanol for the identification of BHCs. Dieldrin is not decomposed in the potassium hydroxide treatment and can thus be detected in the chromatogram of that aliquot. Some common chlorinated hydrocarbon pollutants and the internal standard hexabromobenzene were treated, according to the general procedure described above with sulphuric acid, potassium cyanide and potassium hydroxide. The results of the recovery experiments are shown in Table 9.8. When using packed columns, a pre-column of sodium and potassium hydroxides will give the same effect as the potassium hydroxide treatment described above [74].

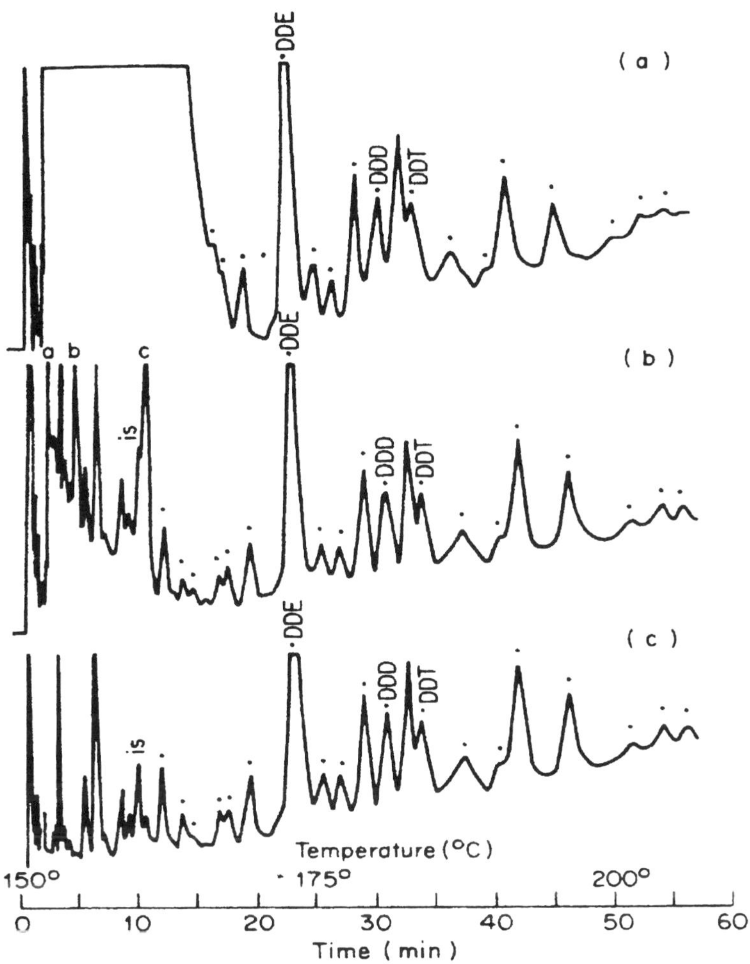

Fig. 9.7 (a) Typical digested sewage sludge chromatogram, severely contaminated with sulphur. (b) The same sample after a normal TBA–sulphite treatment, showing that most of the sulphur has disappeared. A number of peaks (a,b,c) originating from traces of sulphur appear in the BHC–aldrin region. (c) The final chromatogram after additional treatment with sodium sulphite. = • PCB components. IS = internal standard
Source: Reproduced with permission from the American Chemical Society [45]

Jensen *et al.* [45] applied the method described in section 9.1.2.1 for the determination of DDT and polychlorobiphenyls in sulphur containing sediments to the analysis of sludges. The results in Fig. 9.7 show the beneficial effects of pretreatment of the sewage with tetrabutyl ammonium sulphate–sodium sulphite reagent on the recovery of DDT and polychlorobiphenyls from a digested sewage sludge sample.

McIntyre *et al.* [58,59] described a method for the analysis of polychlorobiphenyls and chlorinated insecticides in sewage sludges in which homogenized samples are extracted with hexane, concentrated and cleaned up on an alumina/alumina plus silver nitrate column and eluted with hexane. After concentration of the eluent, polychlorobiphenyl and organochlorine compounds were determined by a silica gel chromatographic procedure and gas chromatography.

McIntyre *et al.* [58,59] found that the silica gel chromatographic method of Holden and Marsden [75] resulted in good reproducible separation of polychlorobiphenyls and organochlorine insecticides. Polychlorobiphenyls and DDE were found to emerge in the first 6ml of hexane eluate, while the remaining organochlorine insecticides emerged in the diethyl ether–hexane eluate. A standard solution containing Aroclor 1260, *p,p'*-DDE, γ-HCH and Dieldrin was carried through this separation procedure in order to assess recovery. The results indicate that recovery from the four compounds was quantitative, with recoveries of 99.8, 99.6, 102.5 and 102.3% for Aroclor 1260, *p,p'*-DDE, γ-HCH, and Dieldrin, and displays little scatter, with relative standard deviations of 1.6, 3.1, 2.9 and 3.1% respectively for the four determinations.

McIntyre *et al.* [58,59] examined the influence of solids concentration in the sewage sample on the extraction of polychlorobiphenyls and organochlorine insecticides and found that in the 0.2–10g L^{-1} solids concentration range, determinations of Aroclor 1260, *p,p'*-DDE, γ-HCH and Dieldrin were not subject to a relative standard and deviation of greater than 7.6 and that the F-test was highly significant for all four determinations. Examination of the percentage recoveries shows that Aroclor 1260 and DDE both exhibited significant increases in percentage recovery from 40.5g L^{-1} total solids to 1.0g L^{-1} total solids, that of Aroclor 1260 increasing from 71.4% to 96.3%, DDE from 52.1% to 61.8%. γ–HCH, again, displayed a different trend, with the increase in percentage recovery over the range of total solids concentrations not being significant due to the degree of scatter of the results, as was the case with Dieldrin. Overall, it can be concluded that the degree of scatter, reflected in the relative standard deviations, was less for both Aroclor 1260 and *p,p'*-DDE than for γ-HCH and Dieldrin and that the solids concentration significantly influenced the percentage recovery of Aroclor 1260 and *p,p'*-DDE from spiked samples, but not those of γ-HCH and Dieldrin.

McIntyre *et al.* [58,59] conclude that the extraction of polychlorobiphenyls and organochlorine insecticides is most efficient at a total solids concentration of 1g L^{-1}, using the extraction procedure described above. The recovery of *p,p'*-DDE from subsamples was always found to be the lowest of the four determinands considered (61.8%), while recoveries of Aroclor 1260, γ-HCH, and Dieldrin from the diluted sample averaged 96.3, 89.4 and 82.9% respectively at 1g L^{-1} total solids.

Garcia-Gutierrez *et al.* [60,76] have shown that the gas chromatographic method is suitable for the determination of Endrin, Aldrin, other chlorinated insecticides and polychlorobiphenyls in sewage sludges.

Maskarinec and Harvey [61] have described a method for the screening of solid wastes and sludges for organic compounds. The technique involves sequential extraction with acidic, basic and neutral media and the extracts are all analysed directly by gas chromatography with fused silica columns. Results achieved using the technique are discussed. The method is compared with traditional Soxhlet extraction.

9.1.4.2 Gas chromatography–mass spectrometry

Erikson and Pellizzari [77] analysed municipal sewage samples in the USA by a gas chromatography–mass spectrometry-computer technique for chlorinated insecticides and polychlorobiphenyls.

In this method the samples (≈300g) were extracted at pH11 six times with a total of 350ml chloroform to remove neutral and basic compounds. The extract was dried with sodium sulphate, vacuum filtered, and concentrated to 2ml using a Kuderna Danish apparatus. In cases where the sample background interfered significantly, an aliquot of the sample was chromatographed on a 1.0 × 30cm silica gel column. Polychlorobiphenyls and related compounds were eluted with 50ml hexane; pesticides and other compounds were eluted with 50ml toluene.

Acidic components of the sludge samples were treated with diazomethane and dimethyl sulphate. Analysis of all samples for polychlorobiphenyls was accomplished using a Finnegan 3300 quadruple gas chromatography–mass spectrometer with a PDP/12 computer. The 180cm × 2mm i.d. glass column, packed with 2% OV–101 on Chromosorb W, was held at 120°C for 3min, programmed to 230°C at 120min $^{-1}$ and held isothermally until all peaks had eluted. Helium flow was 30ml min $^{-1}$. The ionization voltage was nominally 7OeV and multiplier voltages were between 1.8 and 2.2kV. Full scan spectra were obtained from *m/e* 100–500. Polychlorobiphenyls were quantitated by gas chromatography–mass spectrometry-computer using the selected ion monitoring mode to provide maximum sensitivity and precision. Ten ions were selected for monitoring: one from the parent cluster for each of the chlorinated biphenyls ($C_{12}H_9Cl$ through $C_{12}Cl_{10}$). Polychlorobiphenyls were quantitated using anthracene as external standard and a previously determined relative molar response (anthracene parent ion mass 178; 27ng ml $^{-1}$). Anthracene does not interfere with polychlorobiphenyl determination nor do polychlorobiphenyls or their fragment ions interfere with the determination of anthracene.

The retention time results for 35 chlorinated compounds found in sewage sludge are given in Table 9.9. Not all compounds could be identified. A large number of spectra contained what appeared to be chlorine isotope clusters which are not reported. This could be due to interferences, very low levels, or spurious peaks. Although no structure

Table 9.9 Summary of chlorinated compounds found in sewage sludge

Compound[a]	*Retention time (min)*[b]
Dichlorobenzene	0.5
Mol. wt. = 194 Cl_1	0.8
Mol. wt. = 222, Cl_1	0.9
Trichlorobenzene	1.0
Chloroaniline (tent.)	1.2
Dichloroaniline	2.1–4.7 (1.1)
Tetrachlorobenzene	2.2 (0.7)
Mol. wt. 187, Cl_2	2.3–2.7 (0.6)
Mol. wt. = 171, Cl_2	3.0
Mol. wt. = 240, Cl_4[e]	3.0 4.3 (1.1)
Trichloroaniline[d]	3.2, 4.4
Dichloronaphthalene	3.7
Trichlorophenol	3.7
Mol. wt. = 302, Cl_1	4.6
Mol. wt. = 210, Cl_3	5.1
Chlorobiphenyl	6.2
Dichlorobiphenol[f]	6.3–8.2
Trichlorobiphenyl[f]	7.5–9.6
Mol. wt. = 291, Cl_1	7.6
Mol. wt. = 288, Cl_1	8.6–11.1 (5.8)
Tetrachloronaphthalene	8.6
Mol. wt. = 218, Cl_1 (tent.)	9.1
Mol. wt. = 256, Cl_1	9.1
Dichlorobenzophenone	9.2 (7.2)
Mol. wt. = 269, Cl_1	9.9
Mol. wt. = 256, Cl_2	10.1
Pentachlorobiphenyl[f]	10.2–11.1
Mol. wt. 288, Cl_3	10.5
Mol. wt. = 280, Cl_1	10.7
Mol. wt. = 241, Cl_1	10.8
Mol. wt. = 285, Cl_1	12.6
DDE[e] 12.7, 13.2	(8.6)
Mol. wt. = 356, Cl_2	12.8
Mol. wt. = 397, Cl_1	15.0

Source: Reprinted with permission from Springer-Verlag [77]

[a]Unidentified compounds are listed with the apparent molecular weight and number of chlorines. If the identification of a compound is tentative, it is denoted by (tent.).

[b]Retention times are listed for the chromatographic temperature conditions, 12°C for 3min, then 12° min^{-1} to 230°C, then hold. Values in parentheses are for the chromatographic temperature conditions, 150°C for 3min, then 8° min^{-1} to 230°C, then hold.

[d]Differences in retention times possibly indicate different isomers.

[e]Two separate isomers observed in some samples.

[f]Several isomers observed.

could be assigned, the mass spectra indicated possible structures for three compounds in Table 9.9. The compound containing two chlorines with mol. wt. = 187 (RT = 2.3–2.7min) may have the molecular formula $C_8H_7C_{12}$ and could be a dichlorodihydroindole or related compounds. Two distinct compounds were observed with four chlorines and mol. wt. = 240. These compounds appear to be isomers of tetrachloro-styrene ($C_8H_4C_{14}$). The 4,4′-dichlorobenzophenone identification was confirmed by comparison of the retention time with an authentic sample. The two peaks identified as DDE isomers are probably the two common isomers, *o,p*′-DDE and *p,p*′-DDE which generally are separable by gas chromatography.

9.2 Carbamate insecticides

9.2.1 Soil

9.2.1.1 Titration

Singhal *et al.* [78,79] have described a titrimetric method for the determination of low levels of Oxamyl residues in soils. Their investigations revealed that after hydrolysis Oxamyl gave a brown precipitate with carbon disulphide and an alkaline solution of copper(II) sulphate. This reaction suggested a procedure for the determination of Oxamyl by titration with ethylene diamine tetracetic acid of the copper remaining unreacted to the 1-(2 pyridylazo)-2-naphthol end-point indicator). The following stoichiometric reaction appeared to occur between Oxamyl and the reagents:

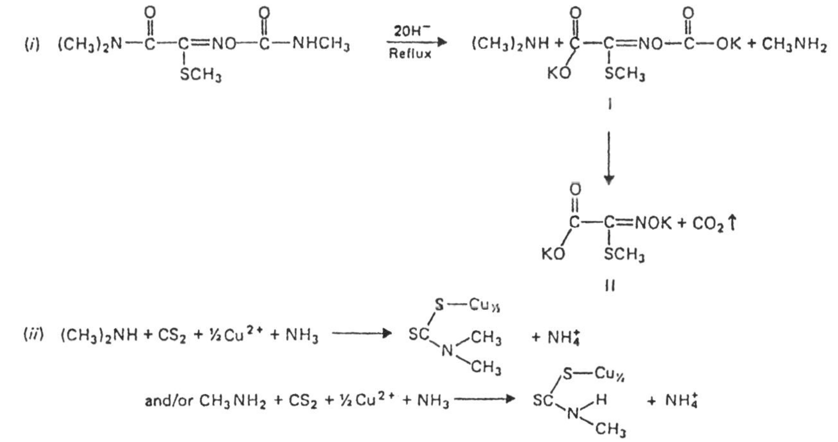

A reproducibility study of two series of 10 solutions containing 1000 and 2500µg of Oxamyl, respectively, gave an arithmetic mean, standard deviation and coefficient of variation of 997µg, 8.0µg and 0.80%, respectively, for the first series and 2489µg, 13.3µg and 0.53%, respectively, for the second. The average recovery varied from 85.5-100.8%.

9.2.1.2 Gas chromatography

Cohen and Wheals [80] used a gas chromatograph equipped with an electron capture detector to determine ten substituted urea and carbamate herbicides in soils in amounts down to 0.001–0.05 1–50μg kg $^{-1}$. The methods are applicable to those urea and carbamate herbicides that can be hydrolysed to yield an aromatic amine. A solution of the herbicide is first spotted on to a silica gel G plate together with herbicide standards (5–10μg) and developed with chloroform or hexane-acetone (5:1). The plate containing the separated herbicide or the free amines is sprayed with 1-fluoro-1,4-dinitrobenzene (4% in acetone) and heated at 190°C for 40min to produce the 2,4-dinitrophenyl derivative of the herbicide amine moiety. Acetone extracts of the areas of interest are subjected to gas chromatography on a column of 1% of XE–60 and 0.1% of Epikote 1001 on Chromosorb G (AW–DCMS) (60–80 mesh) at 215°C.

Westlake *et al.* [82] determined *m*-S-butylphenyl methyl-(phenylthio) carbamate (RE 11775) in soil by a gas chromatographic procedure. The sample is extracted with dichloromethane, chloroform or acetonitrile, followed by clean-up, if necessary, on a column of Florisil, silica gel or alumina. The purified residue is submitted to gas chromatography on either a stainless steel column (90cm × 6mm) packed with 5% OV–225 on Gas–Chrom Q (60–80 mesh) and operated at 242°C, with nitrogen as carrier and a flame photometric detector operated in the S mode, or on a glass column (90cm × 6mm o.d.) with identical packing and operated at 195°C, with hydrogen as carrier gas (100ml min $^{-1}$) and an electrolytic conductivity nitrogen detector. Recoveries of added RE 11775 from soil samples were about 100%. Down to 0.1mg kg $^{-1}$ could be determined in soil.

Reeves and Woodham [83] have described a gas chromatographic method for the determination of Methomyl in soil. The residues were extracted from soil with dichloromethane, and the extracts were purified on a column of Florisil. The residues were extracted from tobacco with dichloromethane–benzene (39:1), and the extracts were purified by a coagulation procedure with ammonium chloride–phosphoric acid. The purified and concentrated extracts were then analysed by gas chromatography on a glass column (180cm × 5mm) packed with 10% DC–200 on Chromosorb WHP (80–100 mesh) and operated at 140°C, with nitrogen as carrier gas (80ml min $^{-1}$) and a 394nm S-interface filter. The limits of detection were 0.05mg kg $^{-1}$ for soil; the recovery was 90.8.

Soils treated with Carbamyl and Carbofuran have been analysed by gas chromatography following conversion of N-methyl carbamates to their pentafluorobenzyl derivatives [84].

Bromilow [85] has described a gas chromatographic procedure for the determination of down to 10μg kg $^{-1}$ of Oxamyl (S-methyl N-N′-thio oxamimidate) without interference by Oxamyl oxime (S-methyl N-N′-dimethyl N-hydroxythiooxamimidate) in soils.

Oxamyl is extracted from the soil with dichloromethane or acetone–dichloromethane and is separated from interfering coextractives by chromatography on a Florasil column.

Extraction

To 50g of soil (as sampled, recoveries of Oxamyl are constant for moisture contents of up to at least 30%) contained in a 450g wide-necked jar fitted with a ground-glass stopper, add 50g of anhydrous sodium sulphate and 200ml of acetone-dichloromethane solvent. Shake orbitally for 4h, ensuring that all the solids are kept in suspension. Allow to settle, and remove a 100ml aliquot of the supernatant liquid by pipette. Evaporate this extract just to dryness in a 250ml round-bottomed flask fitted to a rotary evaporator.

Florasil clean-up

Place a small cotton-wool plug in the bottom of a chromatographic column, and pour in a slurry of 18g of Florisil in the eluting solvent. Add 20mm of anhydrous sodium sulphate on top of the Florisil and allow the solvent to drain until the level reaches the top of the column bed. Transfer the crop or soil extract from the flask on to the column with small portions of the eluting solvent totalling 20ml, swirling the solvent carefully around the flask to dissolve as much of the residue as possible and allowing the solvent level in the column to drain just to the top of the column bed between each addition. Allow a further 150ml eluting solvent to percolate through the column at a rate of about 1 drop s^{-1}; discard the eluate, which contains most of the plant oils and any Oxamyl oxime present in the sample.

Elute the column with 75ml of acetone and collect the eluate in a 100ml round-bottomed flask; this fraction contains the Oxamyl. Using the same conditions as above, evaporate the eluate just to dryness in a rotary evaporator and quantitatively transfer the residue into a 1.0 or 2.0ml calibrated flask with small portions of ethyl acetate. Make up to volume with ethyl acetate and shake thoroughly.

The Oxamyl in this extract is then determined by gas chromatography using on-column reaction with trimethylphenyl ammonium hydroxide, the derivative so formed being determined by a flame photometric detector operated in the sulphur mode. Both Oxamyl and Oxamyl oxime in the soil react with trimethylphenyl ammonium hydroxide to form the same methoxime derivative: $(CH_3)_2NCOC(SCH_3) - NOCH_3$.

Gas–liquid chromatography

Solutions are taken into a 10μl syringe in the following order: ethyl acetate (0.2μl), sample extract or standard Oxamyl solution (2.0μl, containing

0.5–40ng of Oxamyl); and 0.1m trimethylphenylammonium hydroxide in methanol (0.5µl). Inject the contents of the syringe rapidly into the injection port of the gas–liquid chromatograph, and vent the column effluent to the atmosphere for 60–90s after injection. The gas chromatographic column comprised Chromasorb W coated with a mixture of 0.5% Carbowax 20m and 5% SE –30. The methoxime derivative of Oxamyl has a retention time of about 3min. Prepare a calibration graph by plotting log (peak height) against log (amount of Oxamyl injected), and read off the Oxamyl content of the sample extracts by interpolation.

The recovery of Oxamyl from soil samples was checked by adding known volumes of standard solutions of Oxamyl to the sample before extraction (Table 9.10) Mean recoveries of Oxamyl from the samples fortified over the range 0.02–0.4mg kg^{-1} varied from 87–96%. The recovery of Oxamyl was similar at all of the fortification levels tested, except that there was more variation at the very low levels.

Several other sulphur-containing carbamate insecticides/nematicides and their oxidation products give peaks when injected into the gas chromatograph with trimethylphenyl–ammonium hydroxide, some of these peaks having retention times on the 0.5% Carbo-wax 20M – 5% SE–30 column closely similar to that of the methoxime derivative of Oxamyl. To check for possible interference in the analysis for Oxamyl from some of these other pesticides, 50µg amounts each of Aldicarb [2-methyl-2-(methylthio)propionaldehyde O-(methylcarbamoyl)oxime] and Thiofanox [3,3-dimethyl-l-(methylthio)-2-butanone O-(methylcarbamoyl) oxime], and their sulphoxide and sulphone metabolites, were taken separately through the procedure. No detectable peaks were observed, indicating that these potentially interfering compounds are not eluted in the Oxamyl-containing fraction taken from the Florisil column.

Leppert *et al.* [86] have described a procedure for the determination of Carbosulphan and Carbofuran (2,3-dihydro 2,2-dimethylbenzofuran-7-y1-methyl carbamate) utilizing gas chromatography with a nitrogen specific detector. Extraction was carried out using hexane-2-propanol or a methanol buffer.

Bilikova and Kuthan [87] developed a gas chromatographic method for the determination of submicrogram concentrations of Carbofuran (2,3-dihydro-2, 2,-dimethylbenzofuran-7-y1-methyl carbamate) in soils. Soil samples are mixed with methanol-water (80:20) and water samples are extracted with chloroform. After separation of the chloroform, and weak alkaline hydrolysis, derivatization is performed with 1-fluoro-2,4-dinitrobenzene. The ester produced is isolated from benzene and cleaned up with acetone. The acetone extract is used to determine Carbofuran by gas chromatography with a nitrogen-specific detector.

Table 9.10 Recovery of Oxamyl from soil samples

*Sample**	*Extraction solvent*	*Number of samples*	*Oxamyl recovered,† % Range*	*Mean*
Pea loam	Dichloromethane–acetone (1 + 1)	4	83–89	87
Sandy clay loam	Dichloromethane–acetone (1 + 1)	4	87–100	96
Clay loam	Dichloromethane–acetone (1 + 1)	4	89–105	94

Source: Reproduced with permission from the Royal Society of Chemistry [85]

*Oxamyl added at 0.2–0.40mg kg^{-1} level to 50g of sample.
†Duplicate injections into the gas–liquid chromatograph of extract from each sample

9.2.1.3 Gas chromatography–mass spectrometry

Methomyl (S-methyl(-N methyl carbamoxy thioacetimidate or (methylthio)acetald oxime) has been determined in chloroform extracts of soil in amounts down to 1μg kg^{-1} by gas chromatography–mass spectrometry [88].

9.2.1.4 Miscellaneous

Leistra *et al.* [89] have discussed a technique for the measurements of the rate of leaching of Methomyl (S-methyl N-(methyl-carbamoxy)thioacetimidate or S-methyl-N-(methyl carbamoyl)oxy thioacetimidate) from greenhouse soils into watercourses. Its adsorption on and leaching from soil were studied. Adsorption on three typical greenhouse soils was weak to moderate, and the half-life ranged from three to 14 days. These data were used in mathematical models to predict the pesticide's behaviour; only very small amounts would be leached, depending on the rate of transformation in the soil and the amount of irrigation water used.

Concentrations of the pesticide measured in the three soils were higher than estimated, but were less than one μg per litre. Streams in an area where there are a number of greenhouses contained higher concentrations than those measured in drainage water; this was attributed to discharge of surplus spray liquid.

N-methylcarbamate and N,N′-dimethylcarbamates have been determined in soil samples by hydrolyses with sodium bicarbonate and the resulting amines reacted with 4-chloro-7-nitrobenzo-2,1,3-Oxadiazole in isobutyl methyl ketone solution to produce fluorescent derivatives [81]. These derivatives were separated by thin layer chromatography on silica gel G or alumina with tetrahydrofuran–chloroform (1:49) as solvent. The fluorescence is then measured *in situ* (excitation at 436 nm, emission at 528 and 537nm for the derivatives of methylamine and dimethylamine respectively). The method was applied to natural water and to soil samples containing parts-per-10^9 levels of carbamate. The disadvantage of the method is its inability to differentiate between carbamates of any one class.

9.2.2 *Non-saline deposited and suspended sediments*

9.2.2.1 Gas chromatography

Reeves and Woodham [83] have described a gas chromatographic method for the determination of Methomyl (S-methyl-N-(methyl carbamoyl)oxy thioacetimidate) insecticide in sediments. The residues were extracted with dichloromethane, and the extracts were purified on a column of Florisil. The purified and concentrated extracts were then examined by gas chromatography. The limits of detection were 0.05mg kg $^{-1}$ and the recoveries were 91%.

9.2.2.2 Thin layer chromatography

Spengler and Jumar [90] used a spectrophotometric method and thin layer chromatography to determine carbamate and urea herbicide residues in sediments. The sample is extracted with acetone, the extract is evaporated in vacuo at 40°C and the residue is hydrolysed with sulphuric acid. The solution is made alkaline with 15% aqueous sodium hydroxide and the liberated aniline (or substituted aniline) is steam distilled and collected in hydrochloric acid. The amine is diazotized and coupled with thymol, the solution is cleaned up on a column of MN 2100 cellulose power and the azo-dye is determined spectrophotometrically at 440nm (465nm for the dye derived from 3-chloro- or 3.4-dichloroaniline) with correction for the extinction of a reagent blank.

9.3 Organophosphorus insecticides

9.3.1 *Soil*

9.3.1.1 Gas chromatography

Kjolholt [91] determined trace amounts of organophosphorus pesticides and related compounds in soil using capillary gas chromatography and a nitrogen specific detector. Homogenized samples were subjected to Soxhlet extraction with acetane-*n*-hexane. The extract was partitioned between methylene chloride and water, subjected to adsorption chromatography and analysed by gas chromatography. The influences of freeze drying and of the pH on extraction efficiency were studied. Interference by elemental sulphur was examined. Recoveries at the 10μg kg $^{-1}$ level were 54.6–82.4% and detection limits 95–220 g kg $^{-1}$ depending on the type of organophosphorus compound.

Trichlorophon has been determined [92] in acid soil by solvent extraction followed by gas chromatography on a glass column (180cm × 9mm) packed with 16% of XF–1150 on Chromosorb W–AW operated at 125°C with a carrier gas flow of 60ml m^{-1} and a flame photometric detector operated in the phosphorus mode. Average recoveries were 96%, and down to 50µg kg^{-1} of Trichlorophan could be determined in soil.

Fenophos (O-ethyl-S-phenylethyl-phosphorodithioate) insecticide has been determined in soil by gas chromatography. Fenophos is known to degrade to its oxygen analogue (O-ethyl-S-phenylethyl phosphonothioate) in soil but none was found in the soil samples examined [93].

9.3.1.2 Supercritical fluid chromatography

Snyder *et al.* [94] compared supercritical extraction with classical sonication and Soxhlet extraction for the extraction of selected organophosphorus insecticides from soil. Samples extracted with supercritical carbon dioxide modified with 3% methanol at 350atm and 50°C gave a ≥85% recovery of Diazinon (diethyl-2-isopropyl-6-methyl-4-pyrimidinyl phosphorothiodate or 0,0 diethyl-0-(2-isopropyl-6-methyl-4-pyrimidyl) phosphorothioate). Ronnel (or Fenchlorphos) 0,0-dimethyl-0-2,4,5 trichlorophenol phosphorothiodate), Parathion ethyl (diethyl-*p*-nitrophenyl (phosphorothioate), Tetrachlorovinphos (trans,-2-chloro-1-(2,4,5 trichlorophenyl) vinyl (chlorophenyl-O-methylphenyl phosphorothioate) and Methiadathion. Supercritical fluid extraction with methanol modified carbon dioxide has been applied to the determination of organophosphorus insecticides in soil [260].

9.3.1.3 Miscellaneous

Novikova [21] has reviewed the literature (209 references) covering the extraction, clean-up and analysis of organophosphorus insecticides in soil (also food).

9.3.2 Non-saline deposited and suspended sediments

9.3.2.1 Gas chromatography

A gas chromatographic procedure using electron capture detection has been described for the determination of Dursban (O,O-diethyl-O (3, 5, 6-trichloro-2-pyridyl phosphorothioate) in water and silt [95]. In this method, water samples are extracted with dichloromethane, the extract is evaporated, and a solution of the residue is cleaned up on a column of silicic acid, Dursban being eluted with hexane. The eluate is evaporated to dryness under reduced pressure, and a solution of the residue in hexane is subjected to gas chromatography. Down to 0.1µg kg^{-1} of

Dursban in sediment could be determined; average recoveries from sediment was 83%.

Deutsch *et al.* [96] determined Dursban in sediments by an extraction gas chromatographic procedure which was capable of determining down to 0.01mg kg^{-1} Dursban using a 10g sample.

Szeto *et al.* [97] have described a simple gas chromatographic method for the determination of Acephate and Methamidophos residues in sediments.

9.3.2.2 Supercritical fluid chromatography

Supercritical carbon dioxide extraction [98] has been applied with 85% recovery to the organophosphorus insecticides from sediments. Compounds studied include Parathron ethyl, Methiadathion and Tetrachlorovinphos.

Grob *et al.* [47] compared supercritical extraction with classic sonication and including Diazinon, Ronnel, Parathion ethyl, Methiadathion and Tetrachlorovinphos. Samples extracted with supercritical carbon dioxide modified with 3% methyl alcohol at 350°Catm and 50°C gave a recovery of at least 85%.

9.3.3 Saline deposited and suspended sediments

9.3.1.1 Gas chromatography

The capillary column gas chromatographic method [91] discussed in section 9.3.1.1 has been applied to marine sediments.

9.3.4 Sludges

9.3.4.1 Gas chromatography

An official method [99] describes procedure for the analysis of organophosphorus insecticides in sewage sludge. The method includes a primary extraction gas chromatographic procedure for the analysis of sewage sludge and a confirmatory thin layer chromatography procedure for organophosphorus pesticides in sludges, the latter being capable of 80% recovery of Malathion and Ethyl Parathion and 70% recovery of Diazinon, with analytical limits of detection from 4.0 to 8.0ug per kg for these three compounds, are described. In addition a brief note is included on methods of eliminating interference effects when determining pesticides such as Carbophenothion in waters.

9.4 Triazine herbicides

9.4.1 Soil

9.4.1.1 Gas chromatography

Cotterill [100] studied the effect of ammonium nitrate fertilizer on the electron capture or nitrogen specific gas chromatographic determination of Triazine plus other types of herbicide (Atrazine(2-chloro-4-ethylamino-6-isopropylamino, 1,3,5 triazine), Simazine (2-chloro-4.6 *bis* ethyl amino 1,3,5 triazine), Linuron (3,4,-chlorophenyl-1-methoxy-1-methyl urea), Metribuzin, Triallate and Phorate) residues in soil.

The effect of ammonium nitrate concentration on the peak height of a 60ng Atrazine standard solution is shown in Fig. 9.8, the optimum response being at a concentration of approximately 4μg μL $^{-1}$.

Table 9.11 shows the effect of this concentration on the responses of the other pesticides. In every instance the peak height was increased while the peak area remained constant. All of the columns used for this study were aged by repeated sample injections, but had not deteriorated to the point where they would normally be replaced. No values are included in Table 9.11 where the chromatographic system was not suitable for the pesticide concerned.

When the enhancement of peak height was first observed the stainless-steel column packed with UCW 982 was in use. On changing to the glass column packed with OV–17 the effect disappeared. However, after several hundred injections the effect returned. Injections of standards into the chromatograph fitted with an electron-capture detector also showed peak-height enhancement. These observations, together with the constant peak areas, suggested that enhancement was due to an increase in column performance, probably due to the extract. This was confirmed by injecting the column conditioner Silyl 8 (Pierce Chemical Co.) on to an aged OV–17 column. The conditioner restored the column performance and removed the enhancement effect. Theoretical plate measurements were made in order to determine the effect of enhancement. A new OV–17 column gave 2190 plates but after ageing gave only 11240 plates; after enhancement the number of plates was increased to 2070.

To check the enhancement effect in the presence of soil extracts, soil was treated with ammonium nitrate at a rate equivalent to 200kg ha $^{-1}$. When 50g of the soil were extracted with 100ml of methanol using the method of Byast *et al.* [101] the concentration of ammonium nitrate in solution was about 4μg μL $^{-1}$. Injections of the treated soil extract increased the peak height of a subsequent 60ng Atrazine standard by 55–65%. No increase occurred following injections of untreated soil extracts.

An additional complication was that ammonium nitrate-induced enhancement was short-lived. Consecutive injections of an untreated 60ng Atrazine standard decreased in peak height with time after

Table 9.11 Mean peak heights and peak areas of pesticide standards on different columns
Results are means of six determinations; standard deviations are given in parentheses

Pesticide	Column					
	OV–17		SE–30		UCW 982	
	Peak height/ mm	*Peak area, arbitrary units*	*Peak height/ mm*	*Peak area, arbitrary units*	*Peak height/ mm*	*Peak area, arbitrary units*
Atrazine	23 (1.7)	23 (1.7)	8 (0.3)	8 (0.3)	13 (1.2)	15 (0.7)
+ NH_4NO_3	34 (3.1)	24 (1.1)	13 (1.0)	8 (1.0)	20 (2.3)	14 (0.6)
Simazine	33 (0.9)	180 (3.2)	8 (0.4)	7 (1.0)	22 (1.2)	121 (0.8)
+NH_4NO_3	53 (1.0)	186 (3.5)	13 (0.7)	8 (0.2)	34 (2.3)	119 (0.9)
Metribuzin	33 (1.3)	49 (1.9)	–	–	49 (1.9)	44 (1.7)
+NH_4NO_3	52 (2.0)	52 (2.0)	–	–	78 (2.7)	46 (1.8)
Tri-allate	41 (1.9)	62 (4.2)	90 (0.0)	81 (0.0)	27 (1.3)	41 (2.8)
+NH_4NO_3	63 (0.5)	63 (0.5)	122 (2.5)	85 (1.7)	42 (0.3)	42 (0.3)
Linuron	–	–	27 (0.6)	27 (0.6)	–	–
+NH_4NO_3	–	–	42 (1.0)	29 (0.7)	–	–
Phorate	104 (2.0)	125 (2.4)	–	–	27 (2.0)	35 (1.5)
+NH_4NO_3	190 (3.7)	124 (4.4)	–	–	49 (3.6)	34 (2.5)

Source: Reproduced with permission from the Royal Society of Chemistry [100]

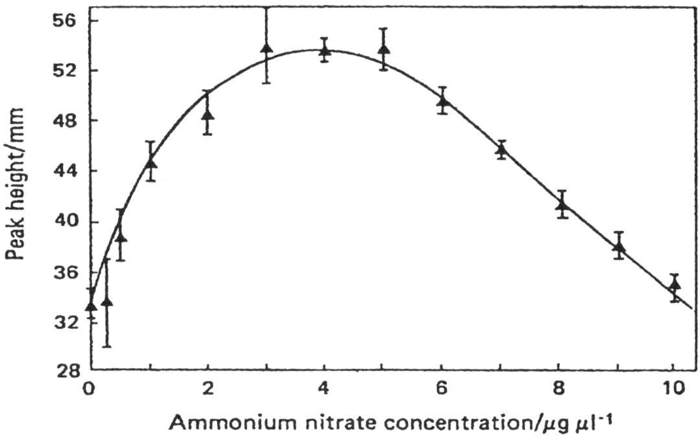

Fig. 9.8 Peak height of 60ng Atrazine standard versus concentration of ammonium nitrate
Source: Reproduced with permission from the Royal Society of Chemistry [100]

enhancement had been induced by injection of ammonium nitrate. The calculated regression line shows a correlation coefficient of 0.95.

Cotterill [100] concluded that the apparent enhancement of response could lead to large errors in the determination of pesticide residues using the peak-height method of measurement. The magnitude could depend on the frequency of injection of standards. Therefore, when a high ammonium nitrate fertiliser concentration is present in pesticide extracts, peak areas are more likely to give accurate values than peak heights. It is advisable that extracts from soils containing high fertiliser levels should be chromatographed using a freshly conditioned column. When ammonium nitrate is present in the soil in sufficient amounts to cause measurement difficulties, its presence in the extract should be avoided by the use of an alternative extraction or partition technique.

9.4.1.2 Gas chromatography–mass spectrometry

The isotope dilution gas chromatography–mass spectrometry method described by Lopez-Avila *et al.* [16] and discussed in section 5.3.1.3 has been applied to the determination of Atrazine in soil. In this method known amounts of labelled Atrazine were specked into soil samples before extraction with acetone–hexane. The ratio of the naturally abundant compound and the stable-labelled isotope was determined by high-resolution gas chromatography–mass spectrometry with the mass spectrometer in the selected ion monitoring mode. Detection limits of 0.1–1.0ppb were achieved. Accuracy was >86% and precision better than 8%.

9.4.1.3 High-performance liquid chromatography

Sanchez-Rasero and Dios [102] have described a high-performance liquid chromatographic method for the determination of Cyanazine in the presence of some normal soil constituents. This method was specific, accurate and precise with a detection limit of 0.253ng Cyanazine, equivalent to 2ul of a 0.1265mg per litre solution. The use of a microbore column, diode-array detector and multichannel integrator was more economical in terms of operating costs and avoided the partial degradation of Cyanazine observed with gas–liquid chromatography. Peat was used in most experiments as the soil constituent which released the greatest quantity of interfering substances in aqueous solutions.

9.4.1.4 Supercritical fluid chromatography

Steinheimer *et al.* [103] used supercritical fluid chromatography to extract Atrazine, diethyl Atrazine and Cyanazine from Canadian cornbelt soils by supercritical fluid extraction with carbon dioxide.

Supercritical fluid extraction with methanol modified carbon dioxide has been applied to the determination of Triazine herbicides in soil [103].

9.4.1.5 Enzyme-based immunoassay

Enzyme-based immunoassay has been applied to the determination of Atrazine (2-chloro-4-ethylamino-6-isopropyl-amino 1,3,5 triazine) residues in soil [104]. In this technique samples of soil (11 different types ranging from clay to sandy) were fortified with 1–80ppb Atrazine and analysed using an enzyme immunoassay (tube system). Reproducibility was good. Percent coefficients of variation ranged from 23.8 to 4.1, with highest values for samples containing 1–2ppb Atrazine. Results were comparable with those obtained using high-performance liquid chromatography. The immunoassay cross-reacted with several other triazine. The reactivity was with the two- and four-position diamine side chains containing the ethyl and isopropyl groups. One advantage of the tube system was its field adaptability.

9.4.1.6 Miscellaneous

Mills and Thurman [105] studied the mixed mode isolation of Triazine metabolites from soils using automated solid phase extraction with methanol:water (4:1u/u) extracts of the sample. Methanol is evaporated from the extract and the metabolites are collected on an octadecyl resin (C_{18}) column. The analytes are eluted with ethyl acetate leaving the impurities on the C_{18} resin column. The detection limit of this method is 0.1µg kg^{-1}.

Studies have been made of the fate of 3-amino 1,2,4 triazole herbicide in soils [106], while adsorption of aminotriazole by clay minerals has been postulated, little is known of the interaction with pure clay minerals, particularly of the montmorillonite group. The importance of such reactions cannot be overemphasized in view of their bearing on the persistence of the herbicide in the soil.

While the high solubility of aminotriazole in water (28g per 100ml at 23°C) suggests ready leaching from whole soil, Russell *et al.* [106] showed if the soil contains a montmorillonite-type mineral, the aminotriazole might be resistant to leaching as a result of adsorption by the montmorillonite.

The 3-aminotriazole molecule is protonated when adsorbed on montmorillonite surfaces to produce the 3-aminotriazolium cation. In the case of montmorillonite saturated with polyvalent cations (Ca^{2+}, Cu^{2+}, Ni^{2+}, Al^{3+}), protonation is believed to be due to the highly polarized water molecules in direct co-ordination to these cations. The decreasing order of extent of protonation (Ca < Mg < Al) reflects the order of decreasing polarizing power of the cations (Table 9.12). Infrared spectra indicate co-ordination of 3-aminotriazole to Ni^{2+} and Cu^{2+} cations. The infrared absorption band at 1696cm $^{-1}$ is assigned to the C=N stretching vibration of the exocyclic C=N $^{+}$HH group. Shifts of the 1696cm $^{-1}$ band to 1683 and 1666cm $^{-1}$ upon dehydration and deuteration, respectively, suggest that the positive charge on the protonated molecule lies on the exocyclic nitrogen. The protonated molecule undergoes normal exchange reactions with other cations.

Table 9.12 3-aminotriazolium ion formed in various cationic saturations of montmorillonite[a]

3-Amino-triazole added	*% Exchange capacity*						
	Exchangeable ion on montmorillonite						
	Na	*NH_4*	*Ca*	*Mg*	*Al*	*Cu*	*Ni*
	3-Aminotriazolium ion formed						
25	6	17			23	3	22
50	6	28	18	29	47	12	46
100	6	48	23	45	~100	32	42
200	6	49	21	42			

Source: Reproduced with permission from the American Chemical Society [106]

[a]Calculation was based on the absorbance of the 1969cm $^{-1}$ band of the 3-aminotriazolium cation using montmorillonite saturated with the cation as the standard. Film weights were standardized either by weight or from the absorbance of a silicate absorption band at 800cm $^{-1}$.

Stransky [107] investigated the possibility of determining the triazine herbicides Atrazine (2-chloro-4-ethylamino-6-isopropylamino, 1,3,5 triazine), Simazine (2-chloro 4,6, bis ethylamino 1,3,5 triazine), Atratone (2-ethyl amino-4-isopropylamino 1,3,5 triazine), Prometryne, Desmetryn and Methoprotryne, also the growth regulator Chlormequat and the quarternery cationic herbicides Paraquat (1,1′ dimethyl-4,4-bipyridinium chloride) and Diquat (1,1′ethylene-2, 2-bipyridinium bromide in soil extracts by capillary isotachophoresis. The more basic triazine could be determined directly using enforced isotachophoresis but very weak triazine bases had to be derivatized by nucleophilic substitution of chlorine by electron-donor or quaternary ammonium groups. Recoveries from soils were 70–80% for quaternary and 90–95% for triazine herbicides. The limit of determination was about 10ug per kg sample.

9.4.2 Non-saline deposited and suspended sediments

9.4.2.1 High-performance liquid chromatography

Recently there has been a growing interest in employing a highly selective analyte-antibody interactions achieved by immunosorbents [108–111]. In the immunosorbent, the antibody is immobilized onto a silica support and used as an affinity ligand to extract the target analyte and other compounds with similar structures from the aqueous sample. In this way, any material not recognized by the antibody is not retained in the immunosorbent while the target analyte remains bound to the antibody, leading to a high selectivity. The development and the evaluation of two immunosorbents for the selective trace solid-phase extraction of phenylurea and triazine herbicides have been presented in previous works [108,109]. Other immunosorbents for the analysis of single pesticides have been described in the literature [112–114].

As an example of this type of application Ferrer *et al.* [115] have described an automated on-line immunosorbent phase extraction method for the analysis of triazine and phenylurea herbicides in sediments. This method is a first application and consists of trace analyte extraction using an immunosorbent column-containing either anti-Atrazine or anti-Chlorotoluron antibodies – combined with a liquid chromatograph by use of an on-line sample preparation system coupled directly to an atmospheric pressure chemical ionization-mass spectrometer in positive mode of operation. After the percolation of 20mL of water through the immunosorbent columns, high recoveries of extraction were obtained for all the compounds with the exception of Deisopropylatrazine and Diflubenzuron. Calibration curves were linear in the range between 0.01 and 0.2μg/L $^{-1}$ in groundwater. The limits of detection ranged from 0.001 to 0.005μg/L $^{-1}$, indicating good sensitivity achieved by both types of immunosorbents. Environmental sediment samples from the Ebre Delta area (Tarragona, Spain) containing several triazines and Linuron were Soxhlet extracted with methanol, and the extracts were brought to a volume of 100mL in water in order to perform the extraction with the immunosorbents. No significant interferences from the sample matrix were noticed, thus indicating a good selectivity of the immunosorbents used.

9.4.2.2 Miscellaneous

Wauchope and Myers [116] studied the adsorption–dispersion kinetics of Atrazine and Linuron in sediment–aqueous slurries. The resulting adsorption or desorption was very rapid, approaching 75% of equilibrium values within 3–6min. Chlorinated adsorption of the herbicide on the sediment was completely reversible after 2h of adsorption.

Mills and Thurmen [105] used a mixed method for the isolation of triazine herbicide metabolites from aquifer sediments using automated solid phases extraction with a mixture of methanol and water (4:1 V/V). Following evaporation of the methanol phases, the metabolites were collected in a column of C18 octadecyl resin. The analytes were then stripped from the column with ethyl acetate leaving impurities on the column. Down to 0.1μg kg^{-1} triazine could be determined.

9.5 Substitute urea herbicides

9.5.1 Soil

Gas chromatography of phenylurea herbicides is difficult because of their ease of decomposition. Procedures have been reported in which careful control of conditions allows these compounds to be chromatographed intact [120,121,127,128]. Alternatively, the urons can be hydrolysed to the corresponding substituted anilines; these compounds are then determined by either gas chromatography directly [122] or as derivatives [123], or colorimetrically [124] after coupling with a suitable chromophore. Methods based on hydrolysis lack specificity and involve lengthy procedures. These disadvantages can be overcome by using liquid chromatography.

Following the earlier work of Kirkland [125] on phenylurea herbicides, Sidwell and Ruzicka [126] applied liquid chromatography to the identification and determination of active ingredient contents of phenylurea herbicide formulations. Smith and Lord [118] have used liquid chromatography for the determination of Chlorotoluron residues in soil, but Diuron and Monuron interfered in their chromatographic system.

9.5.1.1 Thin layer chromatography

Chlorobromuron (3,4 (bromo-3-chloro-phenyl) 1-1 methoxy 1-methylurea)

Katz and Strusz [117] have described gas and thin layer chromatographic methods for the determination of traces of Chlorobromuron and its metabolites. The samples were extracted in a Soxhlet apparatus with 250ml of ethyl acetate for 6h. Each extract was then acidified with 2ml of anhydrous acetic acid and evaporated to 5ml on a steam bath; 15µl portions were applied to the 0.1mm layer of silica gel on an Eastman Chromagram sheet, and the chromatogram was developed twice with hexane–ethyl acetate (15:2). After drying, the chromatogram was developed with chloroform–pyridine (10:1), dried again, and sprayed with Mitchell reagent (20ml of 2-phenoxyethanol added to a solution of 1.7g silver nitrate in 5ml of water, the mixture is diluted with acetone to 200ml, then one drop of 30% aqueous hydrogen peroxide is added), and the spots were located under UV radiation. For gas chromatographic confirmation, standards were treated as above and corresponding unsprayed sample areas were removed, extracted with 5ml of ethyl acetate and evaporated to 1ml; 5µl aliquots were injected on to a glass column (110cm × 6mm) packed with 1.5% of XE–60 on Gas-Chrom Q (80–100 mesh) (previously aged at 240°C for 48h). The column was temperature programmed from 75 to 230°C at 50° per min. Nitrogen was the carrier gas and detection was by flame ionization.

9.5.1.2 Liquid chromatography

Chlorotoluron (3-(3 chlorotoluyl) dimethylurea)

Smith and Lord [118] have used liquid chromatography for the determination of Chlorotoluron residues in soil but report that Diuron and Monuron interfere in the chromatographic system used.

9.5.1.3 Gas chromatography and high-performance liquid chromatography

Diuron (N-(3,4 dichlorophenyl)-N,N dimethyl urea)

Diuron is used for the control of annual weeds in front crops and total weed control in non-crop situations.

Cotterill [119] compared two methods, high-performance liquid chromatography and gas chromatography for the determination of Diuron in soil.

Cotterill [119] used the soil extraction method devised by McKone [120] in which a 25g sample of soil was extracted with 50ml of methanol by shaking on a wrist-action shaker for 1h. The resulting soil slurry was filtered through a Whatman No. 42 filter-paper. For gas chromatography, a 2ml aliquot was evaporated to dryness by gently blowing air and the residue was re-dissolved in 2ml of hexane. For high-performance liquid chromatography a 25ml aliquot was concentrated to about 1ml under reduced pressure while warming in a water-bath at 40°C. The remaining solvent was removed with a gentle stream of dried air and the residue was then re-dissolved in 1ml of the high-performance liquid chromatography eluent.

Gas–liquid chromatography A Pye 104 chromatograph fitted with a nickel-63 electron-capture detector and a 1.5m × 4mm i.d. glass column was used. The conditions employed were as follows: column packing, 5% SE–30 on Chromosorb W HP (80–100 mesh); carrier gas, oxygen-free nitrogen at a flow-rate of 50ml min $^{-1}$; column temperature, 155°C; injector temperature, 250°C; detector temperature, 350°C; attenuation, 10 × 10^2; and pulse width 150µs.

High-performance liquid chromatography A constant-flow pump (HSCP, Bourne End, Buckinghamshire) was connected to a 100mm × 5mm i.d. stainless-steel column packed with Hypersil-ODS (5.5µm mean diameter) (Shandon Southern). Injections were made using a Rheodyne valve and Diuron was measured using a Cecil 212 variable-wavelength ultraviolet monitor set at 250nm and 0.1 absorbance unit for full-scale deflection. Methanol–water (7 + 3) was used as the eluent at a flow-rate of about 0.5ml min $^{-1}$. Standards in the range 5–50ng per 5µl injection gave a linear response, the peak area being determined with a Perkin–Elmer Sigma 10. The optimum absorption wavelength was determined by scanning a Diuron solution between 200 and 300nm prior to chromatography.

Table 9.13 shows the recovery of Diuron from fortified soil using high-performance chromatographic and gas chromatographic determination. There is no practical difference between the means obtained by either method but the variance within the means is greatest for gas chromatographic determination. The results obtained for a Diuron treated field soil showed a mean of 0.13µg g $^{-1}$ and a coefficient of variation of 11.5% for high-performance liquid chromatographic determination and a mean of 0.12µg g $^{-1}$ and a coefficient of variation of 41.7% for gas chromatographic determination. Thus, although the means are almost the same, the variability of the gas chromatographic determination is over three times that of the high-performance liquid chromatographic determination.

Although high-performance liquid chromatography is generally the most reproducible method, gas chromatography has the advantage of being more sensitive. When measuring very low residues in soils with a low organic matter content, gas chromatography could prove to be the better method but the results should be interpreted with caution because of the possible presence of unresolved metabolites. In practice, however,

Table 9.13 Recovery of Diuron from soil

Method	Soil	Diuron added/ μg g⁻¹	Recovery %			Mean recovery %
HPLC	1	1.0	98.8	95.3	102.3	98.8
		0.5	99.9	99.9	92.0	97.2
		0.1	101.4	98.8	96.2	98.8
	2	1.0	101.9	100.0	98.1	100.0
		0.5	101.0	99.4	97.9	99.4
		0.1	104.5	100.9	80.1	94.7
	3	1.0	87.9	102.1	104.9	98.3
		0.5	89.3	92.0	86.6	89.3
		0.1	98.8	93.5	87.3	93.5
GLC	1	1.0	99.6	97.3	94.7	97.2
		0.5	90.1	84.8	96.0	90.3
		0.1	84.7	96.5	96.1	92.4
	2	1.0	102.2	93.2	84.5	93.3
		0.5	96.8	90.5	84.6	90.7
		0.1	73.4	97.5	85.5	85.4
	3	1.0	Not analysable without clean-up			
		0.5				
		0.1				

Some properties of the soils used

Property	Soil 1	Soil 2	Soil 3
Organic carbon %	1.6	4.1	25
pH	7.0	5.1	5.9
Clay %	16	16	56
Silt %	11	16	32
Sand %	73	68	12
Water-holding capacity %	16.6	27.0	38.9

Source: Reproduced with permission from the Royal Society of Chemistry [119]

the limit of detection is set by the signal to background ratio, which is usually similar for both methods, but in the more organic soils high-performance liquid chromatography is favoured. A limit of detection of 0.04μg g⁻¹ can be achieved using either method.

9.5.1.4 Gas chromatography

Miscellaneous urea herbicides (chlorobromuron (3,4(bromo-3-chlorophenyl)-1-methoxy-1-methylurea), Chlorotoluron (3(3 chlorotoluyl) dimethyl urea), Diuron (N-(3,4 chlorophenyl) N,N dimethyl urea), Monolinuron (3(4, chlorophenyl-1-methoxy-1-methylurea), Linuron (3,(3,4 dichlorophenyl)-1-1-methoxy-1-methylurea) Chloroxuron (3,(3,4 dichlorophenyl-1,1 dimethylurea)

Substituted urea herbicide residues have been determined using gas-liquid chromatography by pyrolysis to the phenyl isocyanate in the injection heater [129,130] with electron-capture detection. They have also been determined by high-performance liquid chromatography [126,131–133] and by gas chromatography using a thermionic detector [134]. Lawrence and Laver [135] and Büchert and Lokke [134] have reported methods using gas chromatographic determination after alkylation. Bieser and Grolinmund [137] and Khan *et al.* [121] have devised methods of measuring urea herbicides by gas chromatography without thermal decomposition.

The gas chromatographic procedure described by Cohen and Wheals [80] has been applied to the determination of various substituted urea herbicides in soil in amounts down to 1–50μg kg^{-1}.

Caverly and Denney [138] applied gas chromatography to the determination of a range of substituted urea herbicides in soil. In this method acetone extraction is followed by alkaline hydrolysis, steam distillation and concentration of anilines in toluene, the last three steps being carried out in a single operation using a liquid–liquid extractor. The anilines, after partition into hydrobromic acid, are brominated and determined by gas chromatography with an electron-capture detector. The procedure is sufficiently sensitive for investigations into problems of crop damage and can be applied to a wide variety of soil types. An additional deoxidisation step to remove interference due to the presence of aniline metabolites is also described. The limit of detection is 0.1mg kg^{-1} and recoveries at residual levels are generally better than 80% (Table 9.14).

Fig. 9.9 which was obtained using a Perkin Elmer 452 gas chromatograph shows the separation of 0.5ng amounts of four derivatised herbicides: A, monoLinuron; B, Fenuron; C, Linuron; and D, chlorobromuron.

The separation of Diuron using the Pye 104 chromatographic is shown in Fig. 9.10 and, although greater sensitivity was obtained compared with the much older Perkin–Elmer instrument, both gave satisfactory results with adequate sensitivity. The response due to 0.04ng of Diuron is shown in (a), together with that from a soil extract (b) and a recovery on the same soil (c).

The inclusion of the deoxidisation step effectively removes the anilines without degrading the parent herbicides. When this additional step was

Table 9.14 Recovery of herbicides from soils

Soil type	*pH*	*Organic matter %*	*Herbicide*			
			Name	*Added mg kg^{-1}*	*Found/ mg kg^{-1}*	*Recovery %*
Very fine sandy loam	6.8	2.84	Chlorpropham	0.20	0.17	85
			Chlorpropham	0.20	0.14	70
			Metobromuron	0.20	0.16	80
			Diuron	1.00	1.08	108
			Diuron	0.20	0.16	80
			Chlorburam	1.00	0.85	85
			Propachlor	1.00	0.85	85
			Fenuron	1.00	1.04	104
			Chlorbromuron	1.00	0.91	91
			Monuron	0.20	0.16	80
Silty loam	7.3	2.95	Chlortoluron	1.00	0.99	99
	7.3	4.42	Chlortoluron	1.00	0.84	84
	7.4	4.23	Chlortoluron	1.00	0.84	84
Fine sandy loam	6.3	1.74	Linuron	1.00	0.95	95
	6.3	1.97	Linuron	1.00	0.95	95
	5.3	1.49	Linuron	0.20	0.18	90
Very fine sandy loam	7.7	2.53	Linuron	0.20	0.17	85
	6.7	2.26	Linuron	1.00	1.00	100
Silty loam	7.2	2.00	MonoLinuron	1.00	1.00	100
Calcareous clay loam	7.5	5.35	MonoLinuron	1.00	0.85	85

Source: Reproduced with permission of the Royal Society of Chemistry [138]

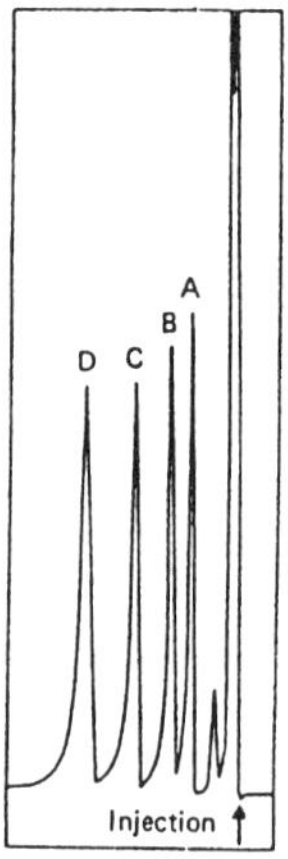

Fig. 9.9 Gas–chromatographic response of a standard mixture of brominated anilines derived from: A monoLinuron; B Fenuron; C, Linuron; and D, chlorobromuron (R_t = 14min). Amount applied, 0.5ng. Perkin–Elmer, Model 452, gas chromatograph
Source: Reproduced with permission of the Royal Society of Chemistry [138]

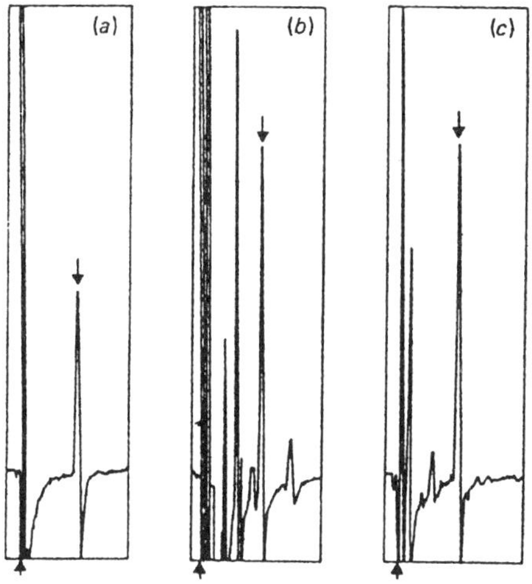

Fig. 9.10 Gas–chromatographic responses of extracts from soils (10g) containing Diuron after hydrolysis and bromination; (a) 0.04ng standard (R_t = 7min); (b) application of 2μl from 10ml of an extract of soil containing 0.04mg kg $^{-1}$; and (c) recovery of Diuron added to soil at 0.8mg kg $^{-1}$, 6μl being applied from a 50-fold dilution of (b). Pye 104 gas chromatography.
Source: Reproduced with permission from the Royal Society of Chemistry [138]

applied to soils from fields treated with these herbicides, only small differences were found compared with the results obtained when it was omitted. Table 9.15 shows the results obtained for Chloropropham. Large differences might be expected in situations that favour reductive anaerobic conditions.

The presence of free anilines or other metabolites in soils and plants has been reported [123,139–143]. Some work has suggested that they are very strongly bound to soil components and the findings of Caverly and Denney [138] are in agreement with these conclusions. The presence in soils of metabolites of Linuron that possess the urea structure have been reported [123,142]; these are produced mainly by microbiological degradation. The dimethyl derivative is considered to be inactive whereas the monomethyl metabolite has phytotoxicity approaching that of the parent herbicide [142]. It is probable that the procedure reported by

Table 9.15 Residues of Chlorpropham found in field-treated soils

Procedure	*Amount found/mg kg* $^{-1}$			
Inclusion of deoxidisation step	0.88	1.04	0.24	1.06
Exclusion of deoxidisation step	0.92	1.09	0.29	1.09

Source: Reproduced with permission from the Royal Society of Chemistry [138]

Caverley and Denney [138] will measure both metabolites. The concentration of the dimethyl compound is unlikely to be large in comparison with that of the parent material. The presence of an aniline as the next product of degradation is, however, effectively removed by introduction of the deoxidisation step.

9.5.1.5 High-performance liquid chromatography

Farrington *et al.* [144] used the method of Sidwell and Ruzicka [126] as the basis for the development of a method for the positive monitoring down to 200μg kg $^{-1}$ of Chlorobromuron, Chlorotoluron, Diuron, Linuron, Monolinuron, Chloroxuron, Monuron and Metobromuron in soils.

The herbicides are extracted from soil samples with methanol. They are chromatographed on microparticulate silica bonded with octadecyltrichlorosilane using a mixture of methanol, water and ammonia as mobile phase and an ultraviolet spectrophotometric detector.

The soil samples are air dried. Transfer 50g into a 500ml flat-bottomed flask and extracted with methanol. Remove the methanol by using a rotary evaporator with a water bath at 55°C. Cool the flask and add 5.0ml of methanol, swirl to dissolve the residue and filter the solution through a Whatman No. 42 filter-paper. Using a flow-rate of 0.6ml min $^{-1}$, inject 5μl of extract into the liquid chromatograph. Calculate the uron content of the sample by comparing the peak height obtained with those obtained from 5μl injections of standard solutions.

The recoveries obtained for urons from samples of soil are shown in Table 9.16. Samples of soil were fortified by adding known volumes of solutions containing (a) Monuron, Metobromuron, Diuron and Chlorbromuron or (b) Monolinuron, Chlorotoluron, Linuron and Chloroxuron.

Table 9.16 Recovery of urons from fortified samples

Five determinations were carried out on each sample. Results given are percentage recoveries

Uron	*Soil, fortified at 2mg kg* $^{-1}$	
	Range	*Mean*
Chlorbromuron	98.0–101.0	99.5
Chlortoluron	100.0–103.0	101.5
Chloroxuron	97.0–103.0	99.0
Diuron	98.0–100.5	97.5
Linuron	94.0–100.5	97.5
Metobromuron	96.5–101.0	99.5
MonoLinuron	100.5–103.0	102.0
Monuron	99.0–102.0	100.5

Source: Reproduced with permission from the Royal Society of Chemistry [144]

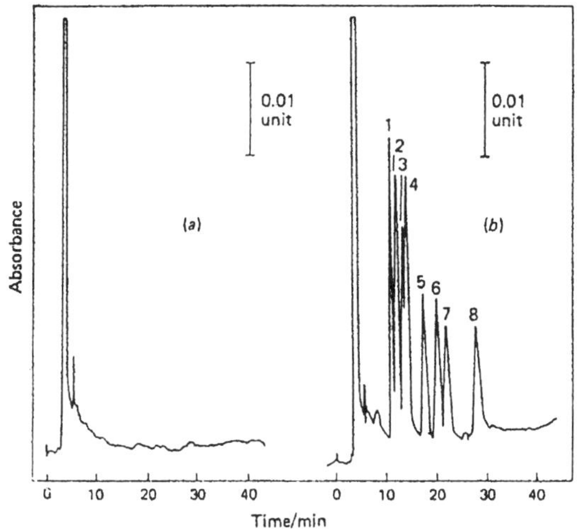

Fig. 9.11 Typical chromatograms obtained from 5μl injections of soil extracts: (a) unfortified; and (b) fortified with uron herbicides at 2mg kg^{-1}. 1, Monuron; 2 monoLinuron; 3 Metobromuron; 4 Chlorotoluron; 5 Diuron; 6 Linuron; 7 Chlorbromuron and 8 Chloroxuron.
Source: Reproduced with permission from Royal Society of Chemistry [144]

Typical chromatograms, obtained from extracts of soil are shown in Fig. 9.11.

9.5.1.6 Miscellaneous

The method based on immunosorbents coupled on-line with liquid chromatography–atmospheric pressure chemical ionization mass spectrometry [109], discussed in section 9.4.2.1, has been applied to the determination of substituted urea type herbicides. Supercritical fluid extraction with methanol modified carbon dioxide has been applied to the determinants of sulfonyl urea herbicides in soil [261].

9.6 Phenoxy acetic acid herbicides

9.6.1 Soil

9.6.1.1 Gas chromatography

MCPA (4-Chloro-2-methylphenoxyacetic acid) and MCPB (4-chloro-2-methylphenoxybutyric acid)

Phenoxyalkanoic acid herbicides are not amenable to direct gas chromatographic determination because of the high polarity or low volatility of the compounds and must be converted to their more volatile derivatives. The sensitivity of the electron capture detector towards alkyl esters of 4-chloro-2-methylphenoxy acetic acid, 4-chloro-2-methylphenoxy butyric acid, etc. is very poor. The methyl ester of 4-chloro-2-methyl phenoxy acetic acid was 100 times less sensitive to electron affinity detection than 2,4-D methyl ester [145].

Chau and Terry [146] reported the formation of penta-fluorobenzyl derivatives of ten herbicidal acids including 4-chloro-2-methyl-phenoxy acetic acid [145]. They found that 5h was an optimum reaction time at room temperature with pentafluorobenzyl bromide in the presence of potassium carbonate solution. Agemian and Chau [147] studied the residue analysis of 4-chloro-2-methyl phenoxy acetic acid and 4-chloro-2-methyl phenoxy butyric acid from water samples by making the pentafluorobenzyl derivatives. Bromination [148], nitrification [149] and esterification with halogenated alcohol [145] have also been used to study the residue analysis of 4-chloro-2-methyl phenoxy acetic acid and 4-chloro-2-methyl phenoxybutyric acid. Recently pentafluorobenzyl derivatives of phenols and carboxylic acids were prepared for detection by electron capture at very low levels [150,151]. Pentafluorobenzyl bromide has also been used for the analytical determination of organophosphorus pesticides [152].

4-chloro-2-methyl phenoxy acetic acid is one of the most effective hormone herbicides. 4-chloro-*o*-cresol (II), 5-chloro-3-methyl catechol (III) and *cis, cis*-4-chloro-α-methyl muconic acid (IV) were first identified as metabolites of 4-chloro-2-methyl-phenoxy acetic acid by Gaunt and Evans [153] (Fig. 9.12).

Renberg [154] has described a method for the determination of 2.4 dichlorophenoxy acetic acid (2.4D) and 2.4.5 tri chlorophenoxy acetic acid (2.4.5T) in soils in which the herbicide is derivativized to its methyl or 2-chloroethyl ester prior to determination by gas chromatography. Recoveries of the two herbicides at the 8–16mg kg^{-1} level in soil were between 70–74% (2.4D) and 84–86% (2.4.5T).

Saltar and Paasivirta [155] have described a method for the analysis in soils of MCPA (4-chloro-2-methyl phenoxy acetic acid) and two of its main metabolites, 4-chloro-*o*-cresol and 6-chloromethyl catechol by gas chromatography of their pentafluorobenzyl derivatives (Fig. 9.12). After derivitization of the residue extract, a clean-up procedure was applied. The best recoveries of compounds from soil were obtained when the extraction was performed by shaking with ether-acetone-heptane-hexane (2:1:1:1) from acidified soil and when the clean-up was done by thin layer chromatography (Table 9.17). Detection limits were in the range 20–25ng absolute.

The mixed solvent ether-acetone–hexane-heptane (2:1.1:1) proved to be an excellent solvent system for the residue analysis of all three compounds with the shaking method. Soxhlet extraction with chloroform, gave quite

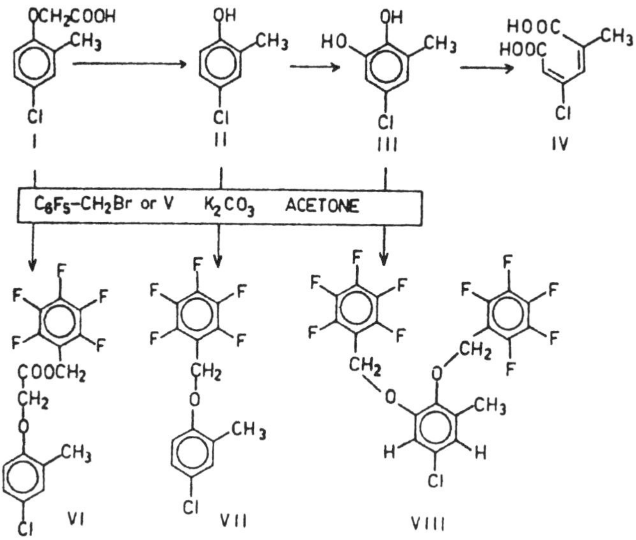

Fig. 9.12 Structures of MCPA (4-chloro-2-methylphenoxyacetic acid (I)), its metabolites 4-chloro-*o*-cresol (II), 5-chloro-3-methyl catechol (III), 4-chloro-2-methyl muconic acid (IV), reagent pentafluorobenzyl bromide (V), and the derivatives VI–VIII from I–III
Source: Reproduced with permission from the American Chemical Society [155]

Table 9.17 Average per cent recoveries of compounds I, II and III mixed (1:1:1) with Finnish (i and ii) and Bangladesh (iii and iv) soils applying different extractions

soil treatment	*solvent*	*extraction*	*compound*	*soil i*	*soil ii*	*soil iii*	*soil iv*
fresh	ether–acetone–hexane–heptane, 2:1:1:1	shaking	I	89.1	95.0	97.2	90.9
			II	82.6	86.3	95.6	89.0
			III	69.2	74.1	80.6	80.1
acidified with H_3PO_4	ether–acetone–hexane–heptane 2:1:1:1	shaking	I	93.1	96.1	100.7	98.5
			II	84.9	91.0	93.5	94.2
			III	73.6	77.8	89.7	86.8
fresh	chloroform	Soxhlet	I	75.1	80.2	85.2	80.9
			II	70.6	78.6	80.5	79.2
			III	60.8	65.6	78.4	75.1
acidified[a] with H_3PO_4	chloroform	Soxhlet	I	92.8	93.6	95.6	93.9
			II	84.1	86.0	90.7	86.7
			III	72.7	73.8	82.1	80.5

Source: Reproduced with permission from the American Chemical Society [155]

[a]Add 3ml IN phosphoric acid to soil sample and leave 15min before solvent extraction

satisfactory results if the soil was acidified with phosphoric acid. This method is most convenient for routine analysis.

Generally, the recoveries were far below 100%. Obviously the extractions were not complete because of the properties of the soils studied. Yip [156] reported that the binding of the soil particles and organic matter with the herbicide residues prevented the complete extraction of them with an organic solvent. Upchurch and Mason [157] found that the extent of adsorption of herbicides is highly dependent on the type of organic matter and clay as well as on the amounts of their constituents in soil. The results (Table 9.17) allow us to compare the possible influence of the characteristics of the four soils studied (Table 9.18) on the recoveries. The results of fresh soil treatments show that the pH has no significant effect in the area 4.6–7.8. Also, one could conclude that the soil texture (especially the differences in clay contents) has no significant influence on the amounts of the residues recovered.

The results support a general conclusion that organic matter is the main factor which influences the fate of herbicides and their analyses in the soil. The sandy and clay loam soils (i, ii) had high but sandy loam and clay soils (iii, iv) very low organic matter contents (Table 9.18). Consequently, significantly higher recoveries of the residues were generally obtained from the latter soil materials than from soils i and ii. In addition, the sandy and clay loam soils (i and ii) gave selectively lower 5-chloro-3-methyl catechol recoveries related to recoveries of 4-chloro-2-methyl phenoxy acetic acid and 4-chloro-*o*-cresol.

Waliszewski and Szynczynski [158] have described a gas chromatographic method for the determination of 4-chloro-2-methyl phenoxy acetic acid (MCPA) and 2.4 chloro phenoxy acetic acid (2.4D) herbicides in soils. The chlorophenoxy acetic acid herbicides were extracted from soil

Table 9.18 Physical and chemical characteristics of Finnish (i and ii, Kemira Co) and Bangladesh (ii and iv) soils

soil types	*sandy clay i*	*clay loam ii*	*sandy loam iii*	*clay iv*
moisture %	13.9	14.8	2.0	2.3
field capacity %	28.1	34.0	12.1	15.5
ignition loss %	33.2	36.0	2.2	2.0
pH	4.6	4.8	6.6	7.5
organic carbon %	10.3	12.2	0.6	0.7
organic matter %	17.8	21.1	1.1	1.2
coarse sand %	19.7	10.8	5.1	4.9
fine sand %	31.1	24.2	20.7	32.7
silt %	30.1	25.3	58.5	18.9
clay %	18.7	39.5	15.7	42.4

Source: Reproduced with permission from the American Chemical Society [155]

samples with dichloromethane. Sulphuric acid was added. The extracts were esterified with 2,3,4,5,6-pentafluorobenzylbromide and extracted with petroleum ether. The levels of MCPA and 2,4–D were determined by gas chromatography with electron capture detection. The detection limits were 0.5µg kg^{-1} and 2µg kg^{-1} 2,4D in soil.

Lopez-Avila *et al.* [159] used isotope dilution gas chromatography–mass spectrometry to determine low levels of µg kg^{-1} 2,4 dichlorophenoxy acetic acid herbicide in soil.

Stable labelled isotopes are spiked into samples before extraction and the ratio of unlabelled compound and stable labelled isotope was used to quantitate the unlabelled compound. Analysis is by high-resolution gas chromatography–mass spectrometry. Fifteen standard water samples and ten standard soil samples containing 2,4–D at known concentrations were analysed. Compound concentrations ranged from 100 to 10 000ug per kg for soil samples. Average recoveries were over 84% and method precision, given as relative standard deviation, was better than 19%.

9.6.1.2 High-performance liquid chromatography

Di Corcia and Marchetti [160] determined chlorinated phenoxy acid and ester type herbicides in amounts down to 1mg kg^{-1} or lower in soil by liquid chromatography combined with particle beam mass spectrometry and ultraviolet absorption spectrometry.

9.6.1.3 High-performance liquid chromatography–mass spectrometry

Kim *et al.* [161] used particle beam mass spectrometry and ultraviolet absorption spectrometry as detectors in a method for the determination of down to 1ppm of chlorinated phenoxy and ester herbicides in soil.

9.6.1.4 Supercritical fluid extraction

Supercritical fluid extraction with methanol modified carbon dioxide has been applied to the determination of acidic herbicides such as chlorophenoxy acetic acids in soil [262].

9.6.2 Non-saline deposited and suspended sediments

9.6.2.1 High-performance liquid chromatography–mass spectrometry

Kim *et al.* [161] determined chlorinated phenoxy and ester herbicides in sediments at the mg k^{-1} level using particle beam mass spectrometry in conjunction with liquid chromatography.

9.6.3 *Sludge*

9.6.3.1 Gas chromatography

Conversion to methyl esters followed by gas chromatography has been used to determine phenoxyacetic acid herbicides in sewage sludge [162].

9.7 Miscellaneous herbicides

9.7.1 *Soil*

Diazinon (diethyl-2-isopropyl-6-methyl-4-pyrimidinyl phosphorothioate or 0,0-diethyl-o-(2-isopropyl-6-methyl-4-pyrimidyl phosphorothioate)

9.7.1.1 Gas chromatography–mass spectrometry

An isotope dilution gas chromatography–mass spectrometric method [16] has been applied to the determination of Diazinon at the 0.1–1μg kg^{-1} level in soils. Accuracy exceeds 86% while precision is better than 8%.

Picloram (4-amino 3,5,6 trichloropicolinic acid)

9.7.1.2 Gas chromatography

Abbott *et al.* [163] described a pyrolysis unit for the determination of Picloram and other herbicides in soil. The determination is effected by electron capture–gas chromatography following thermal decarboxylation of the herbicide. Hall *et al.* [164] reported further on this method. The decarboxylation products are analysed on a column (5mm i.d.) the first 15cm of which is packed with Vycor chips (2–4mm), the next 1.05m with 3% of SE–30 on Chromosorb W (60–80 mesh) and then 0.6m with 10% of DC–200 on Gas Chrom Q (60–80 mesh). The pyrolysis tube, which is packed with Vycor chips, is maintained at 385°C. The column is operated at 165°C with nitrogen as carrier gas (110ml min^{-1}). The method when applied to ethyl ether extracts of soil gives recoveries of 90 ± 5%. Dennis *et al.* [165] have reported on the accumulation and persistence of Picloram in bottom deposits.

Acarol (isopropyl-4,4′ dibromobenzylate)

9.7.1.3 Gas chromatography

Cannizzara *et al.* [166] have carried out gas–liquid radio chromatography of this ^{14}C herbicide present as residues in weathered soil.

Imidazolinones

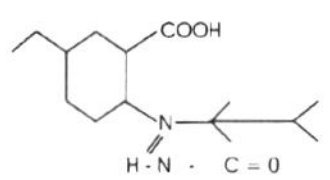

The imidazolinones are a relatively new class of herbicides used to control a wide spectrum of broad-leafed weeds and grasses in a variety of agricultural commodities [167]. These herbicides are very potent weed killers and are used in doses that are substantially lower than those of conventional herbicides. The members of this class of herbicides have similar structural features centred around the imidazolinone ring and an attached aromatic system bearing a carboxylic acid moiety. Imidazolinones have an excellent activity against annual and perennial grasses and broad-leafed weeds when applied either pre- or post-emergence. They function by inhibiting acetohydroxy acid synthase, the feedback enzyme in the biosynthesis of branched-chain essential acids [168–170]. This enzyme is not present in animals. The imidazolinone ring of herbicides is amphoteric and can behave as a weak base or a weak acid. The movement of the acid imidazolinones in the soil can be strongly influenced by many soil properties, the most important of which are pH, organic matter and clay content. Binding of the acid imidazolinones increases as pH decreases. Basic herbicides protonate and are adsorbed on negatively charged soil colloids. Acidic herbicide anions also become protonated as pH decreases, reducing the repulsive forces present when the molecule is dissociated, thus increasing molecular adsorption [171–175]. The typically low application rates used for imidazolinones herbicides make their clinical analysis difficult.

9.7.1.4 Gas chromatography–mass spectrometry

Imazethapyr herbicide has been determined at the μg kg^{-1} level in soil by microwave assisted extraction using electron capture negative chemical ionization mass spectrometry [176]. A soil extract is prepared by extraction with 0.1M ammonium acetate at pH10 or 1M ammonium acetate at pH9.5.

Lagana *et al.* [163] determined imidazolinone herbicides in soil using soil column extraction followed by liquid chromatography with ultraviolet detection or liquid chromatography–electroscope mass spectroscopy. The herbicides studied included Imazapyr, *m*-Imazamethabenz, *p*-Imazamethabenz, *m,p*-Imazamethabenzmethyl, Imazethapyr and Imazaquin. Soil sample analysis utilized combined soil column extraction and off-line solid phase extraction for sample preparation, analyzing with liquid chromatography–electroscope mass spectroscopy under selected ion monitoring. Several different extractants were evaluated for the purpose of soil column extraction optimization. The system that best optimizes the extractibility imidazolines from the soil was found to be the mixture $CH_3OH/(NH_4)_2CO_3$ (0.1M, 50:50 *v/v*). The effect of imidazoline concentration in the matrix on recovery was evaluated. The total recovery of each imidazoline from soil at each of the two levels investigated ranged from 87–95%. Under three ion selected ion monitoring conditions, the limit of detection (S/N = 3) was 0.1–0.05ng/g in soil samples.

Dicamba (2-methoxy 3,6 dichlorobenzoic acid)

9.7.1.5 Gas chromatography–mass spectrometry

Lopez-Avila *et al.* [159] have described a method for the determination of low parts per billion concentrations of Dicamba and 2,4–D in soil. Stable-labelled isotopes are spiked into samples before extraction and the ratio of unlabelled compound and stable-labelled isotope was used to quantitate the unlabelled compound. Analysis is by high-resolution gas chromatography–mass spectrometry. Ten standard soil samples containing Dicamba and 2,4–D at known concentrations were analysed. Compound concentrations ranged from 100 to 10 000ug per kg for soil samples. Average recoveries were over 84% and method precision, given as relative standard deviation, was better than 19%.

2,6 dichlorobenzonitrile

9.7.1.6 Gas chromatography

Herzel [177] has described a procedure for the determination of this herbicide, a component of dichlorophenyl based on extraction of the soil with toluene and analysis of the extract by gas chromatography using an electron capture detector. $\mu g\ kg^{-1}$ levels can be determined with a reproducibility of 92–93%.

Paraquat (1,1′-dimethyl-4,4-bipyridinium chloride) and diquat (1,1′-ethylene-2,2-bipyridinium bromide)

9.7.1.7 Gas chromatography

Calderbank and Yuens [178] and Pope and Benner [179] have described a spectrophotometric method for the determination of paraquat in soil in amounts down to 0. $mg\ kg^{-1}$. Paraquat, Trifluralin and Diphenamid have also been determined gas chromatographically in soil.

To determine Paraquat in soil Payne *et al.* [180] separate the sediment from the sample (2L) by adding calcium chloride to aid flocculation, leaving the mixture overnight in a refrigerator for the sediment to settle, then decanting and filtering through a Whatman No. 42 paper under suction on a Buchner funnel. The wet sediment and soil core samples, are mixed for 4h with dichloromethane in a Soxhlet extractor to remove Trifluralin and Diphenamid. A 1L aliquot of the filtrate is extracted with dichloromethane (100, 50, 50, and 50ml). The dichloromethane extracts are concentrated by evaporation and the Trifluralin and Diphenamid are determined by direct injection, without further purification, on to a glass column (180cm × 6mm o.d.) packed with 10% DC–200 on Gas Chrom Q and operated at 220°C with helium as carrier gas (100ml min^{-1}) and a Coulson electrolytic–conductivity detector (N mode). Paraquat is determined in the dried sediment and core samples, and in a 500 ml aliquot of the filtrate, by a modification of a conventional colorimetric method. Recoveries of the three substances were between 86–96% from soil and 82–95% from water.

Khan [181] has described a method for determining Paraquat and Diquat in soils involving catalytic dehydrogenation of the herbicide followed by gas chromatography and also a pyrolytic method [182].

9.7.1.8 Enzyme-based immunoassay

Niewola *et al.* [183,185] have described a rapid, convenient and accurate method, based upon an enzyme-based immunosorbent assay (ELISA) for the determination of Paraquat residues in soil. Polystyrene plates, coated with paraquat-keyhole limpet haemocyanin (KLH) conjugate, are incubated with the test samples and a known amount of monoclonal antibody. Residual antibody that has not reacted with free Paraquat in the sample combines with paraquat-KLH on the plate. The determination of the fixed antibody is achieved by the addition of peroxidase labelled rabbit anti-mouse immunoglobulin G followed by reaction with a chromogenic substrate. The enzyme activity of the solid phase is determined from the absorbance measurements, which are inversely proportional to the concentration of Paraquat. The method shows high specificity and correlates well with the traditional ion exchange–spectrophotometric method for the determination of Paraquat [178].

In this method the keyhole limpet haemoglobin conjugate was prepared as follows: Keyhole limpet haemocyanin (KLH, Calbiochem, La Jolia, CA) and bovine serum albumin (BSA, BDH Chemicals) were coupled to the adduct (2), derived from 6-bromohexanoic acid and monoquat (3), via a carbodiimide reaction, as reported previously by Niewola *et al.* [184]. The resulting conjugates contained 662mol of Paraquat per mole of KLH and 15mol of Paraquat per mole of 6-bromohexanoic acid. The amount of Paraquat bound to the protein was determined by spectrophotometric dithionite assay for Paraquat and the protein concentration was established by a standard Lowry test.

The preparation of the monoclonal antibody coupled to Paraquat-6-bromohexanoic and has been described by Niewola *et al.* [185] and Kohler *et al.* [186].

Table 9.19 Analysis of fortified soil extracts using the ELISA

	Paraquat in soil/mg kg $^{-1}$								
Added	0	10	20	50	100	150	200	250	300
Found	0	8.5	20	37	90	141	204	247	287

Source: Reproduced with permission from the Royal Society of Chemistry [183]

The suitability of the ELISA for soil analysis was initially tested by assaying a number of control soil samples, fortified after extraction and neutralisation with Paraquat in the range 10–300mg kg $^{-1}$. The results in Table 9.19 were close to the expected values and thus confirmed that natural soil components did not interfere with the determination. These results justified the further refinement of the method for soil analysis.

The limit of detection of the assay for soil extracts is 0.02µg ml $^{-1}$, which corresponds to 0.2mg kg $^{-1}$ of Paraquat in soil for a typical extraction procedure. This level of sensitivity is satisfactory for virtually all determinations of Paraquat in soil and lower limits of detection are rarely required.

Details of the ion exchange spectrometric method [178] for determining paraquat, discussed above, are given below.

A 25g homogeneous representative sample of air-dried soil was refluxed for 5h with 100ml of 6m sulphuric acid. The digest was filtered and percolated through a column of cation-exchange resin (Duolite C225–SRC14), which retained the Paraquat and some of the natural soil constituents. The column was washed successively with 2M hydrochloric acid, 2.5% M/V ammonium chloride solution and water and the Paraquat was finally eluted with saturated ammonium chloride solution. A portion of the column eluate was reacted with sodium dithionite in alkali. The quantitative measurement of Paraquat residues was achieved by differential spectrophotometric measurement of light absorption over the range 430–360nm by the single electron free radical derived from Paraquat, and subsequent comparison with a calibration graph. In this method recovery values were typically in the range 85–100% and where necessary the raw data were corrected, for recovery.

9.7.1.9 Isotachophoresis

Stransky [107] used isotachophoresis to determine Paraquat and Diquat in soils in amounts down to 10µg kg $^{-1}$.

Frenock (sodium 2,2,3,3 tetra fluoro propionate) and Dalapon (sodium 2,2 Dichloropropionate)

9.7.1.10 Mass spectrometry

Tsukioka *et al.* [187] determined these contact herbicides in soil by mass fragmentography. The method is based on the reaction of 1-benzyl-3-*p*-polytriazene with an extract of Frenock and Dalapon from strongly acidified sample solutions to form benzylated species. In the analysis of soil samples, steam distillation was applied prior to extraction. Recoveries were >92% and precision <5%.

Bromacil (5-bromo-3-sec-butyl-6-methyluracil), Lenacil (3-cyclohexyl-5,6-trimethyleneuracil) and Terbacil (5-chloro-3-tert-butyl-6-methyluracil) herbicides

9.7.1.11 Gas chromatography

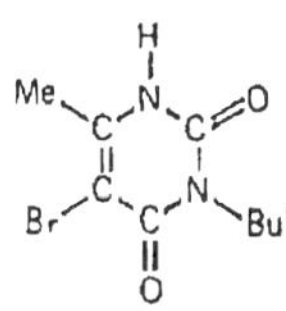

I

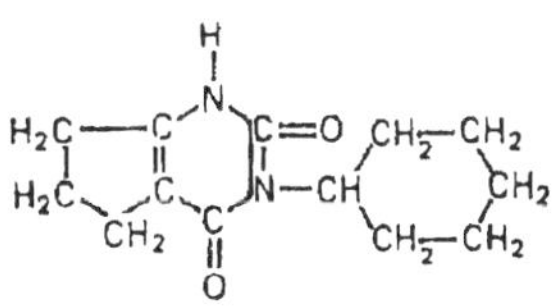

II

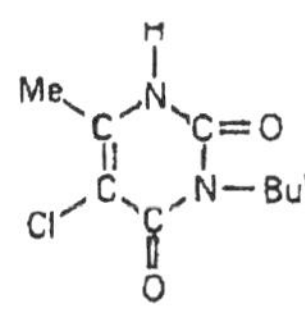

III

Bromacil, Lenacil and Terbacil are widely used herbicides for the control of annual and perennial weeds in fruit and vegetable crops. Procedures [188–191] for the determination of Bromacil and Terbacil residues in soils using electron-capture gas chromatography have been described but these procedures involve time-consuming clean-up techniques to remove interfering co-extractives. Pease [192] described a flame-ionisation gas-chromatographic procedure for the determination of Lenacil residues in soils but this method lacks sensitivity and selectivity at the normal levels found in soils. Micro-coulometric gas chromatography [193,194] has been applied successfully to Bromacil residues alone in soils and fruit, but technique does not possess the simplicity of more standard forms of gas chromatography and has not been applied to residues of Lenacil. Thin-layer techniques [195,196] have been used principally for confirmation of the identity of residues found by gas chromatography; they lack the sensitivity for application to normal soil residue levels. Maier-Bode and Riedmann [197] reported comprehensively on the use of the rubidium bromide thermionic detector for the detection of many nitrogen– and phosphorus-containing pesticides; they did not apply their findings to determination of residues in soils. Jarcjyk [198] has applied the nitrogen-selective detector to the determination of Bromacil residues in soils. The procedure uses relatively large volumes of solvents and extracts require a column chromatographic clean-up.

Calverly and Denny [199] have described a rapid and sensitive procedure for the determination of residue levels of three uracil herbicides in soils. After addition of calcium hydroxide and Celite to the soil the herbicides are eluted from columns with water. After acidification of the eluate and partition into chloroform these herbicides are determined by gas chromatography using a nitrogen-selective detector. Recoveries from a range of soil types are better than 80%, with a sensitivity limit of 20μg kg $^{-1}$.

The recovery of these herbicides was checked by adding known volumes of standard solutions to 10g portions of the ten air-dried soil types followed by removal of solvent by a gentle stream of air. The soils were allowed to stand for 24h and then treated as described above. The results obtained are shown in Table 9.20. Blank determinations carried out on these soils showed that any herbicide present was below the limit of detection.

Diclofop-methyl ($G_5H_{15}C_{12}$) and Diclofop ($C_{15}H_{11}O_4C_{12}$)

9.7.1.12 Gas chromatography

Diclofop-methyl is an early post-emergence herbicide applied to cereals, oil seeds and a variety of other crops for the control of annual grasses. Both Diclofop-methyl and its hydrolysis product, diclofop, are herbicidal [200]; therefore, analysis for Diclofop-methyl in soil includes the determination of both chemicals.

Most laboratory and field studies on Diclofop-methyl persistence in soil have employed radiolabelled chemicals [201–203]. Gas-liquid chromatography has been used in soil analysis to detect Diclofop-methyl directly and diclofop after methylation with diazomethane [204,205]. Although Diclofop-methyl has good chromatographic properties, separate solvent mixtures were recommended for extraction of the ester and acid to reduce interference near the eluting analyte by coextractives [205]. Also, care must be used in the preparation of diazomethane [205], which must be prepared fresh because of its poor storage properties [206]. Pentafluorobenzyl bromide has been used as a derivatising agent for alkylating chemicals, such as carboxylic acids, which have a reactive hydrogen [205]. The response of an electron-capture detector to the resulting derivative is usually enhanced and the retention time increased so that interfering coextractives are less important [206].

Gaynor and MacTavish [207] have described a further sensitive gas chromatographic procedure for the determination in soil of Diclofop-methyl and its hydrolysis product diclofop.

Following extraction of the herbicide and its hydrolysis product from soil with methanol:water:ethyl acetate–acetic acid (40+40+19+1) solution

Table 9.20 Recovery of herbicides added to soil samples

Soil no.	*Bromacil* *Added/ mg kg*$^{-1}$	*Found/ mg kg*$^{-1}$	*Recovery %*	*Lenacil* *Added/ mg kg*$^{-1}$	*Found/ mg kg*$^{-1}$	*Recovery %*	*Terbacil* *Added/ mg kg*$^{-1}$	*Found/ mg kg*$^{-1}$	*Recovery %*
1	5.0	4.3	86	5.0	4.8	96	5.0	4.1	82
	1.0	0.76	76	1.0	0.8	80	1.0	0.7	70
	0.2	0.18	90	0.2	0.2	100	0.2	0.18	90
2	5.0	4.7	94	5.0	5.0	100	5.0	4.8	96
	1.0	0.88	88	1.0	0.96	96	1.0	0.9	90
3	5.0	3.7	74	5.0	4.2	84	5.0	4.0	80
	1.0	1.02	102	1.0	1.10	110	1.0	1.0	100
4	5.0	4.2	84	5.0	4.8	96	5.0	3.7	74
	1.0	0.74	74	1.0	0.8	80	1.0	0.8	80
5	5.0	5.0	100	5.0	4.9	98	5.0	4.4	88
	1.0	0.86	86	1.0	0.8	80	1.0	0.98	98
6	5.0	5.0	100	5.0	5.1	102	5.0	5.0	100
	1.0	0.88	88	1.0	1.0	100	1.0	0.88	88
7	1.0	0.86	86	1.0	0.96	96	1.0	0.96	96
	0.2	0.2	100	0.2	0.21	105	0.2	0.19	95
8	1.0	0.86	86	1.0	0.86	86	1.0	0.86	86
	0.2	0.17	85	0.2	0.17	85	0.2	0.17	85
9	5.0	3.6	72	5.0	4.7	94	5.0	3.6	72
	1.0	0.8	80	1.0	0.96	96	1.0	0.96	96
10	5.0	4.3	86	5.0	5.0	100	5.0	4.7	94
	1.0	0.7	70	1.0	0.8	80	1.0	0.82	82

Characteristics of soils studied

Soil number	*Texture*	*pH*	*Organic matter %*
1	Very fine sandy loam	7.6	2.55
2	Loam	5.1	2.78
3	Calcareous silty loam	7.6	3.57
4	Very fine sandy loam	7.7	4.11
5	Fine sandy loam	6.5	2.37
6	Sandy loam	4.4	2.56
7	Very fine sandy loam	7.2	2.51
8	Clay loam	7.6	5.76
9	Sandy loam	5.7	3.82
10	Very fine sandy loam	7.5	2.24

Source: Reproduced with permission from Royal Society of Chemistry [199]

and back extraction of this phase with 1.5% sodium chloride 5% sodium carbonate aqueous solution, the alkaline phase is acidified, then extracted with hexane.

The next stage is hydrolysis to Diclofop-methyl in the extract to diclofop. This is achieved by heating the residue obtained by removal of hexane with methanoic potassium hydroxide at 60°C, then acidification and extraction with hexane. The hexane extract is treated with pentafluro-benzyl bromide to convert Diclofop to its pentafluorobenzyl derivative. Gas chromatography of the extract was carried out using a Varian 3700 gas chromatograph equipped with a nickel-63 electron-capture detector and a 1.2m × 3mm i.d. glass column packed with 60–80-mesh Gas-Chrom Q coated with a mixture of 5% GE XE–60 and 0.2% Epon Resin 1001. The operating conditions were as follows: inlet temperature, 240°C; column temperature, 240°C; detector temperature, 330°C; and carrier gas, argon–methane (95+5) at a flow-rate of 30ml min^{-1}.

The procedure provides a sensitive derivative for analysis of Diclofop-methyl and of its hydrolysis product by gas–liquid chromatography. Standards of the derivative, stored in hexane at room temperature, were stable for at least one month. The method is rapid and permits the detection of 0.05mg kg^{-1} of diclofop in soil with minimum interference from coextractives. The alkylation product also provides a second method for verification of diclofop by gas–liquid chromatography.

Two soils containing respectively 1% and 3.6% organic matter were spiked with 0.1 and 1.0mg kg^{-1} Diclofop methyl or diclofop. The extraction efficiency for these two soils fortified at 0.1mg kg^{-1} averaged 81 ± 7% and 76 ± 6% for the ester and acid, respectively. For soil fortified at 1mg kg^{-1} the extraction efficiency was 91 ± 4% and 100 ± 2% for the ester and acid, respectively. Chromatograms of untreated and field-treated soils had minimum interference at the retention time of pentafluorobenzyl diclofop, obviating the need for separate extraction solvents for the ester and acid. Chromatograms of soil extracts for the diclofop fraction showed an interference peak (this was relatively small). The nature and intensity of the interference will depend on the properties of the soil being analysed. Interference from coextractives is highest soon after diclofop-methyl application.

Fluazifop-butyl (2-(4-5, trifluoromethyl 2-pyridyloxy)phenoxy)propionate) and Fluazifop

9.7.1.13 High-performance liquid chromatography

Fluazifop-butyl, a selective grass herbicide, and its corresponding acid Fluazifop, have been extracted from soil samples [208]. The compounds were extracted from soil samples using methanol/hydrochloric acid mixtures with and without dichloromethane, and determined by liquid chromatography. The 1:1 *v*/*v* methanol/hydrochloric acid mixtures were not effective in removing Fluazifop-butyl from soil. Both compounds were effectively removed using an extraction mixture of 90% methanol, and 10% of 1m hydrochloric acid to prevent hydrolysis of Fluazifop-butyl. Fluazifop-butyl and Fluazifop had two absorbance maxima at 225nm and 270nm. The latter wavelength was preferred because of the lesser effect of interfering substances in the extracts from soil.

Imugen (N-formyl-N'-3, 4-dichlorophenyltrichloracetaldehydaminal)

9.7.1.14 Miscellaneous

Soil to which Imugen had been applied contained two Imugen metabolites, 3,4-dichloroaniline and 3,3′,4,4-tetrachloroazobenzene [209].

Propanil (3,4-Dichloropropionaniline and 3,4-dichloroaniline)

9.7.1.15 Miscellaneous

Application of either 3,4-dichloroaniline or propanil to soil resulted in production of the metabolite 3,4-dichloroacetanilide, which was identified by infrared spectroscopy and gas chromatography–mass spectrometry [210].

Sencor (Metribuzin, 6-t-butyl 1,2,4-triazine-3-methylthio-2-one)

9.7.1.16 Gas chromatography

This herbicide has been determined in soil by gas chromatography [211].

Trifluralin (2,6-dinitro-N, N-di-n-propyl-α,α,α-trifluoro-p-toluidine) and benefin (N-(n-butyl)-N-ethyl-2, 6-dinitro-α,α,α-trifluoro-p-toluidine)

9.7.1.17 Gas chromatography

Downer *et al.* [212] reported comparable sensitivity with electron capture gas chromatography and specific ion monitoring of characteristic ions of residues of these compounds in soils. For both herbicides, the detection limit was reported to be 50pg, but less clean-up was required for specific ion monitoring than for electron capture gas chromatography.

Cyperquat (1-methyl-4-phenyl pyridinium chloride)

9.7.1.18 Gas chromatography–mass spectrometry

Cyperquat, a post-emergence herbicide has been determined in surface soil by a method involving catalytic hydrogenation to 1-methyl-4-cyclohexypiperdine and analysis by gas chromatography–mass spectrometry. Recovery of cyperquat from fortified soil samples was 77% at the 0.5mg kg $^{-1}$ level and 85% at the 1mg kg $^{-1}$ level [213].

Dacthal

9.7.1.19 Supercritical fluid chromatography

Supercritical fluid extraction is an attractive analytical technique for recovering organic compounds from soils and sediments. Carbon dioxide is currently the fluid of choice, due to its low toxicity and environmental acceptability. The physicochemical properties of supercritical fluids, including low viscosity, variable solvent strength, and high diffusivity, contribute to faster extractions compared to conventional extraction techniques, such as Soxhlet extraction or sonication.

Field *et al.* [221] have described an interesting procedure in which Dacthal and its mono- and diacid-metabolites were sequentially extracted from soils by first performing a supercritical carbon dioxide extraction to recover Dacthal, followed by a subcritical (hot) water extraction step to recover metabolites. Dacthal was recovered from soil in 15min by supercritical carbon dioxide at 150°C and 400bar. The mono- and diacid metabolites were extracted from soil in 10min under the subcritical water conditions of 50°C and 200bar. The metabolites were trapped *in situ* on a strong anion-exchange disk placed over the exit frit of the extraction cell. Metabolites are combined with Dacthal by placing the disk into the gas chromatograph autosampler vial containing the supercritical fluid extract. The metabolites then are simultaneously eluted from the disk and derivatized to their ethyl esters by adding 100µL of ethyl iodide and heating the vial at 100°C for 1h. Using this approach, only a single sample is analyzed, and because the disk-catalyzed alkylation reaction does not transesterify Dacthal, the speciation of Dacthal is maintained. In addition, no sample cleanup steps are required, the use of diazomethane for derivatization is avoided, and the method consumes a total of 5mL of non-chlorinated organic solvent.

9.7.1.20 Miscellaneous

Dacthal is a widely used pre-emergent herbicide that is applied to many crops for the control of annual weeds. Dacthal is typically applied to agricultural soils at 6–14kg ha [214]. In the soil environment, Dacthal transforms to mono- and diacid-metabolites that are more water soluble than the parent herbicide [215–217]. In eastern Oregon, where Dacthal is applied to onions, the diacid metabolite is the principal form of Dacthal detected in groundwater obtained from domestic wells [218,219].

To assess the fate of Dacthal that is applied to soil, both parent and metabolite forms in water and soil should be considered. While rapid methods exist for the determination of Dacthal and its metabolites in water [218,219], quantitative and rapid methods are needed to determine Dacthal and its metabolites in soils, since conventional methods require large volumes of solvent and time to process the extract. For example, the conventional method for extracting Dacthal and its metabolites from soil requires 200mL of 0.4M HCl/acetone to extract a 20g sample and the use of hazardous diazopropane to derivatize the acids to their ester forms [220].

9.7.2 *Non-saline deposited and suspended sediments*

9.7.2.1 Miscellaneous

Goldberg [48] has developed equations relating pesticide concentration in water, pesticide concentration in sediment and surface area of sediment; they can be use to give an approximate estimation of the concentration of pesticides in water at concentrations below those measurable with available analytical equipment.

9.8 Multi insecticide/herbicide mixtures

9.8.1 *Soil*

9.8.1.1 Gas chromatography

Cotterill [222] has developed a procedure, discussed below, in which the herbicides are extracted from soil with saturated calcium hydroxide solution. After clean-up the residues are ethylated using iodoethane and tetrabutylammonium hydrogen sulphate as counter ion. Liquid–liquid partition and the use of a macroreticular resin column were compared as clean-up steps and the reaction conditions for optimum yield of ethyl ester were evaluated.

The herbicides are estimated in the extract by electron capture gas chromatography. Both of the clean-up methods used are capable of giving good results at the 1μg g^{-1} level but XAD–2 is inadequate for smaller amounts of herbicides in soil. This poor clean-up and the higher variability in recovery shows XAD–2 to be unsuitable for residue determinations.

Recoveries in excess of 80% were achieved for 2,4–D, Dicamba, 3,6-dichloropicolinic acid, Dichloroprop, Picloram, 2,4,5–T Fenoprop, 2,3,6–TBA, Bromoxynil and Ioxynil (Table 9.22).

The practical limit of determination for these herbicides is between 0.01 and 0.05μg g^{-1} depending on the background response from the soil extract.

Table 9.21 gives the yields, times to maximum yield, retention times and least detectable amounts of the herbicide esters or ethers prepared using the above method. In no instance was the standard error of the mean yield >2%. The least detectable amount is based on a peak giving a response of twice the background signal.

Gambrell *et al.* [223] have discussed the recovery of DDT, Kepone and Permethrin added to soil suspensions incubated under controlled redox potential and pH conditions to determine the effect of time on the levels of the insecticides and their degradation products. Samples were analysed by gas chromatography, pH and redox potential affected the persistence of pesticides to different degrees. The recovery of DDT was affected by redox potential but not by pH.

Table 9.21 Yields, times to maximum yield, retention times and least detectable amounts of alkylation product

Herbicide	*Yield*	*Time to maximum yield/min*	*Retention time/min*	*Least detectable amount/ng*
2,4–D	95.9	37	3.75	0.05
Dicamba	99.0	56	3.04	0.005
3,6-Dichloropicolinic acid	4.5	60	2.16	0.01
Dichloroprop	100	60	5.52	0.05
Picloram	99.6	15	14.57	0.002
2,4,5–T	90.2	45	5.76	0.005
Fenoprop	94.9	17	6.40	0.005
2,3,6–TBA	92.8	20	3.03	0.005
Bromoxynil	98.8	45	3.88	0.05
Ioxynil	91.2	15	8.52	0.05

Source: Reproduced with permission from the Royal Society of Chemistry [222]

The stability of Kepone (chlorodecone) was not affected by pH or redox potential while Permethrin stability was affected by both parameters.

Kavetskii *et al.* [224] developed a method for the simultaneous determination of pesticides in soil. A combination of thin layer chromatography and gas chromatography was used. The pesticides examined were 4,4′ DDT, 4,4′ DDD, 4,4′ DDE, 2,4′ DDT, γGHCG, αGHCG, Metaphos, Phosphamidon, Phozalone, Atrazine, Prometryne, Simazine and 2,4 dichlorophenoxy acetic acid. Detection limits were in the range 0.5–5μg kg^{-1}.

Acid herbicides such as 2,4 dichlorophenoxy acetic acid, 2,4,5-trichlorophenoxy acetic acid, 3,6-dichloropicolinic acid and other types of herbicides such as Dicamba, Dichloroprop, Picloram, Fenoprop, 2,3,6–TBA, Bromoxynil and Ioxynil are widely used in agriculture and are often formulated as mixtures. They may also be mixed in the spray tank or used in sequence, so it is likely that residues of more than one of these compounds may be present in the soil.

Many methods have been reported for the extraction of these compounds from soil. Khan [225] used acidified acetone followed by methylation with diazomethane for the simultaneous determination of 2,4–D, Dicamba and Mecoprop residues, as did Bache and Lisk [226] for Ioxynil. Abbott *et al.* [227] developed a method for MCPA, MCPB, 2,4–D Dichloroprop and 2,4,5–T in which dilute sulphuric acid and diethyl ether were used for extraction. Byast *et al.* [101] have shown that diethyl ether-chloroform–acetic acid is a suitable extractant for 2,4,5–T, 2,4–D, Dichloroprop and Dicamba [101] and saturated calcium hydroxide solution is efficient for Picloram [228] and 3,6-dichloropicolinic acid [229].

Table 9.22 Recovery of herbicides from soil using alternative clean-ups

Herbicide	*Treatment* $\mu g\ g^{-1}$	*Mean recovery %** *Partition* *Soil 1*	*Soil 2*	*XAD–2†* *Soil 1*	*Soil 2*
2,4–D	1.0	92.9 (2.0)	90.0 (19.6)	76.8 (6.7)	56.3 (9.6)
	0.1	94.7 (2.9)	100.0 (<0.1)	– –	– –
Dicamba	1.0	95.3 (0.1)	100.0 (3.4)	110.4 (3.6)	100.4 (13.7)
	0.1	95.3 (3.2)	97.6 (0)	– –	– –
3,6-Dichloropicolinic acid	1.0	96.7 (5.4)	91.3 (5.2)	66.7 (32.2)	21.6 (23.1)
	0.1	93.2 (5.2)	83.0 (4.7)	– –	– –
Dichloroprop	1.0	80.2 (2.4)	87.2 (2.0)	96.4 (3.6)	77.1 (7.2)
	0.1	88.6 (2.6)	94.0 (2.1)	– –	– –
Picloram	1.0	99.2 (4.4)	99.2 (2.2)	62.7 (6.8)	68.1 (2.4)
	0.1	97.8 (2.2)	98.8 (<0.1)	– –	– –
2,4,5–T	1.0	103.8 (2.7)	100.0 (3.0)	88.9 (6.97)	73.5 (5.9)
	0.1	95.2 (2.5)	98.2 (2.9)	– –	– –
Fenoprop	1.0	90.0 (1.7)	91.3 (1.7)	93.6 (5.7)	92.8 (15.4)
	0.1	84.7 (0.1)	93.4 (1.7)	– –	– –
Bromoxynil	1.0	90.6 (3.2)	88.6 (3.1)	79.3 (6.2)	48.4 (13.6)
	0.1	92.3 (3.1)	93.4 (3.2)	– –	– –
Ioxynil	1.0	98.0 (2.9)	81.1 (2.3)	51.4 (9.7)	38.9 (15.0)
	0.1	99.8 (7.2)	85.5 (6.2)	– –	– –

Source: Reproduced with permission from the Royal Society of Chemistry [222]

*Recovery figures are means of six determinations. Figures in parentheses are standard deviations.
†XAD gave insufficient clean-up to measure recoveries at the 0.1 $\mu g\ g^{-1}$ level

All of the methods discussed above use liquid–liquid partition for clean-up. However, Smith and Hayden [231] and Johnson *et al.* [232] have shown that the macroreticular resin XAD–2 is an efficient adsorber of 2,4–D.

After extraction and clean-up the acid and hydroxybenzonitrile herbicides are either too polar or insufficiently volatile to be determined directly by gas–liquid chromatography, and a suitable derivative must therefore be prepared. Among the methods reported at the preparation of methyl esters using diazomethane [225,226,232–234]. Boron trichloride-methanol reagent [235] or iodomethane with alkali metal carbonate catalysis under anhydrous conditions [236]. Other esters have been prepared by reaction of herbicide acids with the appropriate alcohol [230,237,238]. Some workers have prepared esters with enhanced electron-capturing properties to improve detection limits, such as straight-chain halogenated esters [239] or strongly halogenated aromatic esters such as pentafluorobenzyl [240–244]. Various silyl esters have also been prepared [245,246].

Each of these methods has some disadvantage. Diazomethane is toxic, carcinogenic and explosive. Boron trichloride-methanol will not alkylate hydroxybenzonitriles. The silyl derivatives tend to condense in the electron-capture detector and decrease its sensitivity. The yield of ester from the acid-catalysed reaction of acid and alcohol, although reproducible, is usually of the order of only 80%. Pentafluorobenzylation and alkali metal carbonate-catalysed alkylation require anhydrous conditions. The use of halogenated regents also has the disadvantage of transferring electron-capturing properties to impurities or coextractants that may interfere with measurement of the herbicide. Extractive or ion-pair alkylation is an alternative method that has been reported [247–250].

9.8.1.2 Thin layer chromatography

Abbott and Wagstaff [251] of the Laboratory of the Government Chemist UK have described a thin layer chromatographic method for the identification of 12 acidic herbicides and 19 nitrogenous herbicides (carbamates, substituted ureas and triazine).

Smith and Fitzpatrick [252] have also described a thin layer method for the detection in water and soil of herbicide residues, including Atrazine, Barban, Diuron, Linuron, Monuron, Simazine, Trifluralin, Bromoxynil, Dalapon, Dicamba, MCPB, Mecoprop, Dicloram, 2,4–D, 2,4–DB, Dichloroprop, 2,4,5–T, and 2,3,6-trichlorobenzoic acid.

Neutral and basic herbicides were extracted from water made alkaline with sodium hydroxide or from soil, with chloroform; extracts of soil were cleaned up on basic alumina containing 15% of water. Acidic herbicides were extracted with ethyl ether from water acidified with hydrochloric acid or from an aqueous extract of soil prepared by treatment with 10% aqueous potassium chloride that was 0.05M in sodium hydroxide and filtration into 4M hydrochloric acid. The concentrated chloroform solution of neutral and basic herbicides was applied to a pre-coated silica gel plate containing a fluorescent indicator and a chromatogram was developed two-dimensionally with hexane–acetone (10:3) followed after drying by chloroform–nitromethane (1:1). The spots were detected in UV radiation. Atrazine, Barban, Diuron, Linuron, Monuron, Simazine and Trifluralin were successfully separated and were located as purple spots on a green fluorescent background. The ether extracts were dried over sodium sulphate, concentrated, and applied to a similar plate, which was developed two-dimensionally with chloroform–anhydrous acetic acid (19:1) followed after drying by benzene–hexane–anhydrous acetic acid (5:10:2). The spots were detected by spraying with bromocresol green. Bromoxynil and (as the acids) Dalapon, Dicamba, MCPA, MCPB, Mecoprop, Dicloram, 2,4–D, 2,4–DB, Dichloroprop, 2,4,5–T and 2,3,6-trichlorobenzoic acid were seen as yellow spots on a blue background. The limits of detection were 1ppm in soil and 0.1ppm in natural water.

9.8.1.3 Supercritical fluid chromatography

Snyder *et al.* [253] compared supercritical fluid chromatography with classical sonication procedures and Soxhlet extraction for the determination of selected insecticides in soils and sediments. In this procedure the sample was extracted with carbon dioxide modified with 3% methanol at 350atm and 50°C. An excess of 85% recovery of organochlorine and organophosphorus insecticides was achieved. These included Dichlorvos, Diazinon, (diethyl-2-isopropyl-6-methyl 4-pyrimidinyl phosphorothioate), Ronnel (i.e. Fenchlorphos-O,O dimethyl-O-2,4,5-trichlorophenyl phosphorothioate), Parathion ethyl, Methiadathion, Tetrachlorovinphos (trans-2-chloro-1-(2,4,5 trichlorophenyl) vinylchlorophenyl-O-methyl phenyl phosphoroamidothioate), Endrin, Endrin aldehyde, pp′ DDT, Mirex and decachlorobiphenyl.

9.8.1.4 Miscellaneous

Kearney *et al.* [254] in a study of persistence, binding and metabolism of six dinitroaniline herbicides (Trifluralin, Profluralin, Dinitramine, Butralin, Fluchloralin and Chlornidine) showed that the parent herbicide was the major compound extracted from herbicide-treated soil after three, five and seven months. Trifluralin, Fluchloralin and Profluralin were the most persistent of the six pesticides.

Degradation products of six dinitroaniline herbicides that were added to silt loam were identified by thin layer chromatography, liquid chromatography and mass spectrometry.

9.8.2 Non-saline deposited and suspended sediments

9.8.2.1 Supercritical fluid chromatography

The supercritical fluid extraction procedure [253] discussed in section 9.8.1.3 has been applied to river sediments.

9.9 Growth regulators

9.9.1 Soil

9.9.1.1 Capillary isotachoelectrophoresis

Stransky [255] investigated the determination in amounts down to 10μg kg^{-1} growth regulator Chlormequat in soil by capillary isotachophoresis.

9.10 Fungicides

9.10.1 Soil

Dichloro-1,4-naphthaquinone

9.10.1.1 Spectrophotometric method

A spectrophotometric method has been described for determining down to 2μg of this fungicide in soil based on the formation of a coloured reaction product with aniline [256].

Furaloxyl (methyl N(2,6-dimethyl-phenyl) N(2-furyol)alaninate and metoxyl (methyl N-(2,6, dimethyl phenyl)-N-(2-methoxyacetyl)alaninate

9.10.1.2 Gas chromatography

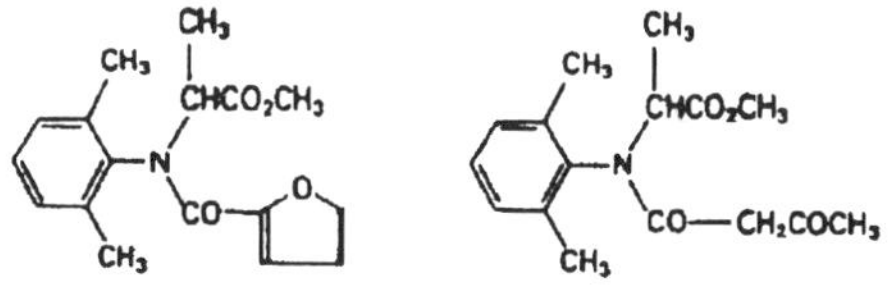

Furaloxyl and Metalaxyl are fungicides with residual and systemic activity against fungi of the order *Peronosphorales*, which attack a wide range of commercial crops. A procedure described by Caverly and Unwin [257] is sensitive and rapid and can be applied to a wide range of agricultural materials. Soils and peat composts are dried and extracted with acetone in a Soxhlet apparatus, and the extract is analysed, without clean-up, by gas chromatography using a nitrogen-selective rubidium chloride thermionic detector and a glass column packed with 5% high vacuum silicone grease on 8–100 mesh Gas–Chrom Q.

Table 9.23 Recovery of fungicides added to samples

Sample	*Furalaxyl* Added/ mg kg $^{-1}$	Found/ mg kg $^{-1}$	Recovery %	*Metalaxyl* Added/ mg kg $^{-1}$	Found/ mg kg $^{-1}$	Recovery %
peat compost	200.00	190.00	95	400.00	313.00	78
	100.00	105.60	106	200.00	200.00	100
	50.00	44.00	88	100.00	88.00	88
	20.00	17.00	85	40.00	40.00	100
	10.00	7.50	7	20.00	21.00	105
	5.00	4.00	80	–	–	–
Soil A (calcareous silty loam; pH7.6; organic matter 3.57%)	200.00	200.00	100	40.00	30.00	75
	100.00	111.00	111	20.00	16.00	80
	50.00	49.30	99	10.00	8.30	83
	20.00	23.00	115	5.00	4.00	80
	10.00	7.50	75	1.00	1.10	110
	5.00	4.30	86	–	–	–
Soil B (sandy loam; pH5.7; organic matter 3.82%)	200.00	147.00	74	20.00	17.00	85
	100.00	80.00	80	10.00	7.70	77
	50.00	39.00	78	5.00	4.50	90
	20.00	16.00	80	1.00	0.70	70
	5.00	4.50	90	–	–	–

Source: Reproduced with permission from the Royal Society of Chemistry [257]

The carrier gas was nitrogen at a flow-rate of 40ml min $^{-1}$. The flow-rate of hydrogen was 40ml min $^{-1}$ and that of air was approximately 250ml min $^{-1}$. The column temperature was 190°C for Metalaxyl and 210°C for Furaloxyl. An almost linear response of peak height and amount of fungicide applied was obtained with 30 and 20ng of Furaloxyl and Metalaxyl, respectively, giving full-scale deflection. Recoveries are generally better than 80% with detection limits of 0.5mg kg $^{-1}$ for peat, 0.1mg kg $^{-1}$ for soils and plants and 0.02mg kg $^{-1}$ for nutrient solutions.

The recovery of the two fungicides from peat composts and soils was measured by the addition of stock solutions to the air-dried materials and, after thorough mixing and standing for 24h, extraction with acetone as detailed above. The mean recoveries of duplicate determinations are given in Table 9.23 and are illustrated in Fig. 9.13 and 9.14.

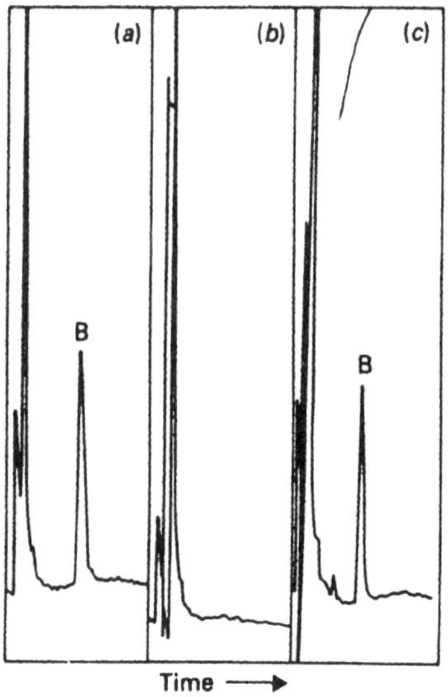

Fig. 9.13 Gas chromatograms of (1) standard containing 6ng of metalaxyl (B); (b) extract from soil A not treated with metalaxyl; and (c) extract of soil A after addition of metalaxyl at 20mg kg $^{-1}$ (6μl of 1g per 20ml). Column temperature 190°C. Detector temperature 250°C. Attenuation 50 × 1.
Source: Reproduced with permission from the Royal Society of Chemistry [257]

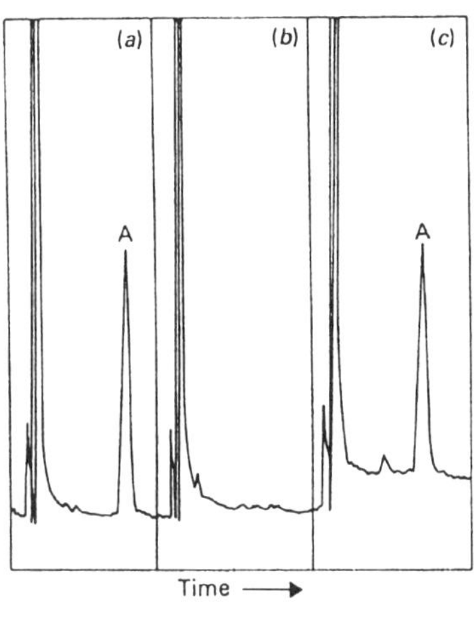

Fig. 9.14 Gas chromatograms of (a) standard containing 8ng of Furaloxyl (A): (b) extract from soil B not treated with Furaloxyl; and (c) extract of soil B after addition of Furaloxyl at 10mg kg $^{-1}$ (4μl of 1g per 5ml). Column temperature 205°C. Detector temperature 250°C. Attenuation 50 × 1.
Source: Reproduced with permission from the Royal Society of Chemistry [257]

2,6-dichloroacetanilide

9.10.1.3 Miscellaneous

Van Alfen and Kosuge [258] incubated this plant disease control fungicide in flooded soil to identify principal metabolites and determine metabolic toxicity to microorganisms; the major soil metabolite of 2,6-dichloro-4-nitroaniline was identified as 4-amino-3,5-dichloroacetanilide and its expected metabolic precursor, 2,6-dichloro-p-phenylenediamine fungi grown on potato-dextrose agar were inhibited by the parent compound but not significantly affected by the two metabolites.

References

1 Deubert, K.H. (1970) *Bulletin of Environmental Contamination and Toxicology*, **5**, 379.
2 Cooke, B.K. and Western, N.M. (1980) *Analyst (London)*, **105**, 490.
3 Goodwin, E.S., Goulden, R. and Reynolds, J.G. (1961) *Analyst (London)*, **86**, 697.
4 Chiba, M. and Morley, H.V. (1968) *Journal of Agriculture and Food Chemistry*, **16**, 916.
5 Stringer, A., Pickard, J.A. and Lyons, C.H. (1974) *Pesticide Science*, **5**, 587.
6 Mills, P.A., Bong, B.A., LaVerna, R.K. and Burke, J.A. (1972) *Journal of Association of Official Analytical Chemists*, **55**, 39.
7 Hesselberg, R.J. and Johnson, J.L. (1972) *Bulletin of Environmental Contamination and Toxicology*, **7**, 115.
8 Woodham, D.W., Loftis, C.D. and Collier, C.W. (1972) *Journal of Agriculture and Food Chemistry*, **20**, 163.
9 Gooding, P.H., Philip, H.G. and Tawnk, H.S. (1972) *Bulletin of Environmental Contamination and Toxicology*, **7**, 288.
10 Mahel'ova, H., Sackmanereva, M., Szokolav, A. and Kovac, J. (1974) *Journal of Chromatography*, **89**, 177.
11 Suzuki, M., Yaomoto, Y. and Wanatabe, Y. (1977) *Environmental Science and Technology*, **11**, 1109.
12 Suzuki, M., Yamoto, Y. and Wanatabe, Y. (1973) *Nippon Nogei. Kagaku Kaishi*, **47**, 1.
13 Mangani, F., Crescentini, G. and Bruner, F. (1981) *Analytical Chemistry*, **53**, 1627.
14 Gambrell, R.P., Reddy, C.N., Collard, V. *et al.* (1984) *Journal of Water Pollution Control*, **56**, 174.
15 Teichman, T., Revenue, A. and Hylin, J.W. (1978) *Journal of Chromatography*, **151**, 155.
16 Lopez-Avila, V., Hirata, P., Kraska, S. *et al.* (1985) *Analytical Chemistry*, **57**, 2797.
17 Gillespie, D.M., Eldridge, J.D. and Thompson, C.E. (1975) *Water Research*, **9**, 817.
18 Brady, B.O., Kao, C.C., Dolley, K.M. *et al.* (1987) *Chemistry Research*, **26**, 261.
19 Von Bavel, B., Jarimo, M., Karlsson, L. and Lindstrom, G. (1996) *Analytical Chemistry*, **68**, 1279.
20 Snyder, J.L., Grobe, R.L., McNally, M.E. and Oostdyk, T.S. (1992) *Analytical Chemistry*, **64**, 1940.
21 Novikova, K.F. (1972) *Zhur Vses Khim Obshch.*, **18**, 562.
22 Johnson, R.E. and Starr, R.I. (1972) *Journal of Agriculture and Food Chemistry*, **20**, 48.
23 Lopez-Avila, V., Young, R. and Beckert, W.F. (1994) *Analytical Chemistry*, **66**, 1097.

24 Miller, J.M. and Singh, J. (1976) *Bulletin of Environmental Contamination and Toxicology*, **16**, 485.
25 Bruner, F., Crescentini, G. and Mangani F. (1979) Paper presented at the XIV International Symposium Advances in Chromatography, Lausanne, Switzerland, September 24–28 1979.
26 Bruner, F., Crescentini, G. and Mangani, F. (1979) Paper presented at the third meeting of the Sezione Marchiglana of the Italian Chemical Society, Urbino, Italy, April 27 1979, *Chim. Industry (Milan)*, **9**, 695.
27 Bacaloni, A., Goretti, G., Lagana, A. *et al.* (1980) *Analytical Chemistry*, **52**,2033.
28 Bruner, F., Cerscentini, B. and Mangani, F. (1980) Paper presented at the II National Congress of Analytical Chemistry of the Italian Chemical Society, Padova, October 2–4 1979. Proceedings of the Congress, pp.146–148. Società Chimica Italiana, Rome.
29 Bertoni, G., Brocco, D., Di Palo, V. *et al.* (1978) *Analytical Chemistry*, **50**, 732.
30 Sackmauereva, M., Pal'usova, O. and Hluchan, E. (1972) *Vodni Hospodarstvi*, **10**, 267.
31 Szokolay, A., Uhnak, J. and Madaric, A. (1971) *Chem. Zvesti*, **25**, 453.
32 Janak, J., Sackmanereva, M., Szokolay, A. and Madaric, A. (1973) *Chem. Zvesti*, **27**, 128.
33 Janak, J., Sackmanereva, M., Szokolay, A. and Pal'usova, O. (1974) *Journal of Chromatography*, **91**, 545.
34 Szokolay, A. and Madaric, A. (1969) *Journal of Chromatography*, **42**, 509.
35 Mills, P.A. (1959) *Journal of Association of Official Analytical Chemists*, **42**, 734.
36 Mills, P.A. (1968) *Journal of Association of Official Analytical Chemists*, **51**, 29.
37 Burke, J.A. and Malone, B. (1960) *Journal of Association of Analytical Chemists*, **49**, 1003.
38 Frank, R., Armstrong, A.F., Boeleus, R.G. *et al.* (1974) *Pesticide Monitoring Journal*, **7**, 165.
39 Langlois, R.E., Stemp, A.R. and Liska, B.J. (1954) *Journal of Milk Food Technology*, **27**, 202.
40 Hamence, J.H., Hall, P.S. and Caverly, D.J. (1965) *Analyst (London)*, **90**, 649.
41 Goerlitz, D.F. and Law, L.J. (1974) *Journal of Association of Official Analytical Chemists*, **57**, 176.
42 Sackmauereva, M., Pal'usova, O. and Szokolay, A. (1977) *Water Research*, **11**, 537.
43 Wegman, R.C.C. and Hofstee, A.W.M. (1982) *Water Research*, **16**, 1265.
44 Janak, J., Sackmauereva, M., Szokolay, A. and Pal'usova, O. (1974) *Journal of Chromatography*, **545**, 91.
45 Jensen, S., Renberg, L. and Reutgard, L. (1977) *Analytical Chemistry*, **49**, 316.
46 Southeast Water Laboratory (1971) Method No. SP8/71. *Sediment extraction procedure*, Athens, Georgia, USA.
47 Grob, R.L., MacNally, M.E. and Oostdyk, T.S. (1992) *Analytical Chemistry*, **64**, 1940.
48 Goldberg, M.C. (1982) *Science of the Total Environment*, **24**, 73.
49 Picer, N., Picer, M. and Strohal, P. (1975) *Bulletin of Environmental Contamination and Toxicology*, **14**, 565.
50 Dube, D.J., Veith, G.D. and Lee, G.F. (1974) *Journal of Water Pollution Control Federation*, **46**, 966.
51 Lawrence, J. and Tosine, H. (1976) *Environmental Science and Technology*, **10**, 381.
52 Shannon, E.E., Ludwig, F.J. and Valdemanis, I. (1976) *Environmental Canada Research Report* No. 49, Ottawa.
53 Bergh, A.K. and Peoples, R.S. (1977) *Science of the Total Environment*, **8**, 197.
54 Harper, D.B., Smith, R.V. and Gotto, D.M. (1977) *Environmental Pollution*, **12**, 223.
55 Lawrence, J. and Tosine, H. (1977) *Bulletin of Environmental Contamination and Toxicology*, **17**, 49.

56 Liur, D., Chawla, V.K. and Chau, A.S.Y. (1975) Proceedings of the 9th Annual Conference on Trace Substances in Environmental Health. University of Missouri, Columbia, Missouri, p.189.
57 Mattson, P.E., Nygren, S. (1976) *Journal of Chromatography*, **124**, 265.
58 McIntyre, A.E., Perry, R. and Lester, J.N. (1980) *Environmental Technology Letters*, **1**, 157.
59 McIntyre, A.E., Lester, J.N. and Perry, R. (1979) Analysis of Organic Substances of Concern in Sewage Sludge, Final Report to the Department of the Environment (UK) for Contracts DGR/480/66 and DGR/480/240. Imperial College, London, UK. pp.42–45.
60 Garcia-Gutierrez, A., McIntyre, A.E. and Lester, J.N. (1982) *Environmental Technology Letters*, **3**, 541.
61 Maskarinec, M.P. and Harvey, R.W. (1982) *International Journal of Environmental Analytical Chemistry*, **11**, 53.
62 Organochlorine insecticides and PCBs in sewages, sludges, muds and fish (1978). Organochlorine insecticides and PCBs in water, an addition (1984). HMSO, London.
63 Pearson, J.R., Aldrich, F.D. and Stone, A.W. (1967) *Journal of Agriculture and Food Chemistry*, **14**, 938.
64 Ashling, B. and Jenson, S. (1970) *Analytical Chemistry*, **42**, 1483.
65 Jenson, S., Johns, A.G., Olsson, M. and Otterlind, G. (1972) *Ambio. Spectroscopy Reports*, **1**, 71.
66 Ahnoff, M., and Josefsson, B. (1975) *Bulletin of Environmental Contamination and Toxicology*, **13**, 159.
67 Goerlitz, D.F. and Law, L.H. (1971) *Bulletin of Environmental Contamination and Toxicology*, **6**, 9.
68 Schulte, A. and Acker, L. (1974) *Analytical Chemistry*, **46**, 260.
69 *Methods of Analysis of the Association of Official Analytical Chemists* (1965). Association of Official Analytical Chemists, Washington D.C., 10th ed., p.393.
70 Erne, K. (1958) *Acta Pharmacology and Toxicology*, **14**, 158.
71 Bartlett, J.K. and Skoog, D.A. (1954) *Analytical Chemistry*, **26**, 1008.
72 Zimmerli, B., Sulser, H. and Marek, B. (1971) *Mitt. Geb. Lebensmittelunters. Hygiene*, **62**, 60.
73 Cochrane, W.P. and Maybury, R.B. (1973) *Journal of Association of Official Analytical Chemists*, **56**, 1324.
74 Miller, G.A. and Wells, C.E. (1969) *Journal of Association of Official Analytical Chemists*, **52**, 548
75 Holden, A.V. and Marsden, K. (1969) *Journal of Chromatography*, **44**, 481.
76 Garcia-Gutierrez, A., McIntyre, A.E. and Lester, J.N. (1980) Water Research Centre Information, UK, No. 35. Abstract No. 2534.
77 Erikson, M. and Pellizzari, E.D. (1979) *Bulletin of Environmental Contamination and Toxicology*, **22**, 688.
78 Singhal, J.P., Khan, S.U. and Bansal, O.P. (1978) *Analyst (London)*, **103**, 872.
79 Singhal, J.P., Khan, S.U. and Bansal, O.P. (1979) *Journal of Agriculture and Food Chemistry*, **25**, 377.
80 Cohen, I.C. and Wheals, B.B. (1969) *Journal of Chromatography*, **43**, 233.
81 Lawrence, J.F. and Frei, R.W. (1972) *Analytical Chemistry*, **44**, 2046.
82 Westlake, W.E., Monika, I. and Gunther, F.A. (1972) *Bulletin of Environmental Contamination and Toxicology*, **8**, 109.
83 Reeves, R.G. and Woodham, D.W. (1974) *Journal of Agriculture and Food Chemistry*, **22**, 76.
84 Coburn, J.A., Ripley, B.D. and Chau, A.S.Y. (1976) *Journal of the Association of Official Analytical Chemists*, **59**, 188.
85 Bromilow, R.H. (1976) *Analyst (London)*, **101**, 982.

86 Leppert, B.C., Markle, J.C., Helt, R.C. and Fujie, G.H. (1983) *Journal of Agriculture and Food Chemistry*, **31**, 220.
87 Bilikova, A. and Kuthan, A. (1983) *Vodni Hospodarstvi. Series B*, **33**, 215.
88 Fung, K.K.H. (1976) *Pesticide Science*, **7**, 571.
89 Leistra, M., Dekker, A. and Van der Bung, A.M.M. (1984) *Water, Air and Soil Pollution*, **23**, 155.
90 Spengler, D. and Jumar, A. (1971) *Pflanzenschutz Nachr. Bayer*, **7**, 151.
91 Kjolholt, J. (1985) *Journal of Chromatography*, **325**, 231.
92 Devine, M. (1973) *Journal of Agriculture and Food Chemistry*, **21**, 1095.
93 Khan, S.U., Hamilton, H.A. and Hogue, E.J. (1976) *Pesticide Science*, **7**, 553.
94 Snyder, J.L., Grobe, R.L., McNally, M.E. and Oostdyk, T.S. (1992) *Analytical Chemistry*, **64**, 1940.
95 Rice, J.R. and Dishberger, H.J.J. (1968) *Journal of Agriculture and Food Chemistry*, **16**, 867.
96 Deutsch, M.E., Westlake, M.E. and Gunther, F.A. (1970) *Journal of Agriculture and Food Chemistry*, **18**, 178.
97 Szeto, S.Y., Yee, J., Brown, M.J. and Olofs, P.C. (1982) *Journal of Chromatography*, **24**, 526.
98 Alford Stevens, A.L., Buddle, W.L. and Bellar, T.A. (1985) *Analytical Chemistry*, **57**, 2452.
99 HMSO (1985) *Organophosphorus pesticides in sewage sludge; organophosphorus pesticides in river and drinking water, an addition.* (1986) *Methods for the Examination of Waters and Associated Materials*, HMSO, London.
100 Cotterill, E.G. (1979) *Analyst (London)*, **104**, 878.
101 Byast, T.H., Cotterill, E.G. and Hance, R.J. (1977) *Methods for the Analysis of Herbicide Residues*, 2nd edn. Technical Report, Agricultural Research Council, Weed Research Organization, Yarnton, UK, No. 15.
102 Sanchez-Rasero, F. and Dios, G.C. (1988) *Journal of Chromatography*, **447**, 426.
103 Steinheimer, T.R., Pfeiffer, R.L. and Scoggins, K.D. (1994) *Analytical Chemistry*, **66**, 645.
104 Bushway, R.J., Perkins, S.B., Savage, S.A. and Ferguson, B.S. (1988) *Bulletin of Environmental Contamination and Toxicology*, **40**, 647.
105 Mills, M.S. and Thurmen, E.M. (1992) *Analytical Chemistry*, **64**, 1985.
106 Russell, J.D., Cruz, M.I. and White, J.L. (1968) *Journal of Agriculture and Food Chemistry*, **16**, 21.
107 Stransky, Z. (1985) *Journal of Chromatography*, **320**, 219.
108 Pichon, V., Chen, L., Hennion, M-C. *et al.* (1995) *Analytical Chemistry*, **67**, 2451.
109 Pichon, V., Chen, L., Durand, N. *et al.* (1996) *Journal of Chromatography*, **725**, 107.
110 Pichon, V., Rogniaux H., Fischer-Durand, N. *et al.* (1997) *Chromatographia*, **45**, 289.
111 Ferrer, I., Pichon, V., Hennion, M-C., and Barcelo, D. (1997) *Journal of Chromatography*, **777**, 91.
112 Rule, G.S., Mordehal, A.V. and Henion, J. (1994) *Analytical Chemistry*, **66**, 230.
113 Shahtaheri, S.J., Katmeh, M., Kwasowski, P. and Stevenson, D. (1995) *Journal of Chromatography*, **697**, 131.
114 Marx, A., Giersch, T. and Hock, B. (1995) *Analytical Letters*, **28**, 267.
115 Ferrer, I., Hennion, M-C. and Barcelo, D. (1997) *Analytical Chemistry*, **69**, 4508.
116 Wauchope, R.D. and Myers, R.S. (1985) *Journal of Environmental Quality*, **14**, 132.
117 Katz, S.E. and Strusz, R.F. (1968) *Bulletin of Environmental Contamination and Toxicology*, **3**, 258.
118 Smith, A.E. and Lord, K.A. (1975) *Journal of Chromatography*, **107**, 407.
119 Cotterill, E.G. (1980) *Analyst (London)*, **105**, 987.
120 McKone, C.E. (1969) *Journal of Chromatography*, **44**, 60.
121 Kahn, S.U., Greenhalgh, R. and Cockrane, W.P. (1975) *Bulletin of Environmental*

Contamination and Toxicology, **13**, 602.
122 Kirkland, J.J. (1962) *Analytical Chemistry*, **34**, 428.
123 Lokke, H. (1974) *Pesticide Science*, **5**, 749.
124 Friestead, H.O. (1974) *Journal of the Association of Official Analytical Chemists*, **57**, 221.
125 Kirkland, J.J. (1969) *Journal of Chromatographic Science*, **7**, 7.
126 Sidwell, J.A. and Ruzicka, J.H.A. (1976) *Analyst (London)*, **101**, 111.
127 Parouchais, C. (1973) *Journal of the Association of Official Analytical Chemists*, 56, **831**.
128 Onley, J.H. and Yip, G. (1971) *Journal of the Association of Official Analytical Chemists*, **54**, 1366.
129 McKane, C.E. (1969) *Journal of Chromatography*, **44**, 60.
130 Spengler, D. and Hamroll, B. (1970) *Journal of Chromatography*, **49**, 1070.
131 Lawrence, J.F. (1976) *Journal of Association of Official Analytical Chemists*, **59**, 1066.
132 Byast, T.H. (1977) *Journal of Chromatography*, **134**, 216.
133 Pribyl, J. and Herzel, F. (1978) *Journal of Chromatography*, **166**, 272.
134 Jarczyk, H.J. (1975) *Pflanzenschutz-Nachr. Bayer*, **28**, 334.
135 Lawrence, J.F. and Laver, G.J. (1975) *Journal of Agriculture and Food Chemistry*, **23**, 1106.
136 Buchert, A. and Lokke, H. (1975) *Journal of Chromatography*, **115**, 682.
137 Bieser, H. and Gronlinmund, K. (1974) *Journal of the Association of Official Analytical Chemists*, **57**, 1294.
138 Caverly, D.J. and Denney, R.C. (1978) *Analyst (London)*, **103**, 368.
139 Grossbard, E. and Marsh, J.A.P. (1974) *Pesticide Science*, **5**, 609.
140 Sheets, T.J. (1964) *Journal of Agriculture and Food Chemistry*, **12**, 30.
141 Bartha, R. (1971) *Journal of Agriculture and Food Chemistry*, **19**, 385.
142 Geissbuhler, H. (1969) in (eds. P.C. Kearney and D.D. Kaufman) *Degradation of Herbicides*, Marcel Dekker, New York, p.79.
143 Burge, W.D. (1972) *Soil Biology and Biochem*istry, **4**, 379.
144 Farrington, D.S., Hopkins. R.G. and Ruzicka, J.H.A. (1977) *Analyst (London)*, **102**, 377.
145 Gutenmann, W.H. and Liks, D.J. (1964) *Journal of the Association of Official Analytical Chemists*, **47**, 353.
146 Chau, A.S.Y. and Terry, K. (1975) *Journal of the Association of Official Analytical Chemists*, **58**, 1294.
147 Agemain, H. and Chau, A.S.Y. (1976) *Analyst (London)*, **101**, 732.
148 Gutenmann, W.H. and Lisk, D.J. (1963) *Journal of the Association of Official Analytical Chemists*, **46**, 859.
149 Bache, C.A., Lisk, D.J. and Loos, M.A. (1964) *Journal of the Association of Official Analytical Chemists*, **47**, 348.
150 Kawahara, F.K. (1968) *Analytical Chemistry*, **40**, 1009.
151 Kawahara, F.K. (1971) *Environmental Science and Technology*, **5**, 235.
152 Coburn, J. and Chau, A.S.Y. (1974) *Journal of the Association of Official Analytical Chemists*, **57**, 1272.
153 Gaunt, J.K. and Evans, W.C. (1971) *Biochemical Journal*, **122**, 519.
154 Renberg, L. (1974) *Analytical Chemistry*, **116**, 459.
155 Sattar, M.A. and Paasivirta, J. (1979) *Analytical Chemistry*, **51**, 598.
156 Yip, G. (1975) *Journal of Chromatographic Science*, **13**, 225.
157 Upchurch, R.P. and Mason, D.D. (1962) *Weeds*, **10**, 9.
158 Waliszewski, S.H. and Szymczynski, G. (1985) *Fresenius Zeitschrift für Analytische Chemie*, **322**, 510.
159 Lopez-Avila, V., Hirta, P., Kraske, S. and Taylor, J.H. (1986) *Journal of Agriculture and Food Chemistry*, **34**, 530.
160 Di Corcia, A. and Marchetti, M. (1991) *Analytical Chemistry*, **63**, 819.

161 Kim, I., Sasinos, F.T., Stephens, R.D. *et al.* (1991) *Analytical Chemistry*, **63**, 819.
162 Yorkshire Water Authority (1971) Determination of Herbicides in Sewage and Trade Effluents (1971). Method YWA 670–01.
163 Abbott, S.D., Hall, R.C. and Giam, G.S. (1969) *Journal of Chromatography*, **45**, 317.
164 Hall, R.C., Giam, G.S. and Merkle, M.G. (1970) *Analytical Chemistry*, **42**, 432.
165 Dennis, D.S., Gillespie, W.H., Moxey, R.A. and Shaw, R. (1977) *Archives of Environmental Contamination and Toxicology*, **6**, 421.
166 Cannizzara, R.D., Cullen, T.E. and Murphey, R.T. (1970) *Journal of Agriculture and Food Chemistry*, **18**, 728.
167 Shaner, D.L. and O'Connor, S.L. (eds) (1991) *The Imidazolinone Herbicides*, CRC Press, Boca Raton, FL.
168 Shaner, D.L., Anderson, P.C. and Stidham, M.A. (1984) *Plant Physiology*, **76**, 545.
169 Anderson, P.C. and Hibert, K.A. (1985) *Weed Science*, **33**, 479.
170 Shaner, D.L. and Millipudi, N.M. (1991) in *The Imidazoline Herbicides*, (eds. D.L. Shaner, and S.L. O'Connor) CRC Press, Boca Raton, FL.
171 Wehtje, G.R., Dickens, R., Wilcut, J.W. and Hajek, B.F. (1987) *Weed Science*, **35**, 858.
172 Stoughaard, R.N., Shea, P.J., Martin, A.R. (1990) *Weed Science*, **38**, 67.
173 Che, M., Loux, M.M., Traina, S.J. and Logan, T.J. (1992) *Journal of Environmental Quality*, **21**, 698.
174 Loux, M.M., Liebl, R.A., Slife, R.W. (1989) *Weed Science*, **37**, 712.
175 Renner, K.A., Meggit, W.F. and Penner, D. (1988) *Weed Science*, **36**, 78.
176 Strout, S.J., da Cuhna, A.R. and Allardice, D.G. (1996) *Analytical Chemistry*, **68**, 653.
177 Herzel, F. (1980) *Journal of Chromatography*, **193**, 320.
178 Calderbank, A. and Yuens, O. (1965) *Analyst (London)*, **90**, 99.
179 Pope, J.D. and Benner, J.E. (1974) *Journal of the Association of Official Analytical Chemists*, **57**, 202.
180 Payne, W.R., Pope, J.D. and Benner, J.E. (1974) *Journal of Agriculture and Food Chemistry*, **22**, 79.
181 Khan, S.L. (1974) *Journal of Agriculture and Food Chemistry*, **22**, 863.
182 Martens, M.A. and Hyndricks, A. (1974) *Journal of Pharmacology, Belgium*, **29**, 449.
183 Niewola, Z., Benner, J.P. and Swaine, H. (1986) *Analyst (London)*, **111**, 399.
184 Niewola, A., Walsh, S.T. and Davies, G.E. (1983) *International Journal of Immunopharmacology*, **5**, 211.
185 Niewola, Z., Hayward, C., Symington, B.A. and Robson, R.T. (1985) *Clinica Chimica Acta*, **148**, 149.
186 Kohler, G. and Milstein, C. (1975) *Nature (London)*, **256**, 495.
187 Tsukioka, T., Shimizu, S. and Hurakami, T. (1985) *Analyst (London)*, **110**, 39.
188 Wheeler, W.B., Thompson, N.P., Ray, B.R. and Wilcox, M. (1971) *Weed Research*, **19**, 307.
189 Jolliffe, U.A., Day, B.E., Jordan, R.S. and Mann, J.D. (1967) *Journal of Agriculture and Food Chemistry*, **15**, 174.
190 Gutamann, W.H. and List, D.J., (1971) *Journal of the Association of Official Analytical Chemists*, **54**, 975.
191 Bevenue, A. and Ogata, J.H. (1970) *Journal of Chromatography*, **46**, 110.
192 Pease, H.L. (1966) *Journal of Science Food Agriculture*, **17**, 121.
193 Pease, H.L. (1966) *Journal of Agriculture and Food Chemistry*, **14**, 94.
194 Gawronski, S. and Skapski, H. (1974) *Zesz. Nauk. Akad., Roln. Warsz., Ogrodnictwo*, **8**, 59.
195 Hamilton, D.J. (1968) *Journal of Agriculture and Food Chemistry*, **16**, 152.
196 Von Stryk, F.G. and Zajacs, G.F. (1969) *Journal of Chromatography*, **41**, 125.
197 Maier-Bode, H. and Riedmann, M. (1975) *Residue Review*, **54**, 113.

198 Jarcjyk, H.J. (1975) *Pflanzeuschutz-Nachr. Bayer*, **27**, 319.
199 Caverly, D.J. and Denney, R.C. (1977) *Analyst (London)*, **102**, 576.
200 Shimabukuro, M.A., Shimabukuro, R.H., Nord, W.S. and Hoerauf, R.A. (1978) *Pestic. Biochem. Physiol.*, **8**, 199.
201 Martens, R. (1978) *Pesticide Science*, **9**, 127.
202 Smith, A.E. (1979) *Journal Agriculture and Food Chemistry*, **27**, 1145.
203 Smith, A.E. (1977) *Journal Agriculture and Food Chemistry*, **25**, 893.
204 Smith, A.E. (1976) *Journal of Chromatography*, **129**, 309.
205 Blau, K. and King, G. (1978) *Handbook of Derivatives for Chromatography*, Heyden, Philadelphia.
206 Gaynor, J.D. and MacTavish, D.C. (1981) *Journal Agriculture and Food Chemistry*, **29**, 626.
207 Gaynor, J.D. and MacTavish, D.C. (1982) *Analyst (London)*, **107**, 700.
208 Negre, M. and Cignetti, A. (1987) *Journal of Chromatography*, **387**, 541.
209 Sotirion, N., Weisgerber, I., Klein, W. and Korte, F. (1976) *Chemosphere*, **1**, 53.
210 Oda, M., Yukimoto, M. and Zasso, O. (1975) *Kenkyu*, **20**, 12.
211 Prestel, D., Weisgerber, I., Klein, W. and Korte, F. (1976) *Chemosphere*, **5**, 137.
212 Downer, G.B., Hall, M. and Mallen, O.N.B. (1976) *Journal of Agriculture and Food Chemistry*, **24**, 1223.
213 Khan, S.U. and Lee, K.S. (1970) *Journal of Agriculture and Food Chemistry*, **24**, 684.
214 Worthing, C.R. (ed.) (1983) *The Pesticide Manual: A World Compendium*. 7th ed. British Crop Protection Council, London, UK.
215 Gershon, H. and McClure, G.W. (1966) *Contrib. Boyce Thompson Institute*, **23**, 291.
216 Miller, J.H., Keeley, P.E., Thullen, R.J. and Carter, C.H. (1976) *Weed Science*, **26**, 20.
217 Ross, L.J., Nicosia, S., McChesney, M.M. *et al.* (1990) *Journal of Environmental Quality*, **19**, 715.
218 Monohan, K., Tinsley, I.J., Shepherd, S.F. and Field, J.A. (1995) *Journal of Agriculture and Food Chemistry*, **43**, 2418.
219 Field, J.A. and Monohan, K. (1995) *Analytical Chemistry*, **67**, 3357.
220 Wettasinghe, A. and Tinsley, I.J. (1993) *Bulletin of Environmental Contamination and Toxicology*, **50**, 226.
221 Field, J.D., Monohan, K. and Reed, R. (1998) *Analytical Chemistry*, **70**, 1956.
222 Cotterill, E.G. (1982) *Analyst (London)*, **107**, 76.
223 Gambrell, R.P., Reddy, C.N., Collard, V. *et al.* (1984) *Journal of Water Pollution Control Federation*, **56**, 174.
224 Kavetskii, V.N., Bublik, L.I. and Fuzik, G.V. (1987) *Journal of Analytical Chemistry of USSR*, **42**, 1037.
225 Khan, S.U. (1975) *Journal of the Association of Official Analytical Chemists*, **58**, 1027.
226 Bache, C.A. and Lisk, D.J. (1966) *Analytical Chemistry*, **38**, 783.
227 Abbott, D.C., Egan, H., Hammond, E.W. and Thomson, J. (1964) *Analyst (London)*, **89**, 480.
228 Fahing, T.M., Paulatis, M.E., Johnson, P.M. and McNally, M.E.P. (1993) *Analytical Chemistry*, **65**, 1462.
229 McKone, C.E. and Cotterill, E.G. (1974) *Bulletin of Environmental Contamination and Toxicology*, **11**, 233.
230 Cotterill, E.G. (1978) *Bulletin of Environmental Contamination and Toxicology*, **19**, 471.
231 Smith, A.E. and Hayden, B.J. (1979) *Journal of Chromatography*, **171**, 482.
232 Johnson, E.R., Yu, T.C. and Mongomery, M.I. (1977) *Bulletin of Environmental Contamination and Toxicology*, **17**, 369.
233 Howard, S.F. and Yip, G. (1971) *Journal of the Association of Official Analytical Chemists*, **54**, 970.
234 St John, L.E. and Lisk, D.J. (1967) *Journal of Dairy Science*, **50**, 582.

234 St John, L.E. and Lisk, D.J. (1967) *Journal of Dairy Science*, **50**, 582.

235 Gutenmann, W.H. and Lisk, D.J. (1963) *Journal of Agriculture and Food Chemistry*, **11**, 301.

236 Thio, A.P., Kornet, M.J., Tan, H.S.I. and Tomkins, D.H. (1979) *Analytical Letters*, **12**, 1009.

237 McKone, C.E. and Hance, R.J. (1972) *Journal of Chromatography*, **69**, 204.

238 Cotterill, E.G. (1975) *Journal of Chromatography*, **106**, 409.

239 Gutenmann, W.H. and Lisk, D.J. (1977) *Journal of the Association of Official Analytical Chemists*, **60**, 1070.

240 Agemian, H. and Chau, A.S.Y. (1977) *Journal of the Association of Official Analytical Chemists*, **60**, 1070.

241 Mierzwa, S. and Witak, S. (1977) *Journal of Chromatography*, **136**, 105.

242 Cotterill, E.G. (1979) *Journal of Chromatography*, **171**, 478.

243 Chau, A.S.Y. and Terry, K. (1976) *Journal of the Association of Official Analytical Chemists*, **59**, 633.

244 Sattar, M.A., Hattula, M.L., Lahtipera, M. and Paasivirta, J. (1977) *Chemosphere*, **11**, 747.

245 Garbrecht, T.P. (1970) *Journal of the Association of Official Analytical Chemists*, **53**, 70.

246 Bache, C.A., St. John, L.E. and Lisk, D.J. (1968) *Analytical Chemistry*, **40**, 1241.

247 Garle, M. and Petters, I. (1977) *Journal of Chromatography*, **140**, 165.

248 Gyllenhaal, O., Naslund, B. and Hartvig, P. (1978) *Journal of Chromatography*, **156**, 330.

249 Gyllenhaal, O., Ehrsson, H. (1975) *Journal of Chromatography*, **107**, 327.

250 Hartvig, P. and Fagerlund, C. (1977) *Journal of Chromatography*, **140**, 170.

251 Abbott, D.C. and Wagstaff, P.J. (1969) *Journal of Chromatography*, **43**, 361.

252 Smith, A.E. and Fitzpatrick, A. (1971) *Journal of Chromatography*, **57**, 303.

253 Snyder, J.L., Grob, R.L., McNally, H.E. and Oosterdyk, T.S. (1992) *Analytical Chemistry*, **64**, 1940.

254 Kearney, P.C., Plimmer, J.R., Wheller, W.B. and Konston, A. (1976) *Pestic. Biochem. Physiol.*, **6**, 229.

255 Stransky, Z. (1985) *Journal of Chromatography*, **320**, 219.

256 Burkat, S.E., Medvedeva, N. Ya. and Ivanov, B.G. (1969) *Z. Analit. Khim*, **24**, 284.

257 Caverly, D.J. and Unwin, J. (1981) *Analyst (London)*, **106**, 389.

258 Van Alfen, N.K. and Kosuge, T. (1976) *Journal of Agriculture and Food Chemistry*, **24**, 584.

259 Wong, J.H., Li, Q.X., Hammock, B.D. and Seiber, J.N. (1991) *Journal of Food and Agricultural Chemistry*, **39**, 1802.

260 Snyder, J.L., Grob, R.L., McNally, M.E. and Oostdyk, T.S. (1993) *Journal of Chromatographic Science*, **31**, 183.

261 McNally, M.E. and Wheeler, J.R. (1985) *Journal of Chromatography*, **435**, 53.

262 Hawthorn, S.B., Miller, D.J., Nivens, D.E. and White, D.C. (1992) *Analytical Chemistry*, **64**, 405.

263 Lagana, P. Private Communication.

Chapter 10

Miscellaneous organic compounds

10.1 Humic and fulvic acids

10.1.1 Soil

10.1.1.1 Spectrofluorimetry

Saar and Weber [1] compared methods based on spectrofluorimetry and ion-selective electrode potentiometry for determining the complexes formed between fulvic acid and heavy metal ions.

The fluorescence properties of two fulvic acids, one derived from the soil and the other from river water were studied. The maximum emission intensity occurred at 445–450nm on excitation at 350nm, and the intensity varied with pH, reaching a maximum at pH 5.0 and decreasing rapidly as pH dropped below 4. Neither oxygen nor electrolyte concentration affected the fluorescence of the fulvic acid derived from the soil. Complexes of fulvic acid with copper, lead, cobalt, nickel and manganese were examined and it was found that bound copper II ions quench fulvic acid fluorescence. Ion-selective electrode potentiometry was used to demonstrate the close relationship between fluorescence quenching and fulvic acid complexation of cupric ions. It is suggested that fluorescence and ion-selective electrode analysis may not be measuring the same complexation phenomenon in the cases of nickel and cobalt complexes with fulvic acid.

10.1.1.2 Nuclear magnetic resonance spectroscopy

Wilson *et al.* [2] carried out a compositional and solid state nuclear magnetic resonance (NMR) spectroscopic study of humic and fulvic acid and fractions present in soil organic matter.

The ^{13}C NMR study utilized cross polarization–magic angle spinning (CP–MAS) with spin counting. The elemental and functional group analyses provided input for a series of analytical constraints calculations that yield an absolute upper limit for the amount of aromatic carbon and most probable estimates of both aromatic and non-carboxyl aliphatic carbon in each sample. Spin counting experiments demonstrate that less than 50% of the carbon in three of the fractions is observed in the NMR experiment, and even after correction for differential relaxation, the amounts of aromatic and non-carboxyl aliphatic carbon determined by ^{13}C CP–MAS NMR are dissimilar to those obtained by calculation. Unambiguous evidence is presented for the predominance of aliphatic carboxyl groups in one of the fulvic acid fractions.

A summary of the results obtained by these workers is given in Table 10.1.

10.1.1.3 Miscellaneous

Weber and Wilson [3] used anion and cation exchange resins to isolate fulvic and humic acids from soil and water.

Ion-selective electrodes have been used to determine the stability constants for the complexation of copper II ions with soil fulvic acids [4]. Two classes of binding sites were found with conditional stability constants of about 1×10^6 and 8×10^3.

10.1.2 Non-saline deposited and suspended sediments

10.1.2.1 Atomic absorption spectrometry

Klenke *et al.* [5] described a technique for extraction of humic and fulvic acids from stream sediments and outlined methods for their determination. By means of flame atomic absorption spectrometry, the levels of environmentally important heavy metals (cadmium, copper, chromium, cobalt, nickel and lead) in the fulvic and humic acid extracts were compared with those in the original sediment samples. The pattern distribution of the respective metals in the two cases showed very close agreement, suggesting that the combined extract of humic and fulvic acids could be used as an indicator of the level of heavy metal pollution in flowing waters.

10.1.2.2 Liquid chromatography

Hayase [6] applied reverse phase liquid chromatography to the examination of molecular weight fractionated sedimentary fulvic acid.

10.1.3 Saline deposited and suspended sediments

10.1.3.1 Spectrofluorimetry

Hayase [7] measured the fluorescence and absorption spectra of humic acid and fulvic acid in sediment collected from Tokyo Bay, at 20°C and pH8. The maximal excitation and emission wavelengths for humic acid were longer than those for fulvic acid, independent of molecular weight,

Table 10.1 Integrated areas of ^{13}C NMR spectra, corrected for variable T_{1p}H's[a]

carbon type	*HA3* N	D	*FA3* N	D	*HA5* N	D	*FA5* N	D
total aliphatic	42	14	51	31	49	19	73	28
alkyl	23	9	16	↑	18	9	18	8
methoxyl	4	↑		25	16	3	tr	tr
		5						
alcoholic	12	↓	27	↓	15	4	39	10
acetal + ketal	3		8	6		3	16	10
total aromatic	42	26	29	23	35	22	12	12
aryl-H,R	↑	↑	↑	↑	24	12	8	8
	42	26	29	23				
aryl-O	↓	↓	↓	↓	11	10	4	4
COOH	15	10	19	15	13	7	18	13
carbonyl	2	2			1	1		

Source: Reproduced with permission from the American Chemical Society [2]

[a]Units in mol %, and tr, N and D are defined as follows: tr, trace, amount detected; N, normal spectra, corrected for effects of T_{1p}H's; D, dipolar dephased spectra (40μs). Arrows imply integrated areas for regions which could not be accurately further subdivided.

and could therefore be used to differentiate between humic acid and fulvic acid in marine deposits. Smaller molecules showed greater fluorescence than larger molecules. Fluorescence intensity per weight concentrations unit increased for humic acid, and decreased for fulvic acid, with increase in absorption coefficient.

10.1.3.2 Liquid chromatography

The Hayase procedure [6] discussed in section 10.1.1.2 has also been applied to marine sediments. Sedimentary fulvic acid exhibited increasing hydrophilic character with increasing molecular weight. The method used was effective for hydrophobichydrophilic characterization of humic substances.

10.1.3.3 Nuclear magnetic resonance spectroscopy

Pontanen and Morris [8] compared the structure of humic acids from marine sediments and degraded diatoms by infrared and C13 and proton NMR spectroscopy. Samples of marine sediments taken from the Peru continental shelf were extracted with water, sodium hydroxide (0.05mol l^{-1}) and sodium pyrophosphate (0.05mol l^{-1}) under an atmosphere of nitrogen and fractionated by ultrafiltration. Humic acids of molecular weight 300 000 and above were examined. Diatoms were collected from ecosystem bags in Loch Ewe, UK. Highly branched aliphatic compounds formed the major fraction of the acids. Carbohydrates, and, to a lesser extent, aromatic compounds, and carbonyl, ether, alcohol and amino groups were found in marine humic acids. The structure of the marine and algal humic acids were similar, and there was evidence that certain sediment humic acids originated from planktonic debris.

10.1.3.4 Miscellaneous

Raspor *et al.* [9] examined humic acids isolated from marine deposits in the Adriatic Sea and the Norwegian Sea, and humic acid and fulvic acid isolated from an estuarine sediment from Borneo. They presented data on their elementary composition, absorption and infrared spectra, distribution of molecular weight, trace metal content and adsorption at a hanging mercury drop electrode. Humic acids from the marine environments had a higher nitrogen content than the estuarine humic acid. The most humified material was that isolated from deep sea sediment in the Norwegian Sea. Fulvic acid had a lower carbon content and a higher oxygen content than the humic acids, and was more hydrophillic.

10.2 Anthropogenic compounds

10.2.1 Non-saline sediments

10.2.1.1 Pyrolysis–gas chromatography–mass spectrometry

de Leeuw *et al.* [10] have screened anthropogenic compounds in sediments by flash evaporation–pyrolysis–gas chromatography–mass spectrometry. Detection limits are in the low ppm range.

10.3 Optical whiteners

10.3.1 Non-saline deposited and suspended sediments

Detergents for laundry washing are mixtures of synthetic chemicals, which are used in very large quantities. Worldwide consumption of a major detergent ingredient, linear alkyl benzenesulphonates, reached 2.8 million tons/year in 1995 [12]. The environmental fate of major detergent components such as surfactants and builders has, therefore been the subject of extensive research. Considerably less attention has been paid to the environmental fate of minor detergent components such as fluorescent whitening agents (Fig. 10.1), which on average contribute only 0.15% of the total mass of laundry detergents [13,14]. The three most important detergent fluorescent whitening agents are sold under the tradenames DAS 1, DSBP and BLS (diaminostilbene, distryrylbiphenyl and bleach

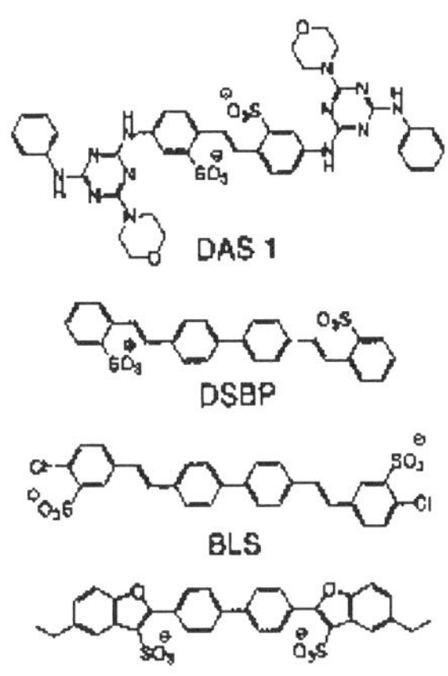

Fig. 10.1 Structures of the fluorescent whitening agents included in this study. Full names: DAS 1, 4,4'-bis[4-anilino-6-morpholino-1,3,5-triazin-2-yl)-amino]stilbene-2,2'-disulphonate; DSBP, 4,4'-bis(2-sulfostyryl)biphenyl; BLS, 4,4'-bis(4-chloro-3-sulfostyryl)biphenyl. Internal standard: 4,4'-bis-5-ethyl-3-sulfobenzofur-2-yl)biphenyl.
Source: Own files

stable, respectively; full names are given in Fig. 10.1). Worldwide production was estimated at 14 000 tons of DAS 1 and 3000 tons of others (predominantly DSBP) for the year 1989 [13]. BLS was used in large-scale laundry facilities (e.g. in hospitals) until quite recently [15].

10.3.1.1 Spectrofluorimetry

Uchiyama [11] has given details of a procedure he developed for the isolation and determination of down to 0.2mg kg^{-1} of fluorescent whitening agents in extracts of bottom deposits. The fluorescent whitening agents were sodium salts of a sulphonated stilbene derivative and this was measured by fluorescence (excitation 370nm, emission 405nm) with the use of tetra-*n*-butyl ammonium hydroxide.

Uchiyama [11] applied this method to the determination of fluorescent whitening agents and alkyl benzenesulphonates and also methylene blue active substances in bottom sediment samples taken in a lake. The muds were filtered off with a suction filter and frozen until analyzed. About 20g of wet bottom mud was extracted three times with a methanol–benzene (1:1) mixture. After the solvent was evaporated using a water bath, the residue was dissolved in hot water and this solution used for analysis. Table 10.2 shows the analytical results for methylene blue active substances (MBAS), alkyl benzene-sulphonate (ABS), and fluorescent whitening agent (FWA) in bottom sediments.

Table 10.2 Analytical results of MBAS, ABS and FWA in bottom sediments

Station no.	*Methylene blue active substances (μg g⁻¹ dry mud)*	*Alkyl benzene–sulphonate (μg g⁻¹ dry mud)*	*Fluorescent whitening agent (μg g⁻¹ dry mud)*
2	377.8	73.2	1.35
3	288.3	96.3	0.83
4	190.3	37.5	0.55
5	281.5	26.9	0.55
6	117.0	28.1	0.25
7	303.5	17.3	0.70
8	107.3	16.9	0.25

Source: Reproduced with permission from Elsevier UK [11]

10.3.1.2 High-performance liquid chromatography

Stroll and Giger [16] have described a reverse phase high-performance liquid chromatographic method for the determination of detergent derived fluorescent whitening agent isomers in lake sediments.

Stereoisomers of the two main laundry detergent fluorescent whitening agents of the diaminostilbene type (DAS 1) (Fig. 10.1) and of the distyryl-biphenyl type (DSBP), as well as total BLS (a compound contained in detergents until a few years ago), were quantitated in sediments and water from Greifensee, a lake in Switzerland. The freeze-dried sediments were extracted in an ultrasonic bath using methanol with tetrabutylammonium hydrogen sulphate as an ion-pairing reagent. Aqueous samples were extracted with C_{18} extraction disks, which were subsequently eluted by methanol with tetrabutylammonium hydrogen sulphate. Extracts from solid samples were analyzed by reversed-phase high-performance liquid chromatography. Fluorescence detection was applied after post-column UV irradiation. Analytical reproducibility ranged from 1 to 12% (relative standard deviation). The limit to quantitation was 1–11g/kg of dry matter. Recoveries ranged from 9.3 to 100% in solid samples. Concentrations of DAS 1 and DSBP ranged from 0.4 to 1.4mg/kg of dry matter in top sediment layers. Concentrations of BLS were between 0.02 and 0.08mg/kg of dry matter in top sediment layers.

10.4 Ethylene diamine tetraacetic acid salts

10.4.1 Soil

10.4.1.1 High-performance liquid chromatography

Nowack *et al.* [17] determined adsorbed iron III and nickel-EDTA species in soil by reverse phase ion-pair high-performance liquid chromatography. Iron III EDTA was found to be the main species present occuring at

30–70% while nickel EDTA species were present in considerably lower amounts (<10%). The adsorbed metal EDTA species were detected in lake sediment and soil cores.

10.5 Mestranol

10.5.1 Soil

10.5.1.1 Gas chromatography

Okuno and Higgins [18] have described a procedure for determining residual levels of mestranol animal damage control chemisterilant and its 3-hydroxy homologue, ethynyloestradiol in soil samples. The lower limits of detection was 0.1ppm. After extraction in acidic medium, samples are cleaned up by Florisil column chromatography. Soil samples are further cleaned up on a gel permeation chromatographic column so that the ethynyloestradiol fraction can be analysed by gas chromatography. The mestranol fraction is again cleaned up by gel separation and a Florisil column. Thin layer chromatography was used to confirm the results obtained by gas chromatography.

In this method 10g of air-dried and pulverized soil sample are weighed with 50ml each of 1.2M hydrochloric acid and acetone, decanted and filtered through a No. 1 Whatman filter paper into another 300ml flask, and rinsed twice with 50ml each of acetone. A three-ball Snyder column is connected to the flask which is placed on a steam bath to evaporate off the acetone. After cooling, 10ml of ethyl ether and 50ml of chloroform are added and the flask shaken for 2min. The contents and two 50ml of chloroform rinses are transferred to a 250ml separatory funnel and shaken for 5min. The lower organic layer is drawn off and passed slowly through 10g of granular anhydrous sodium sulphate into a 300ml flask, then the sodium sulphate is rinsed with 10ml of chloroform and the rinse added to the sample. A Teflon boiling stone is added to the flask, a three-ball Snyder column is attached to the flask which is then evaporated just to dryness on a steam bath.

The sample extract from the soil sample is dissolved in 20ml of chloroform, using an ultrasonic cleaner to facilitate solution. This and a 10ml chloroform rinse are added to a dry 10g Florisil chromatographic column and eluted with an additional 110ml of chloroform. All the eluate is collected in a Kuderna–Danish unit and evaporated just to dryness, then the first 5g Florisil column clean-up procedure is followed.

The concentrate (from the 10g Florisil clean-up of soil samples) is dissolved in 1ml of chloroform and this and a 1ml chloroform rinse are added to a dry 5g Florisil column, then eluated. The remainder is collected in a Kuderna–Danish unit and concentrated just to dryness.

Further clean-up is achieved by gel separation using Biobeads S–X2 gel, 200–400 mesh, Bio Rad Laboratories. Gel permeation chromatography is used to remove extractables of high molecular weight that are not easily removed on the Florisil column. Substances of high molecular weight are eluted first, followed by mestranol and then ethynyloestradiol.

Beckman GC–5 gas chromatograph, equipped with a Model 5671000 Electrometer, flame ionization detector, and a 138cm × 6mm o.d. aluminium column packed with 3% OV–17 on 80–100 mesh Gas Chrom Q were used. The electrometer was operated at a sensitivity of 4×10^{-11}A mV $^{-1}$. Flow rates were about 60ml min $^{-1}$ for hydrogen, 300ml min $^{-1}$ for air, and 60ml min $^{-1}$ for the nitrogen carrier gas. Operating temperatures were 260°C for the column, 275°C for the inlet, and 290°C for the detector. The retention time is about 9min for mestranol and 11.5min for ethynyloestradiol, but at least 40min should be allowed between injection of mestranol sample extracts so that late emerging substances will not interfere with subsequent injections. The ethynyloestradiol fraction is relatively free of these late interfering substances.

Thin layer chromatography was used for qualitative and semi-quantitative confirmation. An appropriate amount of sample solution is spotted on a silica gel thin-layer sheet (Eastman Chromatogram, Type R 30IR2) and developed with chloroform, allowing the solvent front to migrate about 10cm. After drying, the sheet is sprayed with a 1:1 mixture of methanol and sulphuric acid and warmed for 5–10min at 60°C. Mestranol (R_f = 0.75) and ethynyloestradiol (R_f = 0.25) will appear as red spots under visible light and orange under ultraviolet light (154nm). The lower limit of detection is about 100ng for each compound.

Recoveries in soil samples, averaged less than 50% even after corrections. This may have been due to degradation of the compounds by soil micro-organisms or to chemical and physical interactions with the soil. Mestranol recoveries averaged 26–30% from soils at 0.1ppm. Recoveries of ethynyloestradiol were even lower, presumably because of its greater chemical reactivity due to the slightly acidic hydrogen in the 3-hydroxy position.

10.6 Methoxy groups

10.6.1 Soil

10.6.1.1 Gas chromatography

The determination of the methoxy group (OCH_3) content of soils is important in studies concerned with the degree of humification of soil organic matter. Methods used to determine methoxy groups in soils have generally been based on volumetric modifications of the classical Zeisel method, which is a complicated and tedious procedure requiring specialised apparatus.

Magalhâes and Chalk [19] used a combination of Zeisel hydriodic and digestion and gas chromatography to determine methoxy groups in soil.

In this method finely ground soil is boiled with hydriodic acid under reflux for 30min using 1-iodobutane as an internal standard, and the iodomethane formed is extracted with carbon tetrachloride. The gas chromatographic analysis of the extract is rapid with complete resolution of all of the peaks within a total retention time of 1.5min. The method is sensitive and precise, and gives accurate results for pure compounds and pure compounds added to soil. Analysis of a range of soils gave results that were consistent with published data. The method is suitable for the analysis of whole soil, for fractions of soil organic matter or for pure compounds.

In this procedure 1g of dry soil is digested with 5ml of 55%ml hydriodic acid and the mixture refluxed for 30min. 5ml of carbon tetrachloride is then added through the condenser, rinsing with 2 × 5ml of water. 1ml of the solvent layer at the bottom of the flask is removed and transferred to a sealed tube.

Iodoethanes are separated on a glass-lined, stainless-steel column (2m × 3.2mm o.d.) packed with 10% OV–101 on Chromosorb W HP (150–190m), using flow-rates of 90ml min $^{-1}$ for the carrier gas (N_2) and 35 and 350ml min $^{-1}$ for H_2 and air, respectively, for the flame-ionisation detector. Set the injection port, detector and column temperatures at 140, 130 and 60°C, respectively.

This was used to determine the content of methoxy groups in samples of 22 soils from Eastern Australia. A wide range in the methoxy group concentration was observed, which accompanied the wide range in organic concentrations measured (Table 10.3). Positive correlation ($p < 0.001$) was found between the methoxy group and organic C contents of the soils.

Table 10.3 Methoxy and organic carbon contents of surface samples (0–15cm) of 22 soils from Eastern Australia

	Methoxy group content/ μequiv g $^{-1}$ soil	*Organic C, *%*	*Methoxy group content/ organic matter†*
Range	8–128	1.4–14.4	0.29–0.64
Mean±s.d.	38±32	4.5±3.3	0.40±0.09
Median	30	3.5	0.37

Source: Reproduced with permission from the Royal Society of Chemistry [19]

*Determined by Mebius' method
†Assumed soil organic matter = 2 × organic C

10.7 Hexachlorophene

10.7.1 Saline deposited and suspended sediments

10.7.1.1 Miscellaneous

Beller and Simoneit [20] studied the occurrence of hexachlorophene in extracts of estuarine sediments taken from the Hudson River. Hexachlorophene was detected only in the humic acid fractions of the samples, indicating that it could bind strongly to organic matter and was highly resistant to degradation in that form.

10.8 Coprostanol

10.8.1 Sludge

10.8.1.1 Miscellaneous

For the past 50 years the determination of the sanitary quality of water has been based on the enumeration of indicator micro-organisms (e.g. coliform bacteria). The adequacy of coliform enumeration methods for this purpose has been questioned [21]. The current trend of year-round disinfection of waste water effluents and the increasing discharge of both toxic substances and heat from industrial outfalls cast further doubt on the accuracy of biological indicator systems [22].

The use of 5ß-cholestane-3ß-ol (coprostanol) as a molecular marker of faecal pollution of water has been suggested [23–26]. It has been shown that this saturated sterol satisfies the criteria for an indicator of faecal contamination of water [22,27].

5ß-Cholestane-3ß-ol satisfies the generally accepted criteria of a good indicator of faecal pollution. It is believed that the only source of this compound is the faeces of higher animals including man. It is biodegradable and can be removed from domestic sewage by adequate treatment. Furthermore, it has been unambiguously proved that the concentration of coprostanol is highest in the overtly faecal polluted water and there is a progressive decrease in the concentration of this compound in the lesser polluted waters. Since its isolation and identification is unaffected by chlorination or by heat and toxic substances discharged from industrial outfalls the advantage of using a molecular rather than a biological indicator of faecal pollution is further demonstrated. Such a characteristic of coprostanol is especially significant in the current trend promoting disinfection of raw and treated waste water. Because of this unique property, coprostanol might also be a useful indicator in monitoring the source, course and extent of faecal pollution in the ocean or brackish waters where bacteriological evidence is often doubtful.

10.9 Cobalamin

10.9.1 Sludge

10.9.1.1 High-performance liquid chromatography

Beck and Brink [28] have described a sensitive method for the routine assay of cobalamins in activated sewage sludge. The method involves extraction with benzyl alcohol, removal of interfering substances using a combination of gel filtration and chromatography on alumina, concentration of the extract by lyophilization, and direct determination of total cobalamin by high-speed liquid chromatography, in comparison with cobalamin standards.

References

1 Saar, R.A. and Weber, J.H. (1980) *Analytical Chemistry*, **52**, 2095.
2 Wilson, M.A., Vassallo, A.M., Perduc, E.M. and Reuter, J.H. (1987) *Analytical Chemistry*, **59**, 551.
3 Weber, J.H. and Wilson, S.A. (1975) *Water Research*, **9**, 1079
4 Bresnahan, W.T., Grant, C.L. and Weber, J.H. (1978) *Analytical Chemistry*, **50**, 1675.
5 Klenke, T., Oskierski, R.W., Poll, K.G. and Reichel, B. (1986) *GWF Wasser Abwasser*, **127**, 650.
6 Hayase, K. (1984) *Journal of Chromatography*, **295**, 530.
7 Hayase, K. (1985) *Geochemicia and Cosmochimica Acta*, **49**, 159.
8 Pontanen, E.L. and Morris, R.J. (1985) *Marine Chemistry*, **17**, 115.
9 Raspor, B., Nurnberg, H.W., Valentia, P. and Bramica, M. (1984) *Marine Chemistry*, **15**, 217.
10 de Leeuw, J.W., de Leer, E.W.B., Sinnighe Damsté, J.S. and Schnyl, P.J.W. (1986) *Analytical Chemistry*, **58**, 185.
11 Uchiyama, M. (1979) *Water Research*, **13**, 847.
12 Ainsworth, S.J. (1996) *Chemical Engineering News*, **74**, 32.
13 Kramer, J.B. (1992) in *The Handbook of Environmental Chemistry*, (ed. O. Hutzinger), Springer, Berlin, vol. 3, pp.351–366.
14 Anliker, R. (1975) in *Fluorescent Whitening Agents*, (eds. R. Anliker, G. Müller, G.), Georg Thieme Publishers, Stuttgart, Suppl. vol. VI, pp.12–18.
15 Polger, T. (1994) Ph.D. Thesis, ETH Zurich, No. 10832.
16 Stroll, J-M.A. and Giger, W (1997) *Analytical Chemistry*, **69**, 2594.
17 Nowak, B., Kari, F.G., Hilger, S.U. and Sigg, L. (1996) *Analytical Chemistry*, **68**, 561.
18 Okuno, I. and Higgins, A. (1977) *Bulletin of Environmental Contamination and Toxicology*, **18**, 428.
19 Magalhâes, A.M.T. and Chalk, P.M. (1986) *Analyst (London)*, **111**, 77.
20 Beller, H.R. and Simoneit, B.R.T. (1988) *Bulletin of Environmental Contamination and Toxicology*, **41**, 645.
21 Geldreich, E.E. (1971) *Journal of American Water Works Association*, **63**, 225.
22 Kinchmer, C.J. (1971) *5ß Cholestan-3ß-ol: an indication of Fecal Pollution*, PhD Thesis, University of Florida.
23 Dutka, B.J., Chau, A.S.Y. and Coburn, J. (1974) *Water Research*, **8**, 1047.
24 Murtagh, J.J. and Bunch, R.L. (1967) *Journal of Water Pollution Control Federation*, **39**, 404.
25 Smith, L.L. and Gouron, R.E. (1969) *Water Research*, **3**, 141.

26 Taback, H.H., Bloomhuff, R.N. and Bunch, R.C. (1972) *Development in Industrial Biology*. Publication of the Society for Industrial Microbiology, American Institute for Biological Science, Washington, DC, vol. 13. pp.296–307.
27 Gould, R.G. and Cook, R.D. (1958) *Cholesterol, Chemistry, Biochemistry and Pathology*, (ed. R.P. Cook), Academic Press, New York, pp.289–292.
28 Beck, R.A. and Brink, J.J. (1976) *Journal of Environmental Science and Technology*, **10**, 173.

Chapter 11

Mixtures of organic compounds

11.1 Soil

11.1.1 Spectrophotometry

Talsky [1] illustrated the practical use of higher-order derivative spectrophotometry in estimating pollutants in soils by detailed descriptions of the simultaneous estimation of aniline in waste water, phenol in soils and the study of nickel adsorbed on to bentonite powder.

11.1.2 Gas chromatography

Gambrell *et al.* [2] investigated the recovery of DDT, Kepone and Permethrin added to soil suspensions incubated under controlled redox potential and pH conditions.

DDT, Kepone and Permethrin were added to soil and incubated under controlled pH and redox potential conditions to determine the effect with time on the levels of the insecticides and their degradation products. Samples were analysed using gas chromatography, pH and redox potential affected the persistence of pesticides to different degrees. The recovery of DDT was affected by redox potential but not by pH. The stability of Kepone was not affected by pH or redox potential but Permethrin stability was affected by both.

Neumayr [3] has discussed methods for sampling soil atmospheres and gives a detailed account of gas chromatographic methods employing electron capture and flame ionization detectors for detecting and estimating specific components of the soil atmosphere.

Krock and Wilkins [4] have used multidimensional gas chromatography with infrared and mass spectrometric detection to determine organics in soil. Direct acetylation followed by gas chromatography with flame ionization, electron capture and mass spectrometric detectors has been used to determine phenolic residues in soil [5]. Llopart-Visoso *et al.* [6] have used direct acetylation followed by headspace gas chromatography to determine phenolic and cresolic components of soil.

11.1.3 Purge and trap gas chromatography

The use of capillary columns in gas chromatographs coupled with purge and trap concentrators has been accelerated by the publication of methods for the analysis of volatile pollutants using capillary columns by the United States Environmental Protection Agency [7,8]. Analysts using purge and trap gas chromatography in other application areas have long accepted capillary methods. Unfortunately, unique problems are encountered when using a concentrator with a capillary column instead of a packed column. Samples from a concentrator are injected by thermal desorption of an adsorbent trap. This process requires a length of time and minimum flow volume that produce an injection profile of a rather broad nature. Packed columns, due to their comparatively high capacity, can effectively cold trap the sample at the head of the column. Solutes are eluted during temperature programmed runs with the same column efficiency as a normal syringe injection. Due to their reduced capacity, capillary columns cannot trap the analytes efficiently enough at normal operating temperatures to produce good resolution.

A number of methods have been used to refocus purge and trap samples to improve capillary column resolution. All of the methods generally accepted by the analytical community involve cryogenics in one form or another. Non-cryogenic methods that utilize an adsorbent packed refocusing trap exist, but it has been reported [9] that their performance is unsatisfactory unless coupled with a cryogenic gas chromatographic oven. The use of a cryogenic oven to cool the entire column to a point at which the sample is efficiently retrapped has been performed with good results [10]. Unfortunately, this simple method, known as Whole Column Cryofocusing (WCC), suffers from two major shortcomings: the consumption of coolant is excessive and the analysis time is very long. Of these, time is the major factor. The oven takes a fair amount of time to cool to its initial temperature, and then takes more time to heat up to normal temperatures used for separation. These two factors approximately double the time required per sample.

Cryofocusing is a technique in which only a short section of the column or a precolumn is cooled. In its simplest form a section of the column near the inlet is immersed in a flask of coolant during desorb [11]. After desorb the coolant is removed and the column allowed to return to the oven temperature. An automated version of this was introduced in 1983 [12,13]. This system required that a precolumn of uncoated fused silica tubing be attached to the column. Cryofocusing occurred in the precolumn. These systems required the use of a precolumn because sections of the tubing outside of the gas chromatographic oven were not temperature controlled. The presence of a liquid phase in these areas would have led to severe peak distortion and irreproducible retention times. Unfortunately, the use of a precolumn carries two major disadvantages: the precolumn must be connected to the column in a dead-volume free manner, and the cold trap capacity is limited due to the absence of a liquid phase. Zero dead volume connections can be made, yet they require skill and care to properly make the connection. These fittings are then subjected to the temperature cycles of the column oven, and may begin to leak. The cold trap capacity is somewhat limited, and samples with a large amount of organics can cause breakthrough of the trap.

Cryofocusing directly on-column simplifies and improves the focusing process. Since no unions are required, there is no possibility of dead volume or leaks downstream of the cold trap. The stationary phase present significantly improves the trapping efficiency by acting as an adsorbent. Yet trapping on-column requires careful attention to ensure that all parts of the column outside of the oven are carefully temperature controlled. The Tekmar 2000 Series Capillary Interface (Fig. 11.1) uses an advanced cryofocusing trap mounted in a special housing that allows the sample to be refocused on-column with outstanding results [14].

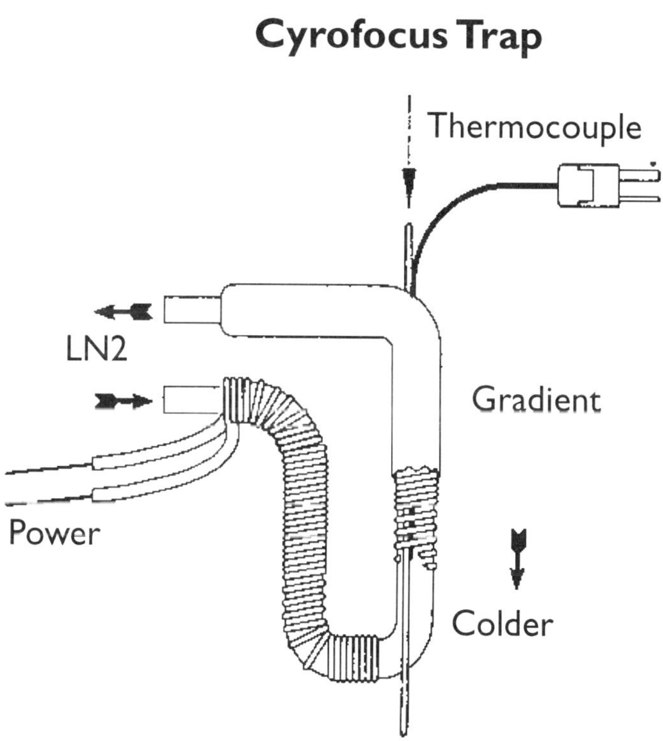

Fig. 11.1 Cut away view of cyrofocus trap.
Source: Own files

The thermal gradient is maintained in all phases of operation. This ensures that in the cyrotrap the sample is always moving from a hotter to a colder region. As they move through the trap the solutes are subjected to an increase in their distribution constants. Solutes at the front of the sample band will move slower than solutes at the rear. This allows the rear portion of the sample band to 'catch up' to the front, effectively compressing the sample band into a very tight slug. With proper establishment of the gradient, heat-up rate of the cyrotrap becomes relatively unimportant. No differences in chromatographic efficiency have been observed for heat-up rates from 100–2000°C/min. Fig. 11.2 illustrates a sample chromatogram obtained on a thin film (0.25micron), narrow bore (0.20mm) column. This therefore represents a 'worst case' scenario, as wider bore and thicker film columns are easier to work with.

Perhaps the greatest limitation of previous generations of cryofocusing traps was the requirement for use of an uncoated precolumn in the trapping zone. Operating in this manner simplified design problems, and made precise temperature control of transfer lines unimportant. However, the absence of any liquid phase limited the trapping mechanism to simple condensation. Highly concentrated or extremely volatile samples often

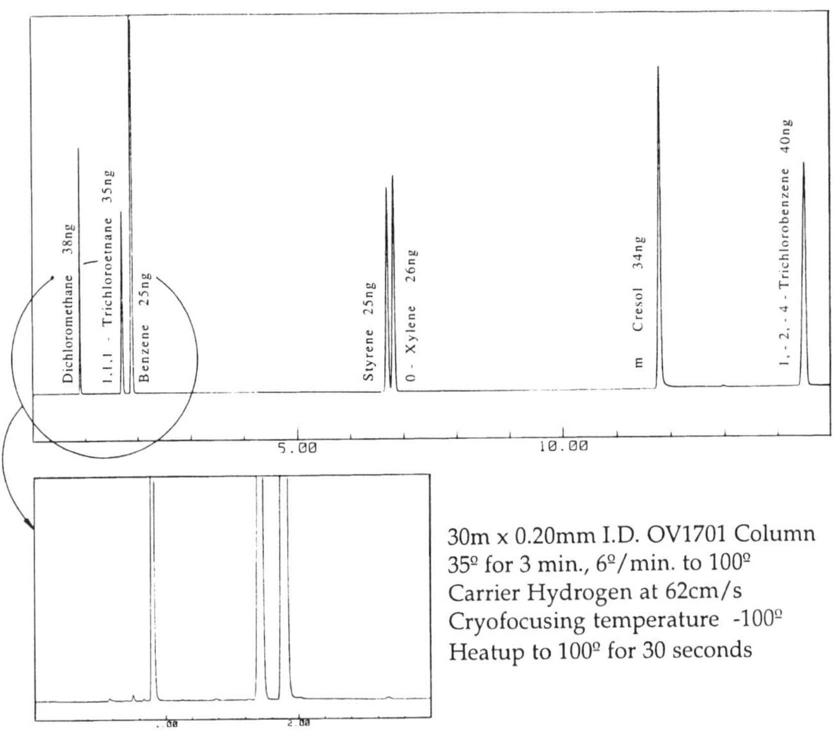

Fig. 11.2 Sample chromatogram obtained using 'worst case' column.
Source: Own files

exceeded the capacity of the cold trap, and the resultant breakthrough produced split peaks and generally poor resolution. By trapping on-column, the capacity of the cyrotrap is approximately equal to the capacity of the column. Breakthrough only occurs with sample loads that would overload the column anyway. The diagram in Fig. 11.3 illustrates the approximate temperature at which breakthrough is experienced on various size columns as a function of solute boiling point. These curves represent hydrocarbon solutes on methyl silicone columns with film thicknesses of 0.25um (0.20mm), 1.0um (0.32mm) and 3.0um (0.53mm).

Cryofocusing traps are often used to interface purge and trap concentrators to gas chromatographs with capillary columns. The enhanced performance characteristics of the design provide a significant improvement over previous systems. The use of a sophisticated cyrotrap with a thermal gradient ensures that the sample will be trapped and injected with high efficiency.

Askari *et al.* [15] have compared purge and trap, methanol immersion and hot solvent extraction methods for the determination of volatile organic compound in aged soil. These workers found that hot solvent extraction is much more effective than the US Environmental Protection Agency approved purge and trap technique [7,8].

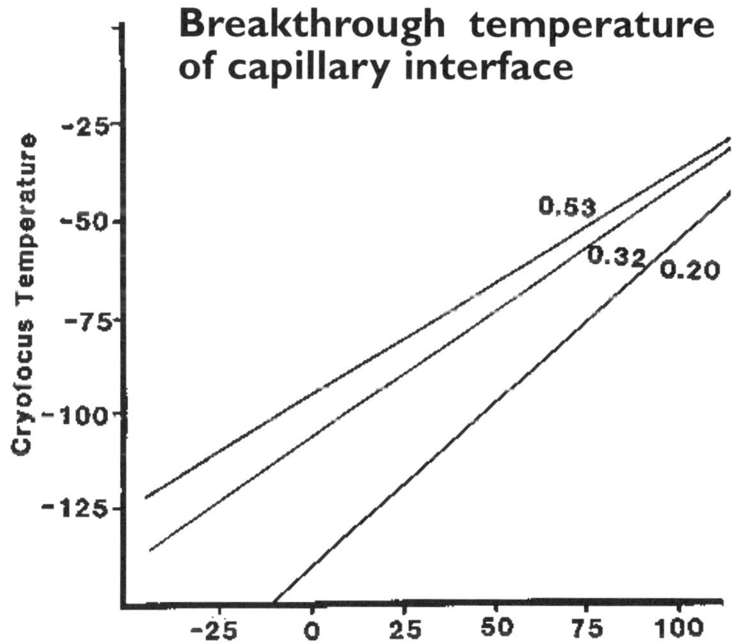

Fig. 11.3 Breakthrough temperature of capillary interface
Source: Own files

11.1.4 Pyrolysis–gas chromatography–mass spectrometry

de Leeuw *et al.* [16] have described a method for the screening in soils and sediments of anthropogenic compounds including polycyclic aromatic hydrocarbons, haloorganics, aliphatic hydrocarbons, heteroaromatics, elemental sulphur and pyrolysis products of synthetic polymers.

Elimination of wet chemical sample preparation enables a complete analysis to be performed and data to be quickly analyzed. The detection limits are in the low part-per-million range using mass spectrometric detection. Alternatively, detection of compounds can be achieved by all common gas chromatography detectors (flame ionization detector, electron capture detector and flame photometric detector), and detection limits are determined by the method of detection employed.

Gas chromatography of the soil extract was carried out on a Varian 3700 instrument equipped with a flame ionization detector and a fused silica capillary column (25in, 0.25mm i.d.) coated with CP–SIL 5. Samples were injected spitlessly, and nitrogen was used as a carrier gas. The oven temperature was programmed from 40°C (5min) to 300°C at 5°C/min.

Both gas chromatographic instruments were connected with a PDP 11/45 computer via an analog-to-digital converter. The peak areas were calculated from the digitalized chromatographic data by means of software developed at Delft University of Technology.

The polyaromatic hydrocarbons in the soil sample were quantitated by using an external standard of anthracene. The results reportedly for a polluted soil and sediment sample indicate that this flash evaporation–pyrolysis technique combined with gas chromatography–mass spectrometry is a valuable tool for rapidly screening polluted samples for virtually all types of anthropogenic contaminants except for heavy metals.

This method allows for the simultaneous detection of highly volatile (e.g. dioxane), volatile (e.g. polyaromatic hydrocarbons) and non-volatile (e.g. polystyrene) substances. The sensitivity of the method depends of course on the detector(s) used. In the gas chromatography–mass spectrometry mode most compounds could be readily detected in the low part-per-million range.

Pyrolysis–gas chromatography has been reported for the analysis of soil samples by several investigators [17–19]. Barrio [17] used pyrolysis gas chromatography to study organic matter evolution in sewage sludge-amended soils by using nitrogen–phosphorus, flame ionization, and mass spectrometric detection. Two methylation reagents and two pyrolysis gas chromatographic techniques have been used by Schulten for organics in soil. Both methylation procedures studied gave valuable additional information on the occurrence of aliphatic and aromatic carboxylic acids, substituted phenols, benzenediols, benzenetriols, phenolic acids and amides in soil organic matter. Pyrolysis products of explosives in soil have been studied by using pyrolysis–gas chromatography with a turnable infrared laser detector [19]. Schnitzer and Schulter [20] have reviewed the analysis of organic matter in soil extracts and whole soils using pyrolysis mass spectrometry.

11.1.5 High-performance liquid chromatography

Reverse phase high-performance liquid chromatography has been used to determine nitrocompounds in soil [21].

11.1.6 High-performance liquid chromatography–mass spectrometry

Supercritical fluid extraction followed by high-performance liquid chromatography–mass spectrometry has been used to determine Atrazine and its metabolites in soil [22].

11.1.7 Supercritical fluid chromatography

Snyder *et al.* [23] compared supercritical fluid extraction with classical sonication and Soxhlet extraction from the analysis of selected insecticides in soils.

Samples extracted with supercritical carbon dioxide modified with 3% methanol at 350atm and 50°C gave greater than 85% recovery of organocholorine and organophosphorus insecticides including Dichlorvos, Diazinon Ronnel, Parathionethyl, Methiadathion, *pp*'DDT, Mirex and decachlorobiphenyl.

Fahing *et al.* [24] studied the modifier effects in the supercritical fluid extraction of organics from soils and clays. Swelling experiments showed that unmodified carbon dioxide did not cause swelling of the soil whereas carbon dioxide modified with water did cause rapid swelling of soil, thereby facilitating extraction of the organics.

Water modified carbon dioxide extraction gave good recoveries of organics from soil. There was a direct correlation between the degree of swelling and extraction efficiency.

11.1.8 Miscellaneous

Lopez-Avila *et al.* [25] studied the microwave assisted extraction of polyaromatic hydrocarbons, phenols and organochlorine insecticides from standard reference soils and sediments.

Solid-phase microextraction has also been used for to determine volatile organic compounds in soil [26]. Target compounds were adsorbed directly from a head-space sample above a soil layer onto a fused-silica fibre. Vacuum distillation coupled with gas chromatography–mass spectrometry [27], head-space membrane extraction with a sorbent interface to multiplex gas chromatography [28], and a prototype near-UV fibre-optic absorption sensor [29] have also been reported for the determination of volatile organic compounds in soil.

Grant *et al.* [30] found that nitramine and nitroaromatic explosive residues in real field soil samples were stable under refrigeration, but nitroaromatics used to fortify samples degraded rapidly, even when samples were refrigerated. Therefore, fortified soils can lead to significant errors.

Soil sampling for volatile organic compounds, including volatile organic compound retention, data quality, sampling devices, storage, preparation and shipping, and sample preparation, have been covered in a recent review [31]. Liikala compared the conventional bulk soil sampling method [32] (i.e. completely filling a container with soil) to placing the soil aliquot in methanol. Results showed large negative biases for aromatic compounds when the conventional bulk sampling method was used, compared to the methanol sampling method.

Grob [32] has produced a well referenced survey of the applications of gas chromatography, liquid chromatography, paper chromatography, thin layer chromatography and ion-exchange chromatography to the determination of organic compounds in soil.

Namiesnik *et al.* [33] have reviewed the analysis of soils and sediments for organic contaminants. They discuss methods of sample preparation and isolation-preconcentration prior to instrumental determination. Compound classes discussed include volatile organic compounds, polychlorobiphenyls, polyaromatic compounds, pesticides and polychlorodibenzo-*p*-dioxins and polychlorodibenzofurans.

Tecator [34] has described an apparatus, the Soxtec System HT6, for the organic solvent extraction of organics from soils preparatory to further analysis.

11.2 Non-saline deposited and suspended sediments

11.2.1 Gas chromatography

The gas chromatographic procedure [2] described in section 11.1.2 for the investigation of the recovery of DDT, Kepone and Permethrin from soils has also been applied to non-saline sediments. Amin and Narang [35] determined volatile organic compounds in sediments at ng g^{-1} concentrations by gas chromatography. Volatile compounds are stripped from sediment samples onto a Porapak N cartridge and then eluted from the cartridge with methanol. Detection limits are ~7ng/g for photoionization-active and 1ng/g for electron-capturing analytes.

11.2.2 Gas chromatography–mass spectrometry

Keith *et al.* [36] and Reijnders *et al.* [37] reviewed applications of gas chromatography–mass spectrometry to sediment analysis. Lopez-Avila *et al.* [38] investigated the efficiency of dichloromethane extraction procedures for the isolation of organic compounds from sediments prior to gas chromatography–mass spectrometry. The compounds investigated were the 51 priority pollutants listed by the Environmental Protection Agency, USA.

11.2.3 High-performance liquid chromatography

Sediments are composed of extremely complex mixtures of compounds belonging to a variety of compound classes. Beside the saturated and aromatic hydrocarbons, heterofunctionalized organic compounds containing nitrogen, oxygen and sulphur (NSO compounds) are important constituents of soluble organic matter from geological materials. A separation of soluble organic matter into compound classes is generally required prior to characterization at the molecular level.

Prior work has been concentrated mostly on the isolation of distinct NSO compound classes from soluble organic matter. Often, selective extraction methods are applied as described for the isolation of, for example, phenols, [39,40] fatty acids [41] and basic nitrogen compounds such benzoquinolines [42], from different matrices. A non-aqueous ion-exchange method has been used for the separation of sediments into acid, base and neutral concentrates [43]. The acid concentrate obtained in this way can be separated into subfractions using *in situ* tetraalkylammonium hydroxide-modified silica high-performance liquid chromatography [44]. Li *et al.* [45] describe a method for the isolation of basic and non-basic azarenes from sediments based on the use of aminocyano-bonded silica high-performance liquid chromatography columns. An overview concerning recent developments in the use of liquid chromatographic techniques is provided by Rowland and Revill [46]. However, in most cases, the described methods are of limited applicability with respect to the analyzable materials and/or the isolable compound classes.

Willish *et al.* [47] have described a method for the rapid fractionation of sediment and rock into compound classes. The method is based on combined polarity/affinity chromatography of soluble organic matter. Five heterocompound fractions are obtained in addition to the conventional saturated and aromatic hydrocarbon fractions. Model compound studies show that those fractions are chemically well-defined. The applicability to a variety of geological materials is demonstrated by analyzing recent lake sediment samples. Generally reproducibilities and linearities are satisfactory. The total recoveries vary between 76 and 57%, depending on the sample type. The method is shown to be well-suited for the bulk compositional characterization of soluble organic matter. Separations can be performed on both the analytical and the semi-preparative scales. Isolated fractions are amenable directly or after derivatization to qualitative and quantitative analysis by gas chromatography and gas chromatography–mass spectrometry.

11.2.4 Gel permeation chromatography

Faure *et al.* [48] have discussed the application of gel chromatography and ultra filtration to the fractionation of organic substances in sediments. Extraction of organic substances from sediments from the River Rhône, France, using caustic soda, and by ion-exchange resins, were compared, and their effect on fractionation by gel chromatography determined. Ultra-filtration and gel chromatography of the extracts were compared and the results are tabulated. The molecular distribution of organic substances in Rhône water samples, concentrated by evaporation, was obtained by using SEPHADEX gel for molecular weights below 700, and ultra-filtration for higher molecular weights. Although the two fractionation methods give comparable results, a combination of the two methods is recommended.

11.2.5 Ultraviolet spectroscopy

Lee *et al.* [49] applied qualitative ultraviolet spectroscopy as an initial guide to source origins of polyaromatic hydrocarbons in river sediments.

11.2.6 Miscellaneous

The flash evaporation pyrolysis gas chromatography method [16] as described in section 11.1.4 for the determination of polycyclic aromatic hydrocarbons, haloorganics, aliphatic hydrocarbons, heteroaromatics, elemental sulphur and pyrolysis products of synthetic polymers in soils has also been applied to non-saline sediments.

Fig. 11.4 shows the total ion current trace and some mass chromatograms obtained by flash evaporation pyrolysis gas chromatography–mass spectrometric analysis of the polluted sediment sample. All compounds present in this complex mixture were not listed. A selection was made to exemplify several aspects of the screening approach. The peak number correspond with the numbers in Table 11.1. Identifications were based on the same criteria as mentioned above. Although several components were shown to be real pyrolysis products, all the compounds are present as such in the sample and resulted from simple thermal extraction from the wire. This was shown in separate analyses using ferromagnetic wires with a Curie temperature of 358°C.

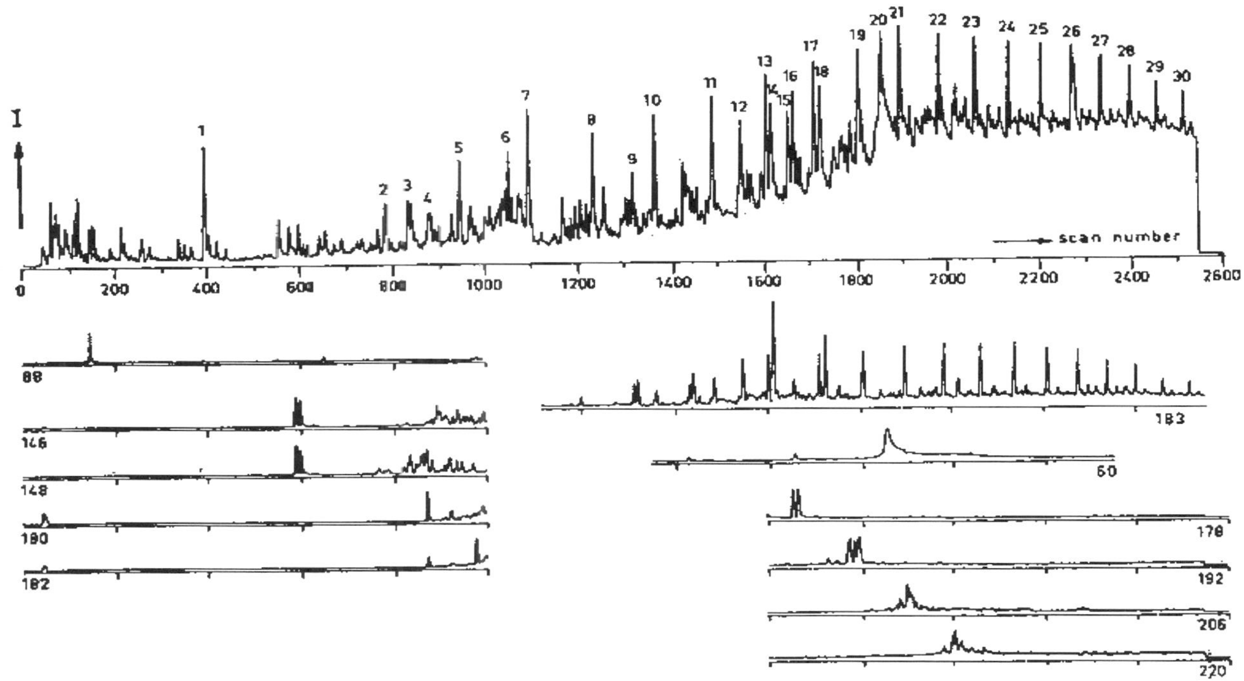

Fig. 11.4 TIC of Ev/Py GS–MS analysis of the polluted sediment sample. The numbers in the mass chromatograms represent the *m/z* values indicative of the following compounds: *m/z* 88 (dioxane); *m/z* 146, 148 (dichlorobenzene); *m/z* 180, 182 (trichlorobenzene); *m/z* 183 (alkanes from C_{14}); *m/z* 60 (hexadecanoic acid); *m/z* 178, 192, 206, 220 (C_0–C_3 anthracenes and phenanthrenes)

Source: Reproduced with permission from the American Chemical Society [16]

Table 11.1 Identified evaporation and pyrolysis products of the sediment sample

peak no.	*compound*	*peak no.*	*compound*
1	styrene	16	anthracene
2	*n*-undecane	17	*n*-octadecane
3	methylidene	18	2,6,10,14-tetramethyl-hexadecane (phytane)
4	naphthalene		
5	*n*-dodecane	19	*n*-nonadecane
6	2-methyl-naphthalene	20	*n*-hexadecanoic acid
7	*n*-tricecane	21	*n*-eicosane
8	*n*-tetradecane	22	*n*-heneicosane
9	2,6,10-trimethyl-tridecane	23	*n*-docosane
10	*n*-pentadecane	24	*n*-tricosane
11	*n*-hexadecane	25	*n*-tetracosane
12	2,6,10-trimethyl-pentadecane (norpristane)	26	*n*-pentacosane
		27	*n*-hexacosane
13	*n*-heptadecane	28	*n*-heptacosane
14	2,6,10,14-tetramethyl-pentadecane (pristane)	29	*n*-octacosane
		30	*n*-nonacosane
15	phenanthrene		

Source: Reproduced with permission from the American Chemical Society [16]

The major contamination found in this sediment sample is derived from mineral oil fractions; a homologous series of *n*-alkanes with no odd-over-even predominance is abundantly present, and other even more specific markers for mineral oil such as the isoprenoid hydrocarbons norpristane, pristane, and phytane are also major components (peaks 12, 14 and 18 in Fig. 11.4). To exemplify this oil pattern, a mass chromatogram of m/z 183 was generated, which almost exclusively reveals the alkane pattern. Phytane and pristane show relatively higher intensities in this mass chromatogram as m/z 183 is enhanced in the mass spectra of these isoprenoid compounds. Mass chromatographic data reduction was further used to visualize a number of other anthropogenic compounds.

The m/z 60 trace shows the presence of stearic acid. A series of mass chromatograms at m/z 178, 192, 206 and 220 selectively revealed the distribution pattern of the unsubstituted and C_1–C_3 substituted phenanthrenes and anthracenes.

Mass chromatography of m/z 146 and 148 and m/z 180 and 182 is shown to be highly selective for di- and trichlorobenzenes. These components are only present in relatively minor amounts. A mass chromatogram at m/z 88 showed the presence of the rather volatile compound dioxane. This sediment sample obviously is heavily polluted with non-biodegraded mineral oil fractions and a number of other components (i.e. stearic acid, chlorinated benzenes), which point to spills of numerous bulk chemicals.

Based on the quantitative determination of pentadecane it was calculated that – with a sample load of about 200μg – alkanes are detected by this screening method if their concentration is 5ppm or more. It is obvious that highly volatile compounds when present as such (e.g. dioxane) cannot be measured quantitatively because considerable losses of such components occur during the evaporation of the suspension liquid from the pyrolysis wire when it is prepared. Quantities measured for such compounds must therefore be considered as minimum values.

Naturally occurring compounds in sediments (lipids, polysaccharides, lignins, etc.) have also been determined by this method. The compounds generated from non-polluted sediments are completely different and easily discriminated from anthropogenic contributions.

The microwave assisted extraction for organic compounds including polyaromatic hydrocarbons, phenols and organochlorine insecticides, described in section 11.1.8 [25] has been applied to sediments. The application of supercritical fluid extraction to the determination of various insecticides in soils described in section 11.1.7 [23] has been applied to river sediments.

Harrison and Young [50] present results of tests to determine the efficiency of, and factors affecting, the extraction of semi-volatile organic priority pollutants from solid samples, using modified Soxhlet procedures. There were marked differences in the efficiency of extraction for different pollutants. The most important factor in obtaining maximal recovery appeared to be the contact period during extraction, but extraction periods longer than 24h resulted in significant errors due to solvent losses.

Nowicki *et al.* [51] point out that in the development of a Soxhlet sample preparation technique for sediment samples, the empty paper Soxhlet thimbles contained organic contaminants which adversely affected results. Glass thimbles were tried and found to be satisfactory. The authors detail the identification of organics solvent-extracted from paper and glass Soxhlet thimbles, and discuss the stability for multiple use of the two materials for trace organic sample preparation.

Workers at the Water Research Centre, UK [52], have drawn up a list of organic substances present in the environment from data contributed by laboratories participating in the COST Project 64b 'Micropollutants', and from the literature, mainly from 1960 onwards. Where possible, the concentration of each substance is given, the contributing laboratory, type of sample, sampling date, method of analysis, and references.

11.3 Saline deposited and suspended sediments

11.3.1 Gas chromatography

Jensen *et al.* [53] have described a method applicable to marine sediments for the determination of polychlorobiphenyls and organochlorine insecticides, in which the sample is extracted with acetone, then n-hexane together with 1% aqueous ammonium chloride. The extracts are then concentrated for purification with concentrated sulphuric acid and aqueous sodium sulphate in the presence of tetrabutyl ammonium sulphate, and finally analysed by gas chromatography.

Ozretich and Schroeder [54] developed an extraction procedure, utilizing sonication with acetonitrile, and cleanup using aminopropyl and/or C–18 bonded phase columns, to prepare marine sediments for priority organic pollutant analysis by gas chromatography. Recoveries from standard reference and interlaboratory comparison sediments and tissue preparations compared favourably to published mean values. Mean recoveries of 22 priority organic pollutants from the sediments ranged from 0% to 84% with a median recovery of 71% and an average percent relative standard deviation (%RSD) of 9%. The effects of sediment type and storage method on the spike recoveries are discussed. Some typical results and recoveries obtained by this procedure are quoted in Table 11.2.

Table 11.2 Standard reference sediment NBS SRM 1645

Compound	*Mean concentration, mg/kg dry, χ (%RSD) n = 3*	*Recovery %*
Naphthalene	0.254(16)	15 ± 6
Hexachlorobutadiene	ND	–
Acenaphthylene	0.092(14)	–
Acenaphthene	0.173 (6)	–
Fluorene	1.93 (4)	69 ± 5
Diethyl phthalate	ND	56 ± 5
Hexachlorobenzene	ND	71 ± 4
Phenanthrene	3.06 (3)	77 ± 2
Anthracene	0.80 (9)	69 ± 5
Di-*n*-butyl phthalate	0.66 (15)	59 ± 8
Fluoranthene	27.5 (5)	–
Pyrene	58.7 (5)	–
n-Butyl benzyl phthalate	0.51 (13)	84 ± 5
Benz (*a*) anthracene plus chrysene	104 (6)	67 ± 4
Bis (2-ethylhexyl) phthalate	0.58 (15)	–
Benzo (*a*) pyrene	75 (7)	179 ± 3
p,p-DDE	0.447 (7)	–
p,p-DDD	0.83 (11)	77 ± 1
p,p-DDT	3.24 (2)	83 ± 1
PCB as Aroclor 1254	1.23 (34)	78 ± 4
Benzo (*k,h*) fluoranthene	31.6 (8)	77 ±10

Source: Reproduced with permission from the American Chemical Society [54]

In this procedure analysis of the extracts was by gas chromatography utilizing either flame ionization or mass spectrometry. The Hewlett-Packard 5880A GC/FID and 5995A GC/MS instruments contained 15m, 0.24mm i.d. fused silica columns with liquid film thicknesses of 1.0μm (DB–5 J&W Scientific Inc., Rancho Cordova, CA). Splitless, 3μL injections with hydrogen carrier gas and a temperature program of 8°C/min from 90–320°C were used for GC/FID analysis. The injection ports were at 250°C.

Japenga *et al.* [56] determined polychlorinated biphenyls and chlorinated insecticides in River Elbe estuary sediments by a procedure in which the sediments were pretreated with acetic acid, mixed with silica and Soxhlet-extracted with benzene/hexane. Humic material and elemental sulphur were removed by passing the extract through a chromatographic column containing basic alumina, on which sodium sulphite and sodium hydroxide were adsorbed. Silica fractionation was followed by gas chromatography to analyse chlorinated pesticides, polychlorinated biphenyls and polyaromatic hydrocarbons. Recovery experiments with standard solutions gave recoveries of 90–102%.

11.3.2 Gas chromatography–mass spectrometry

A method has been described [55] for separating polychlorinated biphenyls from chlorinated insecticides. This procedure involves adsorption chromatography on alumina and charcoal columns, elution with increasing fractional amounts of hexane on alumina columns, and with acetone–diethyl ether on charcoal columns. The polychlorinated biphenyls and chlorinated pesticides are then determined by gas chromatography–mass spectrometry on the separate eluates without interference.

11.4 Sludge

11.4.1 Gas chromatography–mass spectrometry

Warner *et al.* [57] have developed a procedure for the determination of 54 slightly volatile organic contaminants (designated as priority pollutants) in raw and digested sewage sludges at levels down to 0.01ug per g wet weight. The procedure consisted of extraction with methylene chloride or chloroform, followed by clean-up by various chromatographic procedures for specific groups of contaminants, and analysis of the final fractions, in most cases using gas chromatography–mass spectrometry with high-resolution glass capillary columns and specific ion searches. The clean-up steps were regarded as the most critical parts of the sequence, separate techniques being devised for benzidines, phenols, fatty acids and neutral compounds including triglycerides and saturated hydrocarbons. The application of the method to sewage sludge spiked with various compounds is described.

11.4.2 Infrared spectroscopy

Schaumberg *et al.* [58] made a qualitative infrared spectroscopic study of water-soluble compounds extracted from sewage sludge/oil mixtures which were being incubated in the laboratory for 100 weeks at 25°C, and the results are presented. It was found that there was a pattern to the microbial decomposition of anaerobically-digested sewage sludge which involved the disappearance of carbohydrate, protein, sulphonate, and/or sulphate compounds, coupled with the appearance of carboxylates and nitrates.

References

1 Talsky, G. (1983) *International Journal of Environmental Analytical Chemistry*, **14**, 81.
2 Gambrell, R.P., Reddy, C.N., Collard, V. *et al.* (1984) *Journal of Water Pollution Control Federation*, **56**, 174.
3 Neumayr, V. (1986) *Soil and Groundwater Protection* (eds. G. Oklde and R. Leschber), Gustav Fischer Verlag, Stuttgart.
4 Krock, E.A. and Wilkins, C.L. (1996) *Journal of Chromatography A*, **726**, 167.
5 Danis, T.G. and Albanis, T.A. (1996) *Environmental Toxicology and Chemistry*, **53**, 9.
6 Llopart-Visoso, M.P., Lorenzo-Ferreira, R.A. and Cela-Torrifos, R. (1996) *Journal of High Resolution Chromatography and Chromatography Communications*, **19**, 207.
7 United States Environmental Protection Agency (1986) *Volatile Organic Compound in Water by Purge and Trap Capillary Column Gas Chromatography–Mass Spectrometry*, Method 524.2.
8 United States Environmental Protection Agency (1986) *Volatile Organic Compounds in Water by Purge and Trap Capillary Column Gas Chromatography with Photoionization and Electrolytic Detectors in Series*, Method 502.2. September.
9 Pankow, J.W. (1986) *Journal of High Resolution Chromatography and Chromatography Communications*, **9**, 18.
10 Pankow, J.W. and Rosen, M.W. (1984) *Journal of High Resolution Chromatography and Chromatography Communications*, **7**, 504.
11 Trussell, A.R., Moncur, J.G., Lieu, F-Y. and Leong, Y.C. (1983) *Journal of High Resolution Chromatography and Chromatography Communications*, **6**, 292.
12 Capillary Column use in Purge and Trap Gas Chromatography II: Use of the Model 1000 Capillary Interface, Tek/Data B021684, Tekmar Company, Cincinnati, Ohio.
13 Kirshen, N. (1984) *American Laboratory*, **16**, 12.
14 Westendorf, R.G. (1988) *Performance of a Third Generation Cyrofocusing Trap for Purge and Trap Gas Chromatography*. Paper presented at the 39th Pittsburgh Conference, February 1988, New Orleans, LA. Tekmar Company, P.O. Box 371856, Cincinnati, Ohio, 45222–856.
15 Askari, M.D.F., Maskarinec, H.P., Smith, S.M. *et al.* (1996) *Analytical Chemistry*, **68**, 3431.
16 de Leeuw J.W., de Leer, E.W.B., Sinninghe Dansté, J.S. and Schuyl, P.J.W. (1986) *Analytical Chemistry*, **58**, 1852.
17 Barrio, M.E., Lliberia, J. Li., Comellas, L. and Broto-Puig, F. (1996) *Journal of Chromatography*, A, **719**, 131.

18 Schulten, H.R. and Sorge, C. (1995) *Soil Science*, **46**, 567.
19 Wormhoudt, J., Shorter, J.H., McManus, J.B. *et al.* (1996) *Applied Optics*, **35**, 3992.
20 Schnitzer, M. and Schulter, H.R. (1995) *Advanced Agronomy*, **55**, 167.
21 Walsh, M.E., Jenkins, T.F. and Thorne, P.G. (1995) *Journal of Energy and Materials*, **13**, 357.
22 Papillond, S., Haerdi, W. and Chiron, S. (1996) *Environmental Science and Technology*, **30**, 1822.
23 Snyder, J.L., Grob, R.L., McNally, M.E. and Oostdyk, T.S. (1992) *Analytical Chemistry*, **64**, 1940.
24 Fahing, T.M., Paulatic, M.E., Johnson, D.M. and McNally, M.E.P. (1993) *Analytical Chemistry*, **65**, 1462.
25 Lopez-Avila, V., Yong, R. and Beckert, W.F. (1994) *Analytical Chemistry*, **66**, 1097.
26 James, R.J. and Stack, M.A. (1996) *Journal of High Resolution Chromatography and Chromatography Communications*, **19**, 515.
27 Hiatt, M.H. (1995) *Analytical Chemistry*, **67**, 4044.
28 Yong, M.J. and Pawliszya, J. (1996) *Journal of Microcolumn Separation*, **8**, 89.
29 Barber, T.E., Fisher, W.G. and Wachter, E.A. (1995) *Environmental Science and Technology*, **29**, 1576.
30 Grant, C.L., Jenkins, T.F., Myers, K.F. and McCormick, E.F. (1995) *Environmental Toxicology and Chemistry*, **14**, 1865.
31 Liikala, T.L., Olsen, K.B., Teel, S.S. and Lanigan, D.C. (1996) *Environmental Science and Technology*, **30**, 3441.
32 Grob, R.L. (ed.) (1975) *Chromatographic Analysis of the Environment*, Dekker, New York, 744 pp.
33 Namiesnik, J., Zygmunt, B., Biziuk, M. *et al.* (1996) *Environmental Studies*, **5**, 5.
34 Tecator Soxtec System HT6. Tecator A.B., Box 70, 5–26301, Hoganas, Sweden. Tecator Ltd., Cooper Road, Thornbury, Bristol BS12 2UP, UK.
35 Amin, T.A. and Narang, R.S. (1985) *Analytical Chemistry*, **57**, 648.
36 Keith, L.H., Lin, P.H. and Kilpatrick, M.P. (1981) *Water Quality Bulletin*, **6**, 34.
37 Reijnders, H.F.R., Onderdelindin, D., Visser, M.G. and Griepiak, B. (1980) *Water Research*, **14**, 1645.
38 Lopez-Avila, V., Norcutt, R., Oustat, J. and Wiekham, M. (1983) *Analytical Chemistry*, **55**, 881.
39 Ioppolo, M., Alexander, R. and Kagi, R.I. (1992) *Organic Geochemistry*, **18**, 603.
40 Bennett, B., Bowler, B.F.J. and Larter, S. (1996) *Analytical Chemistry*, **68**, 3697.
41 Rezanka, T. (1992) *Journal of Chromatography*, **627**, 241.
42 Yamamoto, M., Taguchi, G. and Sasaki, K. (1991) *Chemical Geology*, **93**, 193.
43 Green, J.B., Hoff, R.J., Woodward, P.W. and Stevens, L.L. (1984) *Fuel*, **63**, 1290.
44 Green, J.B. (1986) *Journal of Chromatography*, **358**, 53.
45 Li, M., Larter, S.R., Stoddart, D. and Bjorøy, M. (1992) *Analytical Chemistry*, **64**, 1337.
46 Rowland, S.J. and Revill, A.T. (1995) in *Chromatography of the Petroleum Industry* (ed. E.R. Adlard), Journal of Chromatography Library, 56, Elsevier, Amsterdam, pp.127–141.
47 Willisch, H., Clegg, H., Horsfield, B. *et al.* (1997) *Analytical Chemistry*, **69**, 4203.
48 Faure, J., Viallet, P. and Picat, P. (1975) *La Tribune Due Cebedeau*, **28**, 439.
49 Lee, H.K., Wright, G.J. and Swallow, W.H. (1988) *Environmental Pollution*, **49**, 167.
50 Harrold, D.E. and Young, J.G. (1982) *Journal of Environmental Engineering Division, ASCE*, **108**, 1211.
51 Nowicki, H.G., Kieda, C.A., Devine, R.F. *et al.* (1979) *Analytical Letters*, **12**, 769.
52 Water Research Centre (Stevenage Laboratory) (1974) COST–Project 646. Commission of the European Communities EUCO/MDU/40/74.

53 Jensen, R., Renberg, L. and Reutergard, L. (1977) *Analytical Chemistry*, **49**, 316.
54 Ozretich, R.J. and Schroeder, W.P. (1986) *Analytical Chemistry*, **58**, 2041.
55 Teichman, J., Benvenue, A. and Hylin, J.W. (1978) *Journal of Chromatography*, **151**, 155.
56 Japenga, J., Wapenaar, N.J., Smedes, F. and Salomons, W. (1987) *Environmental Technology Letters*, **8**, 9.
57 Warner, J.S., Junglcaus, A., Engel, T.M. *et al.* (1980) Report No. EPA/2–80/030, US Environmental Protection Agency, Cincinnati, Ohio.
58 Schaumberg, G.D., Levesque-Madore, C.S., Sposito, G. and Lund, L-J. (1980) *Journal of Environmental Quality*, **9**, 297.

Chapter 12

Non metals and metalloids

12.1 Boron

12.1.1 Soil

12.1.1.1 Spectrophotometry

Spectrophotometric methods have been used to determine water soluble boron in soils. In one method [1] the soil is extracted with boiling water then converted to fluoroborate which is evaluated spectrophotometrically as the methylene blue complex.

Aznarez *et al.* [2] have described a spectrophotometric method using curcumin as chromopore for the determination of boron in soil. Boron is extracted from the soil into methyl isobutyl ketone with 2-methylpentane-2,4-diol. In this method 0.2–1g of finely ground soil is digested with 5ml concentrated nitric-perchloric acid (3 + 1) in a polytetrafluoroethylene lined pressure pump for 2h at 150°C. The filtrate is neutralized with 6M sodium hydroxide and diluted to 100ml with hydrochloric acid 1 + 1. This solution is triple extracted with 10ml of methyl isobutyl ketone to remove iron interference. This solution is then extracted with 10ml 2-methyl pentane-2,4 diol and this extract dried over anhydrous sodium sulphate.

The development of the colour is carried out in the organic phase used for extraction by the addition of curcumin in glacial acetic acid and phosphoric acid as dehydrating agent. Spectrophotometric evaluation is carried out at 510nm.

The effect of those ions most frequently present in soils on the boron determinations is shown in Table 12.1. The interference of iron at concentrations higher than 7×10^{-5}M can be eliminated as the chloro complex by extraction with methyl isobutyl ketone. The total elimination of iron III was not necessary as the phosphoric acid masked the residual iron III in the boric acid–curcumin reaction.

The results of boron determinations in soils by the spectrophotometric method are shown in Table 12.2. Soil samples were provided by Aula Dei, Experimental Station of the Consejo Superior de Investigaciones Cientificas (CSIC), and correspond to alluvial ground or soils with a high limestone content (35–45% of calcium carbonate).

Table 12.1 Effect of foreign ions on boron determination

Determination of 58.3μg of boron by the spectrophotometric method and 2.38μg of boron by the fluorometric method

	Maximum concentration tested without giving interference/m	
Foreign ion	*Spectrophotometric method*	*Fluorometric method*
Cl^-	7.2	7.2
SO_4^{2-}	2.5	2.5
NH_4^+	1.0	0.8
Na^+	0.2	0.2
K^+, Ca^{2+}, Al^{3+}	0.5	0.4
NO_3^-, NO_2^-, Cr^{3+}	0.2	0.2
Mg^{2+}, HCO_3^-, CO_3^{2-}	0.05	0.05
Mn^{2+}	0.05	0.01
F^-	0.02	0.01
Cu^{2+}, Zn^{2+}, Ni^{2+}	4×10^{-3}	4×10^{-3}
Sr^{2+}, PO_4^{3-}, SiO_3^{2-}, Br^-	3×10^{-3}	3×10^{-3}
Fe^{3+}	7×10^{-5}*	7×10^{-5}*
Fe^{3+} (by elimination with three 10ml IBMK washes)	0.1	0.1

Source: Reproduced with permission from the Royal Society of Chemistry [2]

*Tolerated limit (M) as the concentration level at which the interferent causes an error of not more than ±2% (spectrophotometric method) or ±3% (fluorometric method)

Table 12.2 Determination of boron in soils by the spectrophotometric method

Soil	*Source*	*Mean boron content ± standard deviation*/μg g⁻¹*	*Boron added to spiked samples/μg*	*Mean recovery,† %*
1	El Bayo (H_1)	249 ±3.4	250	97.2
2	El Bayo (H_2)	181 ±1.3	200	104.0
3	El Bayo (H_3)	260 ±2.4	250	101.6
4	El Bayo (H_4)	288 ±1.0	300	102.7
5	El Espinal (H_1)	78 ±0.6	80	98.8
6	Aula Dei (H_1)	165 ±0.6	150	100.7
7	Villamayor (H_1)	79 ±0.3	80	98.8
8	Villamayor (H_2)	94 ±0.4	100	99.0

Source: Reproduced with permission from the Royal Society of Chemistry [2]

*Eight determinations

†Three determinations

A further spectrophotometric method [3,4] for water soluble boron in soil, boron is extracted from soil with boiling water. Borate in the extract is converted to fluoroborate by the action of orthophosphoric acid and sodium fluoride. The concentration of fluoroborate is measured spectrophotometrically as the blue complex formed with methylene blue and which is extracted into 1, 2-dichloroethane. Nitrates and nitrites interfere; they are removed by reduction with zinc powder and orthophosphoric acid.

12.1.1.2 Inductively coupled plasma atomic emission spectrometry

Inductively coupled plasma optical emission spectrometry has been applied to the determination of boron in soil extracts in amounts down to 0.5m L^{-1} [5].

To extract boron from the sample a 50g sample of air-dried soil was boiled under reflux with 100ml of water for 10min. After partial cooling, the extract was filtered through an 18.5cm Whatman No. 3 filter paper into a conical beaker and a 40ml aliquot of the filtrate transferred to a silica beaker. The solution was taken to dryness and the residue was oxidized twice with 10ml of 6% hydrogen peroxide solution. The residue was then diluted to 50ml.

Boron has a very simple ICPAES spectrum with the sensitive doublet at 249.7 and 249.8nm being the only useful analytical lines. Between 0.4 and 0.7mg L^{-1} boron was found in soil extracts.

Zarcinas and Cartwright [6] studied the acid dissolution of boron from soils prior to determination by inductively coupled plasma atomic emission spectrometry.

12.1.1.3 Molecular absorption spectrometry

In addition to the spectrophotometric method discussed in section 12.1.1.1 Aznarez *et al.* [2] have described a method based on the molecular fluorescence of boron with dibenzoylmethane. The preliminary soil digestion and extraction procedures are identical to those described earlier. The reactive fluorescence intensity of the boron complex is measured at 400nm with excitation at 390nm and quinine sulphate as reference.

The calibration graph was linear in the range 0.5–5μg of boron in aqueous solution (20–200μg L^{-1} of boron in the final solution to be measured). The detection limit and precision were 1μg L^{-1} of boron and 3% for ten replicate determinations of 1.2μg of boron, respectively. Interference by foreign ions is minimal (see Table 12.1).

12.1.1.4 Miscellaneous

In an interlaboratory comparison of metal determinations in soils the element boron is mentioned [7].

12.1.2 Saline deposited and suspended sediments

Kiss [8] examined various techniques for the efficient separation and preconcentration of boron from marine sediments. Alkaline fusion with potassium carbonate was used to render boron reactive, even in the most resistant silicate minerals. Fusion cakes were extracted with water and borate was isolated by Amberlite XE–243 boron-selective resin. Borate was determined spectrophotometrically, following elution with 2 mol L^{-1} hydrochloric acid. Either the carminic acid complex (620nm), formed in sulphuric acid (94%) or sulphuric:acetic acid (1:4), or the azomethine hydrogen ion association complex (415nm) formed at pH5.2, were used for borate measurement.

12.2 Halogens

12.2.1 Soil

12.2.1.1 Spectrophotometry

Van Vleit *et al.* [9] have described a semiautomatic spectrophotometric method using a Technicon Autoanalyser for the determination of iodine in soil extracts. The method has a coefficient of variation of 2.1% at the 8.6mg L^{-1} iodine level to 6.1% at the 1.4mg L^{-1} iodine level.

In this method 1g of soil is refluxed with 2M sodium hydroxide after centrifuging to remove solids. The clean extract is digested with perchloric acid–nitric acid at 265°C. The iodine content of the extract is determined by the catalytic action of iodine on the oxidation of arsenic III ions with cerium IV ions. Between 96 and 97% recovery of added iodine spikes to soil were obtained by this method.

12.2.1.2 Gas chromatography

Roughan *et al.* [10] have described a gas chromatographic method for the determination of total bromine (and bromide) originating as methyl bromide fumigant in soils (and crops).

In this method 1g dried soil in a silica crucible is mixed with 5ml 0.2N sodium hydroxide and 10ml absolute ethanol and left to evaporate on a hot plate for 12h. The residue is digested with 0.6N sulphuric acid. After treatment with ethylene oxide and acetonitrile the acetonitrile layer is recovered.

The 2-bromoethanol derived from the reaction between the bromide anion and ethylene oxide is determined by electron capture gas chromatography interfaced to the computing integrator. Calibrate with 5μL injections of standard solutions of 2-bromoethanol containing from 1–10ng of bromine using peak height as the measured parameter. Sample solutions should be diluted if necessary to bring them within the calibration range using 5µL injections. The resolution of the 2-bromoethanol peak is good even in the presence of large amounts of extraneous ethylene oxide reaction products. A typical chromatogram is shown in Fig. 12.1; the magnitude of the peaks is variable, depending on the substrate and the characteristics of each individual column.

Excellent recoveries of bromine (bromide) were obtained on soil samples by this procedure (Table 12.3).

12.2.1.3 Neutron activation analysis

Gladney and Perrin [11] used epithermal neutron activation analyses to determine down to 50ppm total bromine in soils. Excellent agreement with recommended values were obtained for a range of Canadian reference soils (Table 12.4).

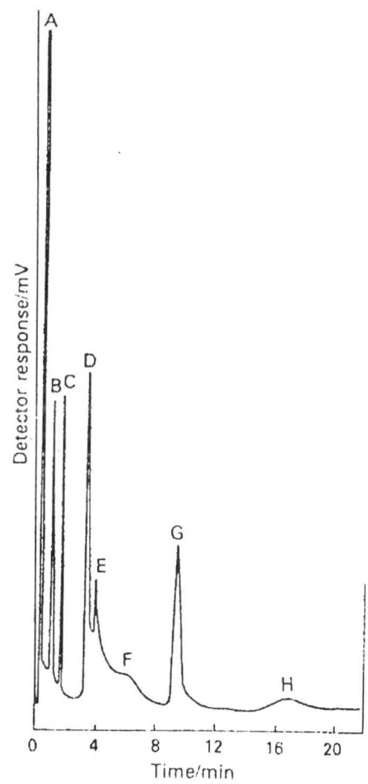

Fig. 12.1 Typical chromatogram containing: A, acetonitrile; B, 2-chloroethanol; C, 4ng of 2 bromoethanol; D, ethane-1,2-diol; E, 2-iodoethanol; F, acetamide; G, unknown and H, 2,2'-dihydroxydiethyl ether.
Source: Reproduced with permission from the Royal Society of Chemistry [10]

Table 12.3 Recoveries from spiked 1g samples of oil

Soil	*Br⁻ added* $\mu g\ g^{-1}$	*Br determined (background corrected) %* $\mu g\ g^{-1}$	*Recovery %*	*Mean recovery %*	*sd %*	*Coefficient of variation %*	*Standard error %*
A	50	40.9	81.8	81.2	1.44	1.77	±0.55
	50	392	78.4				
	50	402	80.4				
	40	41.3	82.6				
	50	40.7	81.4				
	50	40.8	81.6				
B	500	464	92.8	93.8	1.86	1.98	±0.70
	500	460	92.0				
	500	463	92.6				
	500	470	94.0				
	500	481	96.2				
	500	483	96.6				
	600	463	92.6				

Source: Reproduced with permission from the Royal Society of Chemistry [10]

Table 12.4 Bromine concentrations in Canadian certified reference soils

Soil ref no.	*Soil type*	*Bottle no.*	*Br, ppm* $\chi \pm \alpha$	*recommended value*
SO-1	Regosolic	133	1.3 ±0.1	1.4 ±0.2
	clay	711	1.4 ±0.2	
SO–2	Podzolic	97	14.8 ±1.0	15 ±2
	B Horizon	903	15.2 ±1.5	
SO–3	Calcareous	495	5.5 ±0.7	5.2 ±0.8
	C Horizon	1023	4.8 ±0.2	
SO–4	Chernozemic	103	5.5 ±0.3	5.6 ±0.5
	A Horizon	441	5.8 ±0.5	

Source: Reproduced with permission from the American Chemical Society [11]

Randle and Hartman [12] used thermal neutron activation in analysis to investigate total bromine in humic compounds in soil. Bromine was extracted from the soil water with sodium hydroxide or sodium pyrophosphate, then the extract dried prior to analysis.

12.2.2 Non-saline deposited and suspended sediments

12.2.2.1 Neutron activation analysis

Madaro and Moauro [13] as part of an interlaboratory analysis organized by the Italian Association of Oceanology and Limnology, analysed one river sediment and one lake sediment from heavily polluted environments for aluminium, arsenic, barium, bromine, calcium, cerium, cobalt, chromium, caesium, dysprosium, europium, iron, gadolinium, hafnium, potassium, lanthanum, manganese, sodium, neodymium, nickel, rubidium, antimony, scandium, samarium, tantalum, terbium, thorium, uranium, vanadium, tungsten, ytterbium and zinc. Precision and accuracy were determined for Orchard Leaves (National Bureau of Standards (NBS) SRM 1571). Only sodium and zinc differed by more than 20% from NBS certified values. The zinc value was affected by interference from scandium-46. Other interferences were insignificant. Possible errors due to concentration levels being close to detection limits are discussed.

12.2.3 Sludge

12.2.3.1 Ion selective electrode

Rea [14] gives details of a method developed for the determination of fluoride in sludge, using a selective ion electrode with a trisodium citrate buffer and a standard halogen addition procedure. Results obtained on sludges supplied by various water authorities are tabulated; these showed that where specific fluoride-bearing wastes are discharged the fluoride concentration in the sludge is increased, but otherwise the fluoride content of the sludge is little affected by the amount of industry in the area or the extent of fluoridation of water supplies.

12.3 Total organic carbon

12.3.1 Soil

12.3.1.1 Titration

In a standard method [15,19] soil organic matter is almost completely oxidized by boiling gently with a solution of potassium dichromate, sulphuric acid and phosphoric acid. Excess dichromate is determined by titration with standard ferrous sulphate solution.

12.3.1.2 Combustion methods

Carbon in soil and plant materials can be determined by wet- and dry-combustion methods [7,8]. In both instances, soil and plant carbon is converted into carbon dioxide, absorbed in alkali and determined either by titration against a standard acid or by weighing. These methods involve large apparatus, are expensive and time consuming, and therefore cannot be adapted to the routine analysis of a large number of samples.

Dalal [16] has described a wet digestion procedure for the determination of total carbon in soils. The digestion apparatus consisted of a McCartney bottle (i.d. 22mm, capacity 28cm^3) with an aluminum screw-cap fitted with a 2.5mm thick neoprene seal, and a test-tube (i.d. 13mm, capacity 8cm^3). The digestion mixture was prepared by dissolving 25g of chromium trioxide in 100cm^3 of concentrated sulphuric acid–orthophosphoric acid (2 + 1). This mixture was heated to 140–150°C, then cooled and tightly stoppered to prevent the absorption of moisture from the atmosphere.

A suitable amount of soil (particle size less than 0.15mm) containing less than 10mg of carbon, was weighed into a McCartney bottle and 5cm^3 of chromic acid digestion mixture were added rapidly. With the aid of forceps the test-tube, containing 5.0cm^3 of 0.4M sodium hydroxide solution, was lowered into the bottle, which was immediately stoppered tightly. Appropriate blanks were prepared simultaneously. The bottles were autoclaved at 121°C and approximately 105kPa for 1h, then left overnight at room temperature. The test tube was removed from the bottle, its outside washed free of acid and sodium hydroxide solution was transferred into a graduated tube (capacity 36cm^3, length 150mm) fitted with a C19/17 Quickfit or similar stopper. The volume was made up to 10.0cm^3 with carbon dioxide free distilled water and the tube was stoppered.

For the determination of the total carbon in the samples, 1cm^3 of saturated barium chloride solution and 0.05cm^3 of 1% phenolphthalein solution (prepared in ethanol) were added to the remaining 100cm^3 of the sodium hydroxide solution in the tube. The mixture was titrated against 0.10M hydrochloric acid until the colour of the solution changed from red to colourless. By using a magnetic stirrer and titrating under reflected light, the end-point in the titration was a reproducible. From the volume of 0.10M hydrochloric acid used for the titration of the sample and the blank, the amount of organic carbon in the sample can be calculated. 1cm^3 of 0.10M hydrochloric acid ≡0.60mg of carbon

Studies on the effect of the size of the soil sample taken showed that the total carbon recovery was low when the carbon content of the sample did not exceed 1mg or the amount of soil taken exceeded 1g.

Organic carbon can be determined in calcareous soils after the carbonates have been removed by treatment with sulphuric acid–iron(II) sulphate solution and the samples oven-dried at 105°C. However, as in all other wet-combustion methods, chloride ions interfere [7]. Interference from small amounts of chloride ions (up to 4mg of Cl^- as KCl or NaCl) was reduced by adding 2.5% of mercury II oxide or silver I sulphate to the acid digestion mixture.

Good agreement was obtained between this wet combustion method [9] and a method based on dry combustion [17] (Table 12.5).

Charles and Simmons [18] have reviewed methods for determining total, inorganic and organic carbon in soils and sediments, particularly limitations and sources of error associated with dry combustion and wet oxidation techniques. Preferred methods, for greater quantitative

Table 12.5 Carbon contents of soils obtained by the proposed and dry-combustion methods

Sample No.	Carbon content % Proposed method	Dry-combustion method	Sample No.	Carbon content % Proposed method	Dry-combustion method
1	0.353 ± 0.012	0.37 ± 0.02	10	1.272 ± 0.028	1.3 ± 0.05
2	1.049 ± 0.018	1.04 ± 0.02	11	1.673 ± 0.012	1.70 ± 0.03
3	2.297 ± 0.021	2.46 ± 0.07	12	2.155 ± 0.015	2.14 ± 0.04
4	4.102 ± 0.029	4.36 ± 0.13	13	0.859 ± 0.005	0.80 ± 0.03
5	1.502 ± 0.010	1.52 ± 0.04	14	2.196 ± 0.015	2.26 ± 0.04
6	1.728 ± 0.012	1.69 ± 0.02	15	2.940 ± 0.019	3.02 ± 0.06
7	3.319 ± 0.052	3.64 ± 0.07	16	1.386 ± 0.015	1.46 ± 0.07
8	2.237 ± 0.020	2.32 ± 0.04	17	17.148 ± 0.32	16.90 ± 0.55
9	1.428 ± 0.018	1.36 ± 0.02			

Source: Reproduced with permission from the Royal Society of Chemistry [16]

accuracy, were dry combustion methods. Problems associated with cost, convenience, rapidity and applicability had been overcome by the development of microcombustion techniques and commercial analysers. Variations in furnace type and trap type used in combustion processes are outlined. Wet oxidation methods are based on potassium dichromate and gravimetric methods based on carbon dioxide evolution were used. The major sources of error with dichromate oxidations arose from the incomplete oxidation of certain compounds and interference from oxidizable inorganic compounds such as chlorides.

12.3.1.3 Potentiometry

Begheijn [20] determined organic and inorganic carbon in soils by potentiometry. The equivalent amount of carbon dioxide, liberated by a combustion technique, is absorbed and precipitated in a solution containing 2.50mmol of sodium hydroxide and 0.96mmol of barium chloride. The resultant barium carbonate is dissolved by the addition of 0.99 mmol of EDTA (disodium salt). The pH of the final solution is related to the amount of carbon dioxide present by means of a calibration graph. By use of a pH meter with an expanded range of 1pH unit, sensitivity to 10μg of carbon is achieved. Analyses of pure organic compounds, such as benzoic acid and hydroquinone, show recoveries accurate to within 0.1mg of carbon. Results for calcareous soils are in close agreement with values for calcium oxide obtained by use of X-ray fluorescence spectrometry (Table 12.6).

Table 12.6 Results of inorganic carbon analysis of calcareous soils by potentiometry compared with those for calcium oxide by X-ray fluorescence spectrometry

Potentiometric results per 100mg of dry, powdered soil samples; analyses on five batches on five different days

Sample number	*Carbon by potentiometry/mg**	*Mean/mg*	*CaO content if all carbon is in $CaCO_3$, %*	*CaO content by X-ray fluorescence spectrometry, %*
73/230	5.45, 5.55, 5.45	5.48	25.6	26.0
231	5.20, 5.35, 5.20, 5.25	5.25	24.5	25.0
232	5.80, 5.95, 5.80, 5.95	5.88	27.4	26.9
233	5.90, 6.20, 5.85, 6.10, 6,25	6.05	28.3	27.9
234	5.35, 5.55, 5.35	5.42	25.3	25.6
235	3.60, 3.75, 3.50, 3.70, 3.80	3.67	17.1	18.5
236	4.55, 4.70 4.45	4.57	21.3	22.5

Source: Reproduced with permission from the Royal Society of Chemistry [20]

*Readings from graph are accurate to 0.05mg of carbon

Interference from chloride is effectively eliminated by a preliminary evaporation step.

12.3.2 Non-saline deposited and suspended sediments

Various techniques have been used for the determination of organic (and total) carbon in sediments. These include both wet and dry combustion methods which depend on the quantitative conversion of the organic (or total) carbon to carbon dioxide [30–32]. In addition, an approximate assay technique reported by Bremner and Jenkinson has been used [36].

With the exception of instrumental dry combustion methods [32], the techniques referred to above for the analysis of organic (and total) carbon in sediments are time consuming (e.g. 2–3h). An instrumental technique described by Van Hall and Stenger [33] makes use of a non-dispersive infrared detector and measures the carbon dioxide resulting from the combustion of the carbonaceous compounds. Total and inorganic carbon can be differentiated by the use of different combustion columns and temperatures.

The combustion infrared technique has been used for the analysis of diluted sludges [34]. Schaffer *et al.* [34] have used a blender to prepare suspensions of samples of this type. After the sample was homogenized, a microlitre syringe was used to remove a 20µL aliquot from the blender. However, the method described by Schaffer *et al.* [34] does not necessarily allow for the isolation of a representative portion of the sample in a 20µL aliquot [35]. This was found to be particularly true if the sample contained large particles which settled rapidly, and it must be assumed that the precision data presented by Schaffer and coworkers [34] are for samples which have particle sizes of about 20μ or less. Since most sediments are likely to contain a substantial sand fraction, only a small fraction of the particles in sediments are expected to have sizes of less than 20μ. Thus, for determinations involving suspensions of sediments the reductive pyrolysis-flame ionization technique apparently offers the most promising prospects.

12.3.2.1 Spectrophotometry

Lynch *et al.* [21] have described a method for the determination of organic carbon in silty lake sediments. In this method the air-dried and sifted (–250 mesh) sample is leached with 4M nitric acid–0.1M hydrochloric acid for 1.5h at 90–95°C, and the extinction of the cooled, clean solution is measured at 500nm. The extinction correlates well with weight loss (%) on heating the sample between 120 and 800°C. The precision is ±26%. The same leach solution can be used for trace-metal determinations.

12.3.2.2 Combustion methods

McQuaker and Fung [22] used a Model DC–50 Total Organic Carbon Analyser (Dohrmann Envirotech Corp) to determine both total carbon and organic carbon in sediments. Samples for total carbon determination are dispersed in deionized water and injected onto the analyser. Samples for organic carbon determination are dispersed in 0.1M hydrochloric acid (to decompose inorganic carbonates), then injected into the apparatus. Some determinations of total and organic carbon in sediments by this and a more lengthy reference method show an average recovery by the DC–50 TOC Analyser method of 96.8%.

Sample No	*Carbon, mg/g $^{-1}$*		*Organic Carbon, mg/g $^{-1}$*	
	Model DC–50	*Reference*	*Model DC–56*	*Reference*
1	26.8	30.2	26.9	28.1
2	5.5	4.1	3.8	5.3
3	62.0	60.9	55.0	65.6

Various other workers have discussed the determination of organic carbon in river sediments.[18,23,24] Suzuki *et al.* [23] applied wet combustion to the sediment and absorbed the carbon dioxide produced in sodium hydroxide solution, prior to determination in a TOC analyser. Whitfield and McKinley [24] recommended that samples for determination of organic carbon should be filtered immediately and analysed as soon as possible. Charles and Simmons [18] stated that greater accuracy is obtained with dry microcombustion techniques and that wet oxidations with potassium dichromate lead to incomplete oxidation of certain compounds and interference from oxidizable inorganic compounds such as chlorides.

12.3.3 Saline deposited and suspended sediments

12.3.3.1 Spectrophotometry

The Lynch *et al.* [21] procedure discussed in section 12.3.2.1 has been applied to saline sediments.

12.3.3.2 Infrared spectrometry

Mills and Quinn [25] used persulphate oxidation at 130°C with subsequent measurement of carbon dioxide produced by a non-dispersive infrared detector to determine organic carbon in high carbonate oceanic sediments and estuarine sediments. Sediment samples were pre-dried for 4h at 110°C to remove moisture and, if carbonate was present, were pretreated with hydrochloric acid and phosphoric acid to decompose carbonates, then washed and dried. The results of organic carbon determinations in some marine sediments by this method, and by high temperature combustion using the Carlo Erba elemental analyser, shows that there appears to be no significant bias between the methods. The average relative standard deviation obtained by the persulphate method is 11.4%, which is higher than the value of 5.1% obtained by high temperature oxidation. The overall precision of the persulphate method was 6.4%.

12.3.3.3 Combustion method

Microcombustion techniques have been employed to determine organic carbon in estuarine sediments [26].

12.3.3.4 Wet oxidation methods

The persulphate oxidation technique [25] has been applied to estuarine sediments. Organic carbon was determined at 10g kg^{-1} in bay and estuarine sediments with a coefficient of variation of about 10%.

Weliky *et al.* [27] have developed a procedure for determination of both organic and inorganic carbon in a single sample of marine deposit, thus avoiding errors occurring in the method most commonly used. Carbonate carbon is determined from the carbon dioxide evolved by treatment of the sample with phosphoric acid, and the residue is then treated with a concentrated solution of dichromate and sulphuric acid to release carbon dioxide from the organic matter.

de Oliveira *et al.* [28] have also discussed wet oxidation methods for the determination of organic carbon in estuarine sediments.

12.3.3.5 X-ray beam excitation methods

In this method [29] the sample is ground in a ball mill, 25% by weight of finely divided copper is added and the mixture is pressed into a disc against a lead backing plate; standards are prepared from quartz, graphite and calcium carbonate. Calibration graphs are obtained from the intensities of the C Kα lines at $\theta = 44.97$ for calcium carbonate and at $\theta = 45.48°$ for graphite. For samples containing both calcium carbonate and graphite, the ratio of the intensities of the two C Kα lines is measured and a gradient correction is applied to the calibration graph.

12.3.4 Sludge

12.3.4.1 Spectrophotometry

The spectrophotometric method [22] described in section 13.3.2.2 has also been applied to the determination of total organic carbon in sewage sludge.

12.3.4.2 Miscellaneous

Solyom [37] has conducted an intercalibration study. This was carried out on methods for the chemical analysis of sludge using samples of pure biological sludge, biologically-stabilized sludge, and chemical sludge from Orsundsbro, Sweden, sewage works. Methods for the determination of organic carbon, nitrogen, phosphorus, iron, manganese, copper, zinc, chromium, lead, nickel, cadmium and mercury were studied, and recommendations are made for drying and storing samples.

In this study most laboratories employed the chemical oxygen demand procedure for determining organic carbon employing a conversion factor of 0.375 to convert chemical oxygen demand to carbon content.

Waggot and Britcher [38] have discussed experimental considerations in the determination of organic carbon content of sewage effluent. Close attention is paid to the determination of particular classes of organic compounds in sewage including carbohydrates, amino acids, volatiles, steroids, phenols, surface active materials, fluorescent materials, organochlorine pesticides and ethylene diamine tetracetic acid.

12.4 Particulate organic carbon

12.4.1 Non-saline deposited and suspended sediments

For practical purposes, the distinction between dissolved organic carbon and particulate organic carbon is that the latter is retained on filters of known porosity. Membrane filters with 0.25–1.0um pore size and glass fibre Whatman G F/C filters are commonly used. By using the particulate fraction of a sample retained on pre-combusted Whatman G F/C filters for particulate organic carbon measurements and the filtrate for dissolved organic carbon estimations, a single filtration process can yield material for both analyses.

12.4.1.1 Infrared spectroscopy

Baker *et al.* [39] have described a procedure for the measurement of particulate organic carbon in river water and chalk streams. The method is based on the oxidation of organic carbon to carbon dioxide by ultraviolet radiation, the carbon dioxide produced being measured by a non-dispersive infrared gas analyser (Fig. 12.2). The latter was also used in the semi-automatic measurement of particulate organic carbon by a combustion method. Particulate organic carbon concentrations of 0.03–0.04mg L^{-1} were found in chalk spring waters. The relevance of the measurement of dissolved and particulate organic carbon flux to estimates of the energy budgets of stream ecosystems is discussed and published methods for the automatic measurement of dissolved organic carbon are reviewed.

12.4.1.2 Combustion methods

Krambeck *et al.* [40] measured small quantities of particulate carbon in lake waters by an automated furnace combustion infrared procedure. The whole sequence of operations was controlled with the aid of an AIM65 desktop computer. The system was successfully operated for routine analysis of samples of lake water with particulate organic carbon values of 100–300ug L^{-1} carbon; a single analysis takes 8min. The relative standard deviation was about 1%.

12.4.1.3 Wet oxidation

Hulsmann and Hengst [41] used wet oxidation by potassium persulphate in sealed ampoules to estimate particulate organic carbon. Inorganic carbon was first removed from the sample prior to reaction with persulphate.

12.4.1.4 Miscellaneous

Salonen [42] compared different glass fibre and silver metal filters for the determination of particulate organic carbon. He puts forward a view that the varying characteristics of different filters, or pore sizes, appreciably modify the results of particulate organic carbon determinations making comparison of published concentrations unreliable. Solonen compares silver metal filters and different types of glass fibre filters, and seeks to find a filter which would have biologically meaningful cut-off size of

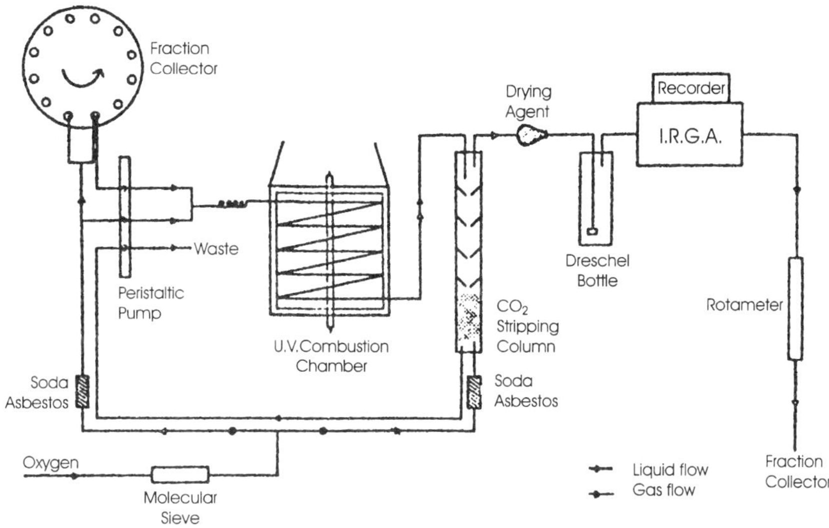

Fig. 12.2 Apparatus layout and flow diagram for dissolved organic carbon measurement
Source: Reproduced with permission from Blackwell Science Publishers Ltd [39]

particles. The most retentive glass fibre filters were able to retain almost all bacteria from the water of an oligotrophic lake; they prove to be quite near to the ideal. Silver filters provide similar retention, but because of their high blank values, price and lower filtration speed and capacity, they are not able to compete with glass fibre filters in practical work.

Whitfield and McKinley [24] studied some of the factors affecting the determination of particulate carbon and nitrogen in river water sediments.

The effect of storage of water samples before determination of nitrogen and carbon associated with particulate matter was investigated. Freezing of the samples, and storage at 5°C, both affected the results obtained, but changes were minimized if the samples were filtered immediately after collection and the particulate matter stored on the filter paper. It is recommended that samples for determination of particulate carbon and nitrogen should be filtered immediately, and analysed as soon as possible.

12.4.2 Saline deposited and suspended sediments

12.4.2.1 Wet digestion methods

Dankers and Laane [43] compared two methods, based on wet oxidation with potassium peroxydisulphate and loss on ignition at 560°C, for the determination of particulate organic carbon in estuarine suspended matter. They found that in estuarine sediments with a high clay content, loss on ignition overestimated organic matter approximately four times. Discrepancies were attributed to incomplete oxidation in the wet digestion method and loss of carbonates and water in the dry ignition method.

12.5 Nitrogen

12.5.1 Soil

12.5.1.1 Titration

Nelson and Sommers [44] have described a Kjeldahl digestion procedure for the determination of total nitrogen in soils in which the sample is digested with sulphuric acid using a selenium catalyst. The digest is steam distilled with sodium hydroxide and ammonia titrated with 3.5mM sulphuric acid. Various other workers have discussed the application of Kjeldahl digestion to the determination of total nitrogen in soils [45–47].

12.5.1.2 Spectrophotometry

Marsh *et al.* [47] have described an apparatus based on an autoanalyser system for the automatic preparation of soil extracts for mineral nitrogen determination. It consists of a reagent adder, which adds the correct volume of extractant for an approximately weighed amount of soil, and a sample preparation unit, which mixes, filters, dilutes and loads samples on to an autoanalyser sampler. A labour saving of 60% is achieved in this method compared to manual method. Examples are given of the determination of nitrate plus nitrate nitrogen and ammonium nitrogen.

Tests were carried out to compare the efficiency of extraction results obtained using a manual weighing and sample preparation method, and the reagent adder and sample preparation unit. Extracts of four replicates of ten soils were prepared by each method and analysed for nitrate- plus nitrite-nitrogen by a diazotisation and coupling reaction with sulphanilic acid and N-(1-naphthyl)ethylenediamine and ammonium-nitrogen by an indophenol method. These methods are described fully by Greaves *et al.* [48].

Results for the nitrate plus nitrite determinations (Table 12.7) show that there was very little difference in the amounts extracted with either technique.

Ammonium-nitrogen (Table 12.8) extracted by the automatic method was slightly higher than that extracted manually, and was significantly higher when less than 2μg of ammonium-nitrogen per gram of dry soil was present. This may be due to the timing for the filtering of the samples and loading the autoanalyzer being shorter than with the manual method. The regression coefficient for results of automatic against manual methods is significantly less than 1 at P = 0.01, the relationship between the two being

$$NH_4^+{}_{Auto} = 0.241 + 0.9285\,(NH_4^+{}_{Manual})$$

Table 12.7 Determination of nitrate- plus nitrate-nitrogen in soil after extraction by manual and automatic techniques

Sample	*Manual preparation*		*Automatic preparation*			*Automatic–manual*			
	Mean of four determinations/ μg g⁻¹ dry soil	*Standard deviation/ μg g⁻¹*	*Coefficient of variation, %*	*Mean of four determinations/ μg g⁻¹ dry soil*	*Standard deviation/ μg g⁻¹*	*Coefficient of variation %*	*Difference/ μg g⁻¹*	*Standard error/μg g⁻¹*	*t*
1	23.15	0.671	2.9	23.26	0.587	2.5	0.11	0.446	0.25
2	14.00	0.303	2.2	13.54	0.220	1.6	–0.46	0.187	2.46
3	8.38	0.120	1.4	8.18	0.160	2.0	–0.20	0.100	2.03
4	38.00	1.281	3.4	36.82	0.913	2.5	–1.17	0.786	1.49
5	19.08	0.426	2.2	19.04	0.323	1.7	–0.04	0.268	0.15
6	4.80	0.293	6.1	4.82	0.223	4.6	0.01	0.184	0.07
7	10.08	0.086	0.9	10.54	0.250	2.4	0.46	0.132	3.49
8	47.45	0.426	0.9	47.66	0.789	1.7	0.20	0.452	0.45
9	14.80	0.172	1.2	15.28	0.215	1.4	0.47	0.138	3.44
10	57.94	1.536	2.7	58.49	0.538	0.9	0.54	0.814	0.67

Source: Reproduced with permission from the Royal Society of Chemistry [47]

Table 12.8 Determination of ammonium-nitrogen in soil after extraction by manual and automatic techniques

Manual preparation			*Automatic preparation*			*Automatic–manual*			
Sample	*Mean of four determinations/ μg g⁻¹ dry soil*	*Standard deviation/ μg g⁻¹*	*Coefficient of variation, %*	*Mean of four determinations/ μg g⁻¹ dry soil*	*Standard deviation/ μg g⁻¹*	*Coefficient of variation %*	*Difference/ μg g⁻¹*	*Standard error/μg g⁻¹*	*t*
1	6.93	0.264	3.8	6.53	0.155	2.4	–0.40	0.153	2.64
2	4.41	0.261	5.9	4.33	0.088	2.0	–0.08	0.138	0.9
3	3.10	0.082	2.6	3.20	0.051	1.6	0.09	0.048	1.93
4	4.28	0.101	2.4	4.37	0.278	6.4	0.09	0.148	0.58
5	2.49	0.113	4.5	2.42	0.118	4.9	–0.07	0.082	0.82
6	0.48	0.024	5.1	0.56	0.025	4.4	0.08	0.018	4.69
7	3.18	0.030	0.9	3.37	0.136	4.0	0.18	0.70	2.65
8	1.46	0.017	1.2	1.63	0.062	3.8	0.17	0.032	5.20
9	0.55	0.024	4.3	0.74	0.025	3.4	0.19	0.017	11.28
10	1.48	0.024	1.7	1.61	0.015	0.9	0.13	0.014	8.89

Source: Reproduced with permission from the Royal Society of Chemistry [47]

Pelts and Belcher [46] have described a semi-automated method for the simultaneous determination of nitrogen and phosphorus in soil. Up to 40 samples are digested with sulphuric acid using a copper sulphate-sodium sulphate catalyst at 370°C for 2.25h in a thermostatically controlled block. Nitrogen is then determined in the 0.2–2.25% range as ammonia by automatic colorimetric analyses of the indophenol blue complex at 630nm.

12.5.1.3 Miscellaneous

Juma *et al.* [49] give details of studies to determine the most suitable equation, and the most appropriate method for calculating the values of the equation parameters, for estimating net nitrogen mineralization in the soil. Data obtained for several different types of soil are examined. Both the hyperbolic equation and a first-order equation gave accurate predictions of the amount of nitrogen mineralized over an incubation period of 14 weeks, but the estimates of potentially mineralizable nitrogen and its half-life depended on the model used.

Brodick *et al.* [50] discuss the use of potassium chloride as an extractant in the determination of exchangeable nitrogen species in soil.

12.5.2 Non-saline deposited and suspended sediments

12.5.2.1 Spectrophotometry

Muhlhauser *et al.* [51] have discussed improvements in the Kjeldahl method for the determination of total nitrogen in sediments. 10mg of sediment and 0.2ml concentrated sulphuric acid are weighed into a thick-walled pyrex tube, the volume made up to 25ml and the mixture evaporated. Heating at 250°C for 4.5h was required, with addition of 0.3–0.4ml hydrogen peroxide. Ammonia was determined by an indophenol blue method. The procedure could not distinguish between nitrite and nitrate nitrogen. Reproducibility was better than 10%. Acid hydrolysable phosphate could also be determined in the reaction mixture.

12.5.2.2 Combustion method

Wong and Kemp [52] have described a rapid automated procedure involving combustion at 2000°C for the determination of total nitrogen in lake sediments.

12.5.2.3 Wet digestion

Potassium persulphate [53], hydrogen peroxide [51], and peroxydisulphate in alkaline medium [59], have all been used to digest river sediments prior to the determination of total nitrogen. Muhlhauser *et al.*

[51] digested 10mg sediment with 25ml water, 0.2ml concentrated sulphuric acid and 0.4ml hydrogen peroxide for 4.5h at 250°C.

Zink-Nielsen [54] showed that higher total nitrogen contents with lower coefficient of variation deviations (7.88 ± CV 3.8mg N g^{-1} dry weight) are obtained when river sediments are digested under reducing Kjeldahl conditions (salicylic acid and sodium thiosulphate) than when digestion was carried out using alkaline potassium peroxydisulphate (6.46 ± CV 8.4mg N g^{-1} dry weight) indicating that peroxydisulphate is not a strong enough oxidising agent to convert all organic nitrogen to nitrate nitrogen.

Smart *et al.* [53] also used persulphate digestion in the determination of total nitrogen in sediments.

12.5.2.4 Miscellaneous

The Kjeldahl digestion–titration technique [44] discussed in section 12.5.1.1 for the determination of total nitrogen in soils has been applied to the determination of total nitrogen in sediments.

Factors affecting the determination of particulate nitrogen in river water have been discussed by Whitfield and McKinley [24].

Ballinger and McKee [55] have discussed the chemical characterization of organic nitrogen and carbon compounds in sediments.

12.5.3 Sludge

12.5.3.1 Wet digestion

The Standing Committee of Analysts (UK) have described a method for the determination of total nitrogen and phosphorus in sewage sludges [56].

The initial preparation of sewage sludges for analysis, followed by a manual method and two semi-automated procedures for determination of total nitrogen and total phosphorus in the digested sample are described. Digestion treatments of three kinds were also used for the these methods, consisting of the application of sulphuric acid together with either mercury/potassium sulphate, copper/sodium sulphate or hydrogen peroxide. By the use of these reagents, the nitrogen compounds were converted to ammonia and the phosphorus compounds to orthophosphate, which were then determined directly. Details in respect of procedures, elimination of interfering substances and interpretation of results are included.

12.5.3.2 Miscellaneous

Solyom [37] has carried out intercalibration studies on methods for the determination of total nitrogen in sludges. Nitrogen was determined by oxidation in alkaline solution with potassium peroxydisulphate, followed by spectrophotometric determination of the nitrate so produced. Alternatively, nitrogen was determined by Kjeldahl digestion. Oxidation with peroxydisulphate was shown to be incomplete (coefficient of variation 32–64%). Kjeldahl digestion gave a lower coefficient of variation of 12–16%.

Ruider and Spatzierer [57] described a simple method for the rapid estimation of nitrogen and phosphorus. They discuss the usefulness of test kits for ammonia nitrogen, nitrate nitrogen and phosphorus.

12.6 Total phosphorus

12.6.1 Soil

12.6.1.1 Spectrophotometry

Matt and Agassiz [58] and Murphy and Riley [59] have described spectrophotometric methods based on the formation of the phosphomolybdate complex for the determination of phosphorus in soils. Up to 500mg L^{-1} iron does not appear to interfere in these procedures. Zink-Nielsen [54] obtained 0.9 ± 5.7mg Pg^{-1} dry weight total phosphorus in a river sediment following digestion with potassium peroxydisulphate and a higher result with a lower standard deviation (0.98 ± 3.4mg Pg^{-1} dry weight) by Kjeldahl digestion followed by spectrophotometric estimation by the molybdate–ascorbic acid method

Aspila *et al.* [60] have described a semi-automated method for the determination of inorganic, organic and total phosphorus in river and lake sediments. Total phosphorus is extracted from sediments with 1M hydrochloric acid after ignition at a high temperature (550°C) (method 1) or by digestion with sulphuric acid–potassium persulphate at 135°C in a sealed PTFE-lined Parr bomb (method 2).

Organic phosphorus is determined by the difference in phosphorus content of the 1M hydrochloric acid extract measured before and after ignition of the dry sediments at 550°C. In all instances the orthophosphate is determined by using standard Technicon AutoAnalyzer II techniques. Silica does not interfere.

The automated spectrophotometric procedure [47] for the determination of total nitrogen in soils discussed in section 12.5.1.2 has also been applied to the determination of 0–0.75% phosphorus in soils. The method is based on the formation of the yellow molybdate–vanadate complex which has an absorption maximum at 420nm.

Bickford and Willett [61] have reported that the filtration of soil extract solutions through a Gelman GA6 0.45µm membrane which contains a wetting agent caused interference in spectrophotometric methods for the determination of phosphate. This was due to the release of some substance from the membrane. It is recommended that 'low extractable' membranes such as Gelman CM–450 are used for this purpose.

Olsen *et al.* [62] have described a method for the determination of pH8.5 sodium bicarbonate extractable phosphorus in soils. The concentration of the blue complex produced by the reduction, with ascorbic acid, of the phosphomolybdate formed when acid ammonium molybdate reacts with phosphate is measured spectrophotometrically at 880 nm [63].

12.6.1.2 Flow-injection analysis

Methods based on flow-injection analyses have been described for the determination of extractable [62,64–66] and available [67] phosphorus in soils.

In the method for extractable phosphorus [62,64–66] the phosphorus is extracted from the soil at 20 ± 1°C with sodium bicarbonate solution at pH8.5. After filtration and release of carbon dioxide the extracts are introduced into a flow-injection system for the determination of phosphate. Phosphate is determined by reaction with vanadomolybdate and the yellow colour evaluated at 410nm. Between 20 and 1000mg kg^{-1} phosphorus in soil has been determined using this method.

In a further method [67] for the determination of ammonium lactate extractable (i.e. available) phosphorus in soils, the sample 5g is extracted with 100ml acidic ammonium lactate and then phosphate determined by flow-injection analysis using the stannous chloride method [68].

12.6.1.3 Gas chromatography

Addison and Ackman [69] have described a direct determination of elementary yellow phosphorus in mud in which the phosphorus is extracted with benzene or isooctane. Gas chromatographic separation is achieved on a 2m × 3mm column packed with 3% OV–1 or SE–30 on Chromosorb W maintained at 100 or 120°C respectively. The carrier gas was helium (80ml m^{-1}). A flame photometric detector with a 526nm filter at 200°C was employed. Down to 1pg of phosphorus could be determined.

12.6.1.4 Inductively coupled plasma atomic emission spectrometry

Que Hee and Boyle [70] analysed soils for total phosphorus using Parr bomb digestion with hydrofluoric–nitric–perchloric acids followed by inductively coupled plasma atomic emission spectrometry.

12.6.1.5 Miscellaneous

Olsen *et al.* [71] have reviewed the determination of inorganic, organic and total phosphorus in soil and sediments. Determination of total phosphorus in aqueous samples commonly involves a hot acid oxidation type digestion procedure, although various other dry-ashing, fusion and

UV irradiation methods have been reported and evaluated e.g. Harwood *et al.* [72] and Osburn *et al.* [73].

Brah and Bishnoi [74] have studied the effect of temperature and soil phosphorus status on the determination of extractable phosphorus by Olsen's method [71].

12.6.2 Non-saline deposited and suspended sediments

12.6.2.1 Spectrophotometry

Aspila *et al.* [60] have described a simple, rapid and semi-automated method for the determination of inorganic, organic and total phosphorus in lake and river sediments. Total phosphorus is extracted from sediments with 1M hydrochloric acid after ignition at a high temperature (550°C) or by digestion with sulphuric acid–potassium persulphate at 135°C in a sealed PTFE-lined Parr bomb. Organic phosphorus is determined by the difference in phosphorus content of the 1M hydrochloric acid extract measured before and after ignition of the dry sediments at 550°C. Orthophosphate is determined by using standard Technicon AutoAnalyzer II techniques. The interferences caused by silica and variable acid concentrations on the determination of phosphorus were studied. Freedom from interferences under the chosen experimental conditions as well as the good results obtained for recovery and precision indicate that the methods are suitable for monitoring inorganic, organic and total phosphorus in sediments.

The determination of the orthophosphate was carried out by using the automated systems described by the Technicon Instruments Corporation. The manifolds used are shown in Fig. 12.3. The procedures referred to below as methods I and II are Technicon industrial methods Nos. 94–70W and 155–71W, respectively. Method I includes ascorbic acid alone for the reduction of the molybdophosphoric acid whereas in method II the mixed reagents ascorbic acid, sulphuric acid, ammonium molybdate and antimony potassium tartrate are used. Method I is intended for use for high levels of phosphorus (up to 10μg ml $^{-1}$) and method II for low levels (less than 0.5μg ml $^{-1}$). The wetting agent (Levor IV) used in order to obtain a smooth bubble pattern, is present in the ascorbic acid reagent line for method I whereas it is added externally Fig. 12.3) in the water line (0.5μg ml $^{-1}$ of Levor) in method II.

Results obtained by method I were found to be linear over the range 0.5–5μg ml $^{-1}$ of phosphorus and by method II from 0.05–1μg ml $^{-1}$. As solution extracts contain 1mg ml $^{-1}$ of sediment, the above concentration ranges allow direct analyses of sediments containing from 100–5000ppm (μg g $^{-1}$) of phosphorus to be made, which encompasses the entire range of sediment phosphate levels expected.

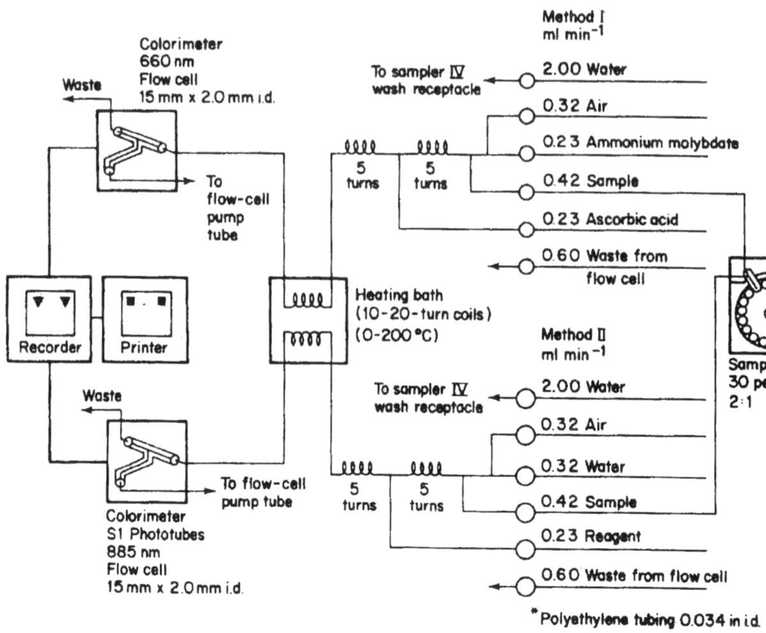

Fig. 12.3 Manifold for phosphorus determination (Technicon)
Source: Reprinted with permission from the Royal Society of Chemistry [60]

Extraction of sediment samples with hydrochloric acid

Aliquots (0.3–0.5g) of dry sediment (passing 100 mesh) were weighed and then transferred into 10cm^3 Coors alumina crucibles. The uncovered crucibles, contained in a suitable tray, were placed into a warm muffle furnace and ignited at 550°C. The samples were maintained at 550°C for 1.5h, then removed, allowed to cool, and transferred into 100ml calibrated flasks; 50ml of 1.0M hydrochloric acid were then added to the flasks. The mixtures were next shaken for 14–18h at about 22°C.

For determinations of 1M hydrochloric acid extractable inorganic phosphate an identical aliquot of sediment was used except that no ignition was performed. After extraction, aliquots of the ignited and non-ignited mixtures were transferred into 15ml test tubes and centrifuged at 2000rev min^{-1} for about 5min. The clarified extracts were finally diluted ten times and analysed by the two automated Technicon procedures described below.

An alternate extraction procedure used by Aspila *et al.* [60] involves ignition of the sediment with potassium persulphate in a PTFE-lined Parr bomb.

Extraction by the bomb method

For the extraction of total phosphate a weighed aliquot of sediment (0.3–0.5g), together with 3 ± 0.1g of potassium persulphate (Analar) and 5.00ml of concentrated sulphuric acid, was added to the bomb, which was then heated in an oven at 135 ± 5°C for 2h. The contents of the bomb were then transferred quantitatively into a 500ml calibrated flask. After dilution to volume with distilled water, the extract (containing 1% *v*/*v* of sulphuric acid) was analysed for total phosphate by the automated procedure outlined below.

Aspila *et al.* [60] found that the concentration of sulphuric acid in the sample solution had an appreciable effect on the response for 1µg L^{-1} of phosphorus, hence the need to carefully control acidity levels during the analysis. Additionally, acidity levels which are too low allow serious interference in the method by silica.

Fig. 12.4 illustrates the complex interactions induced by the presence of silica between reaction temperature, sulphuric acid acidity and the apparent level of phosphorus found. By careful control of acid concentration and reaction temperature, interference by silica can be minimized. Arsenic, germanium and bismuth would interfere in the method but not at the low levels normally encountered in sediment in water samples.

The coefficient of variation obtained for the determination of total phosphorus in sediment at the 1400ppm level was 2.5%. Some 98–100% recovery of inorganic phosphate was obtained in spiking experiments carried out on sediments.

Shulka *et al.* [75] investigated interference by arsenic in the spectrophotometric determination of inorganic phosphate in lake sediments.

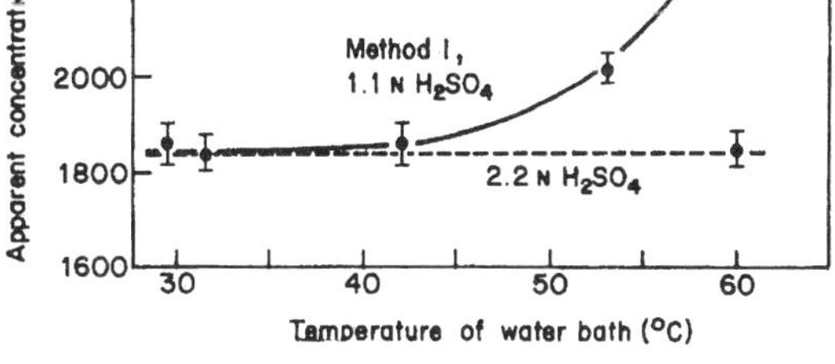

Fig. 12.4 Effect of temperature of heating water bath on the apparent concentration of phosphorus. The 1.1N H_2SO_4 refers to the concentration of sulphuric acid in the ammonium molybdate reagent solution

Source: Reprinted with permission from the Royal Society of Chemistry [60]

Table 12.9 Comparison of results for phosphorus (ppm) obtained by detection with methods I and II

Sediment sample	*Bomb method of extraction*		*Extraction with 1m HCl after ignition*		*Extraction with 1 HCl before ignition,*
	Method I	*Method II*	*Method I*	*Method II*	*Method I*
1	1800	–	1850	1800	1400
	1870	1770	1820*	–	
2	1080	1030	1090	1100	810
	1120	1070	1090*	1180	
3	870	800	860	860	640
	920	860	860*	950	
4	650	(570)	605	520	495
	680	510	755*	815	
5	1060	1000	1060	1030	805
	1120	1070	1080*	1150	
6	1390	1360	1230	1220	820
	1370	1300	1010*	1390	
7	1040	990	1130	1080	840
	1060	1000	1020*	1100	
8	1130	1080	1200	1130	880
	1160	1080	1150*	1070	
9	–	–	855	810	750
	850	780	820*	890	
10	1020	990	980	970	750
	1060	970	930*	1010	
11	970	930	910	900	690
	970	900	930*	1010	
12	690	640	675	660	525
	730	660	675*	730	

Source: Reprinted with permission from the Royal Society of Chemistry [60]

*Sample ignited for 16h at 550°C (normal time is 2h at 550°C)

Muhlhauser *et al.* [51] have discussed methods for the determination of acid hydrolysable phosphorus in sediments.

Table 12.9 compares results for phosphorus determinations in sediments obtained by methods I and II.

12.6.2.2 X-ray diffraction

Dobolyi and Bidlo [76] determined the phosphorus-containing minerals in Balatien lake sediment, and thus the forms in which the phosphorus responsible for the accelerating eutrophication of the lake are present. Samples were subjected to chemical, electron microscope and X-ray analysis. Hydroxylapatite was identified, but no proof of the presence of other phosphorus minerals was obtained.

12.6.2.3 Combustion methods

Anderson [77,78] investigated ignition methods for the determination of total phosphorus in lake sediments and biogenic materials, and compared results with those obtained by perchloric acid digestion. The organic matter is destroyed by ignition. Material remaining after ignition is boiled in hydrochloric acid and orthophosphate determined after dilution. The method generally gave lower recoveries than the perchloric acid digestion method (97.7, 98.7, 94.4, 100.5 and 97.3% for four sediment samples). The reproducibility of the ignition method was slightly less than that observed with the perchloric acid method.

Dry sediment (0.15–0.2g) was ignited in a muffle furnace in a porcelain crucible (550°C for 1h). After cooling, the residue was washed into a 100ml Erlenmeyer flask with 25ml mol L^{-1} hydrochloric acid and boiled for 15min on a hot plate. The sample was diluted to 100ml and orthophosphate was determined as in the perchloric acid method. Standards and blanks were not ignited.

The 95% confidence limits vary from ±0.5 to ±1.4% of the average for the perchloric acid method and from ±1.0 to ±2.1% for the ignition method. Thus, the reproducibility of the two methods is similar, but a little better for the perchloric acid method.

Saunders and Williams [79] have discussed ignition methods for the determination of organic phosphorus in soils. Sodium carbonate fusion has also been used to solubilize phosphate in soil analyses.

12.6.2.4 Wet digestion methods

Persulphate digestion

A rapid method for determination of total phosphorus in water samples by digestion with persulphate was introduced by Koroleff [83], but this method has not been widely used for sediment samples. Preliminary measurements of phosphorus in lake sediments using the persulphate digestion method gave considerably lower values than the perchloric acid method [84].

Perchloric acid digestion

In the Murphy and Riley [85] method 10ml of demineralized water and 2ml of concentrated nitric acid were added to 0.15–0.2g of dry sediment (predried at 103°C) or plant material in a 100ml Erlenmeyer Flask. After a preliminary oxidation by evaporation of water and nitric acid on a hot plate, 2ml of concentrated perchloric acid were added, and the sample was boiled until clear. After cooling, the sample was diluted to 100 ml and an aliquot was withdrawn for orthophosphate determination by the ascorbic acid reduction method of Murphy and Riley [85]. Blanks and standards were treated as samples.

Determination of total phosphorus in lake sediments by ignition of samples in a muffle furnace at 550°C, boiling of the residue from ignition in 1mol L^{-1} hydrochloric acid, and subsequent determination of orthophosphate gave approximately the same values as the perchloric acid digestion.

The American Public Health Authority has published a standard method for the determination of phosphates in sediments [89].

12.6.2.5 Miscellaneous

De Pinto [90] measured the rate at which available phosphorus is released from various types of particulates suspended in lake water. The equipment consists of two culture vessels separated by a thin membrane filter, thus facilitating the separation of two particulate suspensions, while at the same time permitting their interaction by diffusion of solutes through the membrane.

Comparative studies of different methods for determination of total phosphorus in sediment have been made by various workers [80–82,86–88].

Dobolyi and Bidlo [76] have described methods for the determination of phosphorus in lake sediments. Shulka *et al.* [75] investigated the interference by arsenic in the perchloric acid digestion procedure of Murphy and Riley [85] for the determination of phosphorus in sediments. Arsenite concentrations up to 20μg did not interfere but arsenate interfered. Between 1 and 45μg arsenic g^{-1} was extracted from a lake sediment and in all cases the error in the determination of phosphorus due to the presence of arsenic was less than 1%.

Dorich *et al.* [91] estimated algae available phosphorus in extracts of suspended sediments.

12.6.3 Sludge

12.6.3.1 X-ray diffraction

Buchan [92] has carried out a study using electron microscopy combined with energy dispersive X-ray analysis, to determine the location and nature of the stored phosphorus in activated sludge samples from seven different treatment plants in South Africa, which all show enhanced phosphorus removal. The results indicate that there is a biological mechanism of enhanced phosphorus uptake in activated sludge.

12.6.3.2 Miscellaneous

A British standard method [93] has been published for the determination of phosphorus and silicon in sludges.

Mino *et al.* [94] have studied the phosphorus content and metabolism in activated sludges. These workers point out that the Schmidt–Thannhauser–Schneider method estimates the phosphorus composition of organisms. The most important fractions for phosphorus composition were shown to be the same for all sludges. Differences in phosphorus composition among sludges were in the different metal phosphate and low-molecular and high-molecular polyphosphates. A difference was concluded to exist between low- and high-molecular polyphosphates in the phosphorus metabolism of sludges.

The use of test kits to determine phosphorus in biologically treated sewage has been discussed by Ruider and Spatzierer [57].

Solyom has conducted an intercalibration of methods used for the determination of phosphorus in sludges [37]. The methods used to determine phosphorus were that of Koroleff [83] in which the sample is digested with potassium peroxydisulphate and phosphate determined spectro-photometrically. Alternatively a reducing Kjeldahl digestion was used followed by determination of phosphate using molybdate and ascorbic acid. The former method gives somewhat low results. The reducing Kjeldahl method is therefore recommended.

12.7 Particulate phosphorus

12.7.1 Non-deposited and suspended saline sediments

12.7.1.1 Wet digestion methods

Sulphur acid–hydrogen peroxide digestion

Cabrera *et al.* [95] determined total dissolved and suspended phosphorus in natural waters by a method involving digestion with hydrogen peroxide and sulphuric acid, errors may be caused by adsorption of phosphorus on hydrous iron and aluminium oxides formed during neutralization prior to filtration. It is proposed that this can be prevented by adding extra sulphuric acid after neutralization, to dissolve such oxides and release the adsorbed phosphorus into solution.

In this method 50ml of the sample containing sediment is placed in a Kjeldahl flask and 1ml of concentrated sulphuric acid and anti-bumping granules were added. The flasks were heated on a hotplate until fumes of sulphur trioxide appeared. After allowing to cool, 1ml of 30% hydrogen peroxide was added, and the flasks were again gently heated to fumes of sulphur trioxide. If the digest was not clear the hydrogen peroxide treatment was repeated. After the digest is clear, heating was prolonged up to accomplishing the hydrogen peroxide decomposition. When cool, 50ml of distilled water and three drops of phenol–phthalein solution (0.5% in 1:1 water–ethanol) were added, the mixture brought to pink colour with 12% sodium hydroxide solution and back to colourless with 5M sulphuric acid, then a further 0.5ml 1.5M sulphuric acid added, then the solution is made up to 100ml with distilled water.

Phosphorus was determined in the extracts by the spectrophotometric method of Murphy and Riley [63] and Stickland and Parsons [96]. In Table 12.10 results are shown obtained upon applying the procedure to the determination of total phosphorus in solutions of calcium phosphoglyceric acid.

Cabrera *et al.* [95] concluded that the presence of iron and/or aluminium in the sample does not cause interference in the determination of total phosphorus.

Magnesium sulphate digestion

Solazaro and Sharp [181] also described a procedure for separate analyses of total dissolved phosphorus and total particulate phosphorus in natural waters. The method for both procedures involves drying a sample with magnesium sulphate and baking the residue at a high temperature to decompose organic phosphorus compounds. The residue is then treated with hydrochloric acid to hydrolyse polyphosphates and the orthophosphate is measured by the molybdate method. The method gives 100% recovery with refractory phosphorus compounds, is usable on undiluted samples containing up to 18umol L^{-1} phosphorus and has a mid-range precision of ±1%.

To determine particulate phosphorus the insolubles are filtered from the water sample on to a pretreated (baked at 450–500°C for 0.5–1h) Whatman 9F/C glass fibre filter. After filtration the filter is rinsed with 2ml aliquots of 0.17M sodium sulphate and then soaked with 2ml 0.017M magnesium sulphate and dried at 95°C. The disc is then ignited for 2h at 450–500°C. After cooling 5ml 0.2M hydrochloric acid is added and the disc heated to 80°C. After cooling and washings are transferred to a centrifuge tube,

Table 12.10 Results of analysis of phosphoglyceric acid solutions

Solution	*Total P μg*	*Total P μg*
I	3.11 ± 0.21	5.55 ± 0.16
	3.13 ± 0.37	5.74 ± 0.28
II	2.85 ± 0.03	5.30 ± 0.05
	2.96 ± 0.10	5.40 ± 0.18
III	2.77 ± 0.10	5.23 ± 0.06
	2.78 ± 0.16	5.42 ± 0.03

Source: Reproduced with permission from Elsevier Science Ltd [95]

Solution I: C_3O_7H5PCa in water
Solution II: $C_3O_7H_5PCa$ in 100ppm Ca solution
Solution III: $C_3O_7H_5PCa$ in 100ppm Ca, 50ppm Mg, 5ppm K and 5ppm Na solution

chromophoric reagents added, according to the spectrophotometric methods of Murphy and Riley [63] and Stickland and Parsons [96] and the tube spun at 2000rpm for 5m. The optical density of the clear phase is then read at 885nm.

Solazara and Sharp [181] analysed a range of inorganic and organic phosphorus compounds by this procedure and obtained yields of 95–103%.

When magnesium sulphate was omitted from distilled water samples of phosphorus compounds, recovery was variable. Table 12.11 shows yields of a series of standards with and without the magnesium sulphate addition and with and without the final hydrolysis. The magnesium sulphate is used as an acidic solution (after addition to the seawater sample, the pH was about 3) to minimize silicate leaching from the glassware during evaporation. The acid and heating are necessary to hydrolyse any condensed phosphates in the final mixture.

12.8 Sulphur

Sulphur is an important component of both natural and anthropogenic processes. Due to its importance both in the formation of acidic precipitation and as a macronutrient required by all organisms, sulphur's role in atmospheric, aquatic and terrestrial systems has been investigated [100–104]. Sulphur has a vast array of both inorganic and organic chemical species. The understanding of sulphur dynamics has been restricted due to lack of information on the role of specific sulphur constituents in affecting sulphur fluxes and transformations. For example, Mitchell *et al.* [104] have shown the importance of organic sulphur in freshwater

Table 12.11 Yield of phosphate from organic and inorganic salts with ($MgSO_4$) and without (no $MgSO_4$) addition of magnesium sulphate and with hydrol) and without (no hydrol) final acidic hydrolysis step. All yields are percentage of theoretical value; 100 indicates theoretical yield (95–105%)

Compound	*No $MgSO_4$ No hydrol*	*No $MgSO_4$ Hydrol*	*$MgSO_4$ Hydrol*
Sodium phosphate			
monobasic	30	90	100
dibasic	15	58	100
Riboflavin phosphate	29	66	100
Phosphoryl chloride	53	90	100
Dexametasone phosphate	22	59	100
Sodium glycerophosphate	10	100	100
ATP	12	40	100
Disodium phenylphosphate	16	84	100

Source: Reproduced with permission from Elsevier Science Ltd [95]

sediments of lakes in New York and forest soils in the Adirondack Mountains [101]. Previous work on such substances had generally ignored the organic sulphur constituents with most work focusing only on inorganic sulphate or sulphide.

12.8.1 Soil

12.8.1.1 Wet digestion methods

Landers *et al.* [97] and others [98] have described a wet digestion method for the determination of total sulphur in soils. In this method the sample (1.50–500mg) is placed in a digestion flask and heated in a sand bath to dryness at 250°C with 3ml of sodium hypobromite solution. The residue is resuspended with water, neutralized with formic acid, and then hydriodic acid reduction of the sample is followed to quantitatively recover the inorganic sulphate formed by wet oxidation.

These workers [97,99] have also described a dry oxidation procedure for the determination of total sulphur in soils. The samples of dry finely ground soil (1.5–100mg) are combined with about 100mg of a mixed oxidant comprising sodium bicarbonate Ag_2O, 25:1 W/W) in small porcelain crucibles (18mm dia., 12mm h). The sodium bicarbonate in the oxidant should have a sulphur concentration less than 0.001% sulphur or high blanks may result. About 200mg of the mixed oxidant is layered on top of the sample as a trap and the mixture heated in a cold muffle furnace to 550°C and proportional increases in the amount of mixed oxidant. After cooling, the mixture is transferred to the digestion flasks and an hydriodic acid reduction used for determining total sulphur. When the hydriodic acid is added to the digestion flasks containing the solid sodium bicarbonate, considerable liberation of gas may occur. Consequently, to prevent entry of material into the gas import tubes, the nitrogen gas flow should be started before adding the hydriodic acid mixed reagent or, alternatively, up to 2ml of distilled deionized water can be added to decrease the rate of reaction. Beta-casein (0.8% sulphur) is used as a solid standard.

When analyses are complete, a fractionation scheme is used to determine organic and inorganic sulphur constituents (Fig. 12.5).

These procedures are useful for analysing sulphur in soils, sediments and sludges. Examples of these analyses are given in Table 12.12 and show that a wide range of sulphur concentrations can be measured with these procedures. In the sludge, soil and sediments, ester sulphate and carbon-bonded sulphur were the major sulphur constituents. However, the aerobically digested sludge and aerobic lake sediment substrates also had high concentrations of Zn-HCl-S, indicating that inorganic S constituents other than SO_4^{-2} may be important.

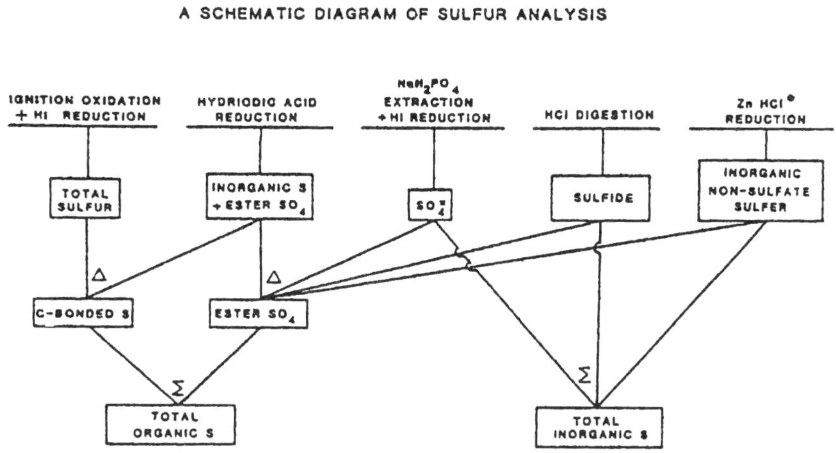

Δ Concentration found by difference
Σ Sum of concentrations
• Analysis not used for lake sediments

Fig. 12.5 Determination of carbon-bonded sulphur, ester sulphate and total organic sulphur from analyses of total sulphur, hydriodic acid reduction, extractable inorganic sulphate, HCl digestion and Zn-HCl reduction.
Source: Reproduced with permission from Gordon AC Breach [97]

12.8.2 Non-saline deposited and suspended sediments

12.8.2.1 Spectrophotometry

Davison and Lishman [105] described a method in which sulphide is released from the sediment using 5.93mol L^{-1} hydrochloric acid, and the resulting solution is separated from the sediment by filtration in a sealed system of syringes. The concentrated sulphide is determined spectrophotometrically at 670nm as ethylene blue. The limit of detection is 2mg kg^{-1} expressed as mass of sulphide in dry mass of sediment. The relative standard deviation was 5% for a sediment containing 118mg kg^{-1} sulphide.

Fig. 12.6 shows values obtained for a sediment core taken from Blelham Tarn in the English Lake District. The core was sampled by inserting syringes fitted with 75cm 21 G stainless steel needles into pre-drilled 3mm diameter holes in the core liner. They were at 1.3cm depth intervals and were sealed with polythene tape. This technique enables a sub-sample to be removed from the core without exposure to oxygen. All measurements were performed in duplicate within 2d of the core being collected. Determinations were in good agreement with those of Rowlett [110] who used a conventional distillation–iodometric procedure for determining sulphide.

Table 12.12 Sulphur constituents of various substrates

Substrate	*n*	*Total S* (*μmol g* $^{-1}$)	*HCl-S* (*dry mass or*	*Zn–HCl–S* *μmol l* $^{-1}$)
Sewage sludge (aerobic) (Waugh and Mitchell) [100]	3	344 ± 1.5	0.0	46.2 ± 3.4
Forest soil Hardwood 01 Horizon (David *et al.*) [101]	4	50.0 ± 9.3	U.D.	0.62 ± 1.1
Forest soil Hardwood B22hir Horizon (David *et al.*) [101]	4	16.5 ± 1.6	U.D.	0.69 ± 0.03
South Lake sediment (unpublished)	12	229 ± 18	1.56 ± 0.36	75.9 ± 9.4

Source: Reproduced with permission from the American Chemical Society [97]

U.D. undetectable

*Not a direct measurement and pyritic S is not subtracted

(*n*) if different from others

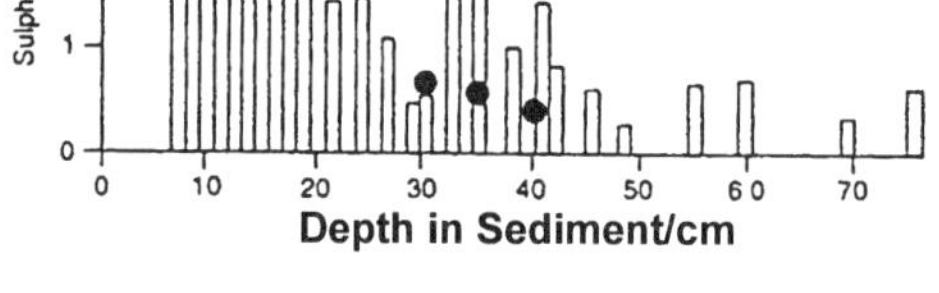

Fig. 12.6 Bar graph of acid volatile sulphide in a sediment core collected from Blelham Tarn on 30 July 1980; • = Rowlatt's mean data for the sediments of this lake (distillation/iodometric procedure) [110]
Source: Reproduced with permission from the Royal Society of Chemistry [105]

12.8.2.2 Wet digestion methods

The method [97] discussed in section 12.8.1.1 for the determination of sulphur in soils has been applied to lake sediments. Landers [97] also determined organic sulphur (carbon bonded sulphur and ester sulphate) and inorganic sulphur (sulphate and sulphide) in lake sediments. Some

HI-S	*SO_4^{-2}–S*	*^{3}C–s*	*^{3}C–O–SO_3*
210 ± 7.5	37 ± 2.5	134 ± 8.4(6)	126 ± 2.5(100)
8.25 ± 0.93	0.48 ± 0.08	41.9	7.16
5.00 ± 0.62	0.72 ± 0.13	11.5	3.59
82.5 ± 12.5	35.3 ± 2.5	146 ± 17.7	5.3 ± 12.4

results obtained by applying these methods to anaerobic lake sediment samples are shown below:

	m mol kg $^{-1}$		
Total S	229	±	18
HCl-S	1.56	±	0.36
Zn-HCl-S	75.9	±	9.4
HI-S	82.5	±	12.5
SO4S	35.3	±	2.5
C-S	146	±	17.7
C-O-SO_3	45.3	±	12.4

Ester sulphate and carbon bonded sulphur are the major sulphur constituents together with a high Zn-HCl reducible sulphur content, indicating that inorganic sulphur constituents other than sulphates may be important.

12.8.3 Saline deposited and suspended sediments

Iron sulphides are ubiquitous in marine and freshwater sediments. They are usually present either as pyrite or as monosulphides, which can be liberated by hydrochloric acid. These acid volatile sulphides give rise to an intense black colour that is characteristic of anoxic sediments. They play an important role in recent diagenetic processes in sediments and the ratio of pyrite to acid volatile sulphides has been used as an historical indicator to determine whether sediments were formed in marine or freshwater conditions. They can be present over a wide range of concentrations. Oxic muds do not have any free sulphide whereas anoxic muds may contain as much as 10mg g^{-1}.

12.8.3.1 Titration method

The traditional method of analysis is based on the method of Kolthoff and Sandell [106] in which 1mol L^{-1} hydrochloric acid is added to a sediment sample and the mixture is boiled. Hydrogen sulphide is trapped as zinc sulphide and the final analysis is performed by iodide titration after re-acidification of the metal sulphide. Gilboa-Garber [107] improved the final analysis step by using a colorimetric procedure based on methylene blue. However, the inherent disadvantage of the method, including the lengthy distillation step and extensive handling of an oxygen-sensitive sample, remained.

Various workers [52,105,108] have modified these procedures to improve precision, by employing zinc sulphide as a standard in contrast to the sodium sulphide solution used in earlier methods. The addition of sodium hydroxide to the hydrogen sulphide absorption solution improves recovery.

Valkov and Zhabina [109] reduced elemental sulphur in marine sediments to hydrogen sulphide using metallic chromium. The liberated hydrogen sulphide was absorbed in cadmium solution and estimated iodometrically.

12.8.3.2 Spectrophotometry

The Davison and Lishman procedure [105] described in section 12.8.2.1 is also applicable to saline sediments.

12.8.3.3 Gas chromatography

Chen *et al.* [111] extracted elemental sulphur from marine sediments with aromatic solvents and determined sulphur in the extract by gas chromatography.

Gas chromatography of the benzene or toluene extracts was carried out on a borosilicate glass column (180cm × 3mm o.d.) packed with 5% DC–200 and 7.5% QF–1 on Chromosorb W (80–100 mesh) operated at 190°C with nitrogen or argon as carrier gas and electron capture detection. Major peaks were obtained for S_4, S_6 and S_8. Down to 1µg L^{-1} of sulphur could be determined in the aromatic solvent extracts.

12.8.3.4 Scanning electron microscopy

Morse and Cornwell [112] investigated methods for determining acid volatile sulphides and pyrites in marine sediments from several typical depositional environments which were permanently or seasonally anoxic. Extractants gave similar results under different acid conditions, with the exception of stannous chloride in hot hydrochloric acid. The use of scanning electron microscopy combined with simultaneous elemental analysis showed that identifiable iron sulphides were almost always pyrite.

12.8.4 Sludge

12.8.4.1 Miscellaneous

Millson [113] investigated components of sewage sludge and found elementary sulphur in the hydrocarbon fractions eluted from liquid adsorption columns. By using a solid adsorbent such as alumina, silica gel, or Florisil, and heptane as eluent, the sulphur could be separated from weakly adsorbed hydrocarbons, e.g. squalene or biphenyl, but not from more strongly adsorbed hydrocarbons such as phenyldodecane.

Kupec *et al.* [180] determined total sulphur in sludge by a method involving magnesium reduction in which the sample is heated with magnesium powder to convert all sulphur compounds into magnesium sulphide. The magnesium sulphide is treated with sulphuric acid and the evolved hydrogen sulphide determined by iodometric titration.

The wet digestion procedure described by Landers *et al.* [97] and discussed in section 12.8.1.1 has also been applied to the determination of total sulphur and various sulphur functions in sludge. A typical analysis of aerobic sewage sludge is shown in Table 12.12.

12.9 Silicon

12.9.1 Soil

12.9.1.1 Inductively coupled plasma atomic emission spectrometry

Que-Hee and Boyle [70] analysed soils for total silicon using Parr bomb digestion with hydrofluoric–nitric–perchloric acid followed by inductively coupled plasma atomic absorption spectrometry.

12.9.1.2 Miscellaneous

Pellenberg [114] analysed soils and river sediment for silicone content by nitrous oxide–acetylene flame atomic absorption spectrophotometry. He showed that total carbon and total carbohydrates both correlate well with silicone content and the correlation between sedimentary silicone and presumed sewage material is good enough to suggest silicone as a totally synthetic, specific tracer for sewage in the aquatic environment.

12.9.2 Sludge

12.9.2.1 Miscellaneous

A standard UK method [93] has been described for the determination of silicon and phosphorus in sludges.

12.10 Arsenic

12.10.1 Soil

The limitations of the Gutzert method for determining arsenic are well known. The spectrophotometric molybdenum blue or silver diethyl/ dithiocarbamate procedures tend to suffer from poor precision and accuracy as shown in collaborative studies [115,116].

12.10.1.1 Atomic absorption spectrometry

The determination of arsenic by atomic absorption spectrometry with thermal atomization and with hydride generation using sodium borohydride has been described by Thompson and Thomerson [117] and it was evident that this method could be modified for the analysis of soil.

Thompson and Thoresby [118] and Wauchaupe [119] undertook an investigation of chemical pretreatment of soils prior to atomic absorption spectrometric analysis for arsenic.

In this method arsenic hydride is generated by the reaction of sodium borohydride with an acidified sample solution and is swept into a heated furnace tube by a current of nitrogen carrier gas. The hydride is thermally decomposed on contact with the hot walls of the tube, yielding a substantial population of ground state atoms for atomic absorption measurement. The principal advantages of this technique are high sensitivity and good precision and accuracy (RSD 11–12% at the 11–150 $\mu g\ g^{-1}$ arsenic level) coupled with speed of analysis (5–100 analyses per day). Blanks and detection limits are low. The detection limit being about $0.001\mu g\ ml^{-1}$ in the test solution. Brief details of the Thompson and Thoresby method [118] are given below.

The soil samples were first heated to fumes with a mixture of nitric and sulphuric acids. This method gave appreciably higher arsenic recoveries (95–102%) than fusion with potassium pyrosulphate and was considerably more rapid than fusion with a mixture of nitric and perchloric acids.

Table 12.13 compares results obtained by this method with those obtained by a molybdenum blue method of spectrophotometry [115,116]. Values obtained by atomic absorption spectrometry are higher than those obtained by the molybdenum blue method and this is believed to reflect the greater inherent accuracy of the former method. A UK standard

Table 12.13 Comparison of results for soils by atomic absorption spectrometry and molybdenum blue methods

Sample	*Arsenic found µg g⁻¹* *atomic absorption*	*molybdenum blue*
1	299	238
2	375	305
3	293	259
4	189	155
5	103	71
6	467	449
7	207	172

Source: Reproduced with permission from the Royal Society of Chemistry [118]

method also discusses the determination of arsenic in soil by atomic absorption spectrometry [120].

12.10.2 Non-saline deposited and suspended sediments

12.10.2.1 Gas chromatography

Cutter *et al.* [121] have described a method for the simultaneous determination of arsenic and antimony species in sediments. This method uses selective hydride generation with gas chromatography using a photoionization detector.

The following species can be separately determined AsIII, AsV, SbIII, SbV. Detection limits are 10p mol L^{-1} (arsenic) and 3.3pmol L^{-1} (antimony).

12.10.2.2 Atomic absorption spectrometry

Cutter [122] used a selective hydride generation procedure as a basis for the differential determination of arsenic and selenium species in sediments. Goulden *et al.* [123] also discuss the determination of arsenic and selenium in sediments by atomic absorption spectrometry.

12.10.2.3 Inductively coupled plasma atomic emission spectrometry

Goulden *et al.* [123] have described a semi-automated system for the determination of arsenic and selenium by hydride generation-industrively coupled plasma atomic-emission spectrometry. Sediments are brought into a solution by fusion with sodium hydroxide.

In this method the sediment is pretreated with 2g solid sodium hydroxide in a zirconium crucible which is then heated to 350°C for 2h. The fused mass is dissolved in 40ml 0.2N hydrochloric acid, then further treated with 20ml of hydrochloric acid to precipitate silica. The clear phase is decanted and is now ready for analysis.

The manifold for hydride generation is shown in Fig. 12.7. The operating conditions are as follows: forward power 1400W, reflected power less than 10W, cooling gas flow 12L m $^{-1}$, plasma gas flow 0.12L m $^{-1}$, injector flow, 0.34L m $^{-1}$. The standard deviation of this procedure was 0.02µL $^{-1}$ arsenic and the detection limit 0.1µg L $^{-1}$. Results obtained on a selection of standard reference sediment samples are quoted in Table 12.14.

Brzezinska-Paudyn *et al.* [124] compared results obtained in determinations of arsenic by conventional atomic emission spectrometry, flow-injection/hydride generation inductively coupled plasma atomic

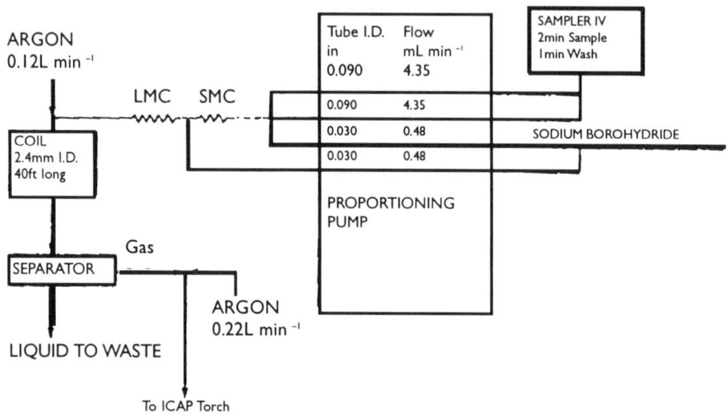

Fig. 12.7 Manifold for hydride generation
Source: Reproduced with permission from the American Chemical Society [123]

Table 12.14 Results in reference sediments

	Arsenic content µg g $^{-1}$ Recommended level	*Found*
Sample[b]		
SO–1	1.9	1.9
SO–2	1.2	1.3
SO–3	2.6	2.6
SO–4	7.1	6.5
IAEA soil	93.9 ± 7.5[a]	95.1
NBS SRM 1645	66	66.4

Source: Reproduced with permission from the American Chemical Society [123]

[a]Certified value; [b]SO, reference soils from Canada Centre for Mineral and Energy Technology; NBS National Bureau of Standards, USA; IAEA International Atomic Energy Agency, Vienna, Austria

emission spectrometry, graphite furnace atomic absorption spectrometry, combined furnace flame atomic absorption and neutron activation analysis. Results obtained show that all these methods can be used for the determination of down to 5μg g^{-1} of arsenic in certified sediments.

These workers used an ARC 34000 inductively coupled plasma emission spectrometer with flow-injection hydride generation. The 189.04nm line (3nd order) was used for arsenic measurement. The flow-injection block and Buckler peristaltic pump, as described by Liversage *et al.* [125] were also used for the determination of arsenic by hydride generation.

Detailed operating conditions for the inductively coupled plasma emission spectrometer have been described by Brzezinska *et al.* [126].

For the determination of arsenic by conventional inductively coupled plasma atomic emission spectrometry, the samples were digested in closed Teflon vessels, similar to the technique described by Brzezinska *et al.* [126].

About 0.1–2.0g of wet sediment was placed into a Teflon vessel and 3mL of concentrated nitric acid, 0.5mL concentrated perchloric acid, and 4mL concentrated hydrofluoric acid were added. The closed vessels were kept at room temperature for 1h. The samples were then placed in a pressure cooker and heated for 1h on a hot plate at a temperature of 300°C. After cooling, the vessels were uncapped and the samples evaporated to 2mL on a hot plate at 250°C. After cooling, 3mL concentrated nitric acid was added. To complex the fluorides, 1g boric acid was added to each sample. The solutions were transferred to 100mL volumetric flasks and adjusted to volume with deionized water. Inorganic arsenic standards, having the same acid content as the samples, were used for calibration.

Inductively coupled plasma atomic emission spectrometric analysis with flow-injection/hydride generation

Samples for inductively coupled plasma atomic emission spectrometric analysis with flow-injection/hydride generation were digested as follows: powdered samples were dried at 80°C for 2h and cooled in a desiccator. About 0.5g of sample was accurately weighed into a 100mL Teflon beaker and 2mL concentrated perchloric acid and 10mL concentrated nitric acid were added. The beakers were covered with Pyrex watch glasses and heated on a hot plate at 120°C for 2–3h. The nitric acid was slowly evaporated until white fumes of perchloric acid were visible. After cooling, the digests were transferred to Teflon dishes. The beakers were then rinsed several times with 3–4mL water each, which was also added to the digests. The final volume at this stage was approximately 25mL. After addition of 5mL concentrated hydrofluoric acid, the dishes were warmed to about 80°C for 15min to decompose and volatilize any siliceous residue. After cooling, the solutions were transferred to 50mL plastic volumetric flasks, 7mL concentrated hydrochloric acid and 1g sodium borohydride were added, and the volume was adjusted with deionized water. The reductant was 0.6% sodium borohydride in alkaline solution. Inorganic arsenic standards containing the same amount of hydrochloric acid and sodium iodide as in the samples were used for calibration.

These methods were used to determine arsenic in certified sediments (Table 12.15). Conventional inductively coupled plasma atomic emission spectrometry is satisfactory for all types of samples, but its usefulness was limited to concentrations of arsenic greater than 5μg g^{-1} dry weight. Better detection limits were achieved using the flow-injection–hydride generation inductively coupled plasma technique in which a coefficient of variation of about 2% for concentrations of 10μg g^{-1} were achieved.

12.10.2.4 Miscellaneous

Laser ablation followed by inductively coupled plasma mass spectrometry [127] has been used to determine down to 0.2μg g^{-1} of arsenic and antimony in solid sediments.

Cheam and Chau [128] used certified Great Lakes reference sediments for the determination of arsenic.

Brannon and Patrick [129] reported on the transformation and fixation of arsenic V in anaerobic sediment, the long term release of natural and added arsenic, and sediment properties which affected the mobilization of arsenic V, arsenic III and organic arsenic. Arsenic in sediments was determined by extraction with various solvents according to conventional methods. Added arsenic was associated with iron and aluminium compounds. Addition of arsenic V prior to anaerobic incubation resulted in accumulation of arsenic III and organic arsenic in the interstitial water and the exchangeable phases of the anaerobic sediments. Mobilization of

Table 12.15 Comparison of arsenic determination in sediments by different analytical methods (μg/g dry weight)

Method	*NRC BESS–1 Marine sediment*	*NRC MESS–1 Marine sediment*	*NBS 1645 River sediment*
Direct ICPAES	11.7 ± 2	8.6 ± 1.6	65 ± 18
ICPAES with flow-injection/ hydride generation	11.9 ± 0.3	10.6 ± 0.2	65 ± 1.2
Graphite furnace	11.5 ± 0.8	10.7 ± 0.7	64 ± 8
Direct introduction of solids into furnace/flame system	12.0 ± 1.6	11.2 ± 1.2	68 ± 8
Neutron activation	12.2 ± 1.1	11.4 ± 1.2	68 ± 10
Certified values	11.1 ± 1.4	10.6 ± 1.2	66

Source: Reproduced with permission from Perkin Elmer Ltd [124]

arsenic occurred over both short term and long term. Short term leaching of arsenic III was correlated with arsenic III concentrations in interstitial water and exchangeable phases, whereas long term release was related to total iron, extractable iron or calcium carbonate equivalent concentration. Arsenic leaching was most likely to be toxic during anaerobic conditions when arsenic III was released.

12.10.3 Saline deposited and suspended sediment

12.10.3.1 Spectrophotometry

Maher [130] has described a procedure for the determination of total arsenic in sediments. Arsenic is converted into arsine using a zinc reductor column, as shown in Fig. 12.8. The evolved arsine is trapped in a potassium iodide–iodine solution and other arsenic determined spectrophotometrically as an arsenomolybdenum blue complex. The detection limit is 0.024μg and the coefficient of variation is 5.1% at the 0.1μg level. The method is free from interferences by other elements at levels normally encountered in sediments. In this method the sediments were freeze-dried and ground (to less than 200μm) before analysis.

A weighed sample (less than 0.5g) was placed in a 30ml Pyrex centrifuge tube, 5ml of concentrated nitric acid were added and the mixture was allowed to stand for at least 12h at room temperature to ensure complete dissolution (this avoids foaming on heating). The tube was then placed in an aluminium heating block and refluxed until the evolution of brown fumes ceased. After cooling, 5ml of a nitric–sulphuric–perchloric acid mixture ($5 + 1 + 3\ V/V$) were added and heating continued until dense fumes of sulphur trioxide appeared. The digest was diluted with 5ml of 1.5M hydrochloric acid, 1ml of reducing solution was added and the solution allowed to stand for 40min to reduce all of the inorganic arsenic to the trivalent form. The solution was then made up to 25ml in a calibrated flask with 1.5M hydrochloric acid. The apparatus was assembled as in Fig. 12.8 with 1.5 ml of the potassium iodide–iodine solution in the centrifuge tube. The nitrogen gas flow-rate was adjusted to 10ml min^{-1}, 1ml of sample was injected on to the zinc column and the evolved arsine collected for 2min. Between each sample, 1ml of 1.5M hydrochloric acid was injected to prevent any accumulation of potentially interfering material on the column. After the arsine had been trapped, the centrifuge tube was removed and 0.5ml of the mixed spectrophotometric reagent added. The solution was mixed by means of a vortex mixer and allowed to stand for 30min to develop the arsenomolybdenum blue. The absorbance was measured at 866nm. Calibration graphs of absorbance versus amount of arsenic (0–1 and 0–5μg) were prepared by using arsenic standards carried through the entire analytical procedure.

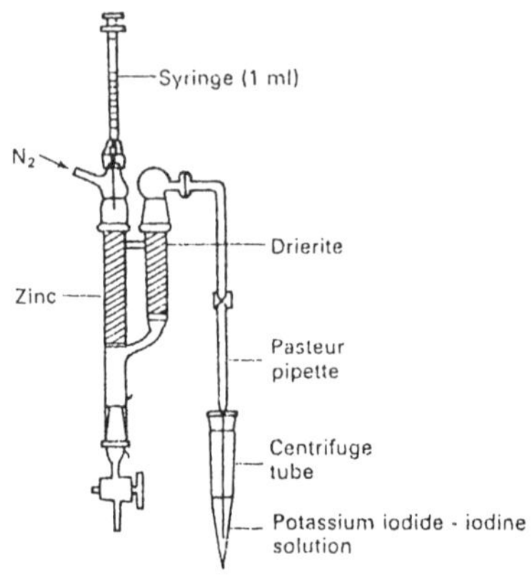

Fig. 12.8 Apparatus to generate, dry and trap arsenic hydride
Source: Reproduced with permission from the Royal Society of Chemistry [130]

Arsenic recoveries from the zinc column in the range 0.1–5μg ml^{-1} arsenic exceeded 97%. The concentrations at which certain elements interfere are shown in Table 12.16. Various other elements [Al III, B III, Ca II, Cd II, Co II, Cr VI, Fe III, K I, Li I, Mg II, Mn II, Na I, Ni II, Pb II, S VI, Sn II and Zn II] showed no significant interference at the 500μg level. Only low senium concentrations in extracts can be tolerated. However, few environmental samples contain appreciable amounts of selenium. As selenium is not reduced to hydrogen selenide on the column, selenium will not interfere in the final determination step, but probably suppresses either arsenic reduction or arsine formation. Selenium appears to suppress arsine generation at high arsenic concentrations but causes a slight enhancement at low arsenic concentrations (around 0.1μg), which could not be traced to arsenic impurities in the selenium standard used.

As shown in Table 12.17, complete recovery of added arsenic was obtained within experimental error for a sediment. The arsenic concentration obtained by replicate analysis of the orchard leaves (9.7 ± 0.3μg g^{-1}) and oyster tissue (13.2 ± 0.4μg g^{-1}) were in agreement with the certified values of 10 ± 2 and 13.4 ± 1.9μg g^{-1}, respectively.

Table 12.16 Effect of inorganic ions on the generation and trapping of arsine

All tests used 0.1μg of As(III) in 1ml of 1.5M hydrochloric acid. Optimised hydride generation and trapping conditions were used

Species	*Cu II*	*Hg II*	*Mo VI*	*Sb III*	*Se IV*	*Si IV*	*V V*
Interference level/μg	50	2.5	300	25	0.1	50	250

Source: Reproduced with permission from the Royal Society of Chemistry [130]

Table 12.17 Recovery of arsenic added to selected biological tissues and a sediment
Arsenic added as inorganic arsenic

Sample	*Arsenic/µg* *Added*	*Found*	*Recovery, %*
Sediment (0.5g)	0	3.1 ± 0.2	
	5	7.9 ± 0.3	96
	10	12.9 ± 0.6	98

Source: Reproduced with permission from the Royal Society of Chemistry [130]

12.10.3.2 Gas chromatography

Siu *et al.* [131] derivitized and determined arsenic in marine sediments using electron capture gas chromatography.

12.10.3.3 Atomic absorption spectrometry

The atomic absorption spectrometric procedure [122] described in section 12.10.2.2 for the determination of arsenic and selenium in non-saline sediments has also been applied to saline sediments.

12.10.3.4 Inductively coupled plasma atomic emission spectrometry

The optimal reaction conditions for the generation of the hydrides can be quite different for the various elements. The type of acid and its concentration in the sample solution often have a marked effect on sensitivity. Additional complications arise because many of the hydride-forming elements exist in two oxidation states which are not equally amenable to borohydride reduction. For example, potassium iodide is often used to pre-reduce AsV and SbV to the 3+ oxidation state for maximum sensitivity, but this can also cause reduction of Se IV to elemental selenium from which no hydride is formed. For this and other reasons Thompson *et al.* [132] found it necessary to develop a separate procedure for the determination of selenium in soils and sediments although arsenic, antimony and bismuth could be determined simultaneously [133]. A method for simultaneous determination of As III, Sb III and Se IV has been reported in which the problem of reduction of Se IV to Se O by potassium iodide was circumvented by adding the potassium iodide after the addition of sodium borohydride [134]. Goulden *et al.* [123] have reported the simultaneous determination of arsenic, antimony, selenium, tin and bismuth, but it appears that in this case the generation of arsine and stibene occurs from the 5+ oxidation state.

de Oliveira *et al.* [28] have described the application of a simple continuous hydride generation system coupled to a low-power (1.4kW) inductively coupled plasma atomic emission spectrometric technique to the determination of trace concentrations of arsenic, antimony and selenium in marine sediments. A variety of sample dissolution procedures and hydride generation reaction conditions were evaluated in an attempt to establish optimal conditions for the simultaneous determination of all three elements. In addition, the effect of the oxidation state of the elements on hydride formation in dilute hydrochloric acid solution was studied.

Four methods of dissolution were evaluated for marine sediments:

1 The acid digestion procedure described above for biological tissues. Crock and Lichte [135] recently described a similar procedure, involving hydrofluoric as well as nitric, perchloric and sulphuric acids, for dissolution of geological materials prior to arsenic and antimony determination by atomic absorption spectrometry.
2 Fusion with sodium hydroxide, as described by Goulden *et al.* [123] but using porcelain or nickel crucibles.
3 Acid digestion with a mixture of nitric, perchloric and hydrofluoric acids in sealed Teflon vessels, as described by McLaren *et al.* [136].
4 Fusion with potassium hydroxide at 500°C.

All four dissolution procedures studied were found to be suitable for arsenic determinations in biological marine samples, but only one (potassium hydroxide fusion) yielded accurate results for antimony in marine sediments and only two (sodium hydroxide fusion or a nitric–perchloric–hydrofluoric acid digestion in sealed Teflon vessels) were appropriate for determination of selenium in marine sediments. Thus, the development of a single procedure for the simultaneous determination of arsenic, antimony and selenium (and perhaps other hydride-forming elements) in marine materials by hydride generation inductively coupled plasma atomic emission spectrometry requires careful consideration not only of the oxidation-reduction chemistry of these elements and its influence on the hydride generation process but also of the chemistry of dissolution of these elements.

The apparatus used by de Oliveira *et al.* [28] consisted of a custom ICP–echelle spectrometer. Hydride generation was accomplished in a continuous mode by using two channels of a four-channel peristaltic pump (Gilson Instrument Co, Minneapolis, Il) to deliver sample and borohydride reagent to a phase separator modified from that of Thompson *et al.* [137]; a schematic diagram of the assembly employed is shown in Fig. 12.9. An air bubble maintained by the surface tension at the junction of the two horizontal arms of the 'T' prevents mixing of the reagent and sample until the two solutions begin to flow down the vertical arm into the phase separator. This results in a smooth and continuous generation of hydrogen

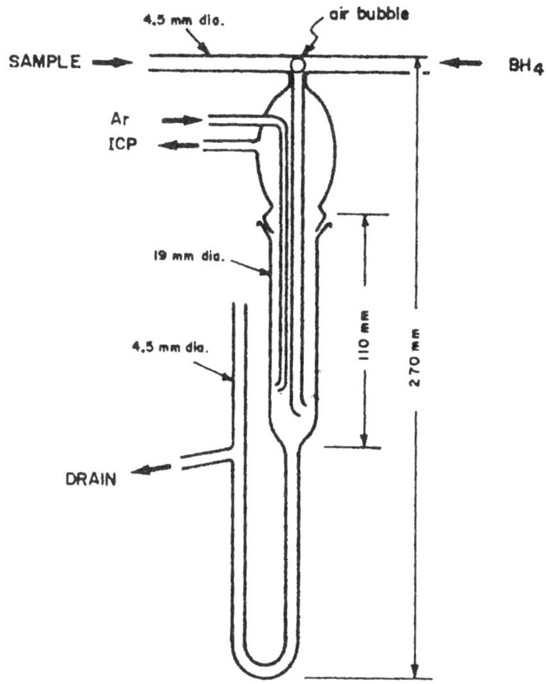

Fig. 12.9 Gas liquid phase separator
Source: Reproduced with permission from the American Chemical Society [28]

which is not significantly disturbed by changeover from one sample to the next, or to the blank. The gaseous hydrides and hydrogen are swept from the phase separator and into the inductively coupled plasma by a continuous flow of argon. In this method, detection limits of 0.5μg L^{-1} for arsenic and 0.5μg L^{-1} for antimony and selenium were achieved.

12.10.4 Sludge

12.10.4.1 Atomic absorption spectrometry

Webster [138] has investigated the determination of arsenic in the concentration range 0.5–10mg kg^{-1} dry solids in sewage sludge. The method involves the use of sodium borohydride to generate arsenic hydrides, and their introduction into a silica furnace, maintained at dull red heat by the air-acetylene flame of an atomic absorption spectrophotometer. Predigestion of sewage sludge with nitric/perchloric acid is recommended. Recoveries, using a standard addition technique, were 80–100% and precision between 16 and 18%, and some interference from heavy metals was observed. Results from a 12-month study of sludges in the Lothian region showed arsenic concentrations between 0.5 and 7mg kg^{-1} dry solids.

Webster [138] made a close study of various parameters influencing results obtained in this procedure such as sample preparation using the nitric perchloric acid method [139]. Hydrochloric acid strength at the reduction stage, concentration and volume of sodium borohydride, volume of nitrogen from purging reduction cell and effect of arsenic valency all have a bearing on results obtained. A UK standard method [120] discusses a hydriodic reduction atomic absorption spectrometric method for determination of arsenic in sewage sludge.

12.10.4.2 Hydride generation inductively coupled plasma atomic emission spectrometry

Atsuya and Akatsuka [140] have described a method for determining trace amounts of arsenic. The technique, which uses capacitively coupled microwave plasma with an arsine generation system, has been used to determine arsenic in sewage sludge.

12.11 Antimony

12.11.1 Non-saline deposited and suspended sediments

12.11.1.1 Spectrophotometry

Abu-Hilal and Riley [182] have investigated a spectrophotometric method for the determination of antimony in sediments and clays. In this procedure 0.5g of the finely ground sample is weighed into a polytetrafluoroethylene baker. This is moistened with a few drops of water and then 10ml of 40% (*w/v*) hydrofluoric acid is cautiously added. The beaker is covered with a polytetrafluoroethylene lid and heated overnight on a boiling water bath. The solution is evaporated to dryness, 10ml of redistilled nitric acid added and the solution evaporated to dryness. A further 5ml of nitric acid is added and the solution evaporated to dryness on a hot plate at low temperature, taking care to avoid baking the residue. The latter is dissolved in 3ml of 6M hydrochloric acid, diluted with water and transferred quantitatively to a 1L Erlenmeyer flask. The solution is diluted to 1L.

If necessary a preconcentration was carried out on this solution to lower the detection limits of the method. Preconcentration was achieved by a method involving co-precipitation of the antimony with hydrous zirconium oxide in which the digest is stirred with 150mg zirconyl chloride and the pH adjusted to 5 with ammonia to coprecipitate antimony and hydrous zirconium oxide. The isolated precipitate is dissolved is 7M hydrochloric acid and 30% sulphuric acid. Antimony is then converted to the pentavalent state by successive treatment with titanium III chloride and sodium nitrite and excess nitrite destroyed by urea.

Antimony is then determined by a spectrophotometric method utilizing crystal violet in which the extract is treated with this chromogenic agent and the coloured complex extracted with benzene. The benzene extract is evaluated spectrophotometrically at 610nm. No inter-ference was exhibited in this method by 10mg of acetate, arsenate, bromide, nitrate, cyanide, fluoride, iodide, nitrate, oxalate, phosphate, sulphate, thiocyanate, thiosulphate and tartrate or 0.5mg of mercury 11 or 50μg of thallium. Gold at concentrations above 0.5μg L^{-1} unfortunately interfered. In this method a clay sample spiked with antimony III gave recoveries in the range 98.5 to 101.5% with an average of 100.5%.

12.11.1.2 Inductively coupled plasma atomic emission spectrometry

The laser ablation inductively coupled plasma atomic emission spectrometry procedure described by Arrowsmith [127] discussed in section 12.10.2.4 has been applied to the determination of down to 0.2μg g^{-1} of antimony in sediments.

12.11.1.3 Miscellaneous

The selective hydride generation–gas chromatographic method [121] using photoionization detection discussed in section 12.10.2.1 for the determination of arsenic III and arsenic V has been applied to the determination of down to 3.3pmol L^{-1} of antimony (Sb III, SbV) in sediments.

Brannon and Patrick [141] give details of studies on the distribution and mobility of antimony in sediments from several sites in rivers, waterways and coastal waters throughout the USA. Most of the naturally-occurring and added antimony in the sediments was associated with relatively immobile iron and aluminium compounds. In sediments containing added antimony, the concentrations of this metal in the interstitial water and of the exchangeable-phase antimony were high. In leaching experiments under anaerobic conditions, the greatest release of antimony occurred early for most of the sediments, suggesting that leaching of antimony from contaminated sediments was most likely to occur during the first few months of interaction between sediment and water. Under aerobic leaching conditions, the antimony moved into a less available sediment phase, thus reducing the possibility of further release. When the sediments were incubated anaerobically, there was evidence of evolution of volatile sulphur compounds.

12.11.2 Saline deposited and suspended sediments

12.11.2.1 Miscellaneous

The inductively coupled plasma atomic emission spectrometric procedure [28] described in section 12.10.3.4 for the determination of arsenic in saline sediments has been applied to the determination of down to 0.5g L^{-1}

antimony in sediments. Of the four sediment digestion procedures discussed only fusion with solid potassium hydroxide at 500°C gives reliable antimony recoveries.

12.12 Bismuth

12.12.1 Non-saline deposited and suspended sediments

12.12.1.1 Atomic absorption spectrometry

Zhe-Ming *et al.* [142] have described a method for the determination of down to 1mg kg $^{-1}$ of bismuth in river sediments by electrothermal atomic absorption spectrometry with low temperature atomization in argon: hydrogen (90:10).

12.2.2 Saline deposited and suspended sediments

12.12.2.1 Atomic absorption spectrometry

Hydride generation and atomic absorption detection provide a very sensitive method for metalloid analyses. Collection of the hydride in a liquid nitrogen trap and then subsequent rapid introduction into an atomizer improved the sensitivities of such elements as arsenic and tin by an order of magnitude compared to the continuous method. However, such a collection method is only applicable to the elements whose hydrides are stable enough to be handled at ambient temperature. Bismuthine is unfortunately too unstable for this technique. Upon warming bismuthine from liquid nitrogen temperatures only 5–15% of the trapped bismuthine was trapped, the remainder decomposed. To overcome this problem Lee [143] used a heated graphite tube to collect the generated bismuthine (Fig. 12.10). Some 72% of the generated bismuthine was collected in the tube reproducibly. In this method the bismuth is reduced in solution by sodium borohydride to bismuthine, stripped with helium gas, and collected *in situ* in a modified carbon rod atomizer (Fig. 12.10). The collected bismuth is subsequently atomised by increasing the atomizer temperature and detected by an atomic absorption spectrophotometer. The absolute detection limit is 3pg of bismuth. The precision of the method is 2.2% for 150pg and 6.7% for 25pg of bismuth.

Down to 6μg kg $^{-1}$ bismuth could be determined in sediments. High concentrations of cobalt, copper, gold, molybdenum, nickel, palladium, platinum, selenium, silver and tellurium interfere in this procedure. Varying amounts of bismuth were found in sea water 0.08–0.63ng L $^{-1}$ in Narragonsett Bay and 0.05ng L $^{-1}$ in the North Pacific and bismuth contents on sediments taken in Narragonsett Bay ranged from 0.27–6.4mg kg $^{-1}$ and from the North Pacific Ocean from 0.10–0.12mg kg $^{-1}$ indicating very high concentration factors.

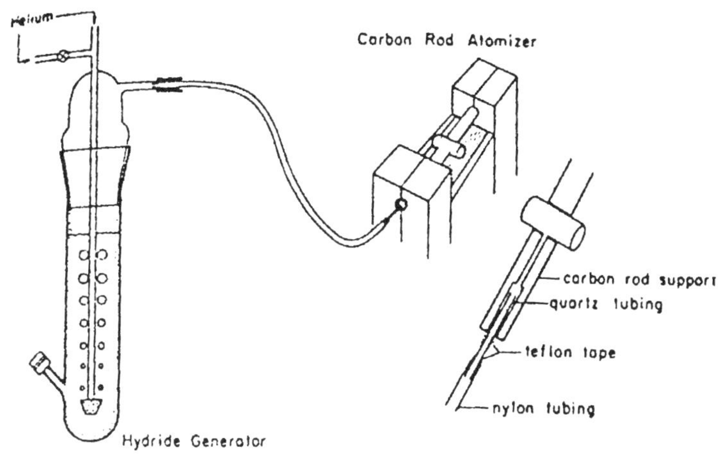

Fig. 12.10 Apparatus for the determination of bismuth by hydride generation
Source: Reproduced with permission from the American Chemical Society [143]

12.13 Selenium

12.13.1 Soil

12.13.1.1 Atomic absorption spectrometry

Atomic absorption spectrometry has found broad application for the determination of selenium in soils and sediments. The most intense resonance line of selenium (196.03 nm) corresponds to a range near to the vacuum ultraviolet. Moreover, the most frequently applied air acetylene flame absorbs about 55% of radiation intensity of the light source. When using electrodeless discharge lamps and air-acetylene flame, a lower detection limit of 0.2μg mL^{-1} can be reached that can be extended down to 0.1μg g^{-1} by application of a deuterium lamp for background correction. The argon-hydrogen flame is often used for augmentation of sensitivity but it increases interferences too. Extraction has also been attempted [150] as a means of improving sensitivity but in selenium determination a re-extraction to a water solution is necessary.

Flameless atomic absorption spectrometric techniques offer a high sensitivity (5×10^{-11}g Se) but are not simple nor free from interference, due to the high volatility of selenium. This technique is suitable specially for direct analysis of samples and its additional advantage lies in possibilities of 'chemical treatment' of samples in the graphite cell in order to diminish chemical interferences.

The addition of nickel enhances significantly the sensitivity for selenium by about 30% and allows higher ashing temperatures (1000°C) without losses [151–153]. Other elements capable of forming selenides (e.g. barium, copper, iron, magnesium and zinc) did not interfere and arsenic interference was minimized. A detection limit of 10–12μg kg $^{-1}$ selenium has been achieved using a graphite electrothermal furnace and background correction with a deuterium lamp [154].

Hydride generation methods are finding increasing favour for the determination of selenium. This method consists of measuring the atomic absorption of selenium hydride formed as a result of reduction of selenium and its compounds with different reducing mixtures such as sodium borohydride or, occasionally zinc-stannous chloride-potassium iodide. Hydride generation techniques are about three orders of magnitude more sensitive for determining selenium than are classical flame ionization techniques, a detection limit of 0.2ng g $^{-1}$ being achievable. They have an additional advantage of separating selenium from the matrix before atomization thus avoiding interferences inherent to the conventional atomic absorption technique. Practical working ranges for selenium are 3–250μg mL $^{-1}$, 0.03–3μg mL $^{-1}$ and up to 0.12μg mL $^{-1}$, respectively, for flame atomic absorption, furnace atomic absorption and vapour generation methods [155–158].

Various workers have discussed the application of atomic absorption spectrometry to the determination of selenium in rocks [159, 160] achieving detection limits of 0.06g g $^{-1}$ [159] and 1.4×10^{-10}g g $^{-1}$ [160] respectively. Hydride generation and measurement of hydride fluorescence has been used to determine selenium [120,161] with a sensitivity of 0.06ug Se mL $^{-1}$ which is 5–30 times than is achieved by conventional atomic absorption spectrometry.

12.13.1.2 Miscellaneous

Neutron activation analysis has been used to determine selenium in soil [144–148]. Nadkarni and Morrison [149] estimated 47 elements in lake sediments and found 0.3–1.01μg selenium per gram using neutron activation analysis. Dong *et al.* [162] used mixtures of phosphoric acid, nitric acid and hydrogen peroxide in the digestion of soils prior to the determination of selenium.

12.13.2 Non-saline deposited and suspended sediments

12.13.2.1 Gas chromatography

de Oliveira [28] digested sediments with a mixture of nitric, perchloric acids and sulphuric acids (section 12.10.3.4) prior to the determination of selenium by a procedure involving co-reaction of selenium with 4-nitro-o-phenylene diamine to produce a volatile product which was determined in amounts down to 100μg kg $^{-1}$ by electron capture gas chromatography [163].

12.13.2.2 Spectrofluorimetry

Wiersma and Lee [164] determined selenium in lake sediments. The sample is digested with 4:1 concentrated nitric acid; 6% perchloric acid and the residue treated with 6M hydrochloric acid then reduced with H_3PO_2. The fluorescing agent used was 2,3 diaminonaphthalene.

12.13.2.3 Atomic absorption spectrometry

The atomic absorption spectrophotometric methods discussed in section 12.10.2.2 [122,124] has been applied to the determination of selenium in sediments. Itoh *et al.* [165] and Cutter [122] (section 12.10.2.2) have used hydrogen generation–atomic absorption spectrophotometric techniques to determine selenium in non-saline sediments. Cutter [122] was able to distinguish between selenite, selenate, total selenium and organic selenium in sediments.

12.13.2.4 Inductively coupled plasma atomic emission spectrometry

The inductively coupled plasma technique [123] discussed in section 12.10.2.3 has been applied to the determination of selenium in non-saline sediments.

12.13.2.5 Miscellaneous

Cheam *et al.* [128] have developed a Great Lakes reference sediment for selenium.

12.3.3 Saline deposited and suspended sediments

12.13.3.1 Spectrophotometric method

Terada *et al.* [166] determined 0.3–1ppm selenium in marine sediments after converting the element to stannic bromide which was distilled off and assayed colorimetrically as piazselenol.

12.13.3.2 Gas chromatography

Siu and Berman [163] determined selenium in marine sediments in amounts down to 0.2pg (or 20ng g^{-1} of sediment) with a precision of 7%. This method is based on the fact that 1,2 diaminobenzene (o-phenylene diamine) and its derivatives react selectively and quantitatively with selenium IV (average accuracy 94 ± 5%) to form piazselenols that are both volatile and stable. Piazselenols can be determined by electron capture gas chromatography. The sediments were digested as follows. A 0.5g sample was placed in a poly(tetrafluoroethylene) pressure decomposition vessel.

A concentrated acid mixture comprising 3mL of nitric, 3mL of hydrochloric, and I mL of perchloric was added. The closed vessel was immersed in boiling water for 2h. The vessel was opened after cooling, and the contents transferred to a poly(tetrafluoroethylene) beaker with doubly distilled water. The solution was taken to dryness overnight on a hot plate at about 90°C. The residue was dissolved in about 20mL of 1M hydrochloric acid and the solution was again taken to dryness at about 90°C. This ensured complete reduction of Se(VI) to Se(IV). The residue was again dissolved and diluted to 50mL with 1M hydrochloric acid. This solution was washed twice with about 20mL of toluene.

One millilitre of sediment solution was allowed to stand with 0.1mL of the 1% 4-nitro-*o*-phenylenediamine hydrochloride solution in a 10 mL glass vial for 2h. One millilitre of toluene was added. The vial was capped and shaken vigorously for 2min. One microlitre of the toluene layer was injected into the gas chromatograph.

A Varian Aerograph Model 1200 gas chromatograph equipped with a Tracor ^{63}Ni electron capture detector was used. The column was a 2m borosilicate tube packed with 3% OV–225 on Chromosorb W, 80/100 mesh. It was normally kept at 200°C. Nitrogen (<10 ppm oxygen), further purified by passage through molecular sieve 5A and a heated oxygen scavenger (Supelco), was used as carrier and detector makeup gas. The flow rates were usually about 25 mL/min. The electron capture detector was heated to 320°C and operated in constant voltage mode at an optimal voltage of usually –13V.

12.13.3.3 Atomic absorption spectrometry

Willie *et al.* [167] applied hydride generation atomic absorption spectrometry with *in situ* concentration in a graphite furnace to the determination of selenium in marine sediments. A custom-made Pyrex cell was used to generate selenium hydride which was carried to a quartz tube and then to a preheated furnace. All the selenium was identified when the furnace was operated at 400°C. Pyrolytic graphic coated tubes could be used and the life of the tubes exceeded 1200 firings. A 10% solution of dimethyldichlorosilane was used to deactivate internal surfaces of the generation cell which could be used for 130 determinations before requiring resilylation. No interferences were found in this study: 1000ug of iron, 6ug of copper, 15ug of nickel, and 2.5ug of arsenic had no effect on the signal from 2ng selenium in 5ml of 0.05M hydrochloric acid solution. An absolute detection limit of 70pg selenium equivalent to 30ng g^{-1} was achieved.

In this method a 0.5µg portion of sediment was decomposed by acid digestion in a PTFE bomb according to the procedure described by Siu and Berman [163] and diluted to 50mL in 1M hydrochloric acid. Total selenium was determined by using 500µL aliquots delivered into the hydride cell containing 5mL of 0.5M hydrochloric acid.

Alternatively 0.5g samples were dry ashed using magnesium nitrate as an ashing aid [168] to prevent volatilization losses of selenium. The solution was diluted to 50mL. Total selenium was determined by using 50μL aliquots diluted to 5mL with 0.5M hydrochloric acid.

Reagent blanks were processed through identical steps for decompositions of sediments.

In the method described by Willie *et al.* [167] atomic absorption measurements were made with a Perkin–Elmer 5000 spectrometer fitted with a Model HGA 500 graphite furnace and Zeeman effect background correction system. Peak absorbance signals were recorded with a Perkin–Elmer PRS–10 printer-sequencer. A selenium electrodeless lamp (Perkin–Elmer Corp.) operated at 6W was used as the source. Absorption was measured at the 196.0nm line. The spectral band-pass was 0.7nm. Standard Perkin–Elmer pyrolytic graphite-coated tubes were used in all studies.

A custom-made Pyrex cell [169] was used to generate selenium hydride. The internal purge gas supply line to the furnace was routed through a stopcock made of Teflon that permitted the operator to select gas flow into either the bottom of the hydride cell or into the furnace, as illustrated in Fig. 12.11.

In this manner an argon flow could be used to strip the generated hydride from solution and carry it out of the top of the cell where it was directed, via a 1mm i.d. × 1.5mm o.d. quartz tube, into the sample introduction port of a preheated furnace.

Prior to use, the cell and transfer line were silylated to deactivate the internal surfaces [170]. A 10% solution of dimethyl-dichlorosilane in toluene was used to rinse the cell, followed by successive rinses in toluene and methanol. The surfaces were then dried at room temperature in a stream of nitrogen.

Sodium borohydride solution was pumped into the cell using a rack-mounted Ismatec peristaltic pump.

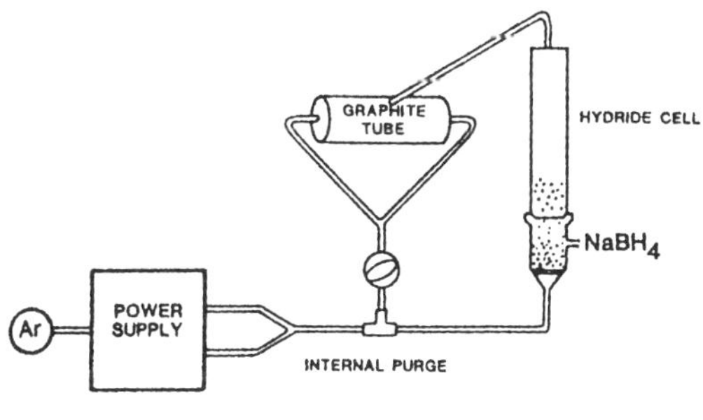

Fig. 12.11 Schematic of gas distribution system
Source: Reproduced with permission from the American Chemical Society [167]

The sequence of operations describing selenium hydride generation collection, and atomization is given below. During collection the stopcock was closed to direct internal purge gas through the hydride cell, and the furnace was preheated for 10s at 600°C. The sodium borohydrite solution was pumped into the cell at a rate of 4mL/min for 30s for 5mL samples (60s for 50mL seawater samples), during which time the selenium hydride was swept, via the generated hydrogen gas stream (≈50mL/min under these conditions), into the furnace where it was trapped. Internal purge gas flow was automatically initiated at the end of sodium borohydride addition, and the cell purged for 120s at a flow rate of 100mL/min for 5mL sample volumes (190–210s for 50mL seawater samples). At the end of this cycle, thermal programming of the furnace was terminated, the quartz transfer line removed from the sample introduction port, and the stopcock opened to permit internal purge gas to flow through the furnace. The sample was then atomized at 2600°C using maximum power heating and internal gas stop, followed by a cleaning cycle at 2700°C with 300mL/min internal purge gas flow. The furnace programme is shown in Table 12.18.

Internal purge gas was again diverted through the reaction cell, which was emptied and rinsed with doubly distilled water. The sodium borohydride solution was withdrawn from the injector tip by reversing the direction of the peristaltic pump. The next sample aliquot was then added to the cell and the measurement process repeated. Replicate measurements could be made every 3–4min.

Application of this method to standard reference sediment samples gave results very close to accepted values as shown in Table 12.19.

12.13.3.4 Inductively coupled plasma atomic emission spectrometry

The procedure described by de Oliveira [28] and discussed in section 12.10.3.4 has been applied to the determination of selenium in potassium hydroxide fusion or nitric–perchloric–hydrofluoric acid digests of marine sediments.

Table 12.18 Furnace programme

	Prog-ramme	*step*	*temp, °C*	*time, s*	*internal purge, mL/min*
generation–collection	1	1	600	10	0
	2	600	30	0	
	3	600	120	100	
atomization	2	1	2600	4	0
	2	2700	2	300	

Source: Reproduced with permission from the American Chemical Society [167]

Table 12.19 Analytical results[a]

Sample	*Determined*	*Accepted value*
NASS–1, ng/g	0.024 ± 0.002	(0.025 ± 0.001)[b]
BCSS–1, µg/g	0.46 ± 0.04	(0.43 ± 0.06)[c]
TORT–1, µg/g	6.74 ± 0.17	6.88 ± 0.47
NBS 1566, µg/g	2.13 ± 0.10	2.1 ± 0.05

Source: Reproduced with permission from the American Chemical Society [167]

[a]Precision expressed as standard deviation based on five determinations. [b]Uncertified value. [c]Uncertified value.

12.13.4 Sludge

12.13.4.1 Atomic absorption spectrometry

A standard UK hydride generation method [171] has been applied to the determination of selenium and arsenic in sludges and soils.

12.14 Oxygen demand parameters

12.14.1 Non-saline deposited and suspended sediments

Bowman and Delfino [172] have reviewed methods for the determination of the sediment oxygen demand of lake sediments. Wang [173] tested several toxicants as a means of distinguishing between the biological and chemical oxygen demand of bottom deposits; phenol was found to be most effective. A method was also developed for separating the chemical demand into ferrous, sulphide and manganous demands. Tests on sediments from lakes in Illinois showed the major component of the oxygen demand to be chemical, and predominantly due to iron.

The determination of the chemical oxygen demand of sediments has been discussed [22,54]. Markert *et al.* [174] have described equipment for the collection of sediment and determination of sediment oxygen demand by means of a flow-through diffuser.

12.14.2 Saline deposited and suspended sediments

Nothlich and Reuter [175] have devised an apparatus for the measurement of oxygen uptake by sediments from coastal waters and estuaries. The specific oxygen uptake per unit surface area of undisturbed sediment was very much smaller (by a factor of about 30) than that of the suspended sediments, which is of considerable significance for an assessment of the oxygen balance in waters transporting dredged solids. In addition, the content of organic degradable matter increases with an increase in the proportion of fines in the suspended sediment, which further enhances the oxygen consumption. Typical silty sediments are characterized by a high oxygen uptake whereas sandy sediments exhibit much lower values.

12.14.3 Sludge

12.14.3.1 Total oxygen demand

The Phillips total oxygen demand meter has been applied to the determination of total oxygen demand of settled sewages and sewage effluents. This instrument utilises two well known electrochemical relationships: Faraday's Law, which relates to the electrochemical transport of elements, and Nernst's Law, which is concerned with electrochemical potentials at boundaries. It is known that, at elevated temperatures and under the influence of an electrical current, oxygen can be transported through specially prepared zirconium oxide and that, in the electrolyte state and in the presence of oxygen zirconium oxide will generate an electrical potential at its surface. Furthermore, when there exists a difference in the oxygen concentration, there will be a measurable potential difference.

In the Philips total oxygen demand meter the zirconium oxide is specially prepared and is in the form of two series connected tubes or cells through which a nitrogen carrier gas is allowed to flow. Each cell is provided with two pairs of annular (internal and external) electrodes and the whole assembly is maintained at about 600°C.

One pair of electrodes conduct an electrical current through the zirconium oxide wall, thereby transporting oxygen from the external atmosphere through the cell wall, into the nitrogen carrier gas. The other pair is used to measure the potential difference over the zirconium oxide.

The transport of oxygen and the potential developed are shown diagrammatically in Fig. 12.12 and 12.13 respectively.

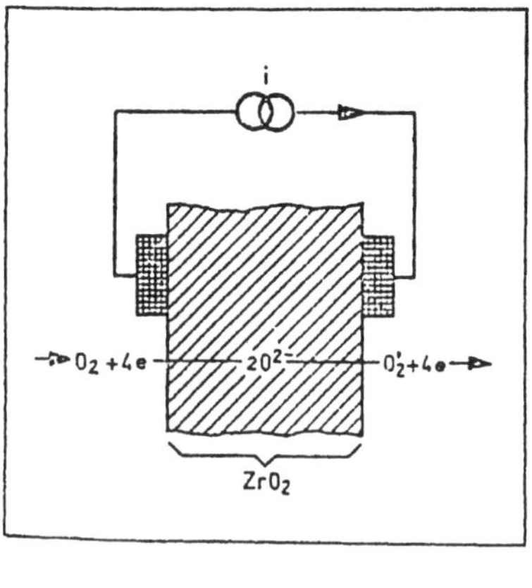

Fig. 12.12 Transport of oxygen
Source: Own files

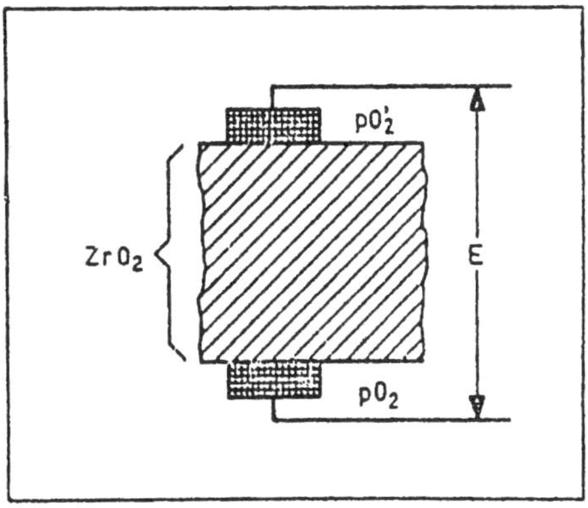

Fig. 12.13 Nerst potential
Source: Own files

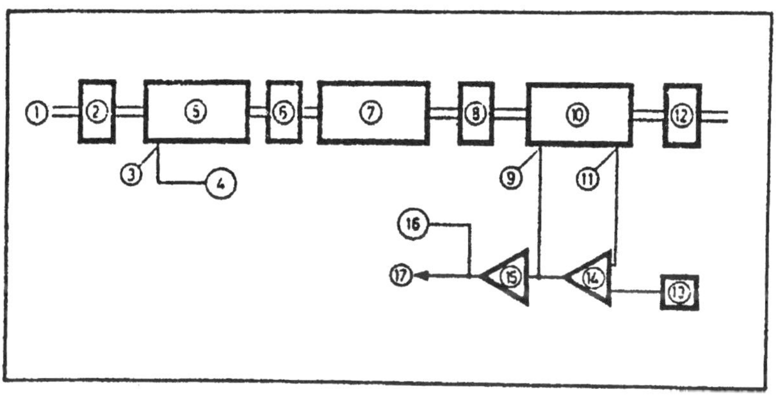

Fig. 12.14 TOD meter schematic: 1, N supply (3 bars); 2, flow controller (50ml N^2 min $^{-12}$); 3, dosing electrodes (dosing cell); 4 and 5, dosing cell (approx. 600°C); 6, sample head (with septum); 7, combustion oven (approx. 900°C); 8, drier; 9, dosing electrodes (measuring cell); 10, measuring cell; 11, measuring electrodes (measuring cell); 12, flow meter; 13, reference voltage memory circuit; 14, amplifier; 15, integrator (gives direct TOD); 16, TOD display meter; 17, recorder output
Source: Own files

Fig. 12.14 shows the operation of the total oxygen demand meter in diagrammatic form. One cell is used as a dosing cell (oxygen transport) and the other as a measuring cell which combines oxygen dosing and oxygen measuring. The cells are located, in the flow sense, prior to and after the combustion oven. A section through a cell is shown in Fig. 12.15.

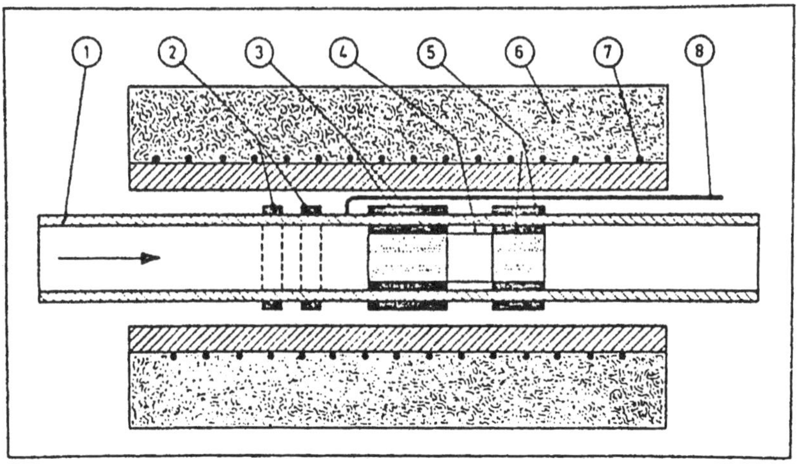

Fig. 12.15 Section through zirconium oxide cell: 1, zirconium oxide tube; 2, NTC resistor; 3, dosing electrodes; 4, ceramic spacer; 5, measuring electrodes; 6, thermal insulation; 7, heating coil; 8, thermocouple
Source: Own files

A preset quantity of oxygen is admitted to the nitrogen carrier gas (whose pressure and flow rate is controlled) via the dosing cell. The quantity of oxygen is controlled by a range selector.

The combined oxygen/nitrogen gas flow, together with the sample, is admitted to the combustion oven which is maintained at about 900°C. The resultant reduction in oxygen content, due to combustion, is measured as the amount of current necessary to maintain the oxygen concentration at the original level, integrated with respect to time and recorded in terms of an electrical output (0–20mA or 1–100mV) which is directly proportional to the mass of oxygen consumed by combustion in the oven and, hence, to the real total oxygen demand.

The electrical output (from the measuring electrode of the dosing cell) measured immediately prior to injection of the sample is stored in a memory and is applied to the input of a differential amplifier; the other input to this amplifier is directly connected to the measuring cycle of the measuring cell itself. During the measuring cycle the two voltages are compared with each other, any difference being amplified and transformed into a current which is then applied to the dosing electrode of the measuring cell. This current, necessary to maintain the oxygen concentration in the carrier gas after combustion, is integrated as described above, and the complete circuit thus operates as a closed-loop feedback control system.

Voorn and Marlow [176] applied this technique to waste water samples. Table 12.20 shows the measured oxidation efficiency of the total oxygen

Table 12.20 Oxidation efficiencies and TOD/COD correlations

Substance	*Conc. g/l*	*Oxidation efficiency measured/theoretical TOD %*	*COD %*	*Correlation TOD/COD*
Glucose	0.093	99	98	1.01
Fumaric acid	2.50	98	99	0.99
Malic acid	2.50	100	100	1.00
Mannose	2.67	99	98	1.01
Sodium citrate	0.046	99	95	1.04
Sodium oxalate	0.670	100	100	1.00
Benzoic acid	0.255	100	100	1.00
Potassium hydrogen phthalate	0.425	100	100	1.00

Source: Own files

demand meter and the excellent correlation with chemical oxygen demand obtained for a range of pure substances containing carbon, hydrogen and oxygen.

In earlier total oxygen demand meters in which large amounts of oxygen are available for combustion any organic nitrogen in the sample is oxidized to nitric oxide. In the Philips meter, because of the controlled oxygen concentrations used in each measuring range, nitrogen is not normally oxidized.

Several nitrogen containing substances were tested by Voorn and Marlow [176] and the results are given in Table 12.21 together with the theoretical values predictable for both the N→N_2 and the N→N0 reactions. The table strongly suggests that hydrogen and carbon were oxidised and nitrogen was not.

Some oxidation of combined nitrogen to nitric oxide may take place when measurements are made low in a range, so that an excess of oxygen is available for combustion.

Although poorer correlation was obtained with organic substances containing nitrogen, where chemical oxygen demand values are lower than total oxygen demand values (with chemical oxygen demand the combined nitrogen is converted to ammonia and many nitrogen containing substances are not oxidised by the dichromate to the theoretical degree), it is firmly believed that the total oxygen demand value measured by the Phillips instrument closely approaches the true oxygen demand due to natural degradation, because nitrogen is the end-product of the natural nitrogen cycle. Sulphur mainly oxidises to sulphur dioxide.

Voorn and Marlow [176] applied this technique to industrial and municipal waste waters. Fig. 12.16 shows TOD/COD ratios obtained from some waste water samples. Where the total oxygen demand value is

Table 12.21 TOD values for nitrogen-containing substances

			TOD				COD	
Substance	*Range used*	*Measure TOD*	*Theory for* $N \rightarrow N^2$	*Measure/ Theory* $N \rightarrow N^2$	*Theory for* $N \rightarrow N0$	*Measure/ Theory* $N \rightarrow N0$	*Theory for* $N \rightarrow NH_3$	*Measured*
EDTA	3000	1040	1070	0.975	1180	0.88	912	772
TRIS	3000	1740	1735	1.00	1900	0.91	1487	1400
Glycine	1000	854	855	1.00	1042	0.82	1382	1170
HMTA	1000	582	585	1.00	695	0.84	1433	2160
Dimethyl formamide	10000	3470	3510	0.99	3923	0.89	2897	1176
Urea	10000	4020	3996	1.01	6660	0.60	0	786

Source: Own files

EDTA: Ethylenediaminetetracetic acid disodium salt
TRIS: Tris(hydroxy)aminomethane
HMTA: Hexamethylenetetramine

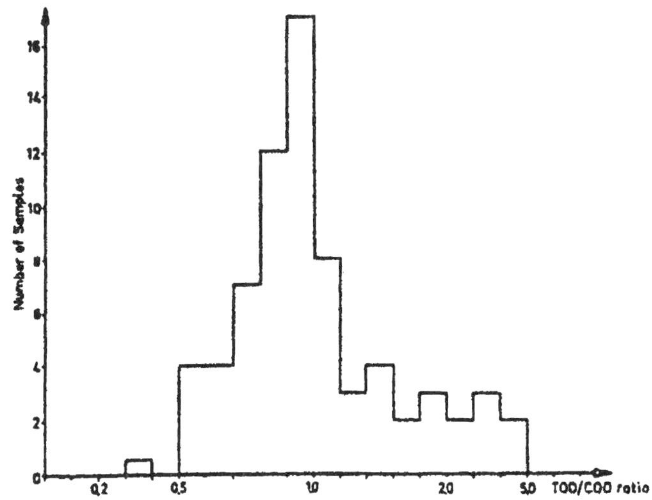

Fig. 12.16 Distribution of TOD/COD ratios for chemical waters
Source: Own files

high with respect to the chemical oxygen demand, this is due to the incomplete oxidation found in the chemical oxygen demand method; low total oxygen demand (with respect to chemical oxygen demand) is due to the presence of nitrates, peroxides and sulphuric acid in the process water.

Analysis of influents to waste water sewage plants showed that the total oxygen demand is higher than the chemical oxygen demand and this is because of the better oxidation efficiency that is achieved with total oxygen demand analysis and also because nitrogen is converted to nitrogen (in the total oxygen demand analysis) and to ammonia in chemical oxygen demand method.

Analysis results from the influents of five communities are given in Table 12.22.

Table 12.22 Analysis of communal influents to sewage waste-water purification plants

Community	*TOD*	*COD*	*BOD_5*	*N*	*TOD/COD*	*TOD/BOD_5*
1	1355	1180	662	42	1.15	2.05
2	920	767	298	71	1.20	3.09
3	352	196	83	37	1.80	4.24
4	160	126	44	18	1.27	3.65
5	848	678	386	76	1.25	2.20

Source: Own files

Marty and Aim [177] described an automatic determination of the total oxygen demand of waste water. The method gives rapid results and correlates well with chemical oxygen demand and biochemical oxygen demand. Measurements of corrections are made for nitrogen compounds which are the principal interferents.

Other workers who have discussed the determination of total oxygen demand include Wells [178] and Ravenscroft [179].

References

1 HMSO (1979) Method 8. Boron, water soluble in soil. *The Analysis of Agricultural Materials*, RB 427, 2nd edn., London.
2 Aznarez, J., Bonilla, A. and Vidal, J.C. (1983) *Analyst (London)*, **108**, 368.
3 Ducret, L. (1957) *Analytica Chimica Acta*, **17**, 213.
4 Pasztor, L., Bode, J.D. and Fernando, Q. (1960) *Analytical Chemistry*, **32**, 277.
5 Manzoori, J.L. (1980) *Talanta*, **27**, 682.
6 Zarcinas, B.A. and Cartwright, B. (1987) *Analyst (London)*, **112**, 1107.
7 Davis, R.D. and Carlton-Smith, C.H. (1983) *Water Pollution Control*, **82**, 290.
8 Kiss, E. (1988) *Analytica Chimica Acta*, **211**, 342.
9 Van Vleit, H., Basson, W.D. and Bohmer, R.G. (1975) *Analyst (London)*, **100**, 405.
10 Rougham, J.A., Rougham, P.A. and Wilkins, J.P.G. (1983) *Analyst (London)*, **108**, 742.
11 Gladney, E.S. and Perrin, D.R. (1979) *Analytical Chemistry*, **51**, 2015.
12 Randle, K. and Hartman, E.H. (1985) *Journal of Radioanalytical and Nuclear Chemistry*, **90**, 309.
13 Madaro, M. and Moauro, A. (1985) *Journal of Radioanalytical and Nuclear Chemistry*, **90**, 129.
14 Rea, P.E. (1979) *Water Pollution Control*, **78**, 139.
15 Tinsley, J. (1950) *4th International Congress on Soil Science*, **1**, 161.
16 Dalal, R.C. (1979) *Analyst (London)*, **104**, 151.
17 Dalal, R.C. (1979) *Analyst (London)*, **104**, 1100.
18 Charles, M.J. and Simmons, M.S. (1986) *Analyst (London)*, **111**, 385.
19 APHA (1979) Method 62. Organic Matter in Soil. *The Analysis of Agricultural Materials*, RB 247, 2nd edn., London.
20 Begheijn, L.T. (1976) *Analyst (London)*, **101**, 710.
21 Lynch, J.J., Garrett, R.G. and Jonasson, I.R. (1973) *Geochemical Exploration*, **2**, 171.
22 McQuaker, N.R. and Fung, T. (1975) *Analytical Chemistry*, **47**, 1435.
23 Suzuki, J., Yokoyama, Y., Unno, Y., Suzuki, S. (1983) *Water Research*, **17**, 431.
24 Whitfield, P.H. and McKinley, J.W. (1981) *Water Resources Bulletin*, **17**, 381.
25 Mills, G.L. and Quinn, L.G. (1979) *Chemical Geology*, **251**, 155.
26 Dankers, N. and Laane, R. (1983) *Environmental Technology Letters*, **4**, 283.
27 Weliky, K., Suess, E., Muller, P.J. and Fischer, K. (1983) *Limnology and Oceanography*, **28**, 1252.
28 de Oliveira, E., McLaren, J.W. and Berman, S.S. (1983) *Analytical Chemistry*, **55**, 2047.
29 Poule, A.B. (1973) *X-ray Spectrometry*, **2**, 165.
30 Allison, L.E. (1960) *Proceedings Soil Science Society America*, **24**, 36.
31 Stewart, B.A., Porter, L.K. and Beard, W.E. (1964) *Proceedings Soil Science Society America*, **28**, 365.
32 Young, J.L. and Lindbeck, M.R. (1964) *Proceedings Soil Science America*, **28**, 377.
33 Van Hall, C.E. and Stenger, V.A. (1967) *Analytical Chemistry*, **39**, 503.

34 Schaffer, R.B., Van Hall, C.E., McDermott, G.N. *et al.* (1965) *Journal of Water Pollution Contr. Fed.*, **37**, 1545.
35 McQuaker, N.R. (1974) Report No. 7416, Chemistry Laboratory, Water Resources Service, Province of British Columbia.
36 Bremner, J.M. and Jenkinson, D.S. (1960) *Journal of Soil Science*, **11**, 394.
37 Solyom, P. (1977) *Vatten*, **33**, 21.
38 Waggott, A., Britcher, H.V. (1976) Technical Report TR29. *Analysis of the Organic Carbon Content of Sewage Effluent: General and Specific Group Analysis*, August 1976. Water Research Centre Medmenham Laboratory, Marlow, Bucks, UK.
39 Baker, C.L., Bartlett, P.D., Far, I.S. and Williams, G.I. (1974) *Freshwater Biology*, **4**, 467.
40 Krambeck, H.J., Lampert, W. and Bredie, H. (1981) *Fachzeitschrift für das Laboratorium*, **25**, 2009.
41 Hulsmann, A. and Hengst, P. (1980) *Hydrobiological Bulletin*, **14**, 135.
42 Salonen, K. (1979) *Hydrobiologia*, **67**, 29.
43 Dankers, N. and Laane, R. (1983) *Environmental Technology Letters*, **4**, 203.
44 Nelson, D.W. and Sommers, L.E. (1972) *Journal of Environmental Quality*, **1**, 423.
45 HMSO (1979) Method 57. Nitrogen in Soil. *The Analysis of Agricultural Materials*. RB 427, 2nd edn., London.
46 Petts, K.W. and Belcher, M. (1980) *Water Pollution Control*, **79**, 399.
47 Marsh, J.A.P., Kibble-White, R. and Stent, C.J. (1979) *Analyst (London)*, **104**, 136.
48 Greaves, M.P., Cooper, S.L., Davies, M.A. *et al.* (1978) Technical Report. Agricultural Research Council Weed Research Organization UK. No. 45, p.55.
49 Juma, N.G., Paul, E.A. and Mary, B. (1984) *Soil Science Society of America Journal*, **48**, 753.
50 Brodick, S., Cullen, P. and Maher, W. (1987) *Bulletin of Environmental Contamination and Toxicology*, **38**, 377.
51 Muhlhauser, H.A., Soto, L. and Zahradnik, P. (1987) *International Journal of Environmental Analytical Chemistry*, **28**, 2115.
52 Wong, H.K.T. and Kemp, A.H.W. (1977) *Soil Science*, **124**, 1.
53 Smart, M.M., Rada, R.G. and Donnermeyer, G.N. (1983) *Water Research*, **17**, 1207.
54 Zink-Nielsen, I. (1977) *Vatten*, **1**, 14.
55 Ballinger, D.G. and McKee, G.D. (1971) *Journal of Water Pollution Control Federation*, **43**, 216.
56 Standing Committee of Analysts (1986) Methods for the examination of waters and associated material. Total nitrogen and total phosphorus in sewage sludge 1985, HMSO, London, 55 pp.
57 Ruider, E. and Spatzierer, G. (1978) *Korrespondenz Abwasser*, **25**, 48.
58 Matt, K.J. and Agassiz, B.C. (1970) *Soil Science*, **109**, 214.
59 Murphy, J. and Riley, J.P. (1963) *Analytical Abstracts*, **10**, 1219.
60 Aspila, K.I., Agemian, H. and Chau, A.S.Y. (1976) *Analyst (London)*, **101**, 187.
61 Bickford, G.P. and Willett, I.R. (1981) *Water Research*, **15**, 511.
62 Olsen, S.R., Cole, C.V., Wanatabe, F.S. and Dean, L.A. (1954) US Department of Agricultural Circular No. 939.
63 Murphy, J. and Riley, J.P. (1962) *Analytica Chemica Acta*, **27**, 31.
64 Tecator Ltd., Sweden. (1984) Application Note AN 73/84. *Determination of Extractable Phosphorus in Soil by Flow Inspection Analysis.*
65 Tecator Ltd., Sweden. (1984) Application Note No. ASN 73–31/84. *Determination of Extractable Phosphorus in Soil by Flow Injection Analysis.*
66 Tecator Ltd., Sweden. (1984) Application Note No. ASN 73–32/54. *Determination of Extractable Phosphorus in Soil by Flow Injection Analysis.*
67 Tecator Ltd., Sweden. (1984) Application Note No. ASTN 9/84. *Determination of Available Phosphorus in Soil by Flow Injection Analysis.*

68 Tecator Ltd., Sweden. (1983) Application Note No. AN 60/83. *Determination of Phosphate Stannous Chloride Method.*
69 Addison, R.F. and Ackman, R.G. (1970) *Journal of Chromatography*, **47**, 421.
70 Que-Hee, S.G. and Boyle, J.R. (1988) *Analytical Chemistry*, **60**, 1033.
71 Olsen, O., Gotterman, H.L. and Elymo, R.S. (eds) (1967) *Environment in the Aqueous Habitats.* Proceedings of the IBP-Symposium in Amsterdam, 10–16 October 1966, NV Noord-Hollandsche Uitgevers, Moat Schappi, Amsterdam.
72 Harwood, J.E., Van Stteerderen, R.A. and Kuhn, A.L. (1969) *Water Research*, **3**, 425.
73 Osburn, Q.W., Lemmel, D.E. and Downey, R.L. (1974) *Environmental Science and Technology*, **8**, 363.
74 Brar, S.P.S. and Bishnoi, S.R. (1987) *Analyst (London)*, **112**, 831.
75 Shulka, S.S., Syers, J.K. and Armstrong, D.E. (1973) *Journal of Environmental Quality*, **1**, 292.
76 Dobolyi, E. and Bidlo, G. (1980) *International Review der Gesamte Hydrobiologie*, **65**, 489.
77 Anderson, J.M. (1975) *Water Research*, **10**, 329.
78 Anderson, J.M. (1974) *Arch. Hydrobiol.* **7**, 528.
79 Saunders, W.M.H. and Williams, E.G. (1955) *Journal of Soil Science*, **6**, 254.
80 Jackson, M.L. (1958) in *Soil Chemical Analysis*, Prentice-Hall Inc., Englewood Cliffs, NJ.
81 Hesse, P.R. (1971) *A Textbook of Soil Chemical Analysis*, Chemical Publishing Co. Inc., New York.
82 Black, C.A. (1965) in *Methods of Soil Analysis Part 2. Chemical and Microbiological Properties*, American Society of Agronomy Inc., Madison, Wisc.
83 Koroleff, F. (1970) *Determination of Total Phosphorus in Natural Waters by Means of Persulfate Oxidation.* International Council for the Exploration of the Sea (ICES), Report No. 3.
84 Singh, N.P., Linsalata, P., Gentry, R. and Wrenn, M.E. (1979) *Analytica Chemica Acta*, **111**, 265.
85 Murphy, J. and Riley, J.P. (1962) *Analytica Chemica Acta*, **12**, 162.
86 Sommers, L.E., Harris, R.F., Williams, J.D.H. *et al.* (1970) *Limnology Oceanography*, **15**, 301.
87 Nordforsk, I. (1974) *Interkalibrering av sedimentkemiska analysmetoder*, Nordforsk, Miljovardssekretariatet.
88 Mehta, O. (1954) *Proceedings Soil Science Society America*, **18**, 443.
89 Standard Methods (1971) *Standard Methods for the Examination of Water and Waste Water*, 13th edn., American Public Health Association, New York, pp.1–874.
90 De Pinto, J.V. (1982) *Water Research*, **16**, 1065.
91 Dorich, R.A., Nelson, D.W. and Sommers, L.E. (1985) *Journal of Environmental Quality*, **14**, 400.
92 Buchan, L. (1981) *Water South Africa*, **7**, 1.
93 Standing Committee of Analysts (1992) *Phosphorus and Silicon in Effluents and Sludges 1992*. 2nd edn. Methods for the Examination of Water and Associated Materials, HMSO, London.
94 Mino, T., Matsuo, T. and Kawakami, T. (1983) *Journal of the Japan Sewage Works Association*, **20**, 28.
95 Cabrera, F., Biaz, E., Toca, C.G. and De Arambarrui, P. (1982) *Water Research*, **16**, 1061.
96 Stickland, J.D. and Parsons, T.R. (1972) *A Practical Handbook of Seawater Analysis.* 2nd edn. Bulletin of the Fisheries Research Board, Canada, 167.
97 Landers, D.R., David, M.B. and Mitchell, M.J. (1981) *International Journal of Environmental Analytical Chemistry*, **14**, 245.

98 Tahatabi, M.A. and Bremner, J.M. (1970) *Proceedings Soil Science Society of America*, **34**, 62.
99 Steinbergs, A., Lismaa, O. and Frency, J.R. (1962) *Analytica Chemica Acta*, **27**, 158.
100 Waugh, J.H. and Mitchell, M.J. (1981) *Pedobiologia*, **22**, 268.
101 David, M.B., Mitchell, M.J. and Nakas, J.P. (1982) *Soil Science Society of America Journal*, **46**, 847.
102 Nriagu, J.O. and Hem, J.D. (1978) in *Sulfur in the Environment Part 2* (ed. J.O. Nriagu), Wiley Interscience, New York, pp.211–270.
103 Shriner, D.S. and Henderson, G.S. (1978) *Journal Environmental Quality*, **7**, 392.
104 Mitchell, M.J., Landers, D.H. and Brodowski, D.F. (1981) *Water, Air and Soil Pollution*, **16**, 351.
105 Davison, W. and Lishman, J.P. (1983) *Analyst (London)*, **108**, 1235.
106 Kolthoff, I.M. and Sandell, E.B. (1952) in *Textbook of Qualitative Inorganic Analysis*, MacMillan, New York.
107 Gilboa-Garber, N. (1971) *Analytical Biochemistry*, **43**, 129.
108 Aspiras, R.B., Keeney, D.R. and Chesters, G. (1972) *Analytical Letters*, **5**, 425.
109 Valkov, I.I. and Zhabina, N.N. (1971) *Zhur Analit Khim.*, **26**, 359.
110 Rowlett, S.M. (1980) PhD Thesis. Geochemical Studies of Recent Sediments from Cumbria, UK, University of Liverpool.
111 Chen, K.W., Monsaavi, M. and Syap, A. (1973) *Environmental Science and Technology*, **7**, 948.
112 Morse, J.W. and Cornwell, J.C. (1987) *Marine Chemistry*, **22**, 55.
113 Millson, M.F. (1970) *Journal of Chromatography*, **50**, 155.
114 Pellenberg, R. (1979) *Marine Pollution Bulletin*, **10**, 267.
115 Analytical Methods Committee (1960) *Analyst (London)*, **85**, 629.
116 Analytical Methods Committee (1975) *Analyst (London)*, **100**, 54.
117 Thompson, K.C. and Thomerson, D.R. (1974) *Analyst (London)*, **99**, 595.
118 Thompson, A.J. and Thoresby, P.A. (1977) *Analyst (London)*, **102**, 9.
119 Wauchaupe, R.D. (1976) *Atomic Absorption Newsletter*, Perkin– Elmer, **15**, 64.
120 HMSO (1987) *Selenium and Arsenic in Sludges Soils and Related Materials. (1985) A note on the use of hydride generator kits*, London.
121 Cutter, L.S., Cutter, G.A. and San Diego-McGlane, M.L.C. (1991) *Analytical Chemistry*, **63**, 1138.
122 Cutter, G.A. (1986) Electric Power Research Institute, Palo Alto, California. Report EPRIEA 4641, Vol. 1 (100pp).
123 Goulden, P.D., Anthony, D.H.J. and Austen, K.D. (1981) *Analytical Chemistry*, **53**, 2027
124 Brzezinska-Paudyn, A., Van Loon, J.C. and Hancock, R. (1986) *Atomic Spectroscopy*, **7**, 72.
125 Liversage, B.R., Van Loon, J.C. and de Andrade, J.C. (1984) *Analytica Chemica Acta*, **161**, 275.
126 Brzezinska, A., Balick, A. and Van Loon, J.C. (1983) *Water, Air and Soil Pollution*, 323.
127 Arrowsmith, P. (1987) *Analytical Chemistry*, **59**, 1437.
128 Cheam, V. and Chau, A.S.Y. (1984) *Analyst (London)*, **109**, 775.
129 Brannon, J.M. and Patrick, W.H. (1987) *Environmental Science and Technology*, **21**, 450.
130 Maher, W.A. (1983) *Analyst (London)*, **108**, 939.
131 Siu, K.W.M., Roberts, S.Y. and Berman, S.S. (1984) *Chromatographia*, **19**, 398.
132 Thompson, M., Pahlavanpour, B. and Pullen, J.H. (1980) *Analyst (London)*, **104**, 274.
133 Pahlavanpour, B., Thompson, M. and Thorne, L. (1980) *Analyst (London)*, **105**, 756.
134 Nygaard, D.D. and Lowry, J.H. (1982) *Analytical Chemistry*, **54**, 803.

135 Crock, J.G. and Lichte, F.E. (1982) *Analytica Chemica Acta*, **144**, 223.
136 McLaren, J.W., Berman, S.S., Boyko, U.J. and Russell, D.S. (1981) *Analytical Chemistry*, **53**, 1802.
137 Thompson, M., Pahlavanpour, B. and Walton, S.J. (1978) *Analyst (London)*, **103**, 568.
138 Webster, T. (1980) *Water Pollution Control*, **79**, 405.
139 Gorsuch, T.T. (1959) *Analyst (London)*, **84**, 135.
140 Atsuya, I. and Akatsuka, K. (1981) *Spectrochimica Acta*, **36B**, 747.
141 Brannon, J.M. and Patrick, W.H. (1985) *Environmental Pollution (Series B)*, **9**, 107.
142 Zhe-Ming, N., Xiao-Chua, L. and Heng-Bin, H. (1986) *Analytica Chemica Acta*, **186**, 147.
143 Lee, D.S. (1982) *Analytical Chemistry*, **54**, 1682.
144 Zmijewska, W. and Semkow, T. (1978) *Chem. Anal. (Warsaw)*, **23**, 583.
145 Kronborg, O.J. and Steinnes, E. (1975) *Analyst (London)*, **100**, 835.
146 Mignosin, E.P. and Roelandts, I. (1975) *Chemical Geology*, **16**, 137.
147 Van der Klugt, N., Poelstra, P. and Zwemmer, E. (1977) *Journal of Radioanalytical Chemistry*, **35**, 109.
148 Baedecker, P.A., Rowe, J.J. and Steinnes, E. (1977) *Journal of Radioanalytical Chemistry*, **40**, 115.
149 Nadkarni, R.A. and Morrison, G.H. (1978) *Analytica Chemica Acta*, **99**, 133.
150 Chambres, J.C. and McClellan, D. (1976) *Analytical Chemistry*, **48**, 2061.
151 Inhat, M. (1976) *Analytica Chemica Acta*, **82**, 292.
152 Henn, E.L. (1975) *Analytical Chemistry*, **47**, 428.
153 Ishizaka, M. (1978) *Talanta*, **25**, 167.
154 Montaser, A. and Mehrabzadeh, A.A. (1978) *Analytical Chemistry*, **50**, 1697.
155 Weltz, B. (1976) *Atomic Absorption Spectrometry*. Verlag Chemie, New York.
156 Pinta, M. (1977) *Atomic Absorption Spectrometry Applications to Chemical Analysis*. P.W.N., Warsaw.
157 Schrenk, W.G. (1975) *Modern Analytical Chemistry. Analytical Atomic Spectroscopy*. Plenum Press, New York.
158 Brodie, K.G. (1979) *International Laboratory*, **40**, July/August.
159 Lavrakas, V., Barry, E. and Golembeski, T. (1975) *Talanta*, **22**, 547.
160 Ohta, K. and Suzuki, M. (1975) *Talanta*, **22**, 465.
161 Thompson, K.C. (1975) *Analyst (London)*, **100**, 307.
162 Dong, A., Rendig, V.V., Buran, R.G. and Besga, G.S. (1987) *Analytical Chemistry*, **59**, 2728.
163 Siu, K.W.M. and Berman, S.S. (1983) *Analytical Chemistry*, **55**, 1603.
164 Wiersma, J.H. and Lee, G.F. (1971) *Environmental Science and Technology*, **5**, 1203.
165 Itoh, K., Chikuma, M. and Tanaka, H. (1988) *Fresenius Zeitschrift für Analytische Chemie*, **330**, 600.
166 Terada, K., Ooba, T. and Kiba, T. (1975) *Talanta*, **22**, 41.
167 Willie, S.N. Sturgeon, R.E. and Berman, S.S. (1986) *Analytical Chemistry*, **58**, 1140.
168 Siu, K.W.M. and Berman, S.S. (1984) *Talanta*, **31**, 1010.
169 Sturgeon, R.E., Willie, S.N. and Berman, S.S. (1985) *Analytical Chemistry*, **57**, 2311.
170 Reamer, D.C., Veillon, G. and Tokousbalides, P.T. (1981) *Analytical Chemistry*, **53**, 245.
171 HMSO (1987) Methods for the Examination of Waters and Associated Materials (40548). *Selenium in Waters: 1984 Selenium and arsenic in sludges, soils and related materials, 1985 a note on the use of hydric generator kits*, London.
172 Bowman, G.T. and Delfino, J.J. (1980) *Water Research*, **14**, 491.
173 Wang, W. (1980) *Water Research*, **14**, 603.
174 Markert, B.E., Tesmer, M.G. and Parker, P.E. (1983) *Water Research*, **17**, 603.

175 Nothlich, L. and Reuter, W. (1982) *Deutsch Gewasserkundliche Mittelungen*, **26**, 162.
176 Voorn, G. and Marlow, J.S. (1977) *Effluent and Water Treatment Journal*, **302**, June.
177 Marty, J.L. and Aim, R.B. (1976) *Techniques et Sciences Municipales*, **71**, 531.
178 Wells, W.N. (1972) *Water Pollution Control Federation*, Deeds and Data D2–D3, February.
179 Ravenscroft, B.D. (1975) *Pollution and Water Chemistry*, **254**, February.
180 Kupec, J., Hasikova, M. and Svancer, J. (1970) *Vod Hospod. Series B*, **20**, 335.
181 Solazaro, A. and Sharp, A. Private communication.
182 Abu-Hilal, A.H. and Riley, J.P. (1981) *Analytica Chemica Acta*, **131**, 175.

Chapter 13

Organometallic compounds

13.1 Organoarsenic compounds

13.1.1 Soil

13.1.1.1 Introduction

Inorganic arsenicals such as arsenic trioxide, sodium arsenite, lead arsenate, calcium arsenate and Paris Green have been used for many years as soil sterilants. Organic arsenical herbicides, in which the organic group is bonded directly to the arsenic atom, have been used extensively for post-emergence control of weeds in cotton. Several of the more important herbicides are sodium cacodylate (monosodium dimethylarsenic acid) and sodium salts of methane arsonic acid. The latter compounds exist in two principal forms: the monosodium salt (MSMA) at pH6.4 and the disodium salt (DSMA) at pH10.2.

In one study of the persistence of disodium methane arsenic acid in soil, the initial and residual phytotoxicity of disodium methane arsenic acid in soil to cotton was measured over a broad range of concentrations in three soils. Toxicity decreased with time, particularly during the first 16 weeks after soil incorporation. Growth of cotton planted immediately after incorporation of disodium methane arsenic acid in Bosket silt loam was reduced significantly by concentrations of 50–80mg kg^{-1} in soil. In the same soils, other plants were shown to have different degrees of susceptibility to disodium methane arsenic acid concentrations. Rice was extremely sensitive to soil concentrations of 5ppm, while corn, cotton and wheat were little affected.

In a further study the oxidation of the methyl carbon of methanearsonate was associated with the oxidation of soil organic matter in a number of soils. Additions of organic matter to a Norfolk loamy sand greatly increased the decomposition of methanearsonate. In three of the soils, there was no evidence of microbiological adaptation to methanearsonate. In Norfolk loamy sand, however, increasing decomposition of methanearsonate relative to soil organic matter occurred with time of incubation.

13.1.1.2 Gas chromatography

Odanake *et al.* [1] have reported the application of gas chromatography with multiple ion detection after hydride generation with sodium borohydride to the determination of mono and dimethyl arsenic compounds, trimethyl arsenic oxide and inorganic arsenic in soil and sediments. Recoveries in spiking experiments were 100–102% (mono and dimethyl arsenic compounds and inorganic arsenic) and 72% (trimethyl arsenic oxide).

Soderquist *et al.* [2] determined hydroxydimethyl arsine oxide in soil by converting it to iododimethylarsine using hydrogen iodide followed determination at 105°C on a column (450cm × 2.8mm) packed with 10% DC–200 on Gas-Chrom Q (60–80 mesh), with nitrogen as carrier gas (20–30min $^{-1}$) and electron capture detection. The recovery of hydroxydimethylane oxide (0.15ppm) added to soil was 91.3 ± 5.1%.

13.1.1.3 Miscellaneous

Because of the wide usage of organic arsenicals, and because little information exists on the fate of these compounds in soils, Von Endt *et al.* [3] studied monosodium methane arsenic acid (MSMA) as a model for studying the metabolism of this class of compounds by soil micro-organisms.

Experiments involving the release of radioactive carbon dioxide from MSMA–14C treated soils were conducted in a system consisting of two test tubes connected in series. One tube contained 5g of treated soil (at 10 and 100ppm of monosodium methane arsenic acid carbon dioxide while a second tube contained a trapping mixture, 2-methoxyethanol and monoethanolamine (7–10, *v*/*v*). Carbon dioxide-free air was passed over the soil and metabolic $^{14}CO_2$ was collected in the trapping solution. The soils studied were Sharkey clay, Hagerstown silty clay loam, Cecil sandy loam, and Dundee silty clay loam. All soils were initially adjusted to field capacity and maintained at 28–30°C; the evolved $^{14}CO_2$ was sampled periodically. Some properties of these soils are shown in Table 13.1.

Table 13.1 Physical properties of soils used in MSMA decomposition studies

Soil	*pH*	*Organic matter %*	*Clay (0.002mm) %*	*Field capacity %*
Sharkey clay	5.2	3.90	67.1	38.4
Hagerstown silty clay loam	6.8	2.50	39.5	25.8
Cecil sandy clay loam	5.3	1.89	20.1	17.8
Dundee silty clay loam	5.0	1.67	29.0	24.3

Source: Reproduced with permission from the American Chemical Society [3]

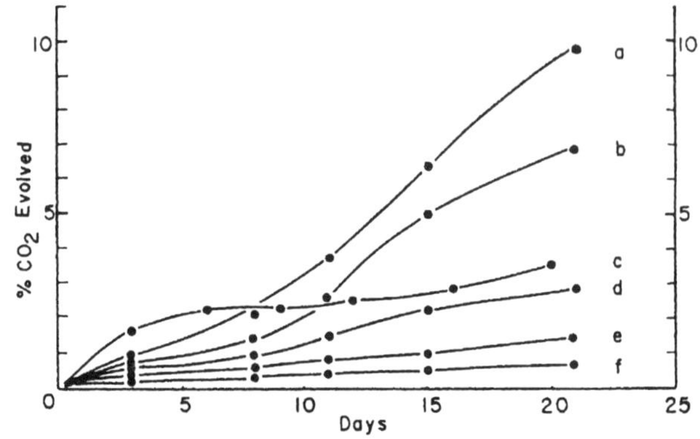

Fig. 13.1 MSMA decomposition: two rates applied to four soils
Source: Reproduced with permission from the American Chemical Society [3]

a Sharkey clay + 10ppm MSMA
b Sharkey clay + 100ppm MSMA
c Hagerstown silty clay loam + 10 or 100ppm MSMA
d Cecil sandy loam + 10 or 100ppm MSMA
e Dundee silty clay loam + 10 or 100ppm MSMA
f Steam-sterilized controls

Preliminary studies indicated that monosodium methane arsenic acid decomposition to carbon dioxide was a slow process that did not involve a lag phase. Hagerstown silty clay loam slowly evolved $^{14}CO_2$ from MSMA–^{14}C applied at a rate of 100ppm. Only 7% decomposition was observed after 60 days. In another experiment involving four soils and two rates of monosodium methane arsenic acid application, the rate of decomposition was again a slow process (Fig. 13.1). After three weeks' incubation, all soils had evolved radioactive carbon dioxide to a degree proportional to the amount of organic matter in the soil. The rate of evolution ranged from 10% decomposition for the highest organic matter soil (Sharkey clay), to 1.7% decomposition in the lowest organic matter soil (Dundee silty clay loam). Only in Sharkey clay was there a difference in carbon dioxide production between the high (100ppm) and low (10ppm) application rates of monosodium methane arsenic acid. The initial rapid rate of $^{14}CO_2$ release from Hagerstown silty clay loam probably occurs because it was a fresh soil (the others were air-dried) with an active microbial population already established. Steam-sterilized soils produced essentially no $^{14}CO_2$; therefore, soil micro-organisms appear to play some role in the decomposition process.

Comparison of evolved $^{14}CO_2$ from four sterile and non-sterile soils 60 days after treatment with MSMA–^{14}C showed that from 1.7 to 10.0% of the MSMA–^{14}C was degraded in non-sterile soil, as compared with 0.7% in steam-sterilized controls. Four soil micro-organisms isolated in pure culture degraded from 3– 20% of the MSMA–^{14}C to $^{14}CO_2$ when grown in liquid culture containing 10ppm of monosodium methane arsenic acid and 1g per litre of yeast extract. Thin layer chromatography on silica gel G-coated plates effected the separation of monosodium methane arsenic acid, arsenate and arsenite. Only arsenate and monosodium methane arsenic acid were detected after thin layer chromatography of extracts from the soil and microbial growth experiments. These data indicate that soil micro-organisms are at least partly responsible for monosodium methane arsenic acid degradation in soil.

Thin layer chromatography was carried out on 20 × 20cm glass plates coated 0.25mm thick with a suitable support and dried overnight. Silica gel G, silica gel H and cellulose were examined as the solid phases for chromatography of methanearsonate, arsenite and arsenate. Several sprays for the visualization of the arsenicals on plates were tested. Three of the more successful reagents and the colour produced with final product are shown in Table 13.2.

13.1.2 Saline deposited and suspended sediments

13.1.2.1 Introduction

The chemical form of arsenic in marine environmental samples is of interest from several standpoints. Marine organisms show widely varying concentrations of arsenic [4–6] and knowledge of the chemical forms in which the element occurs in tissues is relevant to the interpretation of these variable degrees of bioaccumulation and to an understanding of the biochemical mechanisms involved. Different arsenic species have different levels of toxicity [7] and bioavailability [8] and this is important in food chain processes, while physicochemical behaviour in processes such as adsorption onto sediments also varies with the species involved [9]. It has

Table 13.2 Spray reagents used to detect arsenite, arsenate and MSMA on thin layer chromatograms

	Colour and detection limit		
Spray	*Na arsenite*	*Na arsenate*	*MSMA*
2*N* HCl–12%	Yellow		
$(NH_4)_2S$ (1 to 1)	1.5µg		
1% $(NH_4)_6$ Mo_7O_{24}		Blue	Blue
1% $SnCl_2$ in 10% HCl		1.5µg	1.5µg
12% $(NH_4)_2S–H_2O$) (1 to 1)			
1% Et_2NCS_2 in 50%		Orange	Orange
H_2O-acetone		2.5–3.0µg	2.5–3.0µg
0.1% dithizone in benzene			

Source: Reproduced with permission from the American Chemical Society [3]

been shown that inorganic arsenic(III and V), monomethylarsenic and dimethylarsenic acids are present in natural waters [10], biological material [11] and sediments [12]. Thus analytical methods for the separation and measurement of these species are necessary for the study of pathways of accumulation and deposition in marine organisms.

13.1.2.2 Atomic absorption spectrometry

Maher [13] has described a procedure for the determination of inorganic arsenic, monomethylarsenic and dimethylarsenic in marine organisms and estuarine sediments. The arsenic species are isolated by solvent extraction, separated by ion-exchange chromatography and selectively determined by arsine generation. Recoveries of spikes of 5 and 10μg of arsenic taken through the whole procedure were 92–96%.

Sediments were dried at 60–80°C and ground to pass a 340μm sieve. Dried sediments (1g) were shaken with 2 × 10ml of 6m hydrochloric acid for 5–6h in a 50ml polyethylene centrifuge tube, centrifuged and the supernatant solution retained. The residue was further extracted with 20ml of a solution 0.1M in sodium hydroxide and 1M in sodium chloride. This extraction was repeated. The combined alkaline extracted were refluxed for 6h and evaporated to dryness in a rotary evaporator. The residue was dissolved in 15ml of 8.5M hydrochloric acid. This solution, and the supernatant solution from the initial leaching, were analysed separately for arsenic compounds.

An aliquot (10ml) of acid concentrate was transferred to a 50ml glass centrifuge tube, and 1 ml of the reducing solution (1M potassium iodide saturated with ascorbic acid) added. After 30min, the arsenic species were extracted into 5ml of toluene, using a vortex stirrer. Centrifugation was required to separate the organic phase which was transferred to a 100ml separating funnel with a Pasteur pipette. The extraction was repeated with a further 5ml of toluene. The arsenic species were back-extracted from the combined toluene extracts into water (2 × 4ml) and after the addition of 1ml of concentrated hydrochloric acid and 50μl of a 5% potassium dichromate solution, the solution was made up to exactly 10ml.

The arsenic species were separated on a 18 × 1.25cm column of Dowex 50AG–X8 (100–200 mesh) which was washed initially with 50ml of 0.5M HCl. The extract (<1.5ml) was added to the column and eluted sequentially with 0.5M hydrochloric acid (10ml), water (16ml) and 1.5M ammonia (25ml). The fractions contained inorganic arsenic, monomethyl arsenate and dimethyl arsenate.

Determination of arsenic

The arsenic in each fraction was determined by reduction to the corresponding arsine in the zinc reductor column, decomposition of the arsine evolved in a heated carbon tube furnace, and by measurement of atomic absorption of the arsenic at 193.7nm. The principal advantages of using zinc reductor column are its rapidity and freedom from interferences by other elements.

Initial reduction of inorganic arsenic(V), monomethyl arsenate and dimethyl arsenate to the trivalent oxidation state was required for quantitative reduction to the corresponding arsine on the column. Conversion of the first of the above eluates was achieved by the addition of 1.5ml of concentrated hydrochloric acid and 0.5ml of reducing agent. The other eluates were evaporated to dryness after the addition of 2ml of concentrated nitric acid to the monomethylarsenic fractions. The residues were dissolved in 2ml of concentrated hydrochloric acid and 1ml of reducing agent (1M potassium iodide saturated with ascorbic acid) added. After 30min (to allow complete reduction) the solutions were diluted to 10ml.

Before injection of 0.5–1ml of solution into the zinc column, the inert gas flow rate was adjusted to 0.7l min^{-1}, and the furnace (1700°C) and recorder were turned on and allowed to establish a stable baseline (approximately 10s). The solution was injected as quickly as possible using a syringe and the furnace turned off when the recorder signal had returned to the previously established baseline (approximately 20s).

Blanks for the entire procedure were typically less than 4ng ml^{-1} and derived mainly from the hydrochloric acid.

The recoveries of known additions of arsenic in various forms taken through the entire recommended procedure are shown in Table 13.3. With additions in the range 5–10g of arsenic, recoveries of 92–97% were obtained and in all cases recoveries were better than 90%.

In the extraction procedure for sediments an initial leach with 6M hydrochloric acid was used in order to remove the bulk of carbonate and hydrated oxide phases before extraction with sodium hydroxide solution. This procedure gave a recovery of ca. 70% of the total arsenic in the sediment as determined following total decomposition.

Table 13.3 Percentage recoveries* of arsenic species in the ion-exchange separation

Form	*Arsenic added (μg)*		
	0.1	*0.5*	*1.0*
AS(III)	94.0 ± 1.8	–	92.3 ± 0.7
As(V)	96.0 ± 1.6	98.5 ± 0.8	98.0 ± 0.5
MMA	97.0 ± 0.8	96.5 ± 0.5	97.2 ± 0.3
DMA	97.1 ± 1.1	95.2 ± 0.2	96.1 ± 1.3

Source: Reproduced with permission from Elsevier Science Publishers BV [13]

*Values are means of two determinations

13.2 Organolead compounds

13.2.1 Soil

13.2.1.1 Gas chromatography

Blais *et al.* [14] determined alkyl lead salts in soil. They demonstrated that previously published methods gave poor recoveries of lead and the formation of artifacts during the isolation and derivatization procedures. An alternative procedure is described involving a series of selective extractions of tetra-alkylleads, ionic alkylleads, and inorganic ionic lead salts from soils and street dusts. Alkyllead salts were selectively extracted complexometrically from samples containing up to 1000mg ionic lead per kg. The extracts were then butylated and analysed by gas chromatography–atomic absorption spectroscopy. Re-extraction of the sample with methyl isobutyl ketone–dithizone permitted the recovery of ionic lead. In the samples tested, ethyllead salts were detected, but not methyllead salts. Concentrations of these analytes were significantly correlated with levels of extractable ionic lead, but not with total lead.

13.2.2 Non-saline deposited and suspended sediments

13.2.2.1 Gas chromatography

In an early paper Potter *et al.* [15] took samples of sediment in drains and natural water courses from the top 50–100cm, such that a portion of the supernatant water was included, and stored in polypropylene bottles until analysed. Portions (30ml) of the sediment were shaken for 1min with petroleum ether (10ml) and centrifuged. The petroleum ether extract was transferred to a small glass-stoppered flask and stored at 0°C. The extraction procedure was repeated. The extracted sediment was acidified to pH5–6 with dilute nitric acid, allowed to stand until evolution of hydrogen sulphide had ceased, and lead nitrate (5g in 20 ml water) was added and mixed. After standing for 5min the sediment was filtered, the solids were dried and weighed and the filtrate was neutralized with aqueous sodium hydroxide (2M), filtered and made up to 100ml. Samples (10ml) of this filtrate were analysed for alkyllead salts by gas chromatography. The above operations were carried out on the day of sampling. The petroleum ether was analysed for tetraalkyllead compounds.

Potter *et al.* [15] showed that recovery of alkyllead salts obtained from sediments was 90% for Et_3PbCl and Me_3PbCl, 75% for Et_2PbCl_2, and 40% for Me_2PbCl_2. Extraction of Et_2PbCl_2 added to sediment, containing no alkylleads, from a clean and polluted river, from a clean and polluted canal, and from road drainage grids gave recoveries of between 65 and 75%. The lowest detectable concentration of alkyllead salts was 2mg kg^{-1} dry weight of sediment. For direct analysis of a sample of filtered water the lowest detectable concentration was 0.1mg L^{-1}. Extraction of

Table 13.4 Concentration of alkyllead compounds in drainage grid sediments

Size	*Concentration (mg kg $^{-1}$ dry weight of sediment) of*			
	Tetraethyl lead[a]	*Tetraethyl lead*[b]	*Tetramethyl lead*	*Alkyllead salts*
(1) Garage A, near to pumps	96.0[e]	93.0[e]	5.3	6.1
(2) Garage A, near to entrance	4.7[e]	5.1[e]	0.11	
(3) Garage B, near to pumps	54.0[e]	58.0[e]	1.8	4.0
(4) Garage B, near to entrance	25.0[e]	24.0[e]	0.36	
(5) 10m downhill from garage A	0.33	0.33		
(6) 10m downhill from garage A	0.51	0.53		
(7) 100m downhill from garage A	0.20	0.18		
(8) 30m downhill from garage B	0.38	0.35		
(9) 50m downhill from garage B	0.24	0.24		
(10) Aston Expressway	0.25	0.22		
(10)[c] Aston Expressway	0.09	0.07		
(11) Residential area	0.34	0.38		
(12) Residential area	0.36	0.35		
(13)[d] Residential area	0.05	0.05		

Source: Reprinted from Potter *et al.* [15]

[a]Determined on SE–30 column
[b]Determined on TCEP column
[c]Site 10 reasampled after (14)[d]
[d]Solids settlement tank into which water from site 12 passed
[e]Confirmed by t.l.c.

tetraalkyllead (0.01µl) from sediment (30ml) gave recoveries of 36–41% for the initial extraction with petroleum ether and 13–15% for the second. The lowest detectable concentration of tetraalkyllead was 0.02mg kg $^{-1}$ dry weight of sediment. By extracting samples of filtered water with one-tenth their volume of petroleum ether, tetraalkyllead could be detected down to concentrations of 0.002mg L $^{-1}$ in water.

Using these procedures Potter *et al.* [15] found appreciable quantities alkyllead compounds in some samples of drainage grid sediments (Table 13.4). No alkyllead compounds were detected in the filtered water from any of these sediment samples.

Chau *et al.* [16–19] in a series of papers issued between 1979 and 1984 discussed various methods for the determination of alkyllead compounds in sediments.

Chau *et al.* [16] have described a simple and rapid extraction procedure to extract the five tetraalkyllead compounds (Me_4Pb, Me_3EtPb, Me_2Et_2Pb, $MeEt_3Pb$, Et_4Pb) from sediment. The extracted compounds are analysed in their authentic forms by a gas chromatographic–atomic absorption spectrometry system. Other forms of inorganic and organic lead do not interfere. The detection limits for sediment (5g) is 0.01mg kg $^{-1}$. In this method 5g wet sediment and 5mL of EDTA reagent, (0.1M, 37g Na_2EDTA $2H_2O^{-1}$), and 5mL of hexane are placed in a 25mL test tube with a Teflon lined screw cap and shaken for 2h then centrifuged. The extract was analysed by gas chromatography using an atomic absorption spectrometer set at the 217μm lead line and with a silica furnace as a detector. This combination has a detection limit of about 1μg lead, about three orders of magnitude better than can be achieved using a flame ionisation detector.

The system used by these workers consisted of a Microtek 220 gas chromatograph and a Perkin–Elmer 403 atomic absorption spectrophotometer. These instruments were connected by means of a stainless steel tubing (2mm o.d.) connected from the column outlet of the gas chromatograph to the silica furnace of the atomic absorption spectrometer. The silica furnace was set at 1000°C. The gas chromatographic column was packed with 3% OV–1 supported on Chromosorb W. The column was temperature programmed at 15°C h to 150°C.

The furnace was constructed from silica tubing (7mm i.d., 6cm long) with open ends. The lead compounds separated by gas chromatography were introduced to the centre of the furnace through a side-arm. Hydrogen gas was introduced at the same point at a flow rate of 1.35mL min^{-1}, the burning of the hydrogen improved the sensitivity. The silica furnace was mounted on top of the atomic absorption spectrometer burner and aligned to the light path. When the absorbances were plotted against lead concentrations, each of the five tetraalkyl compounds gave similar calibration curves; the response was linear up to at least 200ng Pb, above which over-lapping of the peaks occurred.

To digest sediment samples, 5g of sediment and EDTA solution are extracted with 5ml hexane and the resulting solution gas chromatographed. Concentrations found in a marine sediment ranged from 8.3mg kg^{-1}, (tetramethyllead and methyltriethyllead) to 12mg kg (dimethyldiethyllead and tetraethyllead) and recoveries in spiking experiments were between 81 and 84%.

Determination of the ionic forms of alkyllead compounds is difficult because of the incomplete extraction of the dimethyl and trimethyl species from sample matrices. Recently a chelation extraction method followed by derivatization to their butyl homologues has overcome all the previous difficulties to achieve quantitative extraction of the dialkyl- and trialkyllead (R = Me, Et) from water samples at nanogram levels [18].

The application of a combination of gas chromatography and atomic absorption spectrometry to the determination of tetraalkyllead compounds has been studied by Chau *et al.* [17] and by Segar [20]. In these methods the gas chromatography flame combination showed a detection limit of about 0.1μg Pb. Chau *et al.* [17,18] have applied the silica furnace in the atomic absorption unit and have shown that the sensitivity limit for the detection of lead can be enhanced by three orders of magnitude. They applied the method to the determination of tetramethyllead in sediment systems.

The system used by these workers consisted of a Microtek 220 gas chromatograph and a Perkin–Elmer 403 atomic absorption spectrophotometer. These instruments were connected by means of stainless steel tubing (2mm o.d.) connected from the column outlet of the gas chromatograph to the silica furnace of the a.a.s. (Fig. 13.2). A four-way valve was installed between the carrier gas inlet and the column injection port so that a sample trap could be mounted, and the sample could be swept into the gas chromatographic column by the carrier gas. The recorder (10mV) was equipped with an electronic integrator to measure the peak areas, and was simultaneously actuated with the sample introduction so that the retention time of each component could be used for identification of peaks.

The furnace was constructed from silica tubing (7mm i.d., 6cm long) with open ends (Fig. 13.3). The lead compounds separated by gas chromatography were introduced to the centre of the furnace through a side-arm. Hydrogen gas was introduced at the same point at a flow rate of 1.35ml min $^{-1}$; the burning of hydrogen improved the sensitivity. The furnace was wound with 26-gauge Chromel wire to give a resistance of about 5ohms. The voltage applied to the furnace was about 20Va.c. regulated by a variable transformer so that the furnace temperature with hydrogen burning was about 1000°C. The silica furnace was mounted on top of the a.a.s. burner and aligned to the light path.

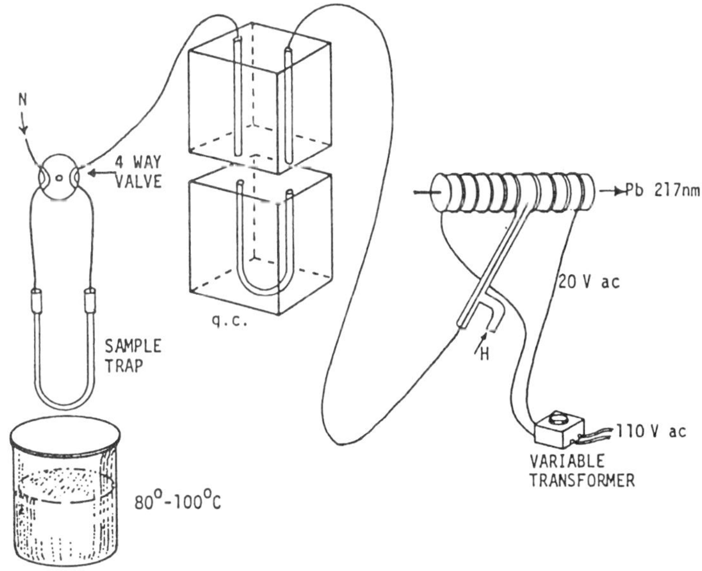

Fig. 13.2 Schematic diagram showing the interfacing of the g.c.–a.a.s. system
Source: Reproduced with permission from Preston Publications Ltd. [17]

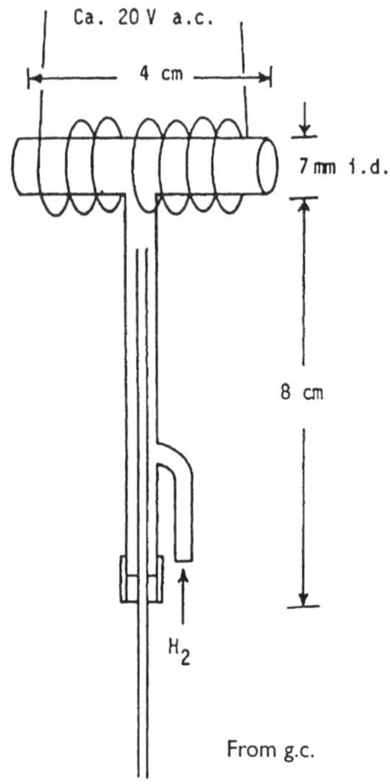

Fig. 13.3 Silica furnace
Source: Reproduced with permission from Preston Publications Ltd. [17]

The sample trap was a glass U-tube (6mm dia., 26cm long) packed with 3% OV–1 on Chromosorb W, which was immersed in a dry ice–methanol bath at ca –70°C as described by Chau *et al.*[17]. A known amount of gaseous sample was drawn through the trap by a peristaltic pump operated at 130–150ml min $^{-1}$. After sampling, the trap was mounted to the four-way valve and heated to ca. 80–100°C by a beaker of hot water, and the adsorbed compounds were swept into the gas chromatographic column.

Liquid samples can be directly injected to the column through the injection port, without a sample trap.

Instrument parameters

Glass column, 1.8m long, 6mm dia., packed with 3% OV–1 on Chromosorb W 80–100 mesh; carrier gas, 70ml nitrogen min $^{-1}$; injection port temperature, 150°C; temperature programme, initial 50°C for 2min, programmed at 15°C min $^{-1}$, until 150°C; sample trap temperature, 80–100°C.

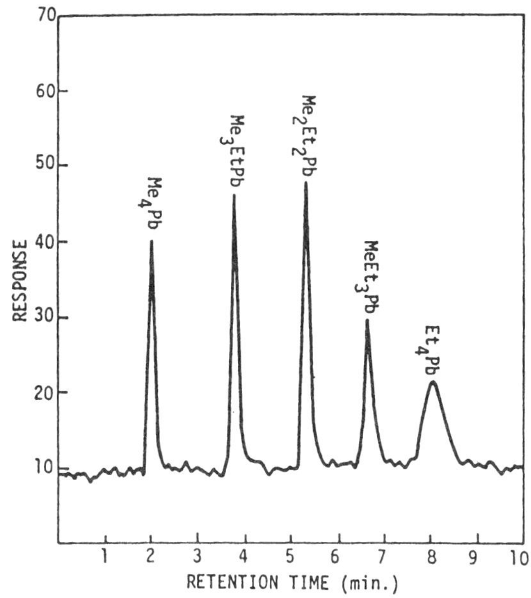

Fig. 13.4 Recorder tracings for a mixture of five tetralkyllead compounds. Each peak represents ca. 5ng of the compound expressed as lead
Source: Reproduced with permission from Preston Publications Ltd. [17]

Lead line, 217nm; lamp current, 8mA; spectral band width, 0.7nm; scale expansion 4× (0.25A full scale); furnace gas, 135ml H_2 min $^{-1}$. The deuterium background corrector was used.

The relative standard deviation was in the range 10–15% at the 5ng level (as Pb). When the absorbances were plotted against lead concentrations, each of the five tetraalkyl compounds gave similar calibration curves; the response was linear up to at least 200ng Pb, above which overlapping of the peaks occurred. If only one compound was present (e.g. tetramethyllead), the plot was linear up to at least 2000ng. For determinations at the microgram level, the flame atomic absorption spectrometric technique [17] is more suitable. Fig. 13.4 illustrates a typical recorder tracing of a mixture of the five tetraalkyllead compounds.

Solvents such as chloroform, carbon tetrachloride, hexane and benzene gave absorption signals because of their non-specific absorption at the lead resonance line. Although these solvent peaks generally emerged well before the lead compounds, the use of the background corrector is recommended to eliminate these potential interferences.

Chau *et al.* [19] have described the optimum conditions for extraction of alkyllead compounds from sediments originating in non-saline waters and in saline waters [16]. Analyses of some environmental samples revealed for the first time the occurrence of dialkyl- and trialkyllead in sediments in areas of lead contamination.

The various alkyllead species and lead(II) are isolated quantitatively by chelation extraction with sodium diethyldithiocarbamate, followed by *n*-butylation to their corresponding tetraalkyl forms, $R_nPbBu_{(4-n)}$, and Bu_4Pb, respectively (R = Me, Et) all of which can be determined by a gas chromatograph using an atomic absorption detector. The method determines simultaneously the following species in one sample: tetra-alkyllead (Me_4Pb, Me_3EtPb, Me_2Et_2Pb, $MeEt_3Pb$, Et_4Pb); ionic alkyllead (Me_2Pb^{2+}, Et_2Pb^{2+}; Me_3Pb^+, Et_3Pb^+); Pb^{2+}. Detection limits expressed for Pb were 15μg kg $^{-1}$ for sediment samples.

In this method, the sediment (1–2g) sample was extracted for 2h in a capped vial with 3mL of benzene after addition of 10mL of water, 6g of sodium chloride, 1g of potassium iodide, 2g of sodium benzoate, 3mL of sodium diethyldithiocarbamate and 2g of coarse glass beads (20–40 mesh). After centrifugation of the mixture, a measured aliquot (1mL) of the benzene was butylated using 0.2ML *n*-butyl magnesium chloride with occasional mixing for 10min. The mixture was washed with 2mL sulphuric acid (0.5M) to destroy excess Grignard reagent. The organic layer was separated in a capped vial and dried with anhydrous sodium sulphate. Suitable aliquots were injected into the gas chromatograph.

Chau *et al.* [19] found that in spiking experiments on sediments both the diethyl and triethyl species were recovered at satisfactory levels (Table 13.5).

Table 13.5 Recovery and reproducibility of alkyllead and lead(II) compounds from sediment

Amount of Pb added (μg)	*Recovery[b] (%)*				
	Me_3Pb	*Et_3Pb*	*Me_2Pb*	*Et_2Pb*	*Pb(II)*
1	113(9)	73(14)	103(6)	104(15)	
5	111(3)	86(2)	116(2)	94(5)	
10	122(4)	106(1)	114(3)	85(1)	
20	99(1)	111(4)	118(2)	89(3)	
av	111	94	113	93	
% rel std dev	4	4	14	15	9[c]
(*n* = 6) at 5mg kg $^{-1}$ level					

Source: Reproduced with permission from the American Chemical Society [19]

[a]Sediment 1g; spiked compounds expressed as Pb.

[b]Average of two results with average deviation in parentheses.

[c]The sediment contained 71mg kg $^{-1}$ of Pb(II) which was used to evaluate the reproducibility. No Pb(II) was added to sample.

13.2.2.2 Miscellaneous

Biomethylation of organolead compounds

Reisinger *et al.* [21] used the gas chromatographic–atomic absorption spectrometric technique to demonstrate that biomethylation of inorganic lead does not account for the presence of organolead compounds in sediments. Sulphide induced chemical conversion of organic lead(IV) salts into alkyl lead compounds is, however, possible.

Wong [22], on the other hand, claims that the conversion of inorganic lead to tetramethyllead in river and marine sediments is purely a micro-organism induced biological process. These workers demonstrated that incubation of some lead-containing sediments generates tetramethyllead; that Me_3Pb^+ salts are readily converted to tetra methyllead by micro-organisms in lake water or nutrient medium, with or without the sediment, and in the presence or the absence of light; that conversion of inorganic lead (such as lead nitrate or lead chloride) to tetramethyllead occurred on several occasions in the presence of certain sediments; and that the conversion is purely a biological process.

13.2.3 Saline deposited and suspended sediments

13.2.3.1 Gas chromatography

Chau *et al.* [16] described a hexane extraction procedure to extract tetramethyllead, trimethylethyllead, methyltriethyllead, dimethyldiethyllead and tetraethyllead from marine sediments. The extracted compounds were analysed in their authentic forms by a gas chromatographic–atomic absorption spectrometric system. Other forms of organic and inorganic lead do not interfere. The detection limit was 0.01mg kg $^{-1}$ as lead.

To digest sediment samples, 5g of sediment and EDTA solution are extracted with 5ml hexane and the resulting solution gas chromatographed. Concentrations found in a marine sediment ranged from 8.3mg kg $^{-1}$ (tetramethyllead and methyltriethyllead) to 12mg kg (dimethyldiethyllead and tetraethyllead) and recoveries in spiking experiments were between 81 and 84%. Down to 0.01mg kg $^{-1}$ organolead can be determined by this method.

The procedure [16] discussed in section 13.2.2.1 for the determination of alkyllead compounds in non-saline sediments has also been applied to marine sediments. Recoveries from sediments ranged from 94% for triethyllead to 111% for trimethyllead in the range 1–20μg alkyllead spiked to 1g of sediment. An average standard deviation of 4% for trimethyllead and triethyllead and 15% for dialkylead compounds was obtained.

13.3 Organomercury compounds

13.3.1 Soil

13.3.1.1 Introduction

In lakes, streams and rivers, mercury can collect in the bottom sediments and soils where it may remain for a long time. It is difficult to release this mercury from the matrixes for analysis. Mercury is also found in soil as a result of applications of mercury containing compounds, or sewage contaminated with traces of mercury. Much concern has been expressed in recent years concerning the contamination of the environment by mercury compounds, both organic mercury originating in industrial effluents and organic mercury originating as fungicides, seed dressings, etc.

Decomposition of mercurial fungicides in contact with soil has long been known [23–25]. The escape of metallic and organic mercury vapours and the amount of organic mercurial remaining in soil, however, have not been investigated except through indirect biological techniques, because of inadequacy of chemical methods. Booer [23], basing his conclusions on biological phytotoxicity experiments, postulated a mechanism for organic mercury decomposition in soil. He suggested that organic mercury compounds reacted with the clay micelle in soil to form an intermediate which subsequently gave a dialkylmercury or diphenylmercury and a mercury-clay compound. Based on this hypothesis, the dialkylmercury compounds would escape into the atmosphere while diphenylmercury would accumulate in the soil. Metallic mercury would result from the further degradation of the mercury–clay compound. However, repeated attempts to detect the disubstituted organic mercury compounds formed in soil through degradation failed, indicating that decomposition was not by Booer's mechanism.

Work has also been done on the absorption and inactivation of organomercurials by micro-organisms that tolerate and even thrive on mercurials [26,27]. It has been postulated that inactivation occurred by the uptake of fungicide by micro-organisms, followed by metabolic break-down and by possible utilization of portions of the byproducts. However, whether or not biological inactivation and mercury evolution occur together has not been determined.

13.3.1.2 Spectrophotometric methods

Kimura and Miller [28] have also studied the decomposition of organic fungicides in soil to mercury vapour and to methyl- or ethylmercury compounds and devised methods for the determination of these compounds in the vapours liberated from the soil sample. The mixed vapours of mercury and organomercury compounds is passed successively through bubblers containing a carbonate-phosphate solution to absorb organic

Table 13.6 Retention and distribution of metallic mercury vapour in an aeration train composed of one carbonate-phosphate absorber and two acid permanganate absorbers (Mercury in μg)

	First absorber carbonate-phosphate		*Second absorber acid permanganate*		*Third absorber acid permanganate*	
Trial	*Oxidation by H_2O_2 method*	*Acid dichromate washings*	*Without further oxidation*	*Further oxidation by H_2O_2 method*	*Without further oxidation*	*Further oxidation by H_2O_2 method*
I	0	0	1.860	1.960	0	4
II	0	0	1.780	1.780	0	3
III	0	0	1.730	1.700	0	4
IV	1	0	1.660	1.670	2	7

Source: Reproduced with permission from the American Chemical Society [28]

mercury and through an acidic potassium permanganate solution to absorb inorganic mercury vapour. In both cases the mercury in the scrubber solution is determined photometrically at 605nm with dithizone. The method is capable of determining 10μg or more of organic mercury/ 1000L air in the presence of mercury vapour.

In this procedure the separation of metallic mercury and organic mercury vapour is based on the passage of the metallic mercury vapours through a solution of sodium carbonate and dibasic sodium phosphate and its quantitative capture in acid permanganate solution (Table 13.6). Vapours of ethyl- and methylmercury chloride, in the 100–1000μg range, are 95–99% retained in a single carbonate–phosphate absorber. This retention is attributed to the formation of the extremely water soluble methyl- and ethylmercury hydroxide and phosphate, and the foaming characteristics of the carbonate solution.

Kimura and Miller [28] collected and determined vapours produced by the decomposition in soil of phenyl alkylmercury compounds. They found that the air above soil containing phenylmercury acetate contained mercury vapour and traces of phenylmercury acetate.

Ethylmercury produced about equal amounts of mercury vapour and an uncharacterized volatile ethylmercury compound, while methylmercury chloride and methylmercury dicyanamide both produced an uncharacterized methylmercury compound plus some mercury vapour.

Kimura and Miller [29] have described a procedure for the determination of organomercury (methylmercury, ethylmercury and phenylmercury compounds) and inorganic mercury in soil. In this method the sample is digested in a steam bath with sulphuric acid (0.9M) containing hydroxy ammonium sulphate, sodium chloride and, if high concentrations of organic matter are present, potassium dichromate solution. Then,

50% hydrogen peroxide is added in portions with vigorous mixing. Allow sufficient time for the peroxide to decompose after each addition. The decomposition of hydrogen peroxide being exothermic, the temperature will rise gradually to about 150°C. The addition of peroxide is discontinued after the solution turns blue green, or in the absence of chromium, light yellow, and residual peroxide is allowed to decompose with the use of low heat after the condenser has been washed down with water. When the decomposition of peroxide has apparently ceased, add 5% permanganate slowly to give a 5ml excess while the temperature is maintained. The permanganate is added in 5ml portions, or less, until the mixed colour persists for 15min. Cool the sample and add 20ml of reducing solution consisting of hydroxy ammonium sulphate dissolved in sodium chloride.

Air is then passed through this solution in the aerator apparatus and the volatilized elemental mercury collected in an absorber solution consisting potassium permanganate in dilute sulphuric acid, followed by stannous chloride solution. To the absorber solution is added a reducing solution consisting of hydroxy ammonium sulphate dissolved in sodium chloride. To determine mercury in the absorber solution transfer a suitable aliquot or the entire sample containing not more than 10μg mercury to a separatory funnel and dilute to 50ml with the 0.9M sulphuric acid solution. Add 3.5ml of dithizone solution (containing 11mg L^{-1}) and shake for 1min. Transfer the chloroform phase to a 13 × 100mm test-tube, allowing any residual water to adhere to its walls. Then transfer to a $1cm^2$ cuvette and measure the excess of dithizone at 605mμ. Read as soon as possible after shaking with the mercury solution.

Kimura and Miller [29] demonstrated (Table 13.7) that mercury in several organic forms can be digested and aerated from unfiltered soil digests. For samples of 10g of soil cores containing 5μg mercury or less, the standard deviations of a single determination were 0.12, 0.15 and 0.23μg, respectively, using 2cm cylindrical optical cells.

Kimura and Miller [30] also described the following methods for the determination in soil samples of extractable organic mercury, total mercury and extractable ionic mercury.

Extractable phenyl- and alkylmercury compounds

Phenyl and alkylmercury compounds are extracted from about 1g soil by shaking for 2h with 25mL 0.1M phosphate pH8 buffer containing 6mg thiomalic acid, added just prior to use, and analysed after dilution of a 5mL aliquot of the centrifuged extract with 5mL water, and acidification with 5mL 9M hydrochloric acid containing 150mg hydroxylammonium chloride. The final determination is made by the dithizone microprocedure of Polley and Miller [31]. Diphenyl- and dialkylmercury compounds are extracted from 1g soil by shaking for 2h with 10mL chloroform and

Table 13.7 Mercury recovery

Sample	*Hg added (μg)*	*Total Hg found (μg)*	*Added Hg recovered (μg)*
	(Added as $HgCl_2$)		
10g Puyallup sandy loam soil (air dried) and screened, organic matter 6% air-dry basis)	0.00	1.66	–
	0.56	2.17	0.51
	7.77	9.41	7.75
	111.0	111.1	110.1
	(Added as methylmercury chloride)		
	0.40	2.15	0.49
	4.00	5.55	3.89
	(Added as ethylmercury chloride)		
	0.38	2.08	0.41
	7.57	9.19	7.53
	(Added as $HgCl_2$)		
1.9 × 2.4cm Turf cores (18–22g as sampled)	0.00	1.35	–
	1.00	2.18	0.83
	5.00	6.36	5.11
	100.0	97.45	96.1

Source: Reproduced with permission from the American Chemical Society [29]

analysed by cleaving the disubstituted mercurial to give an aryl- or alkylmercury salt, using 9M or 12M hydrochloric acid, followed by the dithizone microprocedure [31].

Vapours of mercury, phenyl, and alkylmercury compounds were collected and measured as described except that the dichromate absorption solution was analyzed by the mercury reduction technique [28] after treatment with an excess of chloride hydroxylammonium chloride solution.

Ionic mercury is extracted from about 1g soil by shaking for 2h with each of two 25mL portions of 2M sodium chloride. The combined centrifuged and filtered (using 1M sodium chloride for washing) extract is analysed by the procedure of Polley and Miller [31].

Total mercury is determined in soils containing phenylmercury acetate and or ethylmercury acetate using the method described by Polley and Miller [31]. Total mercury is determined in soils containing methylmercury chloride and methylmercury dicyanamide by the method described by Kimura and Miller [32]. Kimura and Miller [30] present chemical data on the nature of residual mercurials in soil and in the atmosphere surrounding the treated soil to further elucidate the phenomena of degradation in soil.

They showed that metallic mercury vapour and trace amounts of phenylmercury acetate were present in the air surrounding phenylmercury acetate-treated soil. About equal amounts of the vapours of

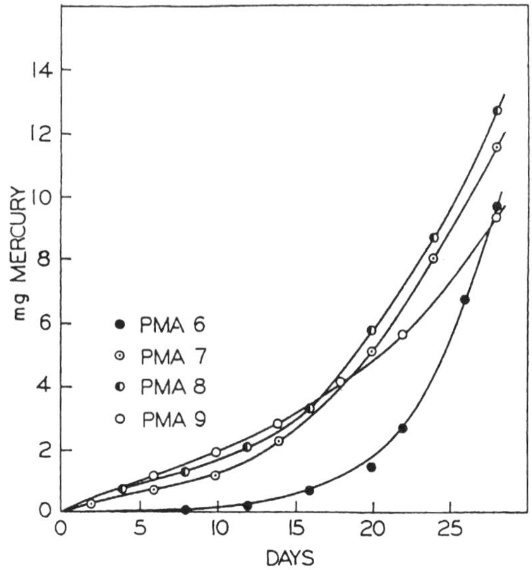

Fig. 13.5 Cumulative mercury vapour from PMA-treated soil in relation to elapsed days
Soil in PMA pot 6 was autoclaved before addition of PMA; others (PMA pots 7, 8, 9), containing unautoclaved treated soil, varied in initial moisture
Source: Reproduced with permission from the American Chemical Society [30]

metallic mercury and a volatile ethylmercury compound were present when ethylmercury acetate was used. With the use of methylmercury compounds, methylmercury vapour was present with trace amounts of mercury vapour. The chloride was about twice as volatile as the dicyandiamide. A large portion of the organic mercurial applied to the soil was found to be in the organomercury form after the lapse of 30–50 days. Moisture in soil decreased the amount of escaping organic mercury vapour.

In an experiment phenyl mercury acetate was introduced into soils containing various initial levels of water between 6.5% and 22.5%. The accumulative values for mercury vapour evolution from these pots are shown in Fig. 13.5. For the first two weeks of aeration, mercury vapour captured was increasingly greater with increasing initial moisture. The pot containing autoclaved soil (PMA–6) gave very little mercury during this period in comparison to a similar non-autoclaved pot (PMA–8), suggesting a microbiological mode of degradation. The increase in mercury vapour with increasing moisture where moisture should hinder volatilization of metallic mercury, suggests a mechanism of degradation requiring water. This is also consistent with microbiological degradation. While autoclaving decreased the rate of mercury evolution, as indicated by the slope of the cumulative curves, during the first several weeks, autoclaving increased the mercury evolution rate over that of comparable non-autoclaved pot after three weeks. If biological degradation were assumed as the only cause of mercury evolution, this could be interpreted as the building up of a PMA-degrading microbial population beyond that in the comparable non-autoclaved pot.

Traces of phenylmercury acetate were found in the carbonate absorbers. Cumulative phenylmercury acetate at the end of a 28-day period was 0.3, 0.2 and 0.1mg as Hg for PMA–7, 8, 9 in the order of increasing initial moisture and 0.2mg Hg for PMA-6. Analyses of the pots by 2.5cm layers indicated the greatest loss generally to be from the surface (Table 13.8). Where unglazed clay pots were employed, as in the phenyl mercury acetate and ethyl mercury acetate series, the next greatest loss was from the bottom, indicating diffusion through the unglazed clay pots. The zone of heaviest concentration of the residual mercurial was at the 2.5–5cm depth. In the methylmercury chloride and methylmercury dicyanamide series, where glass pots were used, concentration of mercurials below the 2.5cm depth was very close to the original. Table 13.9 shows that 14 to 16% of the original phenyl mercury acetate mercury applied was lost as mercury vapour. Of the original mercury, 60 to 70% was extractable as the intact phenylmercury compound at the end of the experiment. About 20% of the original mercury, although recovered as part of final total mercury, was not characterized. This portion may consist of irreversibly bound or physically unavailable phenyl mercury acetate, small amounts of unvolatilized metallic mercury and sulphide.

In a similar procedure [32] the sediment is wet oxidised with dilute sulphuric acid and nitric acids in an apparatus in which the vapour from the digestion is condensed into a reservoir from which it can be collected or returned to the digestion flask as required. The combined oxidised residue and condensate are diluted until the acid concentration is 1N and nitrate is removed by addition of hydroxylammonium chloride with boiling. Fat is removed from the cooled solution with carbon tetrachlo-dithizone in carbon tetrachloride. The extract is shaken with 0.1M hydrochloric acid and sodium nitrite solution and, after treatment of the separated aqueous layer with hydroxylammonium chloride a solution of urea and then EDTA solution are added to prevent subsequent extraction of copper. The liquid is then extracted with a 0.01% solution of dithizone in carbon tetrachloride and mercury estimated in the extract spectrophotometrically at 485nm.

A disadvantage of all the above procedures is that the lowest concentration of mercury that can be determined in the soil or sediment samples is of the order of 0.05–1mg kg^{-1}. These high detection limits are in part due to high blanks caused by the multiplicity of digestion reagents used in the procedures. Several investigators have liberated mercury from soil and sediment samples by application of heat to the samples and collection of the released mercury on gold surfaces. The mercury was then

Table 13.8 Pot mercury analyses by soil depth

	Pot No.	*Depth, Inches*	*Total Hg, PPM* 0 days	n[a] days	*Extractable organic mercury as PPM Hg* 0 days	n[a] days
PMA	6	0–1	113	78	113	59
	6	1–2	113	111	113	83
	6	2–B[b]	113	95	113	75
	7	0–1	125	116	121	102
	7	1–2	125	112	121	82
	7	2–B	125	101	121	73
	8	0–1	125	105	122	74
	8	1–2	126	113	122	77
	8	2–B	126	105	122	76
	9	0–1	112	94	114	60
	9	1–2	112	112	114	75
	9	2–B	112	105	114	79
EMA	10	0–1	77	46	63	28
	10	1–2	77	63	63	45
	10	2–B	77	66	63	42
	11	0–1	84	50	69	29
	11	1–2	84	61	69	38
	11	2–B	84	60	69	36
	12	0–1	84	46	69	29
	12	1–2	84	67	69	47
	12	2–B	84	65	69	45
	13	0–1	102	78	83	56
	13	1–2	102	91	83	66
	13	2–B	102	71	83	49
MMD	19	0–1	84	74	73	59
	19	1–B	84	83	73	67
	20	0–1	86	72	74	58
	20	1–B	86	82	74	68
	21	0–1	80	63	67	51
	21	1–B	80	80	67	66
MMC	22	0–1	76	59	63	47
	22	1–B	76	70	63	59

Source: Reproduced with permission from the American Chemical Society [30]

[a]PMA, 28 days; EMA, 53 days; MMD and MMC, 35 days
[b]Two inch depth to bottom

released from the gold by application of heat or by absorption in a solution containing oxidising agents [33].

Bretthauer *et al.* [34] and Anderson *et al.* [35] described a method in which samples were ignited in a high-pressure oxygen-filled bomb. After ignition, the mercury was absorbed in a nitric acid solution. Pillay *et al.* [36] used a wet-ashing procedure with sulphuric acid and perchloric acid

Table 13.9 Analyses of phenylmercury acetate (PMA)-treated soil

Pot No.	*H_2O in Soil %*	*Total Hg in soil*	*Extractable PMA in soil*	*Extractable Hg^{+2} in soil*	*Uncharacterized Hg in Soil (by difference)*	*Carbonate absorber PMA in air*	*$KMnO_4$ absorber HG a in air*
Initial analyses				Expressed as Mg Hg per pot (dry basis)			
PMA–6	14.1	65.4	67.6 (102%)[a]	0.7 (1.1%)			
PMA–7	6.5	75.8	73.3 (96.7%)	0.6 (0.8%)			
PMA–8	13.6	75.2	73.8 (96.8%)	0.8 (1.1%)			
PMA–9	22.5	65.0	66.2 (102%)	0.0 (...)			
Final analyses,[b] after 28 days							
PMA–6	7.4	55.1 (83.0%)	42.3 (63.7%)	1.1 (1.7%)	13.2 (19.9%)	0.2 (0.3%)	9.6 (14.5%)
PMA–	4.5	67.5 (89.1%)	52.8 (69.7%)	2.0 (2.6%)	9.2 (12.1%)	0.3 (0.4%)	1.5 (15.2%)
PMA–8	7.1	65.2 (85.6%)	46.1 (60.5%)	2.2 (2.9%)	15.1 (19.8%)	0.2 (0.3%)	12.6 (16.5%)
PMA–9	12.0	60.2 (92.6%)	39.5 (60.8%)	1.6 (2.5%)	14.5 (22.3%)	0.1 (0.2%)	9.3 (14.3%)

Source: Reproduced with permission from the American Chemical Society [30]

[a]All percentages are based on the tidal mercury of the initial analyses
[b]Values calculated as describec under Experimental

to digest samples. The released mercury was precipitated as the sulphide. The precipitate was then redigested using aqua regia.

Feldman [37] digested solid samples with potassium dichromate, nitric acid, perchloric and sulphuric acid. Bishop *et al.* [38] used aqua regia and potassium permanganate for digestion.

The approved US Environmental Protection Agency [39] digestion procedure requires aqua regia and potassium permanganate as oxidants. These digestion procedures are slow and often hazardous because of the combination of strong oxidizing agents and high temperatures. In some of the methods, mercuric sulphide is not adequately recovered. The oxidising reagents, especially the potassium permanganate, are commonly contaminated with mercury, which prevents accurate results at low concentrations.

13.3.1.3 Gas chromatography

Longbottom *et al.* [40] have described gas chromatographic methods for the determination of alkylmercury in soils and sediments.

13.3.1.4 Atomic absorption spectrometry

Earlier work on the determination of total mercury in river sediments also include that of Iskander *et al.* [41]. Iskander applied flameless atomic absorption to a sulphuric acid nitric acid digest of the sample following reduction with potassium permanganate, potassium persulphate and stannous chloride. A detection limit of one part in 10^9 is claimed for this somewhat laborious method.

As Umezaki and Iwamoto [42] have reported that organic mercury can be reduced directly with stannous chloride in the presence of sodium hydroxide and copper II, the determination of organic mercury can be simplified, particularly if the reagent used for back-extraction does not interfere with the reduction of organic mercury. Matsunga and Takahasi [43] found that extraction with an ammoniacal glutathione solution was satisfactory.

In their proposed method, contamination only from the ammoniacal glutathione solution is expected. However, any inorganic mercury in this solution will be adsorbed on the glass container walls with a half-life about 2d, i.e. the blank value becomes zero if the solution is left to stand for more than a week. This method for mercury in sediments does not distinguish between the different forms of organomercury. Results are calculated as methylmercury.

In the method 10ml 2M hydrochloric acid is added to 10.2g sediment and the mixture left two days [44]. Filter the samples through a glass filter and wash with 10ml of 2M hydrochloric acid. Extract organic mercury from 20ml of the filtrate into 40ml of benzene by shaking for 3min, and discard the aqueous layer. Add 20ml of 3–10 $^{-4}$% glutathione in 0.1M ammonia solution, and back-extract organic mercury into the aqueous solution by shaking for 2min.

To a gas washing bottle, add 150ml of water, 10ml of 10M sodium hydroxide, 2ml of 1000ppm copper solution and 5ml of 5% tin(II) chloride dihydrate solution. Pass nitrogen gas at a flow rate of 1.41min $^{-1}$ for 6min to eliminate any mercury in the reagent solutions. Then add the aqueous extract from the sample. Concentrate the mercury on 1.5g of gold granules (about 1mm diameter) packed in a glass tube (4mm i.d.) by passing nitrogen gas for 6min. Heat the gold granules in a boat to 500°C in a furnace for 2min, and measure the absorbence at 253.7um using an atomic absorption spectrophotometer with a 1.5cm dia × 20cm quartz tube by passing nitrogen gas at a flow rate of 1.21min $^{-1}$.

In this method, the relative standard deviations were 4.1 and 10.2% at the 50 and 5ng absolute level of methylmercury(II) added to 20ml of hydrochloric acid. The final recovery of methylmercury was 86%.

This method was applied to the determination of organic mercury in polluted sediments. Organic mercury in two sediment samples was 0.22 ± 0.1 and 0.43 ± 0.03ng Hg g $^{-1}$ (dry weight).

Langmyhr *et al.* [45] have applied cold vapour atomic absorption spectrometry to the determination of organomercury compounds in soils and sediments.

13.3.1.5 Miscellaneous

Bretthaur *et al.* [34] described a method in which samples were ignited in a high-pressure oxygen-filled bomb. After ignition, the mercury was absorbed in a nitric acid solution. Pillay *et al.* [36] used a wet-ashing procedure with sulphuric acid and perchloric acid to digest samples. The released mercury was precipitated as the sulphide. The precipitate was then redigested using aqua regia.

Feldman digested solid samples with potassium dichromate, nitric acid, perchloric acid and sulphuric acid [46]. Bishop *et al.* [47] used aqua regia and potassium permanganate for digestion. Jacobs and Keeney oxidized sediment samples using aqua regia, potassium permanganate and potassium persulphate [48]. The approved US Environmental Protection Agency digestion procedure requires aqua regia and potassium permanganate as oxidants [39].

These digestion procedures are slow and often hazardous because of the combination of strong oxidizing agents and high temperatures. In some of the methods, mercuric sulphide is not adequately recovered. The oxidizing reagents, especially the potassium permanganate, are commonly contaminated with mercury, which prevents accurate results at low concentrations.

13.3.2 Non-saline deposited and suspended sediments

13.3.2.1 Introduction

In lakes and streams, mercury can collect in the bottom sediments, where it may remain for long periods of time. It is difficult to release the mercury from these matrices for analysis. Several investigators have liberated mercury from soil and sediment samples by the application of heat to the samples and the collection of the released mercury on gold surfaces. The mercury was then released from the gold by application of heat or by absorption in a solution containing oxidizing agents as discussed below [35,49].

13.3.2.2 Spectrophotometric method

Jurka and Carter [50] have described an automated determination of down to 0.1µg L^{-1} mercury in river sediment samples. This method is based on the automated procedure of El-Awady [51] for the determination of total mercury in waters and waste waters in which potassium persulphate and sulphuric acid were used to digest samples for analysis by the cold vapour technique. These workers proved that the use of potassium permanganate as an additional oxidizing agent was unnecessary.

Sediment samples were passed through a No. 10 polypropylene sieve to remove large debris. If necessary, the samples were blended using a Waring blender. Approximately 1g of wet sediment was accurately weighed into a 360ml polyethylene bottle. Five ml of nitric acid–potassium dichromate preservative solution were added to drive off or oxidize any free sulphides as well as to preserve the sample. If the potassium dichromate was entirely reduced, as indicated by a green colour, additional preservative solution was added. Then 245ml of distilled water was added, and the aqueous samples were blended. The samples were allowed to stand overnight. Additional preservative solution was added if the dichromate was entirely reduced after standing. The aqueous samples were then analysed, using the modified automated analytical system in the manner described by El-Awady [51] (Fig. 13.6).

Table 13.10 contains the results of analyses for organic and inorganic mercury standards which were spiked with sulphide. There was no significant interference due to sulphide in the solutions containing 10mg sulphide L^{-1}. However, a negative interference was observed for both organic and inorganic standards containing 100mg sulphide L^{-1} which is equivalent to 25 000mg sulphide kg^{-1} in the sediment. The spiked blank also resulted in a small negative interference.

It is interesting to note that exactly the same interference occurred for both organic and inorganic mercury standards, since methyl mercuric chloride does not directly react with sodium sulphide to form mercuric sulphide. Therefore the interference could not be the result of incomplete digestion of mercuric sulphide or CH_3Hg^+.

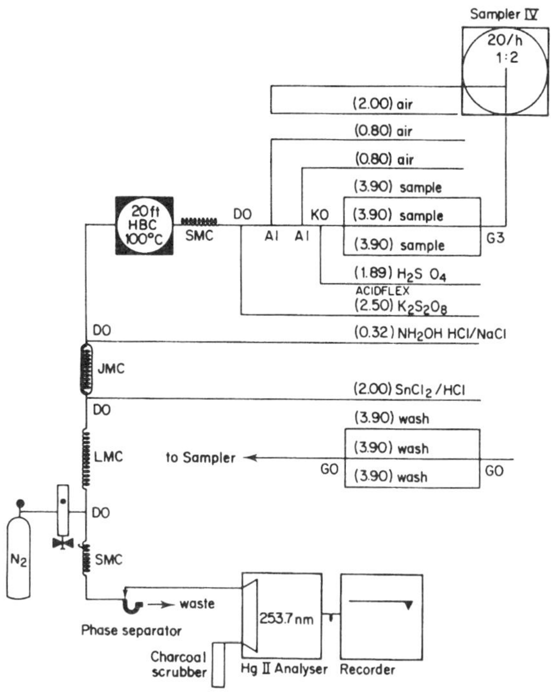

Fig. 13.6 Modified automated total mercury manifold. Numbers in parentheses correspond to the flow rate of the pump tubes in ml min $^{-1}$. Numbers adjacent to glass coils and fittings are Technicon Corp. part numbers.
Source: Reproduced with permission from the American Chemical Society [50]

Table 13.10 Recovery of mercury standards spiked with sulphide

Standard solution	*S^{2-} spike observed (mg L $^{-1}$)*	*Hg concentration (µg L $^{-1}$)*
Blank	–	0.0
1 mg Hg L $^{-1}$ ($HgCl_2$)	–	1.0
1 µg Hg L $^{-1}$ (CH_3HgCl)	–	1.0
1 µg Hg L $^{-1}$ ($HgCl_2$)	10	1.0
1 µg Hg L $^{-1}$ (CH_3HgCl)	10	1.0
Blank	10	0.0
1 µg Hg L $^{-1}$ ($HgCl_2$)	100	0.4*
1 µg Hg L $^{-1}$ (CH_3HgCl)	100	0.4*
Blank	100	–0.1
1 µg Hg L $^{-1}$ ($HgCl_2$) + 4 times normal preservative	100	1.1
1 µg Hg L $^{-1}$ (CH_3HgCl) + 4 times normal preservative	100	1.1
Blank + 4 times normal preservative	100	0.0

Source: Reproduced with permission from the American Chemical Society [50]

* Odd shaped peak

Sulphur, ozone and hydrogen sulphide were investigated as possible causes for the interference. No interference was observed. When the automated method was used, the interferences which were observed for the standards and blank spiked with 100mg sulphide L^{-1}, occurred when an excess of dichromate did not exist in the solutions.

Aromatic organic compounds such as benzene, which are not oxidized in the digestion, absorb at the same wavelength as mercury. This represents a positive interference in all cold vapour methods for the determination of mercury. For samples containing aromatics, i.e. those contaminated by some industrial wastes, a blank analysis must be performed, and the blank results must be subtracted from the sample results. The blank analysis is accomplished by replacing the potassium persulphate reagent and the stannous chloride reagent with distilled water, and reanalysing the sample.

This automated procedure was estimated to have a precision of 0.13–0.21mg Hg kg^{-1} at the 1mg Hg kg^{-1} level with standard deviations varying from 0.011–0.02mg Hg kg^{-1}, i.e. relative standard deviations of 8.4–12% at the 17.2–32.3mg Hg kg^{-1} level in sediments. Recoveries in methyl mercuric chloride spiking studies were between 85 and 125%. The detection limit for the automated method is dependent upon the weight of sample taken for analysis. It is 0.1μg Hg L^{-1} in the aqueous samples. The results for the automated method are routinely reported to a lower limit of 0.1mg kg^{-1} which corresponds to a dry sample weight of 0.25g.

13.3.2.3 Gas chromatography

Jensen and Jernelou [52] reported that both mono and dimethylmercury (CH_3Hg^+ and $(CH_3)_2Hg$) can be produced in lake sediments. The gases evolved from incubated sediment samples were analysed for monomethyl mercury by conversion to methylmercury halide by means of gas chromatography, using electron capture and mass spectrometric detection.

Ealy *et al.* [53] determined methyl-, ethyl- and methoxyethylmercury compounds in sediments by leaching the sample with sodium iodide for 24h and then extracting the alkylmercury iodides into benzene. These iodides are then determined by gas chromatography of the benzene extract on a glass column packed with 5% of cyclohexylenedimethanolsuccinate on Anakrom ABS (70–80 mesh) and operated at 200°C with nitrogen (56ml min^{-1}) as carrier gas and electron capture detection (^3H foil). Good separation of chromatographic peaks is obtained for the mercury compounds as either chlorides, bromides, or iodides.

Longbottom *et al.* [40] have described gas chromatographic methods for the determination of alkyl mercury in soils and sediments.

Andren and Harris [54] have reported a methylmercury concentration of 0.02–0.1ng Hg g^{-1} in unpolluted sediments by using a gas chromatograph with an electron capture detector.

Table 13.11 Typical GLC retention times for organomercury dithizonates. (1) 2% polyethyleneglycol succinate on Chromosorb G (acid-washed, DMCS-treated, 60–80 mesh) in glass column 1.5m long, 3mm I.D.; carrier gas, nitrogen

Dithizonate	*Column temperature (°C)* 140	150	160	170	180
Methylmercury	3.8	2.8	2.2	1.6	1.2
Ethylmercury	6.6	4.6	3.6	2.7	2.0
Ethoxyethylmercury	17.0	11.6	8.7	6.2	4.9
Methoxyethylmercury	17.4	12.0	8.7	6.2	4.9
Tolymercury	–	–	–	29.0	19.5
Phenylmercury	–	–	–	42.0	27.0

Source: Reproduced with permission from McMillan Magazines [55]

Bartlett *et al.* [55] used the method of Uthe *et al.* [56–58] for determining methylmercury. Sediment samples of 2–5g were extracted with toluene after treatment with copper sulphate and an acidic solution of potassium bromide. Methylmercury was then back-extracted into aqueous sodium thiosulphate. This was then treated with acidic potassium bromide and copper sulphate following which the methylmercury was extracted onto pesticide grade benzene containing approximately 100μg L^{-1} of ethyl mercuric chloride as an internal standard. The extract was analysed by electron capture gas chromatography using a Pye 104 chromatography equipped with ^{65}Ni detector. The glass column (1m × 0.4cm) was packed with 5% neopentyl glycol adipate on Chromosorb G (AW–DMCS). Methylmercury was measured by comparing the peak heights with standards of methylmercuric chloride made up in the ethylmercury benzene solution (see Table 13.11). The detection limit was 1–2μg kg^{-1}.

Cappon and Crispin-Smith [59] have described a method for the extraction, clean-up and gas chromatographic determination of alkyl and aryl mercury compounds in sediments. The organomercury compounds are converted to their chloroderivatives and solvent extracted. Inorganic mercury is then isolated as methylmercury upon reaction with tetramethyltin. The initial extract is subjected to a thiosulphate clean-up and the organomercury species are isolated as their bromoderivatives. Total mercury recovery was in the range 75–90% and down to 1μg kg^{-1} of specific compounds can be determined.

13.3.2.4 Atomic absorption spectrometry

Numerous [34,36,41–46,49,60–65] workers have discussed the application of atomic absorption spectrometry to the determination of organomercury compounds in river and lake sediments. Methods [41–44] discussed in section 13.3.1.4 for the determination of organomercury compounds in soil are also applicable to non-saline sediments.

A method [62] has been described for the determination of down to 2.5µg kg $^{-1}$ alkylmercury compounds and inorganic mercury in river sediments. This method uses steam distillation to separate methylmercury in the distillate and inorganic mercury in the residue. The methylmercury is then determined by flameless atomic absorption spectrophotometry and the inorganic mercury by the same technique after wet digestion with nitric acid and potassium permanganate [63]. The well known adsorptive properties of clays for alkylmercury compounds does not cause a problem in the above method. The presence of humic acid in the sediment did not depress the recovery of alkylmercury compounds by more than 20%. In the presence of metallic sulphides in the sediment sample the recovery of alkylmercury compounds decreased when more than 1mg of sulphur was present in the distillate. The addition of 4M hydrochloric acid, instead of 2M hydrochloric acid before distillation completely, eliminated this effect giving a recovery of 90–100%.

This excellent method was sufficiently sensitive to determine 0.02mg kg $^{-1}$ methylmercury and 9mg kg $^{-1}$ inorganic mercury in river sediment samples.

This method was applied to river sediment samples spiked with between zero and 0.06g g $^{-1}$ methylmercury and 0.6µg g $^{-1}$ mercuric chloride. Results indicated the presence in the original sediment of about 0.02µg g $^{-1}$ methylmercury and 9µg g $^{-1}$ inorganic mercury.

Umezaki and Iwamoto [42] have reported that organic mercury can be reduced directly with stannous chloride in the presence of sodium hydroxide and copper II. The determination of organic mercury can be simplified, particularly if the reagent used for back-extraction does not interfere with the reduction of organic mercury.

Matsumaya and Takahasi [43] found that back-extraction with an ammoniacal glutathione solution was satisfactory. In this method, contamination only from the animoniacal glutathione solution is expected. However, any inorganic mercury in this solution will be adsorbed on the glass container walls with a half-life about 2d (i.e. the blank value becomes zero if the solution is left to stand for more than a week). This method for mercury in sediments does not distinguish between the different forms of organomercury. Down to 0.2µg kg $^{-1}$ mercury in sediments can be determined by this method with a standard deviation of 0.03µg kg $^{-1}$. In this method, a large weight sample (10–20g) is extracted with 2M hydrochloric acid for two days and organic mercury then extracted from the filtrate with benzene. Mercury is back-extracted from the benzene with aqueous ammoniacal glutathione. This extract is then added to aqueous solution containing sodium hydroxide, cupric sulphate and stannous chloride and the elemental mercury released is swept off with nitrogen and, in a further concentration step is collected on gold granules. Finally, the granules are heated at 500°C to release mercury which is determined by flameless atomic absorption spectrophotometry at 253.7nm.

Jurka and Carter [50] have described an automated determination of down to 0.1mg kg^{-1} total mercury in river sediment samples with a precision of 0.13–0.21µg Hg kg^{-1} at the 1mg Hg kg^{-1} level and with standard deviations varying from 0.011–0.02mg Hg kg^{-1} (i.e. relative standard deviations of 8.4 to 12%). At the 17.2–32.3mg Hg kg^{-1} level in sediments recoveries in methyl mercuric chloride spiking studies were between 85 and 125%. This method is based on the automated procedure of El-Awady *et al.* [51] for the determination of total mercury in waters and waste waters in which potassium persulphate and sulphuric acid were used to digest samples for analysis by the cold-vapour technique. These workers proved that the use of potassium permanganate as an additional oxidising agent was unnecessary.

Aromatic organic compounds such as benzene, which are not oxidised in the digestion, absorb at the same wavelength as mercury. This represents a positive interference in all cold vapour methods for the determination of mercury. For samples containing aromatics (i.e. those contaminated by some industrial wastes), a blank analysis must be performed and the blank results must be subtracted from the sample results. The blank analysis is accomplished by replacing the potassium persulphate reagent and the stannous chloride reagent with distilled water and reanalysing the sample.

A method [62] has been described for the determination of down to 2.5ppb alkylmercury compounds and inorganic mercury in river sediments. This method uses steam distillation to separate methylmercury in the distillate and inorganic mercury in the residue. The methylmercury is then determined by flameless atomic absorption spectrophotometry and the inorganic mercury by the same technique after wet digestion with nitric acid and potassium permanganate [63]. These workers considered the possible interference effects of clay, humic acids, and sulphides, all possible components of river sediment samples on the determination of alkylmercury compounds and inorganic mercury.

In this method for determining alkylmercury compounds 5–10g of the sediment were weighed accurately in a conical beaker, and 50ml of 2M hydrochloric acid solution and 10g of sodium chloride were added. The mixture was stirred with a glass rod and was left for about 1h, and then transferred into a distilling flask with a small amount of distilled–deionized water. The steam distillation was continued to make the final amount of 200ml of the distillate. The distillate was collected in the flask containing 10ml of 2M hydrochloric acid solution. Fifty millilitres of the distillate were mixed with 5ml of saturated solution of sodium hydroxide, 3ml of 1% cupric sulphate ($5H_2O$) and 2ml of 10% stannous chloride solution. Then the vessel was closed with the lid and slightly shaken. The vessel was coupled to a measuring cell for the determination of alkyl mercury by flameless atomic absorption spectrophotometry.

A method for the determination of inorganic mercury is also described. The well known absorptive properties of clays for alkylmercury compounds does not cause a problem in the above method. The presence of humic acid in the sediment did not depress the recovery of alkylmercury compounds by more than 20%. In the presence of metallic sulphides in the sediment sample the recovery of alkylmercury compounds decreased when more than 1mg of sulphur was present in the distillate. The addition of 4M hydrochloric acid instead of 2M hydrochloric acid before distillation completely eliminated this effect giving a recovery of 90–100%.

Fig. 13.7 shows the results of applying the method described above to river sediment samples spiked with between zero and 0.06µg g $^{-1}$ methylmercury and 0–6µg g $^{-1}$ mercuric chloride. Smooth curves were obtained indicating the presence in the original sediment of about 0.02µg g $^{-1}$ methylmercury and 0µg g $^{-1}$ inorganic mercury.

Workers at the Department of the Environment, UK [47], have described a procedure for the determination of methylmercury compounds in soils and sediments which involves extraction with a carbon tetrachloride solution of dithizone, reduction to elemental mercury then analysis by atomic absorption spectrometry.

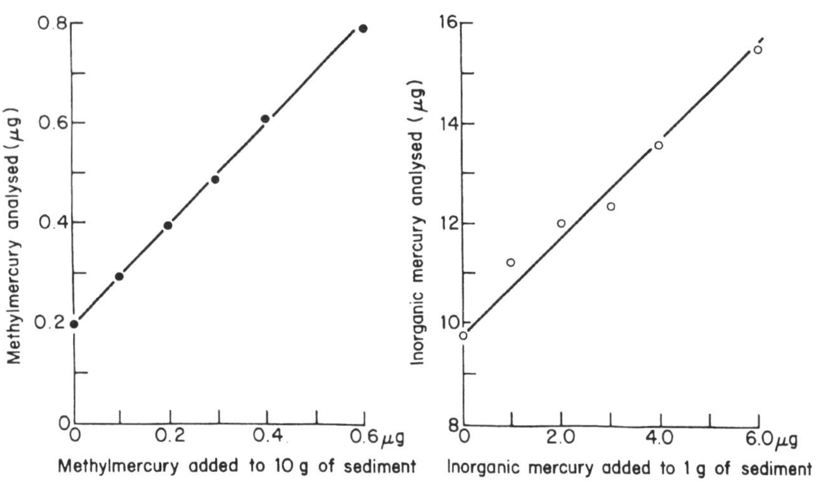

Fig. 13.7 Inorganic mercury and methylmercury recovery test
Source: Reproduced with permission from the Association of Official Analytical Chemists [62]

13.3.2.5 Nuclear magnetic resonance spectroscopy

Robert and Robenstein [66] carried out indirect determination of Hg^{119} NMR spectra of methylmercury complexes, e.g. CH_3Hg^{II} thiol ligands in sediment samples.

13.3.2.6 Miscellaneous

Feldman digested solid samples with potassium dichromate, nitric acid, perchloric acid and sulphuric acid [46]. Bishop *et al.* [47] used aqua regia and potassium permanganate for digestion of organomercury compounds. Jacobs and Keeney oxidized sediment samples using aqua regia, potassium permanganate and potassium persulphate [67]. The approved US Environmental Protection Agency digestion procedure requires aqua regia and potassium permanganate as oxidants [39].

These digestion procedures are slow and often hazardous because of the combination of strong oxidizing agents and high temperatures. In some of the methods, mercuric sulphide is not adequately recovered. The oxidizing reagents, especially the potassium permanganate, are commonly contaminated with mercury, which prevents accurate results at low concentrations.

In lakes and streams, mercury can collect in the bottom sediments, where it may remain for long periods of time. It is difficult to release the mercury from these matrices for analysis. Several investigators have liberated mercury from soil and sediment samples by the application of heat to the samples and the collection of the released mercury on gold surfaces. The mercury was then released from the gold by application of heat or by absorption in a solution containing oxidizing agents [49, 61].

Batti *et al.* [68] determined methylmercury in river sediments from industrial and mining areas.

Bretthauer *et al.* [34] described a method in which samples were ignited in a high-pressure oxygen-filled bomb. After ignition, the mercury was absorbed in a nitric acid solution. Pillay *et al.* [36] used a wet-ashing procedure with sulphuric acid and perchloric acid to digest samples. The released mercury was precipitated as the sulphide. The precipitate was then redigested using aqua regia.

Feldman digested solid samples with potassium dichromate, nitric acid, perchloric acid and sulphuric acid [46]. Bishop *et al.* [47] used aqua regia and potassium permanganate for digestion. Jacobs and Keeney oxidized sediment samples using aqua regia, potassium permanganate, and potassium persulphate [69]. The approved US Environmental Protection Agency digestion procedure requires aqua regia and potassium permanganate and oxidants [39].

These digestion procedures are slow and often hazardous because of the combination of strong oxidizing agents and high temperatures. In some of the methods, mercuric sulphide is not adequately recovered. The oxidizing reagents, especially the potassium permanganate, are commonly contaminated with mercury, which prevents accurate results at low concentrations.

13.3.3 Saline deposited and suspended sediments

13.3.3.1 Gas chromatography

Bartlett *et al.* [55] observed unexpected behaviour of methylmercury containing River Mersey sediments during storage. They experienced difficulty in obtaining consistent methylmercury values; supposedly identical samples analysed at intervals of a few days gave markedly different results. They followed the levels of methylmercury in selected sediments over a period, to determine if any change was occurring on storage. They found that the amounts of methylmercury observed in the stored sediments did not remain constant; initially there was a rise in the amount of methylmercury observed, and then, after about ten days, the amount present began to decline to levels which in general only approximate those originally present. They have observed this phenomenon in nearly all of the Mersey sediment samples they examined.

It was noted that sediments sterilized, normally by autoclaving at approximately 120°C, did not produce methylmercury on incubation with inorganic mercury, suggesting a microbiological origin for the methylmercury. A control experiment was carried out in which identical samples were collected and homogenized. Some of the samples were sterilized by treatment with an approximate 4% *w/w* solution of formaldehyde. Several samples of both sterilized and unsterilized sediments were analysed at intervals and all of the samples were stored at ambient room temperature (18°C) in the laboratory. It can be seen from Fig. 13.8 that there is a difference in behaviour between the sterilized and unsterilized samples. Some of the samples were separately inoculated into various growth media to test for microbiological activity.

This work suggests that the application of laboratory-derived results directly to natural conditions could, in these cases, be misleading: analytical results for day ten if extrapolated directly might lead to the conclusion that natural methylmercury levels and rates of methylation are much greater than in fact they really are. Work in this area with model or laboratory systems needs to be interpreted with particular caution.

Bartlett *et al.* [55] used the method of Uthe *et al.* [70] for determining methylmercury. Sediment samples of 2–5g were extracted with toluene after treatment with copper sulphate and an acidic solution of potassium bromide. Methylmercury was then back extracted into aqueous sodium thiosulphate. This was then treated with acidic potassium bromide and copper sulphate following which the methylmercury was extracted into

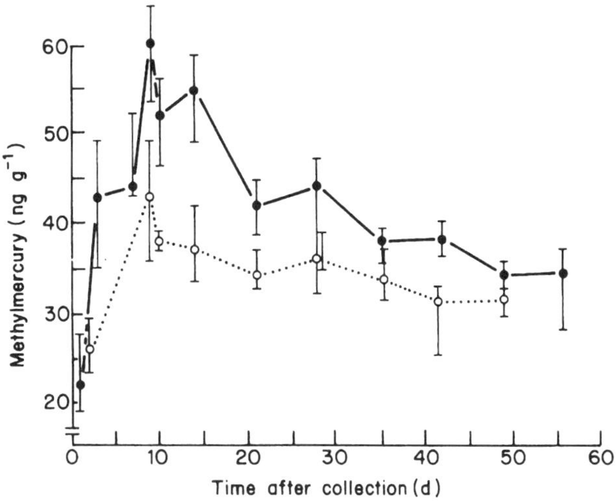

Fig. 13.8 Analysis of sterilized and unsterilized sediments from Hale Point, for methylmercury. Total mercury is 7.24 μg g^{-1}. Results up to day 25 are the mean of eight determinations; results beyond day 25 are the mean of four determinations. Error bars represent range limits for each analysis series. The samples were stored at room temperature (18°C), • untreated; o sterilized samples
Source: Reproduced with permission from McMillan Magazines [55]

pesticide grade benzene containing approximately 100μg dl^{-1} of ethyl mercuric chloride as an internal standard. The extract was analysed by electron capture gas chromatography using a Pye 104 chromatograph equipped with a nickel 65 detector. The glass column (1m × 0.4cm) was packed with 5% neopentyl glycol adipate on Chromosorb G (AW–DMCS). Methylmercury was measured by comparing the peak heights with standards of methyl mercuric chloride made up in the ethylmercury-benzene solution. The results were calculated as nanograms of methyl-mercury per gram of dry sediment. The detection limit was 1–2ng g^{-1}.

Craig and Morton [64] found 2.2μg L^{-1} mean total mercury level in 136 samples of bottom deposits from the Mersey Estuary.

13.4 Organotin compounds

The preparation of volatile derivatives makes the ionic organotin compounds amenable to evaporative separation techniques (purge and trap or gas chromatography). Hydride formation in dilute aqueous solutions is becoming a routine method for determination of methyltins [101,104,105,109,110], methyl- and butyltins [100,111,112], and phenyl-

and various other organotin compounds [77,113,114] to form the volatile hydrides (stannanes), which are analyzed either by purging and atomic absorption spectrometry or flame photometric detection or by liquid–liquid extraction with subsequent gas chromatographic analysis.

Unfortunately, stannanes are rather labile thus preventing further cleanup steps. Therefore, alkylation is often preferred over hydride formation, as the resulting tetrasubstituted organotin compounds can easily be purified and concentrated, which is necessary for low-level samples and complex matrices such as animal tissue or sewage sludge. A Grignard reagent or an alkyllithium compound is used to convert the ionic mono-, di-, or triorganotin compounds into the corresponding non-polar tetrasubstituted compound. The reaction has to be carried out in aprotic solvents and thus requires extraction of aqueous samples prior to derivatization. Procedures have been described for the analysis of methyltins [115], butyltins [116,117], mixed methylbutyltins [118], various alkyltins [119], cyclohexyltins [121] and phenyltins [120]. Alkylation also offers the possibility for selection of the volatility range of the derivatives, which are in most cases analyzed by gas chromatography. However, there are few methods for the sensitive determination of a broad range of organotin compounds in environmental samples. Recently, the sensitive determination of butyltin residues in sediment and surface water was described on the basis of extraction/methylation and high-resolution gas chromatography with flame photometric detection [122].

13.4.1 Soil

13.4.1.1 Gas chromatography

Sinex *et al.* [71] have described a method for the determination of methyltin compounds based on reaction with sodium borohydride to form tin hydrides then purge and trap analysis followed by gas chromatography with mass spectrometric detection. Down to 3–5pg absolute (as tin) of methyltin compounds equivalent to the sub μg kg $^{-1}$ range can be determined by this procedure.

Lobinski *et al.* [72] optimized conditions for the comprehensive speciation of organotin compounds in soils and sediments. They used capillary gas chromatography coupled to helium microwave induced plasma emission spectrometry to determine mono-, di-, tri- and some tetraalkylated tin compounds. Ionic organotin compounds were extracted with pentane from the sample as the organotin-diethyldithiocarbamate complexes then converted to their pentabromo derivatives prior to gas chromatography. The absolute detection limit was 0.5pg as tin; equivalent to 10–30μg kg $^{-1}$.

13.4.1.2 Supercritical fluid chromatography

In situ derivativization and supercritical fluid extraction has been used for the simultaneous extraction and determination of butyl tin and phenyltin compounds in soils and sediments [73].

13.4.1.3 Atomic absorption spectrometry

To determine methyltin, butyltin and inorganic tin in Great Bay estuary soils and sediments, Randall *et al.* [74] extracted the freeze dried sediment with 2.5mol L^{-1} calcium chloride and 2.5mol L^{-1} hydrochloric acid and analysed by hydride generation atomic absorption spectrometry. Detection limits for inorganic tin and tributyltin were 2.2ng kg^{-1} and 0.6ng kg^{-1} respectively. Recoveries of methyltin and butyltin species from spiking experiments were greater than 70 ± 10%. Tributyltin was found in all sampled sites, probably originating from tributyltin based anti-fouling paints.

Berling Gong and Matsumoto [75] used tributyl phosphate as a sensitivity enhancing solvent for the determination of organotin compounds in soil by carbon furnace atomic absorption spectrometry. Addition of tributyl phosphate and utilizing a temperature of 1000°C improves the sensitivity of detection of organotin to that achieved for inorganic tin. At 1000°C the organotin is converted to SnP_2O_7 and $Sn_2P_2O_7$.

13.4.2 Non-saline deposited and suspended sediments

13.4.2.1 Gas chromatography

Muller [76] has described a gas chromatographic method for the determination of tributyltin compounds in sediments. The tributyltin compounds are first converted to tributylmethyltin by reaction with ethyl magnesium bromide, and then analysed using capillary gas chromatography with flame photometric detection and gas chromatography–mass spectrometry. Tributyltin was found in samples of sediment and these results demonstrated that the technique has detection limits of less than 0.5pg L^{-1}.

Hattori [77] extracted alkyl and alkyltin compounds from sediments with methanoic hydrochloric acid and then, following mixture with sodium chloride and water, the mixture was extracted with benzene and converted to hydrides with sodium borohydride and analysed by gas chromatography using an electron capture detector. Down to 0.02mg kg^{-1} organotin compounds in sediments could be determined with a recovery of 70–95%.

Lobinski *et al.* [78] speciated organotin compounds in sediment samples by capillary gas chromatography using helium microwave induced plasma emission spectrometry as a detector. They used the procedure to determine mono-, di- and tri- and some tetraalkylated tin compounds in sediments. The ionic tin compounds were extracted as diethyldithiocarbamates into pentane then converted to pentyl magnesium bromide derivatives prior to gas chromatography. The absolute detection limit was 0.05pg tin, equivalent to 10–30ng kg^{-1}.

Arakawa *et al.* [96] have pointed out that methyltin compounds may be extracted from complex matrices and analysed by conventional gas chromatography. However, the procedure is lengthy, involving multiple steps where speciation may be altered and vessel adsorption effects may be large. Detection limits achievable with a flame ionization detector are 10–100µg [97].

Muller [79] has described a comprehensive method for determining trace analysis of mono-, di-, tri- and some tetrasubstituted organotin compounds in sewage sludge. The ionic compounds are extracted from diluted aqueous solutions as chlorides by using a Tropolon–C_{18} silica cartridge and from sewage sludge or sediments by using an ethereal Tropolon solution. The extracted organotin compounds ethylated by a Grignard reagent and analyzed by using high-resolution gas chromatography with flame photometric detection. Gas chromatography–mass spectrometry was used for confirmation. The extraction behaviour, gas chromatographic retention, and photometric response of a series of organotin compounds are described, and the identification via electron impact (El) and chemical ionization (CI) mass spectrometry is discussed. The main organotin compounds detected in various samples are butyltins; cyclohexyl- and phenyltins were identified in some of the sediment and sewage sludge samples. Methylbutyltins and tetrabutyltin were not detected. Concentrations were found to be in the low mg kg^{-1} range in sewage sludge.

In this method the sample (1–20g) is weighed in a flask, internal standard is added, and the sample acidified with hydrochloric acid to pH2–3. When the evolution of gas has ceased, the slurry is extracted with three portions (10, 5 and 5mL) of a 0.25% ethereal Tropolone solution. After being shaken vigorously and centrifuged, the organic phases are combined, filtered through anhydrous calcium chloride, and reduced in volume to about 2mL at room temperature by a rotary evaporator. The extract is then ready for derivatization.

Small portions (0.5mL) of the Grignard reagent, ethylmagnesium bromide (2M in tetrahydrofuran), are carefully added to the extracts prepared as described above. Reagent is dropwise and carefully added until an excess (indicated by a steady evolution of ethane) is present. The mixture is allowed to stand at least for 10min and the excess of reagent is destroyed by careful, dropwise addition of about 3mL of 2M hydrochloric acid. The organic layer, containing the compounds of interest, is dried over anhydrous sodium sulphate, reduced in volume to about 0.5mL, and purified by adsorption chromatography on silica gel (0.5g of silica gel in a Pasteur pipet, using elution with 10mL of 10% diethyl ether/hexane).

A Carlo Erba 2010 gas chromatograph fitted with a split/splitless injector, glass capillary column (30m in length, 0.3mm i.d., coated with either a 0.15μm film of Pluronic L 64 or a 0.5μm film of PS 255) and a Carlo Erba flame photometric detector SD 250 were used. The detector was operated without a filter and with a hydrogen-rich flame. Injector and detector temperature were set at 250°C and 225°C, respectively. Hydrogen (0.6 bar) served as carrier gas. Sample aliquots of 2μL were injected at room temperature in the splitless mode (45s); the compounds of interest were eluted with a temperature programme of 4°C/min.

13.4.2.2 Gas chromatography–mass spectrometry

Muller [79] has described a comprehensive method for determining traces of mono-, di-, tri- and some tetrasubstituted organotin compounds in lake sediments. The ionic compounds are extracted from acidified sediment as chlorides using ethereal Tropalone solution. The extracted organotin compounds are ethylated using a Grignard reagent (EtMgBr) and analysed by high-resolution gas chromatography with flame photometric and mass spectrometric detection. Ethylation using ethyl magnesium bromide was chosen for conversion of the various mono-, di- and tri-substituted organotin compounds in sediments into tetrasubstituted ones. Ethylation was preferred over either methylation or alkyllation using a larger alkyl group because methylation of tin (IV) and butyltin species seems to occur in the environment leading to methyltins and mixed methylalkyltins. Further methylation of these environmental metabolites in the derivatization step would exclude the possibility of determining these conversion and degradation products. The ethylation reaction of these compounds leads to a series of tetrabutyltin compounds as shown in Table 13.12. Environmental methylation is easily recognized, as the methylated products show typical relative retention time shifts compared to their ethylbutyltin analogues (Fig. 13.9). Furthermore, ethylation facilitates identification of organotin derivatives in the gas chromatogram, as the order of elution follows increasing degrees of substitution, which is not the case for hexylated products.

Table 13.12 Composition of mixed methyl ethylbutyltin compounds

	Butyl groups				
Methyl groups	*0*	*1*	*2*	*3*	*4*
0	Et_4	$BuEt_3$	Bu_2Et_2	Bu_3Et	Bu_4
1	$MeEt_3$	MeE_2Bu	$MeEtBu_2$	$MeBu_3$	
2	Me_2Et_2	Me_2EtBu	Me_3Bu_2		
3	Me_3Et	Me_3Bu			
4	Me_4				

Source: Reproduced with permission from the American Chemical Society [79]

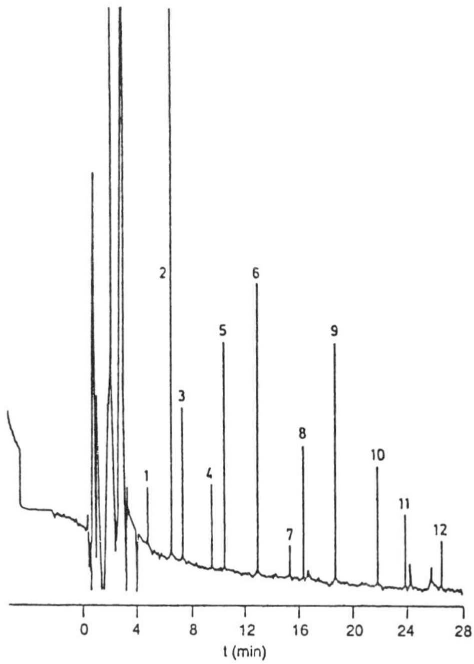

Fig. 13.9 GC/FPD (column B) of a solution containing all methyethylbutyltin compounds. Peak identification: 1 = Me_2Et_2Sn; 2 = Me_3BuSn; 3 = $MeEt_3Sn$; 4 = $Me_2EtBuSn$; 5 = Et_4Sn; 6 = $MeEt_2BuSn$; 7 = Me_2Bu_2Sn; 8 = $BuEt_3Sn$; 9 = $MeEtBu_2Sn$; 10 = Bu_2Et_2Sn; 11 = $MeBu_3Sn$; 12 = Bu_3EtSn
Source: Reproduced with permission from the American Chemical Society [79]

A recent sediment (period 1980–1984) and a sediment from the late 19th century (1880–1885) taken from Lake Zurich were investigated. No organotin compounds could be detected in the 1880 sediment, but a series of organotin compounds ranging from 280μg kg^{-1} tributyltin to 10mg kg^{-1} dicyclohexyltin were present in the recent sediment. The main components were again the butyltin compounds, indicating their frequent use, persistence, and bioaccumulative power. $Cyhex_2Sn^{2+}$ and $Cyhex_3Sn^{+}$ were also identified, reflecting the use of the parent compound, tricyclohexyltin, as a miticide in the region around Lake Zurich. The absence of organotin residues in the 1880 sediment is explained by the fact that technical use of these compounds started after 1936. The residues found in the recent sediment are considerably higher than those detected in surface sediment from the lower basin. As the sedimentation near the effluent of Lake Zurich in the shallow and oxygen-saturated water is dominated by processes leading to resuspension and oxidation of the fine, carbon-containing particles, the sediment taken at the deepest (and anoxic) part of the lake accumulates higher organotin residues and is therefore more representative of the overall situation in the lake sediment.

Szpunar *et al.* [80] and Prange and Jansen [81] have determined organotin species by using gas chromatography with indirectly coupled plasma mass spectrometric detection. In one study, butyltin species were extracted from sediments and biomaterials in only 1–5min by using microwave digestion [80]. In another investigation, detection limits of ~50 (Sn), 100 (Pb) and 120 fg (Hg) were achieved [81].

13.4.2.3 Purge and trap gas chromatography

Sinex *et al.* [71] have described a method for the determination of methyltin compounds based on reaction with sodium borohydride to form tin hydrides then purge and trap analysis followed by gas chromatography with mass spectrometric detection. Down to 3–5pg absolute (as tin) of methyltin compounds equivalent to the sub-μg kg^{-1} range can be determined by this procedure.

13.4.2.4 High-performance liquid chromatography

Epler *et al.* [82] used laser enhanced ionisation as a selective detector for the liquid chromatographic determination of alkyltin compounds in sediments. The analysis was performed on a 1-butanol extract of the sediment.

Yang *et al.* [83] accomplished speciation of organotin compounds using reverse-phase liquid chromatography with inductively coupled plasma mass spectrometric detection. The separation was complete in 6min and detection limits were in the range 2.8–16pg of tin for various species.

High-performance liquid chromatography coupled with fluorescence detection [106,107] or ion-exchange high-performance liquid chromatography with detection by graphite furnace atomic absorption spectroscopy [108] proved to be sensitive methods, but may lack from limitations in separation power and ease of identification of unknown products.

13.4.2.5 Atomic absorption spectrometry

Butylation of methyl tin species before solvent extraction and the use of atomic absorption spectrometry shortens the extraction procedure and reduces detection limits to about 0.1ng [102].

Stephenson and Smith [84] used graphite furnace atomic absorption spectrometry to determine tributyltin in sediments. Recoveries from spiked samples ranged to from 72–111%. The detection limit was 2.5mg kg^{-1} of sample.

13.4.2.6 Supercritical fluid chromatography

Cal *et al.* [85] carried *in situ* derivativisation and supercritical fluid extraction for the simultaneous determination of butyltin and phenyltin compounds in sediments.

13.4.2.7 Miscellaneous

Electrochemical and fluorometric methods of determining alkyl tin compounds have been discussed [104,105].

Unger *et al.* [87] have studied the sorption behaviour of tributyltin on estuarine sediments.

Ebdon *et al.* [88] have discussed a programme to improve the quality of analytical results in the environmental monitoring of organotin compounds. They discuss the evolution of a sensitive, reliable and robust analytical method for the determination of tributyltin, with emphasis on the difficulties of determining it at the ng per litre levels at which it was usually encountered, more especially as other forms of tin frequently occurred together at similar levels. The preparation of a standard reference sample, for use in inter-laboratory comparative determinations, under the aegis of the Bureau of Community Reference of the EU is described, and plans for subsequent distributions of blank, artificially spiked, and genuinely affected sediments are sketched.

Gilmour *et al.* [89] converted methyltin compounds in sediments to tin hydrides with sodium borohydride prior to ion monitoring by mass spectrometry.

Methyltin compounds were determined in complex samples by a sensitive purge and trap method at detection limits of 3–5pg with 6–18% precision. The method is selective and specific and free from most interferences.

Rapsomankis *et al.* [86] have studied the biological methylation of inorganic tin in soils and sediments.

Three groups have independently reported on the methylation of both inorganic tin and organotin substrates by the mixed populations of microbial flora present in sediments collected from a Canadian fresh water lake [90], and from estuarine sites in San Francisco Bay and Chesapeake Bay [91–93]. Biogenesis of Me_4Sn was seen only to occur with additions of Me_3Sn^+ to incubated sediments [90,94], but the redistribution reactions of intermediate methyltins to form Me_4Sn by non-biological pathways must be noted as competitive events in such experiments [94,95]. The concentrations of tin compounds added to incubated sediments were consistent with values found in polluted sediments. The influence of other bioactive pollutant metals also commonly found in such sediments was not investigated.

13.4.3 Saline deposited and suspended sediments

13.4.3.1 Gas chromatography

The procedures described in section 13.4.2.1 [76–78] have been applied to the determination of alkyltin compounds in marine sediments.

13.4.3.2 Purge and trap chromatography

Chromatographic methods have been applied with hydridization. Jackson *et al.* [98] used a commercial purge and trap apparatus fitted to a packed gas chromatographic column and flame photometric detector to achieve a 0.1ng detection. Purge and trap procedures followed by boiling point separations and detection by spectrophotometric methods yield detection limits in water of between 0.01 and 1ng. Detection of SnH emission by flame emission gives the greatest sensitivity.

Sinex *et al.* [71] determined methyltin compounds in amounts down to 3–5pg (as Sn absolute), i.e. the sub-µg kg^{-1} level, in marine sediments by a procedure involving reaction with sodium borohydride to produce tin hydrides, followed by purge and trap analysis then gas chromatography with mass spectrometric detection.

Gilmour *et al.* [89] have developed an extremely sensitive purge and trap method for the determination of methyltin compounds as methylstannanes in marine sediments. Hydride derivatives were prepared with sodium borohydride in a closed, flow through system consisting of a purge vessel, gas chromatograph and mass spectrometer. Borate buffer added to samples generated hydrogen from sodium borohydride, resulting in high purge efficiencies for mono-, di-, and trimethyltin. Selected ion mode monitoring with the mass spectrometer gave detection limits for methyltins of 3–5pg as tin. The concentration detection limits for a 5g sediment sample were in the sub-µg kg^{-1} level, with a standard deviation of 6–18%, depending upon the methyltin species and sample type. The method is both selective and specific, eliminating most interferences while permitting positive identification of individual methyltin species. Sample weights were typically 5g and these were treated directly with the borate buffered sodium borohydride reagent.

The recovery of methyltins from sediments was tested by using anoxic, sulphitic clay sediments from a mid-salinity region of Chesapeake Bay. Monomethyltin (11.2ng) and dimethyltin (11.5ng) were completely recovered from sediment. However, recovery of 10ng of trimethyltin chloride from sediment was only about 70%.

In this procedure methylstannanes were generated directly in buffered sample, purged and cryogenically trapped on the head of a chromatographic column. Gas flow was diverted from the hydride generator while the purge tube was filled with up to 10mL of fluid, usually consisting of 5mL of sample and 5mL of buffer (saturated sodium borate, adjusted to pH8.0 with sodium hydroxide). The purge tube was then sealed to the Teflon head. Samples were spiked with known amounts of diethyltin dichloride, which served as an internal standard. The gas chromatographic column was held at –40°C for 2min before 1mL of sodium borohydride (4% sodium borohydride, 1% sodium hydroxide in water) solution was injected into the purge vessel through a butyl rubber septum on the vessel sidearm. Helium flow was then directed through the hydride generator for 3min. After the purge was completed, carrier gas was again diverted away from the sample and the column oven temperature then increased at 15°C/min to 120°C, then 30°C/min to 150°C. Mass spectrometer operation began 6 min after the purge cycle was complete and operated 7.5min.

Identification and quantitation of methyltin hydrides were achieved by selected ion monitoring [99]. Major ions representative of the three methylstannanes and the internal standard (diethyltin) were nominal, *m/e* 116, 118, 120, 133, 135, 148, 150 and 165.

13.4.3.3 Atomic absorption spectrometry

Andreae and Byrd [101] have pointed out that methylstannanes produced by hydridization of methyltin compounds is both stable and volatile with boiling points from 0–59°C.

Hodge *et al.* [100] determined nanogram quantities of the halides of methyltin, dimethyltin, trimethyltin, diethyltin, triethyltin, *n*-butyltin, di-*n*-butyltin, tri-*n*-butyltin, phenyltin and inorganic tin(IV) in marine sediments by a procedure involving reaction with sodium borohydride to convert to tin hydrides, which are then detected by atomic absorption spectrometry. The compounds are separated on the basis of their differing boiling points, which range from 1.4°C (CH_3SnH_3) to 280°C ($(n\text{-}C_4H_4)_3SnH$). Detection limits range from 0.4μg kg $^{-1}$ (SnIV) to 2μg kg $^{-1}$ (tri-*n*-butyltin chloride). Fig. 13.10 shows that stannane and the organotin hydrides evolve from the hydride trap in such a manner that they can be identified by a 'retention time'. Tin levels in core samples taken in Narragonsett Bay, USA (expressed as total tin) ranged from 1mg kg $^{-1}$ (pre 1900) to 20mg kg $^{-1}$ in present-day samples.

In this method approximately 19 samples of marine sediment were oven dried at 110°C then digested with nitric acid–perchloric acid and hydrofluoric acid–hydrochloric acid. The digested solution is made up to 50ml of an equal volume mixture of 6M hydrochloric acid and 2M nitric acid. 0.1ml or less of the digest was pipetted into the hydride generator, followed by 1ml 2M acetic acid, diluted to 100ml with double distilled water and reacted with sodium borohydride.

Randall *et al.* [74] determined down to 0.6×10^{-6}mg kg $^{-1}$ of methyl and butyltin compounds in estuary water sediments.

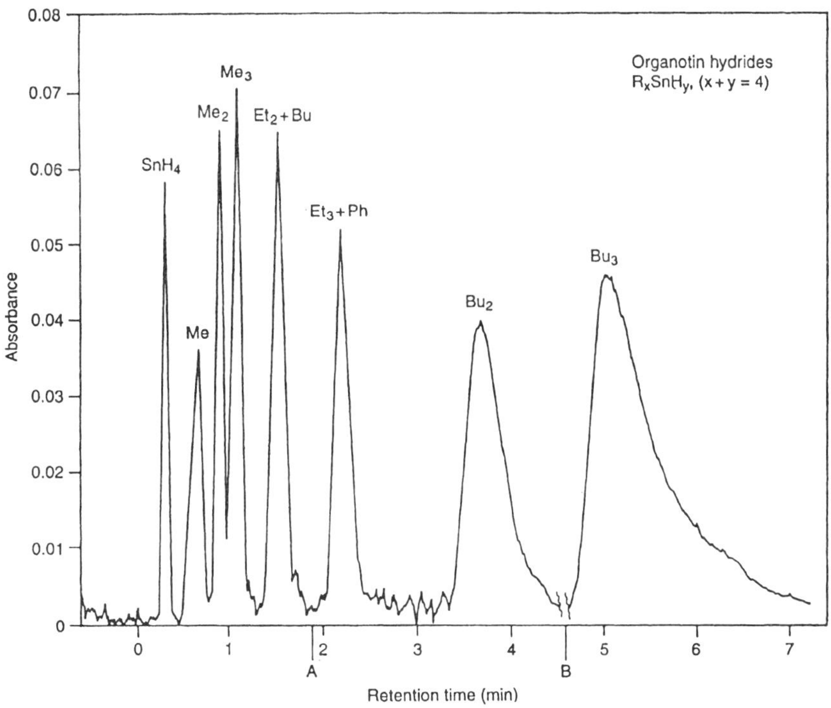

Fig. 13.10 Chromatogram of hydrides generated from a known mixture of Sn(IV) and nine organotin halides
Source: Reproduced by permission of American Chemical Society [100]

Me = methyl; Et = ethyl; Bu = *n*-butyl; Ph = phenyl. Et_2SnH_2 and $BuSnH_3$ have the same retention time, and so do EtSnH and $PhSnH_3$. Approximate concentrations of the reactants that produced this chromatogram are: Sn(IV) 6 ng; $MeSnCl_3$ 14ng; Me_3SNCl 18ng; Et_2SnCl_2 22ng; Et_3SnB_4 or $PhSnCl_3$ 40ng; Bu_2SnCl_2 110ng, and BU_3SnCl 470ng. Chromatogram terminated at 'A' with water trap immersed in dry ice/2-propanol. Tri-*n*-butyltin hydride is released from the hydride trap after di-*n*-butyltin dihydride it trap immersed in 80°C water bath at B.

To determine methyltin, butyltin and inorganic tin in Great Bay estuary soils and sediments, Randall *et al.* [74] extracted the freeze dried sediment with 1.5mol L^{-1} calcium chloride and 2.5mol L^{-1} hydrochloric acid and analysed by hydride generation atomic absorption spectrometry. Detection limits for inorganic tin and tributyltin were 2.2ng kg^{-1} and 0.6ng kg^{-1} respectively. Recoveries of methyltin and butyltin species from spiking experiments were greater than 70 ± 10%. Tributyltin was found in all sampled sites, probably originating from tributyltin based anti-fouling plants.

Table 13.13 Selection of methods of analysis for organometallic compounds

Type of sample	*Type of organo-metallic*	*Method*	*LD mg kg $^{-1}$ (unless otherwise stated)*	*Reference*	*Conc range found in environmental samples mg kg $^{-1}$*
Marine sediments	Organo arsenic	Ion-exchange chromatography–gas	<5	[13]	0.05–0.3 (rivers)
River sediment	Alkyltin, aryltin	Benzene extraction; conversion to hydride with $NaBH_4$, glc with electron capture detector	0.02	[77]	0.0001–0.08 (coastal and marine)
River sediment	Alkyltin	Reverse phase hplc with ICP mass spectrometric detector	3–16 pg	[83]	
River sediment	Bu_3Sn	Digestion–gas	2.5	[84]	
River sediment	MeSn	Purge and trap analysis	3–5 pg	[89]	
River sediment	Bu_3Sn	Reaction with EtMgBe, glc with flame photometric detector	0.5pg L $^{-1}$	[76]	
River sediment	MeSn	Extraction glc	10–100ng	[96]	
River sediment	Mono to tetraalkyl tins	Adsorption on to G8 silica, tropalone extraction, reaction with Grignard reagent, glc with flame photometric detector	low mg kg $^{-1}$	[79]	
River sediment	Alklytin	Microwave digestion, glc with ICP mass spectrometric detector	50µg	[81]	
River and marine sediments, soils	MeSn	Reduction to hydride with $NaBH_4$ and cold trap controlled evaporation, glc with mass spectrometric detection	0.001 (0.3–0.5pg)	[71]	
River and marine sediments, soils	MeSu BuSn	ditto, detection with aas	0.0000006	[74]	
River and marine sediments, soils	Bu_3Sn Bu_2Sn BuSn	Capillary glc–helium microwave induced plasma emission spectrometric detection	0.0001–0.0003	[72]	

Saline sediments	Alkyl tin	Purge and trap analysis	0.01–1ng	[98]	
River and marine sediments, soils	Mono-tri methyltin mono-tri ethyl tin butyl phenyltin	$NaBH_4$ reduction, aas	0.002 (nBu_3Sn)	[48]	
River sediments	MeHg	Steam distillation cold vapour aas	0.0025	[62]	0.0001–0.004 (river)
River sediments	MeHg	Cold vapour aas	0.0002	[81,21,125]	0.0001–0.0004 (coastal and marine)
River sediments	Alkyl Hg Ch_3 HgCl	Sulphuric acid–perchloric acid digestion–spectrophotometry	0.1	[50]	
River sediments	MeHg	Toluene extraction, bromo derivitization, glc	0.001	[55]	
River and marine sediments	MeHg	Cold vapour aas	0.002 (0.1μg L^{-1})	[43,45,124]	
Soil	Alkyl Hg aryl Hg	Spectrophotometric	0.05–1	[28,31,32]	
Soil	Alkyl Hg	Nitric acid–sulphuric acid digestion–aas	0.001	[41]	
River	MePb Et Pb	Conversion to tetra-alkyl lead compounds using nBu MgCl then gas chromatography with atomic absorption spectrometric detector	0.015	[19]	0.0001 (river) 0.00001 (coastal and marine)
River	$PbEt_4$ Et_3PbCl Me_2PbCl Et_2PbCl_2 Me_2PbCl_2	Petroleum ether extraction, glc	0.02	[15]	
River and marine sediments	Me_4Pb Me_3EtPb Me_2Et_2Pb $MeEt_3Pb$ Et_4Pb	Hexane extraction glc with aas detector	0.01	[16]	

Source: Own files

13.4.4 Sludge

13.4.4.1 Gas chromatography

The derivatization gas chromatographic procedure described in section 13.4.2.2 [79] has been applied to the determination of mg kg $^{-1}$ levels of organotin compounds in sewage sludge. Butyltin compounds were found in many of the sewage sludge samples analysed. Some samples also contained phenyltin compounds, but cyclohexyltin and mixed butylmethyltins were absent.

13.4.4.2 Miscellaneous

A standard UK procedure discusses the determination of organic, inorganic and total tin in sludges and sediments [103].

13.5 Organosilicon compounds

13.5.1 Non-saline deposited and suspended sediments

13.5.1.1 Atomic absorption spectrometry

Pellenberg [123] analysed river sediment for silicone content by nitrous oxide-acetylene flame atomic absorption spectrophotometry. He showed that total carbon and total carbohydrates both correlate well with silicone content and the correlation between sedimentary silicone and presumed sewage material is good enough to suggest silicone as a totally synthetic, specific tracer for sewage in the aquatic environment.

13.5.1.2 Inductively coupled plasma atomic emission spectrometry

Wanatabe *et al.* [57] have described a method for the separation and determination of siloxanes in sediment, using inductively coupled plasma emission spectrometry. The organosilicon extract with petroleum ether is evaporated to dryness. The damp residue is dissolved in methyl isobutyl ketone, aspirated into the plasma. The detection limit is 0.01mg kg $^{-1}$. Recoveries are about 50% with a coefficient of variation of about 11%.

13.5.1.3 Miscellaneous

Van der Post [124] has described a method for the determination of silanols in water based on their ability to reduce nitrite or nitrate to ammonia at normal temperature. Individual silanols are identified by mass spectrometry.

13.6 Selection of appropriate analytical methods

The analytical method to be used will depend on the type or organometallic compound, its concentration range in the sample, the detection limit required and the type of sample. This information is summarised in Table 13.13.

First, it must be explained that not all published procedures give details of the detection limits achievable. Where this data is available it is quoted in Table 13.13, and compared to the range of concentrations of organometallic compounds actually encountered in environmental samples.

This table highlights some, but not all, of the available methods that have sufficiently low detection limits to render them applicable to even the lowest concentrations of organometallic compounds encountered in various types of environmental samples. Thus, Table 13.13 is a useful starting point for the selection for any particular type of any organometallic compound and sample type of an appropriate analytical method that will cover the range of concentrations likely to be encountered in the sample including very low concentrations. The situation will depend on the type of sample, viz. sediments, soils or sludges.

References

1 Odanake, Y., Tsuchlya, N. and Goto, S. (1983) *Analytical Chemistry*, **55**, 929.
2 Soderquist, C.J., Crosby, D.G. and Bowers, J.B. (1974) *Analytical Chemistry*, **46**, 155.
3 Von Endt, D.W., Kearney, P.C. and Kaufman, D.D. (1968) *Journal of Agriculture and Food Chemistry*, **16**, 17.
4 Leatherland, T.M. and Burton, J.D. (1974) *Journal Marine Biology Association UK*, **54**, 457.
5 Lunde, G. (1977) *Environmental Health Perspectives*, **19**, 47.
6 Greig, R.A., Wenzloff, D.R. and Pearce, J.B. (1976) *Marine Pollution Bulletin*, **7**, 185.
7 Dean Luh, M., Baker R.A. and Henley, D.E. (1973) *Journal of Science of the Total Environment*, **2**, 1.
8 Coulson, E.J., Remington, R.E. and Lynch, K.M. (1935) *Journal of Nutrition*, **19**, 255.
9 Jacobs, L.W., Syers, J.K. and Keeney, D.R. (1970) *Proceedings of the Soil Science of America*, **34**, 750
10 Andreae, M.O. (1977) *Analytical Chemistry*, **49**, 820.
11 Johnson, D.L. and Braman, R.S. (1975) *Deep Sea Research*, **22**, 503.
12 Inverson, D.G., Anderson, M.A., Holm, T.R. and Stanforth, R.R. (1979) *Environmental Science and Technology*, **13**, 1491.
13 Maher, W.A. (1981) *Analytica Chemica Acta*, **126**, 157.
14 Blais, J.S., Momplasir, G.M. and Marshall, W.P. (1990) *Analytical Chemistry*, **62**, 1611.
15 Potter, H.R., Jarview, A.W.P. and Markell, R.N. (1977) *Water Pollution Control*, **76**, 123.
16 Chau, Y.K., Wong, P.T.S., Bengert, G.A., Kramar, O. (1979) *Analytical Chemistry*, **51**, 186.
17 Chau, Y.K., Wong, P.T.S. and Saitoh, H. (1976) *Journal of Chromatographic Science*, **162**, 14.

18 Chau, Y.K., Wong, P.T.S. and Kramer, O. (1983) *Analytica Chemica*, **146**, 211.
19 Chau, Y.K., Wong, P.T.S., Bengert, G.A. and Dunn, J.L. (1984) *Analytical Chemistry*, **56**, 271.
20 Segar, D.A. (1974) *Analytical Letters (London)*, **7**, 89.
21 Reisinger, K., Stoeppler, M. and Nurnberg, H.W. (1981) *Nature (London)*, **291**, 228.
22 Wong, P.T.S., Chan, Y.K. and Luxon, P.L. (1975) *Nature (London)*, **253**, 26.
23 Booer, J.R. (1944) *Annals Applied Biology*, **31**, 34.
24 Hitchcock, A.E. and Zimmerman, P.W. (1957) *Annals New York Academy of Science*, **65**, 474.
25 Zimmerman, P.W. and Crocker, W. (1934) *Contribution Boyce Thompson Institute*, **6**, 167.
26 Kiessling, H. (1961) *Svensk Papperstid.*, **64**, 689.
27 Spanis, W.C., Munnecke, D.E. and Solberg, R.A. (1962) *Phytopathology*, **52**, 455.
28 Kimura, Y. and Miller, V.L. (1960) *Analytical Chemistry*, **32**, 420.
29 Kimura, Y. and Miller, V.L. (1962) *Analytical Chemistry*, **34**, 325.
30 Kimura, Y. and Miller, V.L. (1964) *Journal of Agriculture and Food Chemistry*, **15**, 253.
31 Polley, D. and Miller, V.L. (1955) *Analytical Chemistry*, **27**, 1162.
32 Subcommittee (of the Analytical Methods Committee) of the Society for Analytical Chemistry (London) (1965) *Metallic Impurities in Organic Matter.* Report by *Analyst*, **90**, 515.
33 Leong, P.C. and Ong, H.O. (1971) *Analytical Chemistry*, **43**, 940.
34 Bretthauer, E.W., Moghissi, A.A., Snyder, S.S. and Matthews, N.W. (1974) *Analytical Chemistry*, **46**, 445.
35 Anderson, D.H., Evans, J.H., Murphy, J.J. and White, W.W. (1971) *Analytical Chemistry*, **43**, 1151.
36 Pillay, K.K.S., Thomas, C.C., Sonde, C.J.A. and Hyone, C.M. (1971) *Analytical Chemistry*, **43**, 1419.
37 Feldman, C. (1974) *Analytical Chemistry*, **46**, 99.
38 Bishop, J.N., Taylor, L.A. and Nearby, B.O. (1975) *The Determination of Mercury in Environment.* US Environmental Protection Agency, Cincinnati, Ohio, p.120.
39 Environmental Protection Agency (1974) *Methods for Chemical Analysis of Water and Wastes.* US Environmental Protection Agency, Cincinnati, Ohio, pp.134–138.
40 Longbottom, J.E., Dressman, R.C. and Lichtenberg, J.J. (1973) *Journal of Association of Official Analytical Chemists*, **56**, 1297.
41 Iskander, I.K., Syers, J.K., Jakobs, L.W. *et al.* (1972) *Analyst (London)*, **97**, 388.
42 Umezaki, Y. and Iwamoto, K. (1971) *Japan Analyst*, **20**, 173.
43 Matsunaga, K. and Takahashi, S. (1976) *Analytica Chemica Acta*, **87**, 487.
44 Irukayama, K., Fukiki, M., Tajima, S. and Omori, S. (1972) *Japan Journal Public Health*, **19**, 25.
45 Langmyhr, F.J. and Aamodt, J. (1976) *Analytica Chemica Acta*, **87**, 483.
46 Feldman, C. (1974) *Analytical Chemistry*, **46**, 1606.
47 Bishop, J.N., Taylor, L.A. and Neary, B.P. (1973) in *The Determination of Mercury in Environment Samples*, Ministry of the Environment, Ontario, Canada.
48 Jacobs, L.W. and Keeney, D.R. (1976) *Environmental Science and Technology*, **8**, 267.
49 Leong, P.C. and Ong, H.P. (1971) *Analytical Chemistry*, **43**, 940.
50 Jurka, A.M. and Carter, M.J. (1978) *Analytical Chemistry*, **50**, 91.
51 El-Awady, A.A., Miller, R.B. and Carter, M.J. (1976) *Analytical Chemistry*, **48**, 110.
52 Jensen, S. and Jernelou, O. (1969) *Nature (London)*, **223**, 753.

53 Ealy, J.D., Schulz, W.D. and Dean, J.A. (1973) *Analytica Chemica Acta*, **54**, 235.
54 Audren, A.W. and Harris, R.C. (1973) *Nature (London)*, **245**, 256.
55 Bartlett, P.D., Craig, P.J. and Morton, S.F. (1977) *Nature (London)*, **267**, 606.
56 Uthe, J.F., Armstrong, F.A.J. and Stainton, M.P. (1979) *Journal Fisheries Research Board, Canada*, **27**, 805.
57 Wanatabe, N., Yasuda, Y., Kato, K. *et al.* (1984) *Science of the Total Environment*, **34**, 169.
58 Uthe, J.F., Armstrong, F.A.J. and Tam, K.C. (1972) *Journal of Association of Official Analytical Chemists*, **54**, 866.
59 Cappon, C.J. and Crispin-Smith, V. (1977) *Journal of Analytical Chemistry*, **49**, 365.
60 Jensen, S. and Jernelov, M.P. (1969) *Nature (London)*, **223**, 753.
61 Anderson, D.H., Evans, J.H., Murphy, J.J. and White, W.W. (1971) *Analytical Chemistry*, **43**, 1511.
62 AOAC (1979) *Official Methods of the Association of Official Analytical Chemists*, 11th ed, p.418.
63 Nagese, H., Sato, T., Ishikawa, T. and Mitani, K. (1980) *International Journal of Environmental Analytical Chemistry*, **7**, 261.
64 Craig, P.J. and Morton, S.F. (1976) *Nature (London)*, **261**, 125.
65 Department of the Environment and National Water Council, UK (1984) *Mercury in Waters, Effluents, Soils and Sediments*, Additional Methods (PE–22) SGENW. HMSO London, 1985.
66 Robert, J.M. and Robenstein, D.L. (1991) *Analytical Chemistry*, **53**, 2074.
67 Jacobs, L.W. and Keeney, D.R. (1976) *Environmental Science and Technology*, **8**, 267.
68 Batti, R., Magnaval, R. and Lanzala, E. (1975) *Chemosphere*, **4**, 13.
69 Jacobs, L.W. and Keeney, D.R. (1976) *Environmental Science and Technology*, **8**, 267.
70 Uthe, J.F., Solomon, J. and Grift, B. (1972) *Journal of Association of Official Analytical Chemists*, **55**, 583.
71 Sinex, S.A., Cantillo, A.Y. and Helz, G.R. (1980) *Analytical Chemistry*, **52**, 2342.
72 Lobinski, R., Dirlex, W.M.R. and Adams, F.C. (1992) *Analytical Chemistry*, **64**, 159.
73 Cal, Y., Aizaga, R. and Bayona, J.M. (1994) *Analytical Chemistry*, **66**, 1161.
74 Randall, L., Hans, J.S. and Ucher, J.H. (1988) *Environmental Technology Letters*, **7**, 471.
75 Berling Gong H. Li and Matsumoto, K. (1996) *Analytical Chemistry*, **68**, 2277.
76 Muller, M.D. (1984) *Fresenius Zeitschrift für Analytische Chemie*, **317**, 32.
77 Hattori, Y., Kabayashi, A., Takemoto S. *et al.* (1984) *Journal of Chromatography*, **315**, 341.
78 Lobinski, R., Dirlex, W.M.R. and Adams, F.C. (1992) *Analytical Chemistry*, **64**, 159.
79 Muller, M.D. (1987) *Analytical Chemistry*, **59**, 617.
80 Szpunar, J., Schmitt, V.O., Lobinski, R. and Moned, J.L. (1996) *Journal of Analytical Atomic Spectroscopy*, **11**, 193.
81 Prange, A. and Jansen, E. (1995) *Journal of Analytical Atomic Spectroscopy*, **10**, 105.
82 Epler, K.S., O'Haver, T.C., Turk, G.C. and MacCrehan, W.A. (1988) *Analytical Chemistry*, **60**, 2062.
83 Yang, H.J., Jiang, S.J., Yang, Y. and Hwang, C. (1995) *Analytica Chemica Acta*, **312**, 141.
84 Stephenson, M.D. and Smith, D.R. (1988) *Analytical Chemistry*, **60**, 696.
85 Cal, Y., Aizaga, R. and Bayona, J.M. (1994) *Analytical Chemistry*, **66**, 1164.

86 Rapsomankis, S., Donard, O.F. and Weber, H. (1987) *Applied Organometallic Chemistry*, **1**, 115.
87 Unger, M.A., MacIntyre, W.G. and Huggett, R.J. (1988) *Environmental Toxicology and Chemistry*, **7**, 907.
88 Ebdon, L., Hill, S. and Griepink, B. (1988) *Environmental Technology Letters*, **9**, 965.
89 Gilmour, C.C., Tuttle, J.H. and Means, J.C. (1986) *Analytical Chemistry*, **58**, 1848.
90 Chau, Y.K. (1980) Biological methylation of tin compounds in the aquatic environment. 3rd International Conference Organometal Coordinating Chem. Germanium, Tin, Lead, July 1980, University of Dortmund, West Germany.
91 Hallas, L.E. (1981) PhD Dissertation. University of Maryland.
92 Hallas, L.E., Mearns, J.C. and Cooney, J.J. (1982) *Science*, **213**, 1505.
93 Hallas, L.E. and Cooney, J.J. (1981) *Applied and Environmental Microbiology*, **41**, 466.
94 Guard, H.E., Cobet, A.B. and Coleman, W.M. (1981) *Science*, **213**, 770.
95 Craig, P.J. (1980) *Environmental Technology Letters*, **1**, 225.
96 Arakawa, Y., Wada, O., Yu, T.H. and Iwai, H. (1981) *Journal of Chromatography*, **216**, 209.
97 Tam, G.K., Lacroix, G. and Lawrence, J.F. (1983) *Journal of Chromatography*, **259**, 350.
98 Jackson, J.A., Blair, W.R., Brinckman, F.E. and Iveson, W.P. (1982) *Environmental Science and Technology*, **16**, 110.
99 Means, J.C. and Hulebak, K.L. (1983) *Neurotoxicology*, **4**, 37.
100 Hodge, V.F., Snider, S.L. and Goldberg, D. (1979) *Analytical Chemistry*, **51**, 1256.
101 Andreae, M.O. and Byrd, J.T. (1984) *Analytica Chemica Acta*, **156**, 147.
102 Chau, Y.K., Wong, P.T.S. and Bengert, G.A. (1982) *Analytical Chemistry*, **54**, 946.
103 HMSO (1992) *Determination of organic, inorganic and total tin compounds in waters, sediments and biota*, London..
104 Kenis, P. and Zirino, A. (1983) *Analytica Chemica Acta*, **149**, 157.
105 Arakawa, Y., Wadd, O. and Wanatabe, M. (1983) *Analytical Chemistry*, **55**, 1901.
106 Langseth, W. (1984) *Talanta*, **31**, 975.
107 Yu, T.-H. and Arakawa, Y. (1983) *Journal of Chromatography*, **258**, 189.
108 Jewett, K.L. and Brinckman, F.E. (1981) *Journal of Chromatographic Science*, **19**, 583.
109 Tugrul, S., Balkas, T.I. and Goldberg, E.G. (1983) *Marine Pollution Bulletin*, **14**, 297.
110 Jackson, J.A., Blair, W.R, Brinckman, F.E. and Iverson, W.P. (1982) *Environmental Science and Technology*, **16**, 110.
111 Donard, O.F.X., Rapsomanikis, S. and Weber, J.H. (1986) *Analytical Chemistry*, **54**, 772.
112 Matthias, C.L., Bellama, J.M., Olson, G.J. and Brinckman, F.E. (1986) *Environmental Science and Technology*, **20**, 609.
113 Woollins, A. and Cullen, W.R. (1984) *Analyst (London)*, **109**, 1527.
114 Soderquist, C.J. and Crosby, D.G. (1978) *Analytical Chemistry*, **50**, 1435.
115 Chau, Y.K., Wong, P.T.S. and Bengert, G.A. (1982) *Analytical Chemistry*, **54**, 246.
116 Meinema, H.A., Burger-Wiersma, T., Versluis-de Haan, G. and Gevers, E.G. (1978) *Environmental Science and Technology*, **12**, 288.
117 Zimmerli, B. and Zimmermann, H. (1980) *Fresenius Zeitschrift für Analytische Chemie*, **304**, 23.
118 Maguire, R.J. (1984) *Environmental Science and Technology*, **18**, 291.
119 Maguire, R.J., Tkacz, R.J., Chau, Y.K. *et al.* (1986) *Chemosphere*, **15**, 253.
120 Möllhoff, E. (1977) *Pflanzenschutz-Nachr*, **30**, 249.

121 Wright, W.B., Lee, M.L. and Booth, G.M. (1979) *HRC CC. Journal of High Resolution Chromatography and Chromatography Communications*, **1**, 189.
122 Muller, M.D. (1984) *Fresenius Zeitschrift für Analytische Chemie*, **317**, 32.
123 Pellenberg, R. (1979) *Marine Pollution Bulletin*, **10**, 267.
124 Van der Post, D.C. (1978) *Water Pollution Control*, **77**, 520.
125 Birnie, S.E. and Hodge, D.J. (1981) *Environmental Technology Letters*, **2**, 433.

Chapter 14

Sampling procedures

14.1 Introduction

Sampling procedures are extremely important in the analysis of soils, sediments and sludges. It is essential to ensure that the composition of the portion of the sample being analysed is representative of the material being analysed. This fact is even more evident when it is conceded that the size of the portion of sample being analysed is in many modern methods of analysis extremely small. It is therefore essential to ensure before the analysis is commenced that correct statistically validated sampling procedures are used to ensure as far as is possible that the portion of the sample being analysed is representative of the bulk of material from which the sample was taken.

The collection and handling of samples prior to analysis has been discussed by various workers and organizations, including Smith and James [1], Kratochvil *et al.* [2,3], Gy [4], Woodget and Cooper [5], Harrison [6], Walton and Hoffman [7], Laitinen [8,9], Ingamells and Pitard [10], Kratochvil and Taylor [11], Kratochvil [12], Wallace and Kratochvil [13], Ministry of Agriculture, Fisheries and Food [14] and HMSO [15]. Other bodies who have discussed sampling procedures include the American Society for Testing Materials, the US Environmental Protection Agency, the American Public Health Association, the British Standards Institution, [16] etc.

The principal step of the sampling process is the taking of the sample. Here we intend to deal only with the risk of contaminating the sample during its collection, storage and processing, since any subsequent separation is applied only after the sample has been brought into solution.

By knowing the history of the sample it is possible to act correctly during all of the sampling steps, in order to avoid contamination of the material either from the utensils used to collect the sample, or from the reagents, the laboratory atmosphere and even the laboratory personnel. The contamination risk is greatest in analysis for trace components. Trace analysis requires the use of specially acclimatized, sometimes over-pressurized laboratories, of very pure reagents and sensitive instruments, and of specialized personnel, who possess a broad range of knowledge both in basic and analytical chemistry. In such cases, the 'method-man-instrument' correlation has to be correctly applied, since any failure of the system would result in unreliable analytical information being obtained.

In analysis of solid materials [17,18], the determination of trace elements requires knowledge of the exact manner in which these traces were introduced or pre-existed in the samples. The most complex problem occurs whenever the trace component to be determined has properties which are very similar to those of a major component of the sample to be analysed. To set up an efficient extraction process when using these techniques, it is necessary to resort once again to a knowledge of basic chemistry. For selecting the separation methods, besides knowing the history of the sample, a decision must be made about the choice of separation methods, a decision which will involve knowledge of the solvents and the ion-exchangers used, the kinetics of the processes and the nature and mechanism of the extraction equilibria and the ion-exchange. Once these problems have been solved, the analytical separation process can be set up.

The 'art' of the analytical chemist consists of knowing the history of the sample, and of choosing the simplest possible analytical procedure.

14.2 Sample homogeneity

The problem of the homogeneity of the sample is closely related to the problem of the history of the sample.

Whereas the literature reports many specific sampling situations, there are few papers which consider the fundamental aspects of the sampling process and its implications for the general analytical process.

The aspects which relate to the homogeneity of the sample have to be considered in the context of the nature of the analytical process. Also the nature of the analytical process is determined by the characteristics of the sample to be analysed. Hence the two groups of analytical methods, destructive and non-destructive, should be considered separately. Although both groups have wide applicability, the analytical chemist has a tendency to prefer the non-destructive methods. Since such methods act directly upon the sample, they have the advantage of partly – and with some precautions, totally – eliminating the risk of contamination of the sample.

The choice of one or other of the methods depends on the nature of the sample. Generally, the non-destructive methods are applied to samples with relatively simple composition. However, for more complex samples or when the determination of major and minor components is required, preliminary separation of the components and concentration of the minor or trace components are necessary, before the actual determination may be performed. It follows therefore, that the analytical chemist must resort in many cases, willingly or not, to destructive methods of analysis.

The homogeneity of the sample depends on the physical state of the material. Because of natural diffusion processes, liquid and gaseous samples are much more homogeneous than solid ones. Solid samples are often heterogeneous, and have first to be homogenized by mechanical means (grinding, ball-milling, etc.) before specimens are selected. The lack of homogeneity of solid samples is the main factor which renders their processing difficult. The problem of sample homogeneity will be discussed under two headings, depending on the type of analysis to be performed.

14.3 Destructive analysis

'Destroying' a sample means to bring it into a homogeneous form as a solution, normally in an aqueous or a partially non-aqueous medium. There are two means of bringing solid samples such as soils, sediments and sludges into solution, either by dissolving them, or by decomposing or disintegrating them in dry form by means of fluxes. These supplementary operations not only increase the duration of the analysis proper, but also introduce the risk of contamination of the samples by reagents and working techniques.

The dissolution agents used for soil and sediment samples are very diverse, and the analytical chemist must understand thoroughly the chemistry underlying the dissolution process when using a particular reagent. Although the most common dissolution agent is water, there are many situations where water may be unsatisfactory.

To dissolve certain solid samples, acids or mixtures of acids may be used. However, besides dissolving the sample, the acids may interfere with the subsequent analysis either by converting some components of the sample into extremely stable complexes or by creating volatile components which may be lost partly or even totally during the dissolution process. Hence, before dissolving a sample, it is necessary to acquire some knowledge of the nature of its components and their relative proportions, i.e. some knowledge of the origin of the sample, and its history.

Some wet dissolution/decomposition reagents such as hydrofluoric or hydrochloric acid may have strong completing action. Very often, the complex formed may prevent the determination proper from being performed, because it is kinetically or thermodynamically very stable. In many cases the dissolution/decomposition reagents are used to destroy an organic substrate. For example, the use of nitric–perchloric or nitric–sulphuric–perchloric acid mixtures is well known.

The art of the analytical chemist consists in choosing the most suitable dissolution/decomposition system for a given sample, so that the resulting solution contains the components in a form directly usable in the subsequent concentration and separation processes.

When the aim of the analytical chemist is to determine trace components after a wet decomposition with water or acids, care must be taken to use clean vessels and pure reagents in order to minimize contamination risks. Trace analysis presupposes an appropriate sampling procedure and dedicated high-purity reagents. The water to be used must be purified by ion-exchange and then distilled, and stored in polyethylene vessels. The acids and other reagents used for decomposing samples must be of suitable purity, e.g. so-called 'electronic' or 'semiconductor' grade.

An interesting example of the contamination risks which may be caused by a laboratory vessel is that of boron. Determination of very low boron concentrations, involves a prior separation by distillation and subsequent analysis by spectrometry, with a suitable reagent such a curcumin or carminic acid. The use of laboratory vessels made of borosilicate glass (such as Duran or Pyrex) could lead to very large errors in the boron content found. Such errors are caused by sample contamination from the boron present in the glassware.

For samples which are virtually insoluble in water or acids, either cold or hot, so-called dry decomposition may be used. This system is more tedious than wet decomposition since it involves two independent operations, the decomposition proper and the succeeding dissolution of the product. For dry decomposition various fluxes can be used, such as Na_2CO_3+K_2CO_3, Na_2CO_3+borax, or Na_2CO_3+S (Freiberger decomposition). To transfer the sample completely into solution, it must first be perfectly homogenized with the decomposition agents. These decomposition systems normally require the use of expensive laboratory vessels, for instance platinum crucibles. However, if one is unaware of the history of the sample, such vessels might be damaged or even destroyed (e.g. whenever the samples contain sulphur, phosphorus, arsenic, antimony, etc.).

The fact that 'classical' systems of dry decomposition still persist in today's analytical chemistry is due to the traditional thinking of the analytical chemist, who still believes that the most favourable agent for speeding up the process is temperature. For this reason, dry decomposition has now become the greatest drawback in sampling; on the one hand, it has led to the lengthening of the analysis time and, on the other, to increased contamination risks due to the decomposition agents used. Unlike acids, which can now be obtained in a high degree of purity, solid reagents are often of insufficient purity for trace analysis. It is this aspect of trace analysis which has led to the development of some non-contaminant decomposition systems. The simplest way of achieving faster (and non-contaminating) decomposition has been to resort to an additional physical parameter, namely pressure, coupled with an adequate decomposition temperature. As discussed later, the use of high-pressure decomposition vessels requires much lower temperatures for decomposing a sample, than those necessary for dry decomposition at atmospheric pressure. The appearance of the high-pressure decomposition vessels (bombs) is a direct result of the availability of a chemically inert plastic, namely Teflon. Teflon exhibits good thermal stability and offers minimal contamination risks. High-pressure decompositions involve the use of some decomposition agents which can be prepared easily in a high degree of purity (e.g. hydrochloric acid). The great advantage offered by these disintegration systems is that they make use of relatively cheap laboratory apparatus and avoid expensive materials such as platinum. These high-pressure decomposition systems have now become commonplace in the laboratory.

In analysing solid samples, regardless of the chosen decomposition system, a preliminary and extremely important step is granulometry, which plays a decisive role when preparing solid samples for chemical analysis. Although many studies have been written on granulometry, these studies could also be considered as part of the general sampling process. Wolfson and Belyaev [19] reviewed current work in this field and discussed the role and importance of granulometry for the general sampling process. This work underlined the necessity of granulometric control of the composition of a sample, during its preparation before chemical analysis. Vulfson and Belyaev [19] examined the modern methods of fine grinding and granulometric analysis and attention was given to problems of the influence of the granulometric composition of the dispersed substance on the chemical analysis results and sampling errors.

A great number of separation processes are based on solvent extraction, especially since this is also a concentration technique. For these reasons, solvent extraction will be considered, both from the point of view of the sampling process and from that of the general analytical process. Solvent extraction is ultimately a process of partitioning between two immiscible solvents, and for its optimization it is necessary to know first of all the operational parameters of the system.

The technique of solvent extraction has long been used in organic chemistry for concentrating and purifying some substances. In the case of organic compounds, the separation process is simple, in many cases being based only on differences in the solubility of the compounds in different solvents.

Many attempts at classifying solvent extraction systems have been made. Thus Diamond and Tuck [20] have described a classification of the solutes that can be separated by solvent extraction.

A number of conclusions may be drawn from this discussion of the destructive analysis of samples.

1 Destructive analysis will continue to be necessary, for many types of samples, owing to the nature of certain samples. The complexity of the composition of some samples often imposes use of a separation step prior to the analysis itself. Furthermore, the classical techniques

are usually the only ones suitable for validation of the major components of reference materials.

2 Destructive analysis is also necessary when there is a need to concentrate the components of a given sample, the components being present at very low concentrations (or in traces), often in a very complex matrix.

3 The techniques of destructive analysis involve taking some special precautions concerning the sampling in general, and the dissolution–decomposition process in particular, in order to avoid the risk of contamination.

4 The techniques of destructive analysis may lead to reliable analytical results only if the analytical process is correctly planned, by taking into account the interdependence of all the operational parameters of the process.

5 The concentration–separation process is a necessary step used in most destructive analyses. Regardless of the actual concentration–separation procedures used, this process belongs to the general sampling process. For this reason, terms such as chromatographic analysis, perpetuated by long routine use, should be replaced by terms such as 'chromatographic separation'. In the framework of the chromatographic separation systems, the detector would constitute an independent, well-defined entity of the chromatograph, the 'analyser' proper.

6 Owing to its complexity, in most cases destructive analysis, including the separation methods, resorts not only to the theoretical and basic knowledge of the analytical chemist, but also to the analyst's 'art' of optimizing the analytical process step-by-step, taking account of the whole series of factors which may interfere with the sampling process. In this connection much experimental work might be saved by using simplex optimization [41] in the exploratory research.

Considering these conclusions, it is apparent that, destructive analysis still has a place in the analysis of soils and sediments – and for this very reason, we should correlate the necessary knowledge so as to simplify as much as possible the analytical process. Such an action is necessary in order to shorten the analysis time.

14.4 Analysis of soils and sediments

The analytical chemist is frequently requested to determine trace elements in solid samples.

The sample when it arrives in the laboratory is usually in a form unsuitable for analysis, e.g. a river sediment or sewage sludge suspended in water. Table 14.1 shows steps that may then be required to convert the sample into a form suitable for analysis.

Table 14.1 Conversion of sample into form suitable for analysis

Sediments, sludges
Comminute sieve Particle-size measurement possibly Digest sample with acid

Source: Own files

Note if determinations of certain volatile elements such as mercury or selenium are required it is necessary to carry out these analyses on the wet sample as received (to avoid loss of element by drying at 105°C). The dry weight of material in the sample is obtained by determining moisture in a separate position of the sample and applying a correction to the sample weight used in metals determination.

As an alternative to drying at 105°C microwave drying has been used to remove moisture from aqueous slurries [21].

14.4.1 Comminution of samples

Various comminution devices (Table 14.2 A–E) are available for handling these types of samples.

Table 14.2 Laboratory homogenizers and comminution equipment supplied by Fritsch

Homogenizers		*Description*	*Sample type*
Comminution			
A	Vibrating cup mill	Pulverisette 9	A gate graining to 20μm of dry or wet sediments, fish, crustacea or plant material
B	Laboratory desk mill	Pulverisette 13	Grinding 0.1mm of dried sediments, soils, sewage sludge, hydrological sediments and drilling cores
C	Mortar grinder	Pulverisette 2	Grinding to 10μm of sediments and soils
D	Centrifugal mill	Pulverisette 6	Grinding to 1μm of sediments and soils
E	Planetary mill	Pulverisette 5	Grinding to 0.1μm of sediments and soils
F	Sieving devices Vibratory sieve Shaker for micro-precision sieving	Analysette 3	Dry and wet sieving (25mm to 20μm) or micro-precision sieving (5–100μm)
G	Rotary sieve shaker	Analysette 18	Satisfied ASTM E–11–190 BS 410 1969; AFNOR NFX 11–501 and DIN 53477/1 separation of coarse grain material

Source: Own files

Grinding elements are offered in various non-contaminating materials such as corundum (Al_2O_3), agate (SiO_2), or zirconium oxide (ZrO_2).

14.4.2 Sieving analysis of samples

Having comminuted the sample it may now be required to carry out a sieving analysis in order to obtain different size fractions for chemical analysis. Fritsch supply a range of devices for sieving analysers (Table 14.2).

A practical scheme for sieving a soil sample is discussed below.

14.4.2.1 Field or moist soil

Remove the sample from its container and either chop the soil with a knife or rub the soil through a 5.6mm mesh wire sieve. If an additive has been used to prevent nitrification during transit, gloves should be worn. Thoroughly mix the chopped or sieved sample and commence the analysis without delay.

14.4.2.2 Air-dried soil

Transfer the soil sample to a suitable metal tray to form a thin layer and, as far as possible, remove any stones present. With very heavy soils it is necessary to break any clods between the fingers. Dry the soil by placing the tray in a current of air at a temperature not exceeding 30°C. With large numbers of samples it is convenient to place the trays on a series of metal racks over which air may be blown from thermostatically controlled fan heaters. Continue the process until the soil feels quite dry. If the soil appears to contain moisture after grinding, return it to the drying rack.

14.4.3 Grinding of samples

Grind the air-dried soil until the whole of the sample, excluding stones, any fibrous material from roots etc. passes through a 2mm mesh sieve. There is a limited range of apparatus available for grinding soil but the Rukuhia-type soil grinding machine is suitable. (This is obtainable from D. Mackay, 85 East Road, Cambridge, CBI 1BY.) The apparatus consists of a number of cylinders into which the samples and metal pestles are placed. The cylinders, which have walls of 2mm mesh perforated steel, are rotated horizontally by means of electrically driven rollers. As the cylinders rotate, the soil is ground by the pestle and falls through the mesh into a tray below. When grinding samples containing soft rock, the action of the pestle should be cushioned by encasing it in a nylon or polythene tube. With very heavy soils it may be better to grind the sample while still slightly moist and to complete the air-drying after grinding.

14.4.4 Particle-size distribution measurement

A complete particle-size analysis can require the use of various analysis technologies. A microscopic examination may be performed before the sieve analysis, which in turn can be followed by a sedimentation analysis or the recording and the evaluation of a diffraction pattern.

The working ranges of the analysis methods overlap and can be subdivided as shown in Table 14.3, which also details equipment suppliers.

14.4.4.1 Sieving methods (5μm to 63mm)

Sieving methods have been discussed in the previous section.

14.4.4.2 Gravitational sedimentation, 0.5–500μm

An optical measuring system is used in sedimentation analysis, whereby a concentrated beam of light is deflected horizontally through the lower section of a measuring vessel onto a photoelectric cell. The amount of light absorbed by the sedimenting particles decreases with time as the

Table 14.3 Suppliers and working ranges of particle-size distribution methods

Method	*Particle-size range*	*Equipment supplier*	*Model*
Dry sieving	63um–63mm	Fritsch	Analysette 3, (20μm–25mm)
Wet sieving	20μm–200μm	Fritsch	Analysette 18
Microsieving	5μm–100μm	Fritsch	
Sedimentation in gravitational field	0.5μm–500μm	Fritsch	Analysette 20
Laser diffraction	0.1μm–1100μm	Fritsch	Analysette 22
Electrical zone sensing	0.4μm–1200μm	Coulter	Model ZM, Coulter multisizer
Electron microscopy	0.5μ m–100μm	–	–
Photocorrelation spectroscopy	0.5μm–5μm	–	–
Sedimentation in centrifugal field	0.5μm–10μm	Fritsch	Analysette 21 (Anderson Pipette centrifuge)
Diffraction spectroscopy	1μm–1mm		
Optical microscale	0.5μm–1mm		
Projection microscopy	0.05μm–1mm		
Image analysis systems	0.8μm–150μm down to 0.5μn	Joyce–Leebl Leitz, Karl Zeiss, Cambridge Instruments	Magiscan and Magiscan P Autoscope P Videoplan II Quantimet 520

Source: Own files

number of particles passing the measuring beam increases. The increase in the photoelectric current as a function of time is then a measure of the particle size.

A major step on the road to reducing the measuring time is provided by the 'Analysette 20' scanning photo sedimentograph (Table 14.2). Using this device, the measuring time is considerably reduced by a continuous movement of the light beam towards the direction of fall of the particle.

14.4.4.3 Centrifugal sedimentation (0.05–10µm)

The Andereasen pipette (Fritsch Analysette 21) is extremely well suited to this type of analysis. The measuring radius is determined by six rotating capillaries of equal length in a centrifuge drum. At certain predetermined times, samples are drawn from this radius using a pipette and the solid content of these samples mathematically evaluated to determine the particle-size distribution for the whole sample. The volume of material remaining in the centrifuge is reduced with each sampling and the distance between the surface of the sample liquid and the measuring plane is also reduced, thus reducing the sedimentation time for the smallest particles without the accuracy of the measurement being affected.

14.4.4.4 Laser diffraction (0.1–1100µm)

This is a universally applicable instrument for determining particle-size distributions of all kinds of solids which can be analysed either in suspension in a measuring cell or dry by feeding through a solid particle feeder. In the Fritsch Analysette 22 laser diffraction apparatus the measured particle-size distribution is displayed on the monitor in various forms, either as a frequency distribution, as a summary curve or in tabular form and can be subsequently recorded on a plotter, stored on hard disk or transferred to a central computer via an interface. The time required for one measurement is approximately 2min.

14.4.4.5 Electrical zone sensing (0.4–1200µm)

This is the classical method of carrying out particle-size analysis. Coulter supply two instruments – the Model ZM (video display optical) and the top-of-the-range multisizer – the latter having built-in video display of results.

The Coulter method of sizing and counting is based on measurable changes in electrical resistance produced by non-conductive particles suspended in an electrolyte.

By means of the Coulter channelizer 256 module an optional extra on the model ZM but built-in on the multisizer, enables biological cell-size distributions to be measured. This provides an ability to measure

suspension concentration and distributions of populations against size with a choice of 64-, 128- or 256-channel resolution over a range approximately 3:1 diameter. Size differences as small as 0.05μm^3 (fL) are detected.
A data management system is also available for the model ZM.

14.4.5 Digestion of solid samples preparatory to chemical analysis

Having as necessary, dried, homogenized or comminuted the samples, they must now be digested in a suitable reagent to extract elements in a suitable form for chemical analysis. In many organizations we have reached the point where the analyses pass from the hands of the person who took the sample to those of the analytical chemist. In the author's experience, however, it must be emphasized that to ensure best-quality results the whole procedure from, for example, statistically sampling a sediment to the final chemical analysis, should be handled by the same person.

14.4.5.1 Wet ashing

Digestion of the sample with hydrochloric acid, hydrofluoric and (if silicaceous material present) nitric acid and aqua regia have all been used. Aqua regia will dissolve most metals. Nitric acid provides an oxidizing attack for organic materials which are usually present at very high concentrations in soil, sediment and biological specimens. Perchloric acid is a very strong oxidizing agent, especially when used in conjunction with nitric acid, but its use is favoured by all chemists and certainly it must not be used in the pressure dissolution technique discussed below.

14.4.5.2 Fusion

Fusion with a flux such as sodium hydroxide, potassium bifluoride potassium pyrosulphate has been used extensively in the water industry.

14.4.5.3 Dry ashing

This is often used to remove organic material from the sample. The sample is weighed into a suitable container such as a ceramic or metal crucible, heated in a muffle furnace and the residue dissolved in an appropriate acid. It is not suitable for the analysis of volatile elements such as mercury and arsenic, since they may volatize during the ashing process. Magnesium nitrate has been used as an ashing agent to prevent volatilization or arsenic during dry ashing. Dry ashing has been used in the analysis of municipal waste [22,23].

14.4.5.4 Pressure dissolution

Pressure dissolution and digestion bombs have been used to dissolve samples for which wet digestion is unsuitable. In this technique the sample is placed in a pressure dissolution vessel with a suitable mixture of acids and the combination of temperature and pressure effects dissolution of the sample. This technique is particularly useful for the analysis of volatile elements which may be lost in an open digestion [24].

14.4.5.5 Microwave dissolution

More recently, microwave ovens have been used for sample dissolution. The sample is sealed in a Teflon bottle or a specially designed microwave digestion vessel with a mixture of suitable acids. The high-frequency microwave, temperature (ca. 100–250°C) and increased pressure have a role to play in the success of this technique. An added advantage is the significant reduction in sample dissolution time [25,26].

14.4.5.6 Equipment for sample digestions

Pressure dissolution acid digestion bombs

Inorganic and organic materials can be dissolved rapidly in Parr acid digestion bombs with Teflon liners and using strong mineral acids, usually nitric and/or aqua regia and, occasionally, hydrofluoric acid. Perchloric acid must not be used in these bombs due to the high risk of explosion.

Table 14.4 contains temperature and pressure data obtained while using microwave heating with a single closed vessel for two different acids. For nitric acid, 200°C (80°C over the atmospheric boiling point) and 7kg cm^{-2} was achieved in 12min and for hydrochloric acid 153°C (43°C over the atmospheric boiling point) and 7kg cm^{-2} was obtained in 5min.

At such elevated temperatures these and other acids become more corrosive. Materials that digest slowly or will not digest at the atmospheric boiling points of the acids become more soluble so dissolution times are greatly reduced. The aggressive digestion action

Table 14.4 Temperature pressure data for acids heated in a 120ml closed vessel

Acid (wt%)		*Temperature (°C)*	*Pressure (kg cm^{-2})*
HNO_3	70	200	8.5
HCl	37	153	8.5
HNO_3	70	193	7.1
HCl	37	30	7.1

Source: Own files

produced at the higher temperatures and pressures generated in these bombs result in remarkably short digestion times, with many materials requiring less than 1min to obtain a complete dissolution, i.e. considerably quicker than open-tube wet-ashing or acid-digestion procedures (Table 14.5).

Several manufacturers supply microwave ovens and digestion bombs (Tables 14.6 and 14.7). CFM Corporation state that their solid PTFE bombs are suitable for the digestion of soils and sediments.

14.4.5.7 Oxygen combustion bombs (Tables 14.6 and 14.7)

Combustion with oxygen in a sealed Parr bomb has been accepted for many years as a standard method for converting solid and liquid combustible samples into soluble forms for chemical analysis. It is a reliable method whose effectiveness stems from its ability to treat samples quickly and conveniently within a closed system without losing any of the sample or its combustion products. Sulphur compounds are converted to soluble forms and absorbed in a small amount of water placed in the bomb. Organic chlorine compounds are converted to hydrochloric acid or chlorides. Any mineral constituents remain as ash but other inorganic elements such as arsenic, boron, mercury, phosphorus and nitrogen and all of the halogens are recovered with the bomb washings. In recent years the list of applications has been expanded to include metals such as chromium, iron, nickel, manganese, beryllium, cadmium, copper, lead, vanadium and zinc by using a quartz liner to eliminate interference from trace amounts of heavy metals leached from the bomb walls and electrodes [27, 28].

14.4.6 Elemental analysis of sample digests

Once the sample is in solution in the acid and the digest made up to a standard volume the determination of metals is completed by standard procedures such as atomic absorption spectrometry or inductively coupled plasma optical emission spectrometry.

Table 14.5 Single-vessel dissolution of inorganic sample using $HF:HNO_3:H_2O$

Sample size	*Acid volume*	*Microwave digestion time*	*Hot-plate digestion time*	*Time saved*
1g	36ml [1]	1h	5h	4h

Source: Own files

[1] $HF:HNO_3:H_2O$, 1:1:1
HF–48wt %
HNO_3–70wt %

Table 14.6 Microwave oven digestion bombs

Supplier	*Oven part no.*	*Bomb part no.*	*Comments*
Acid digester types			
Parr Instruments	Not supplied	4781 4782	See Table 14.7 for metals determination in organic material
CEM Corporation	MD581D	Solid PTFE	See Table 14.7 for metals determination in organic material
Prolabo	Microdigest 300 Microdigest A300	Solid PTFE	See Table 14.7 for metals determination in organic material
Oxygen combustion types			
Parr instruments	Not supplied	1108	For sulphur, chlorine, etc. determination

Source: Own files

Table 14.7 Microwave digestion bombs supplied by Parr Instruments

Catalogue No.	*4781*	*4782*
Maximum charge of:		
inorganic sample	1.0	1.0
organic sample	0.1	0.2
Maximum temperature (°C)	250	250
Cup seal	Teflon-o-ring	Teflon-o-ring
Overpressure protection	Compressible relief disc	Compressible relief disc
Closure style	Band tighten	Band tighten
Bomb dimensions, cm:		
height overall	112	14.3
max. o.d.	7.8	7.8
Cup dimensions, cm:		
Inner diameter	3.1	3.1
Inner depth	3.0	6.1
weight, g	515	625

Source: Own files

If the sample matrix is complex, it may be necessary to determine if there are any interference effects from the matrix, on the analyte response. This is usually done by spiking the sample with a known amount of analyte. Two equal portions of sample are taken and an appropriate quantity of analyte is added to one to effectively double the absorbance. A similar quantity of analyte is added to water to make a 'spike-alone' solution. Readings are taken for sample, sample-plus-spike and spike-alone solutions and the amount of interference calculated as a percentage enhancement or suppression of the response. The interference can then be corrected or preferably removed by use of a separation technique.

It is advisable to include in the sample run standard materials of a type similar to the samples being examined. Standard biological materials and river sediments are available from the National Bureau of Standards USA.

Table 14.8 shows results obtained in the digestion in closed vessels of 1g samples of NBS SRM 1645 river sediment samples, digested (a) in 20ml of 1:1 nitric acid water and (b) in 5ml concentrated nitric acid and 3ml 30% hydrogen peroxide. In the former, at a power input of 450W, the temperature and pressure rose to 180°C and 7kg cm^{-2}. At that point, microwave power was reduced to maintain the temperature and pressure at those values for an additional 50min. In the latter case, 1g samples were open-vessel digested in 1:1 nitric acid:water for 10min at 180W. After cooling to room temperature, 5ml of concentrated nitric acid and 3ml of 30% hydrogen peroxide were added to each. The vessels were then sealed and power was applied for 15min at 180W followed by 15min at 300W power. As can be seen, the temperature rose to 11°C after the first 152°C at 2.8kg cm^{-2} after the final 15min of heating. With both reagent systems element recoveries are in good agreement with the certified values obtained using a hot plate total sample digestion technique which typically requires 4–6h.

Table 14.9 demonstrates the fact that in the case of sewage sludge the use of closed vessels in combination with microwave heating can speed up sludge digestion significantly. To demonstrate this, three sets of duplicate samples of the same standard EPA sludge sample were digested using different methods. The first set was microwave digested in closed vessels using 70% nitric acid and 30% hydrogen peroxide. The second set was also microwave digested in closed vessels but 1:1 nitric acid:water was used. The third set of samples was digested in glass beakers on a hot plate following EPA SW–846 procedures. The microwave digestions required 40min for the nitric acid:hydrogen peroxide dissolution and 60min for the 1:1 nitric acid:water dissolution. The hot-plate dissolution required 10h. Agreement on element recoveries among the three digestion procedures was very good for selenium and other metals and, except for arsenic, they agree well with the average sample reference values.

Table 14.8 SRM 1645 river sediment microwave digested in 1:1 HNO_3:H_2O

(a) in 1:1 HNO_3:H_2O $_2$ Element	Amount recovered (%)	*(b) in 5:3 HNO_3:H_2O_2* Amount recovered (%)	*Certified value (%)*
As	0.0060, 0.0060	0.0076 0,0070	0.0012 ± 0.00015
Se	0.0001, 0.0001	0.0001, 0.0001,	0.0001 0.172 ± 0.017

Source: Own file

Table 14.9 Element recovery for closed vessel microwave digestion versus open vessel hot plate digestion of EPA sludge sample (all values are μg g^{-1})

Digestion	*As*	*Se*
Microwave closed vessel		
10ml HNO_3	4	4
3ml H_2O_2	3	4
40min		
Microwave closed vessel	4	4
20ml 1:1 HNO_3:H_2O	4	4
60min		
Open vessel hot plate		
10ml HNO_3	3	4
4ml H_2O_2	3	4
10h		
EPA χ value	17.0[1]	not reported

Source: Own file

[1]Range of reported values was 0–89μg g^{-1}

Meriwether reported development of a coring sampler made from standard plumbing parts [29]. It was found especially useful for sampling and maintaining the depth profile of the soft sediments underlying relatively shallow waters. A rapid sampling method using passive sampling devices for soil contaminant characterization can provide a more thorough site assessment [30]. Analysis of the overall analytical error showed that, independent of the grain size fraction, sampling was the main source of variance for the determination of metals in river sediments [31]. Two approaches were described to determine the number of samples required for representative sampling. Hewitt found that volatile organic compounds are readily lost from soil samples unless care is taken to limit surface area exposure and to ensure subsample isolation [32]. Volatile organic carbon losses were found to be most abundant during field collection and storage. Hewitt reported that fortified soils held in sealed glass ampoules at 4°C, or dispersed in methanol and held at 22°C, showed no significant losses over 20 and 98 days, respectively [33].

Hunt [34] has described a simple method of filtering soil extracts that eliminates the need for filter-funnels and receivers. It therefore reduces the risk of contamination and speeds up the procedure. It also offers a convenient means of obtaining filtrates in the field for subsequent analysis.

After shaking the soil suspension in the extraction bottle, a tube of filter-paper folded about the centre to form a V with the open ends uppermost is inserted into the bottle. Clear filtrate collects inside the paper tube and aliquots are removed with a pipette.

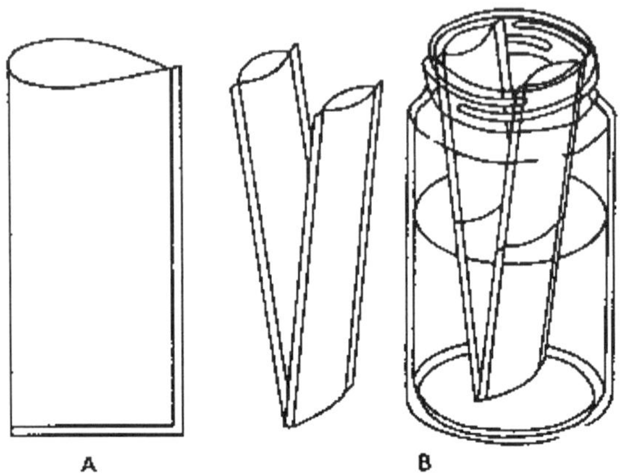

Fig. 14.1 Filtration apparatus
Source: Reproduced with permission from the Royal Society of Chemistry [34]

Fig. 14.1 shows two types of tube that were satisfactory. Type B is preferred because it is easier to produce. Type A is formed from a piece of filter-paper of dimensions 92 × 85mm with two edges glued together with clear impact adhesive. Type B is made from a piece of filter-paper of dimensions 200 × 60mm glued along the long edge with a 4mm overlap (shown folded). The adhesive did not produce any contamination in soil extracts.

Bates *et al.* [35] collected suspended particulate matter from river water and wastewater effluents using high speed continuous flow centrifugation, and analysed the isolated solids for hydrocarbons. The results were compared with those obtained on samples obtained by glass filter filtration. It was concluded that the use of a continuance flow centrifuge allows the concentration of organic associated with suspended particulate matter to be estimated more accurately.

14.5 Non-destructive analysis of solid samples

14.5.1 Introduction

There are methods which are capable of showing the distribution of elements on the surface of solid samples such as soils and sediments. As such they enable one to ascertain the homogeneity of distribution of elements on the surface of and, presumably, within the portion of sample analysed.

We have at our disposal a large number of methods for analysing solid materials without altering the sample in any way all of which enable us to characterize them qualitatively, quantitatively and sometimes structurally, by the direct action of a 'reagent' upon a previously prepared surface of the sample. Although there are a number of techniques which involve the destruction of at least a fraction of the sample, either because of the width of the 'reagent' beam used or the sensitivity of the determination (e.g. laser-source emission spectrometry or spark-source mass spectrometry). In the following discussion we will refer mainly to those techniques for analysis of solid materials which maintain the sample almost intact after impact of the 'reagent'. These are the so-called surface-analysis techniques or, more correctly, beam-analysis techniques. The general scheme of beam analysis is given in Fig. 14.2, from which it may be concluded that beam-analysis systems can be practically unlimited, and even multibeam systems, such as electron spectrometry for chemical analysis (ESCA) are available: these techniques provide various data concerning the sample analysed.

As mentioned already, many surface-analysis techniques are available nowadays. In the opinion of some specialists in this field [36,37], four of these are greater in importance: X-ray photoelectron spectrometry (ESCA), Auger electron spectrometry (AES), secondary-ion mass spectrometry (SIMS), and low-energy ion scattering spectrometry (ISS).

The importance of these surface-analysis techniques has resulted in the development of a range of highly automated instruments. In the effort to obtain multiple analytical data, a trend has occurred during the last ten years to build combined instruments, that is apparatus which will permit measurements by several techniques, in a single vacuum system. In this way, greater utilization of the complex instrumentation involved and a more economic use of the functional parameters of the instruments are ensured.

There is no such thing as a completely non-destructive analysis. Upon interaction of the beam with the sample, a series of surface-disruption phenomena can occur. This fact is illustrated in Table 14.10 for the four major surface analysis techniques mentioned above. The least surface disruption occurs in ESCA and the measurements are characteristic of the surface. However, when charged particles such as electrons and ions are

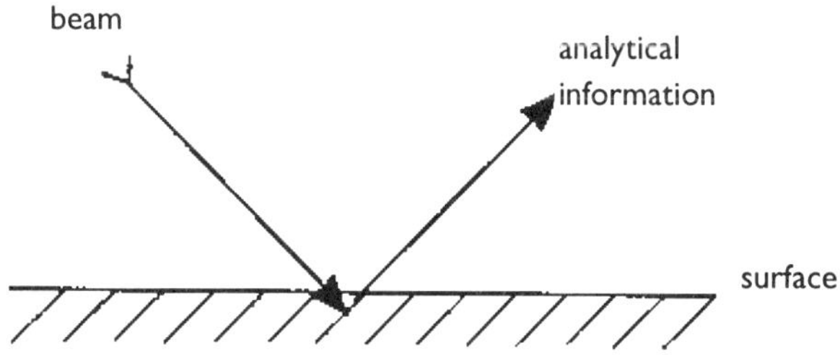

Fig. 14.2 Schematic view of beam analysis
Source: Own files

used, this is not necessarily true, as ions may cause considerable sputtering of any material. Although they induce less sputtering, electrons are chemically active and thus may cause chemical effects, which are usually most severe in insulators and least severe in conductors. ESCA measures electrons photo-ejected from a surface by soft X-rays. The technique is applicable to all elements with an atomic number greater than two. The information obtained by this technique is valuable – for instance we can gather data on oxidation states, structural effects, etc.

14.5.2 X-ray fluorescence spectroscopy

This technique is extremely useful for determining the surface concentration and distribution of elements e.g. chlorine, arsenic, etc. in soils and sediments.

X-ray fluorescence spectrometry was the first non-destructive technique for analysing surfaces and produced some remarkable results. The Water Research Association, UK, has been investigating the application of X-ray fluorescence spectroscopy to solid samples. Some advantages of non-destructive methods are no risk of loss of elements during sample handling operations, the absence of contamination from reagents, etc. and the avoidance of capital outlay on expensive instruments and highly trained staff.

A wide variety of X-ray fluorescence spectrometers may be used, depending on the nature and complexity of the sample, and on the number of samples to be analysed. To prove this and to indicate the substantial influence which the sample has on the choice of measuring instrument, let us consider some of the main characteristics of some X-ray fluorescence instruments used today [38]. These are shown in Table 14.11.

Table 14.10 Surface disruption caused by several surface analysis techniques

ESCA	*SIMS and ISS*
least destructive 95% of samples OK charging effects can be compensated	sputtering occurs in both for all materials sputtering rate depends on beam current desorption effects severe
AES more destructive than ESCA particularly bad for organic materials desorption and diffusion altered by electrons charging effects usually can be compensated all effects: insulators >> >> semiconductors > conductors	charging effects usually can be compensated

Source: Reproduced with permission from the American Chemical Society [36]

Table 14.11 Types of X-ray spectrometers used for X-ray fluorescence analysis

Category	*Type*	*Typical source*	*No. of elements measured*	*Speed*	*Cost (×$1000)*
Wavelength-dispersive	Single-channel	High-powered X-ray tube, 2–4kW	One at a time	Slow	60–100
	Multi-channel	High-powered X-ray tube, 3–4kW	20–30	Very fast	120–180
Energy-dispersive	Bremsstrahlung source	Low-powered (pulsed) X-ray tube, 0.5–1kW transmission target	In principle, all	Fast	30–70
	Secondary fluorescer	High-powered bremsstrahlung source, 2–4kW, used to generate monochromatic radiation from a selected fluorescer	Groups of 1–10	Relatively fast	60–90

Source: Reproduced with permission from the American Chemical Society [38]

14.5.3 Electron probe X-ray microanalysis

Even though X-ray fluorescence is now widely used to analyse a large variety of samples, it does have some drawbacks. For instance the X-ray beam used is wide, and this is of no great use for analysing tiny inclusions present in samples, and also does not allow point-by-point analysis on surfaces ('scanning analysis'). The first electron probe X-ray primary emission spectrometer was built in 1949. No doubt this encouraged the use of surface analyses, by allowing samples of very small dimensions to be studied. This was possible because the electron beam had a diameter of only about 1μm. Although the small size of this beam permitted the analysis of some micro-inclusions in samples, and also multiple analyses by scanning, the main problem, which still remains unsolved, is that of the microhomogeneity and microtopography of the samples. Thus, whereas polishing the solid samples with a 30–100μm grade abrasive is usually satisfactory for X-ray fluorescence spectrometry, a 0.25μm grade abrasive or finer may be required for electron-probe microanalysis.

In principle, the difference between X-ray fluorescence spectrometry and electron-probe microanalysis lies in the fact that the analytical information is provided, in the first case, by secondary, fluorescence X-rays, and in the second by primary X-rays, emitted as a result of the impact of the electron beam on the sample's electrons.

Owing to the small size of the electron beam on the one hand, and to the high sensitivity of the method on the other (a sensitivity which can go down to detection of 10^{-16}g), electron-probe microanalysis has found applications in many fields.

Some of the disadvantages of the electron-probe method may be overcome, as in other methods, by the use of complementary techniques. Such techniques can complete the results obtained by electron microprobe. For instance, the introduction of a proton microprobe [39], which is much more sensitive (by two orders of magnitude) than the electron microprobe, and may be used with very good results in geochemical and cosmochemical studies.

14.5.4 Auger electron spectrometry

Auger electron spectrometry (AES), reported by Auger in 1923 [40], is also a valuable technique for analysing surfaces. The technique is somewhat similar to ESCA, measuring electrons emitted from a surface as a result of electron bombardment. In both cases, the sampling depth is ca. 20A. Coupling this technique with scanning electron microscopy (SEM) produced a tandem (AES–SEM) technique which has proved extremely productive.

14.5.5 Secondary ion mass spectrometry

In secondary ion mass spectrometry (SIMS), a primary ion beam bombards the surface and a mass spectrometer analyses the ions sputtered from the surface by the primary bombardment. This extremely sensitive technique provides both elemental and structural information.

14.5.6 Ion scattering spectrometry

Ion scattering spectrometry (ISS) is also a technique which is sensitive for all elements with an atomic number greater than 2, and measures the energy change of the bombarding ions, caused by elastic collisions with surface atoms. Like SIMS, it has limited spatial capabilities.

References

1 Smith, R. and James, G.V. (1981) *The Sampling of Bulk Materials*, The Royal Society of Chemistry, London.
2 Kratochvil, B., Wallace, D. and Taylor, J.K. (1984) *Analytical Chemistry*, **56**, 113R.
3 Kratochvil, B. and Taylor, J.K. (1982) *A Survey of Recent Literature on Sampling for Chemical Analysis*, NBS. Technical Note 1153. US Department of Commerce, Washington, DC, January.
4 Gy, P.M. (1979) *Sampling of Particulate Mixtures: Theory and Practice*, Elsevier, New York.
5 Woodget, B.W. and Cooper, D. (1987) *Samples and Standards*, Wiley, Basingstoke.
6 Harrison. T.S. (1979) *Handbook of Control of Iron and Steel Production*, Harward, Chichester.
7 Walton, W.W. and Hoffman, J.I. (1969) in *Treatise on Analytical Chemistry*, Part I, vol. 1, 1st edn (eds I.M. Kolthoff and P.J. Elving), Wiley-Interscience, New York, pp.67–97.

8 Laitnen, H.A. (1960) *Chemical Analyses*, 1st edn. McGraw-Hill, New York.
9 Laitnen H.A. and Harris, W.E. (1975) *Chemical Analysis*, 2nd edn., McGraw-Hill, New York.
10 Ingamells, C.O. and Pitard, F.F. (1986) *Applied Geochemical Analysis*, Wiley-Interscience, New York.
11 Kratochvil, B. and Taylor, J.K. (1981) *Analytical Chemistry*, **58**, 924A.
12 Kratochvil, B. (1989) *Samples for Microanalysis: Theories and Strategies*. Paper presented at the 11th International Symposium of Microchemical Techniques, Wiesbaden, 28th August–1st September 1989.
13 Wallace, D. and Kratochvil, B. (1987) *Analytical Chemistry*, 59, 226.
14 Ministry of Agriculture, Fisheries and Food. (1979) *The Analysis of Agricultural Materials* R.B. 427, HMSO, London.
15 HMSO (1977) *Sampling and Initial Preparation of Sewage and Waterworks Sludges, Soils, Sediments and Plant Materials*, London.
16 British Standards Institution (1971) *Sampling Procedures*, London.
17 Veillon, C. (1986) *Analytical Chemistry*, **58**, 851A.
18 Versieck, J., Barbier, F., Gornelis, R. and Hoste, J. (1982) *Talanta*, **29**, 973.
19 Wulfson, E.K. and Belyaev, Yu I. (1985) *Zhur Analit Khim*, **40**, 1364.
20 Diamond, R.M. and Tuck, D.G. (1960) in *Progress in Inorganic Chemistry*, vol. 2. (ed. F.A. Cotton), Wiley-Interscience, New York, pp.109–192.
21 Kuchn, D.G., Brandvig, R.L., Lunden, D.C. and Jefferson, R.H. (1986) *International Laboratory*, 82, September.
22 Haynes, W. (1978) *Perkin–Elmer Atomic Absorption Newsletter*, **17**, 49.
23 Dalton, E.F. and Melanoski, A.J. (1969) *Journal of Association of Official Analytical Chemists*, **52**, 1035.
24 Adrian, W.A. (1971) *Perkin–Elmer Atomic Absorption Newsletter*, **10**, 96.
25 Reverz, R. and Hasty, E. (1987) *Recovery study using an elevated pressure temperature microwave dissolution technique*. Paper presented at the Pittsberg Conference and Exposition on Analytical Chemistry and Applied Spectroscopy, March 1987.
26 Nadkarni, R.A. (1984) *Analytical Chemistry*, **56**, 2233.
27 Nadkarni, R.A. (1981) *American Laboratory*, **13**, 2 August.
28 Parr Manual (1974) 207M Parr Instrument Co., 211 53 Rd St. Moine, Illinois 61265.
29 Meriwether, J.R., Shew, W.J., Hardway, C. and Beck, J.N. (1996) *Microchem. Journal*, **53**, 201.
30 Johnson, K.A., Naddy, R.B. and Weisskopf, C.P. (1995) *Toxicology and Environmental Chemistry*, **51**, 31.
31 Truckenbrodt, D. and Einax, J. (1995) *Fresenius Journal of Analytical Chemistry*, **352**, 437.
32 Hewitt, A.D. (1996) *ASTM Special Publication*, **1261**, 170. (Volatile Organic Compounds in the Environment).
33 Hewitt, A.D. (1996) *Special Technical Publication*, **1261**, 181. (Volatile Organic Compounds in the Environment).
34 Hunt, J. (1981) *Analyst (London)*, **106**, 374.
35 Bates, T.S., Hamilton, S.E. and Cline, J.D. (1983) *Estuarine, Coastal and Shelf Science*, **16**, 107.
36 Hercules, D.M. (1978) *Analytical Chemistry*, **50**, 734A.
37 Hercules, D.M. (1986) *Analytical Chemistry*, **58**, 1177A.
38 Jenkins, R (1984) *Analytical Chemistry*, **56**, 1099A.
39 Bosch, F., El Goresy, A. Martin B. *et al.* (1978) *Science*, *199*, 765.
40 Auger, P. (1923) *Compt. Rend.*, **177**, 169.
41 Betteridge, D., Wade, A.P. and Howard, A.G. (1985) *Talanta*, **32**, 709, 723.

Chapter 15

Accumulation processes in sediments

15.1 Introduction

It has been observed that sediments in rivers and the oceans have the property of adsorbing some types of dissolved substances in the overlying water so that the concentration in the sediment (in $\mu g\ kg^{-1}$) is appreciably greater than that in the water (in $\mu g\ L^{-1}$) with which the solid is in contact.

A convenient method of expressing this phenomena is by calculating a concentration factor (C) expressed by:

$$C = \frac{\text{Concentration of substance in sediment } (\mu g\ kg^{-1})}{\text{Concentration of substance in water } (\mu g\ L^{-1})}$$

The concentration of toxicants present in sediments is a measure of its concentration in the water over a period of time and is therefore a measure of the risk to creatures. In the case of bottom feeding creatures there is the additional risk of direct ingestion of contaminated sediments in the gills and mouth with consequent adverse effects.

Monitoring of bioaccumulation of fresh and tidal waters as trends in spacial monitoring has two purposes:

1 Macroscale, i.e. the identification of potentially unknown areas of elevated contamination and assessment of the extents of the zone of contamination.
2 Monitoring of bioaccumulation in fresh and tidal waters as trends in time. These need to be monitored to identify trends in contamination, especially near effluent discharges, in order to identify stability, improvement or deterioration in contamination levels.

Spatial and time monitoring programmes of the type discussed above will also give information needed to assess the risk to top predators including man in a particular ecosystem.

In this respect it is also interesting to note that in bioaccumulation it is found that the concentration of toxicant in the tissues and particularly in some of its organs such as kidney, liver and opercle, increases with both concentration in the water and exposure time. Measurement of toxicant levels in tissues or organs provide an indication of the amount of exposure to toxicants that the creature has suffered over a period of time. Only limited information is yet available relating concentrations of toxicant in tissues and the onset of ill health or mortality. This is clearly an area where much further work remains to be done.

15.2 Accumulation of organic compounds

Stroll and Giger [1] have recently studied the bioaccumulation of three optical whitening agents in lake sediments obtained in Lake Griefensee, Switzerland. The compounds studied were:

- DAS.1 4′4′-bis[(4 anilino-6-morpholino-1,3,5 triazin-2yl)amino] stilbenzene-2, 21 dilsulphonate
- DSBP 4,4′-bis (2-sulphostyryl)biphenyl
- BLS 4,4′-bis (4-chloro-3-sulphostyryl) biphenyl

Analysis of the lake sediments was performed by reverse phase high-performance liquid chromatography.

The concentrations of these substances found in the sediments and the overlying water were as follows.

	DAS1 and DSBP			BLS		
	max	min	mean	max	min	mean
sediment (μg kg $^{-1}$)	1400	400	900	80	20	50
water (μg L $^{-1}$)	0.098	0.012	0.055	<0.0002	<0.0002	<0.0002
bioaccumulation factor	117,000	4082	16,353	>400,000	>100,000	>250,000

Thus it can be seen that the mean concentration of DAS1 plus DSBP (μg kg $^{-1}$) in sediments is some 16 000 times greater than it is in the overlying water layer. In the case of BLS this factor exceeds 250000.

Similar large factors have been observed in the case of chlorinated insecticides in river waters i.e. bioaccumulation factors of the order of 104.

15.3 Accumulation of organometallic compounds

Observed concentration factors for a range of organometallic compounds and the corresponding metal ions in different types of water are tabulated in Table 15.1. Where the concentration factor is appreciably greater than unity the dissolved phase shows a tendency to be adsorbed by the

Table 15.1 Concentration factors for organometallics and inorganic ions between sediments and liquid phases in water

	in water ($\mu g\ L^{-1}$)	in sediment ($\mu g\ kg^{-1}$)		Factor = Sediment ($\mu g\ kg^{-1}$) / Water ($\mu g\ L^{-1}$) Minimum	Maximum	Mean
Alkylmercury	0.06	<0.01	Coastal	–	–	<0.17
	0.006–1.15 (mean 0.57)	<0.01	Rivers	<0.0086	<1.66	<0.02
Inorganic mercury	0.009–13.0 (mean 6.5)	910–46,800 (mean 23 850)	River	3600	101 110	3669
Alkyl lead	<0.00001	<0.01	Coastal	–	–	1000
	50–530 (mean 290)	<0.01	Rivers	0.000019	0.0002	<0.00003
Inorganic lead	0.02–200 (mean 100)	23 000–38 200 (mean 30,600)	Coastal	191	1.15×10^6	306
	0.13–60 (mean 30)	110–506 000 (mean 253 000)	River	846	8430	8433
Monobutyltin	0.035–0.050 (mean 0.042)	280	River	5600	8000	6660
	<0.0001–0.3 (mean 0.15)	1–8 (mean 4.5)	Sea	26.6	>10 000	30
Dibutyltin	0.010–0.040 (mean 0.025)	140	River	3500	14 000	5600
	<0.001–1.6 (mean 0.8)	0.1–1.0 (mean 0.55)	Sea	0.62	>100	0.69
Tributyltin	0.005–0.015 (mean 0.010)	55	River	3667	11,000	5500
	0.06–0.78 (mean 0.42)	0.01–0.3 (mean 0.15)	Sea	0.17	0.38	0.36

Inorganic tin	<0.0001	1000–20,000 (mean 10,500)	Coastal	>10^7	>2×10^8	>1.05×10^8
Alkylarsenic	2.5–2.6 (mean 2.55)	<0.01	Coastal	<0.0038	<0.004	<0.0039
Inorganic arsenic	1.00–1.04 (mean 1.02)	1600–117 000 (mean 59,300)	Coastal	1600	112 500	58 140
	0.42–490 (mean 245)	220–28,000 (mean 14,100)	River	57.1	523	57.6
Inorganic copper	0.069–9.7 (mean 4.85)	5400–84 800 (mean 45,100)	Coastal	8742	78 260	9298
	0.11–200 (mean 100)	70–244 000 (mean 122,000)	River	636	1220	1220
Inorganic nickel	0.2–15.0 (mean 7.6)	30 000–57 000 (mean 43,500)	Coastal	3800	1.5×10^5	5723
	1.5–4.5 (mean 3.0)	1000–238 000 (mean 119,500)	River	666	52 888	39 883
Inorganic selenium	<0.01–0.08 (mean 0.04)	1500–9000 (mean 5250)	Coastal	112 500	150 000	131 250
Inorganic antimony	0.30–0.32 (mean 0.56)	6200–134 000 (mean 9800)	Coastal	20 ,666	163 414	17 500
	0.08–0.42 (mean 0.25)	10–2900 (mean 1455)	River	125	6905	5820
Inorganic manganese	0.35–250 (mean 125)	21 800–750 000 (mean 386,000)	Coastal	3000	52 285	3088

Source: Own file

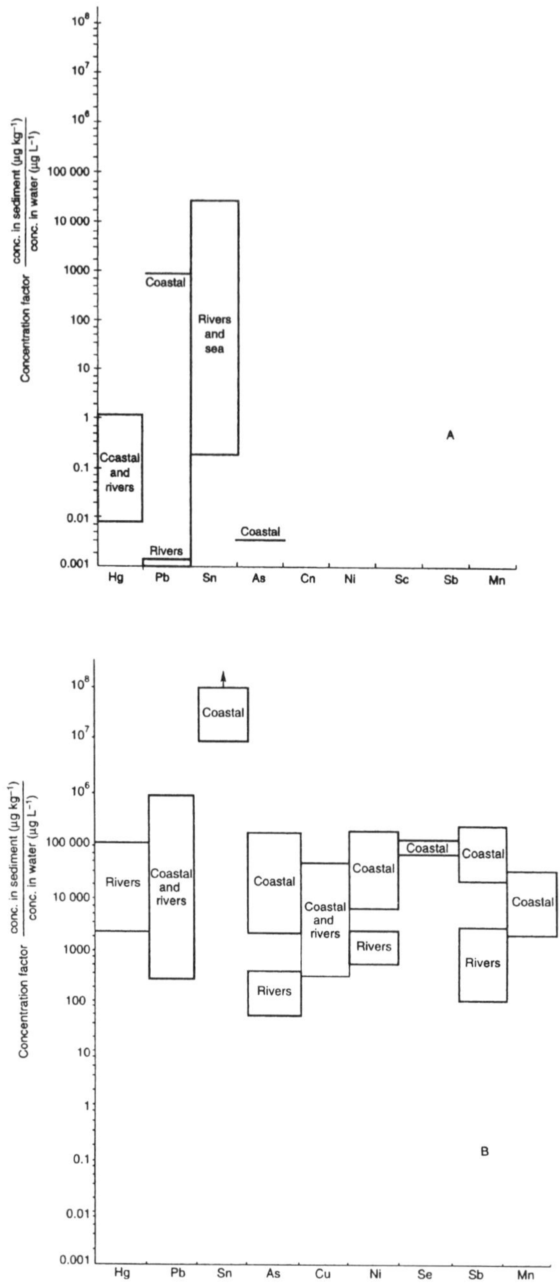

Fig. 15.1 Concentration factor of organometallic compounds and corresponding inorganic metal ions in various matrices. A: Organometallic compounds. B: Inorganic metal ions
Source: Own files

sediment. Examination of the data in Table 15.1 allows the following conclusions to be drawn, see also Fig. 15.1.

1 All the inorganic metal ions listed are strongly adsorbed on to sediments.
2 Organotin compounds are strongly adsorbed on to river water sediments.

The data suggests that organotin compounds may not be as strongly adsorbed on to sediment in saline water, i.e. coastal and seawaters as they are in non-saline waters.

3 Organocompounds of mercury and arsenic are not adsorbed on to sediments in either non-saline or saline waters.
4 The concentration factors for organolead compounds range from very low values, i.e. no concentration in sediments, to 100, i.e. some adsorption.

15.4 Accumulation of metalloids

Lee [2] using a hydride generation atomic absorption spectrometric method has investigated the bioaccumulation of bismuth on marine sediment samples collected in Narragensett Bay and the North Pacific Ocean.

The concentrations of bismuth found in the sediment samples and in the overlying sea water are tabulated below:

	Narragensett Bay			North Pacific		
	Max	Min	Mean	Max	Min	Mean
Sediment μg kg $^{-1}$	640	270	455	120	100	110
Water μg L $^{-1}$	0.00063	0.000083	0.000356	0.00005	0.00005	0.00005
Bioaccumulation factor	7.71×10^6	4.28×10^5	1.28×10^6	2.4×10^6	2.0×10^6	2.2×10^6

Thus it is seen that the concentrations of bismuth in the sediment (μg kg $^{-1}$) are some one to two million times greater than they are in the overlying water (in μg L $^{-1}$).

References

1 Stroll, J.M.A. and Giger, W. (1997) *Analytical Chemistry*, **69**, 2594.
2 Lee, D.S. (1982) *Analytical Chemistry*, **54**, 1682.

Chapter 16

Disposal of wastes to land

16.1 Introduction

In all developed countries large quantities of agricultural, industrial and municipal wastes are generated. As authorities move to protect the environment by regulating waste disposal practices, environmentally sound methods of waste disposal are being sought. In particular, land application of wastes as a means of disposal also nutrient recycling and water conservation are becoming increasingly popular [1].

Before discussing the question of disposal of wastes to land, other means of waste disposal currently in use will be briefly discussed below.

At present land filling and disposal to water are the most common means of disposing municipal and industrial wastes. Other methods which are growing in importance are incineration and discharge to land as discussed below.

In Australia, about 96% of the waste (>12mt) is disposed of by landfilling. In the US and most OECD countries in Europe, about 60–70% of the municipal and industrial wastes are disposed of in this way [2].

16.2 Disposal of waste by landfilling

Many old landfills were located on permeable ground, where leachates were allowed to percolate and attenuate through the porous material. Table 16.1 shows that the leachates percolating out of landfill sites contain significant concentrations of nutrients, heavy metals and organics; this uncontrolled means of landfilling has resulted in the contamination of aquifers at many landfill sites around the world [45].

In particular it is interesting to note the levels of phenols, organochlorine and organophosphorus pesticides and polychlorobiphenyls present in leachates. These leachates are likely to percolate through the layer and enter the groundwater. Leachates need to be controlled (e.g. by capping landfills with low permeability material). Organics have been detected in leachates from landfill sites or groundwater beneath [26–29].

A major attraction of landfill disposal of wastes has been its low cost. However, the increasingly stringent regulatory requirements have

Table 16.1 Typical compositions of leachates (g/m^3) from landfill sites

Species	*UK*	*NZ*	*Species*	*UK*	*NZ*
pH	8–8.5	6–8	Iron	10	50–400
COD	850–1350	500–5000	Nickel	0.04	0.1–1
BOD	80–250	100–4000	Copper	0.9	0.05–0.5
Volatile acids	20	20–200	Zinc	0.1	1–100
Ammonium N	200–600	100–1000	Cadmium	0.02	0.01–6
Chloride	3400	200–2000	Lead	0.1	0.05–0.5
Sulphate	340	200	Monocyclic phenols	0.01	n.d.
Sodium	2185	200–2000	Total cyanide	0.01	n.d.
Potassium	888	100–1000	Organochlorine pesticides	0.01	n.d.
Magnesium	214	50–400			
Calcium	88	100–1000	Organophosphorus pesticides	0.05	n.d.
Chromium	0.05	0.05–0.5			
Manganese	0.5	5–50	PCBs	0.05	n.d.

Source: Own files

resulted in sharp increases in the cost of landfilling in recent years. A serious problem facing many cities is that existing landfill sites are near saturation, and it is increasingly difficult to initiate new landfills due to community opposition and lack of suitable sites. Waste prevention, minimisation, recovery and recycling are strongly encouraged. New technologies are sought for the disposal of wastes.

16.3 Disposal of waste by incineration

Incineration is a controlled flame combustion process for the decomposition of wastes. It can effectively reduce the waste volume by up to 90%. The process is particularly suitable for organic and other combustible wastes with a high energy content. Energy generated by incineration can be harvested for generating electricity. The process is not suitable for non-combustible wastes, including metals, or for stable compounds such as polychlorinated biphenols and dioxins. In Australia, about 143 000t of wastes are incinerated annually, accounting for about 1% of total waste disposal. In Japan and Switzerland more than 75% of municipal waste is incinerated [2,3], and in the US the figure is about 16%. Although incineration is a versatile method for the destruction of many types of wastes, the process itself also creates environmental problems. The issues of concern relate mainly to the emissions of pollutants to the atmosphere. Pollutants emitted to the atmosphere include heavy metals, acidic gases (e.g. hydrogen sulphide, sulphur dioxide), and trace organic (e.g. polychlorinated dibenzo-*p*-dioxins) and polychlorinated dibenzofurans (PCDFs).

16.4 Disposal of waste to the oceans

Discharge of sewage waters, sludges and other wastes (e.g. dredged spoils, hazardous wastes) into the marine environment (rivers, lakes and sea) is practised in many countries [2]. However, even after secondary treatment the treated effluents still retain high concentrations of organic matter, suspended solids, nutrients (particularly N and P), and other contaminants. It is estimated that Australia produces about 100 000t of N and 10 000t of P in sewage effluent annually and much of this is discharged to coastal waters. In low river-flow conditions, sewage effluent may be the major source of nutrients for many rivers, leading to eutrophication and depletion of dissolved oxygen, chemical toxicity and salinity. Organic compounds such as PCBs, DDT, polycyclic aromatic hydrocarbons and organochlorine pesticides and heavy metals such as mercury, have all been found at elevated concentrations in marine organisms, particularly near effluent discharge sites [5–7].

World, regional and national organisations have imposed, or are imposing, increasingly strict regulations on the discharge of wastes to the sea. The 1972 London Dumping Convention specifies the ban of sea dumping of certain hazardous wastes unless it is proven that the hazardous substance is in trace amounts and would be made harmless in the sea [2]. In the EU, discharge of untreated sewage to the sea will be phased out in the next few years; except in special circumstances, the sewage is required to be secondary-treated, and in areas sensitive to eutrophication, the sewage is required to be tertiary-treated, and wherever possible the sewage should be reused.

16.5 Disposal of waste to land

The management of waste application on land is a challenging task and requires rigorous scientific input. Sludges and effluents contain significant concentrations of plant nutrients, particularly nitrogen, phosphorus and organic matter. Their application on land has been shown, in many cases, to result in significant increases in plant yields and improvements in soil physical conditions and chemical fertility. The constraints with some wastes, particularly those of industrial and municipal origin, are that they contain undesirable constituents, e.g. heavy metals, toxic organic, pathogens and salts, or have extremely high or low pH. High concentrations of nitrate and phosphate derived from wastes are also of concern for ground and surface water contamination. The processes that control the fate of wastes in the soil are complex and many of them are poorly understood, e.g. rate of release of nutrients and other chemicals; leaching of nutrients, metals and organics through macropores and as suspended solids; emission of greenhouse gases; impact of solvents, surfactants and sludge organic matter on the sorption, degradation, and leaching of hydrophobic organics; and the long-term bioavailability and fate of metals and organic fixed by soil organic matter.

Land application of wastes is becoming more widespread as regulatory authorities move to protect water quality by restricting waste disposal into rivers, lakes and the marine environment. It is not clear, however, that soil is in fact an appropriate dumping ground for all our wastes.

The pressure to dispose of wastes onto land rather than into water often results in engineers being forced to design land treatment systems with little rigorous scientific information to guide them. One of the main problems is that there is such a wide range of waste materials with different physical, chemical and biological characteristics that it is inappropriate and indeed risky to transfer guidelines from one waste disposal system to another.

In addition to the phasing out of waste disposal into waterways and the ocean, the renewed interest in land application of wastes in the past 20 years is partly to use them because of the need to conserve water and nutrient resources and to use them efficiently and because of the high costs of incineration and landfilling.

In many parts of the world, land application of organic waste is not only an economic imperative but also a management necessity in order to stem the degradation and erosion of soil.

Although the nutrient content of wastes makes them attractive as fertilisers, the application of many industrial wastes and sewage is constrained by the presence of heavy metals, hazardous organic chemicals, salts and extreme pH values.

An example of the type of wastes that might be applied to land is given in Table 16.2.

Most agricultural wastes contain valuable nutrients that could be recycled back onto the land in order to improve soil fertility and increase the sustainability of farming systems. For example, in South Australia there are over 400 000 pigs that produce about 2400ML of waste annually and this waste contains enough nitrogen to fertilise about 200 000ha of wheat or barley.

The composition of industrial wastes varies depending on the industrial structure of a country or region. It consists of general rubbish, packaging, food wastes, acids, alkalis, oils, solvents, resins, paints, mine spoils and sludges. A proportion of the industrial waste is classified as hazardous waste because it contains materials that are presently or potentially hazardous to humans and other living organisms.

Of particular interest to land application are waste waters from factories. Most factory waste waters are treated before disposal; however, many nutrients, metals, organic chemicals remain in significant concentrations in the treated sludges and effluents. While the nutrients contained in these wastes, e.g. N and P, make wastes attractive as fertilisers, their application on land may be constrained by the presence of toxic metals, toxic organics,

Table 16.2 Composition of sludges or effluents from a few selected waste sources

Units are g/m^3 except for pulp and paper sludges (mg/kg). Dashes, not determined

Parameter	*Milk powder/ butter factory wastewater*	*Meat processing secondary effluent*	*Tannery secondary effluent*[A]	*Pulp and paper secondary sludges*[B]	*Dairy shed effluent*	*Piggery effluent*
Suspended solids	–	20–100	120	–	–	–
BOD5	1500	20–100	30	–	–	–
COD	–	80–400	410	–	–	–
pH	10–12	–	7.6	–	–	–
Total N	70	40–200	130	32 000	190	1300
Total P	35	5–30	1.6	8075	30	600
Fat	400	0–30	–	–	–	–
Na	560	50–250	2700	4586	50	–
K	13	20–150	–	2905	220	500
Ca	8	3–250	340	17 000	110	–
Mg	1	3–10	36	2000	30	–

Source: Own files

A Tannery secondary effluent contains chromium (1), chloride (3430) and sulphate (2410g/m^2)

B Pulp and paper secondary sludges also contain iron (2842), manganese (483), aluminium (18,000), tin (15), arsenic 0.17), cadmium (4.5), chromium (20), copper (206), lead (42), nickel (35), mercury (0.3) and zinc (513mg/kg)

excessive concentrations of salt and extreme pH. For example, waste waters from dairy, tannery and pulp factories contain high concentrations of sodium ions. Tannery waste waters contain undesirable constituents, e.g. chromium, aluminium, polyphenolics and aldehydes. Wastes from pulp and paper mills contain metals and a range of toxic organic compounds (Table 16.2).

Sewage sludge contains significant concentrations of nutrients, particularly N and phosphorus and is recognised as having considerable potential as a fertiliser material and soil conditioner. Concentrations of nutrients vary considerably between sludges. Sommers [4], reported ranges of <0.1–17–6% for total N, <0.1–14.3% for total P, and 0.02–2.64% for total potassium.

Unfortunately, in addition to nutrients, sewage sludge can also contain a range of potentially toxic contaminants including heavy metals, pesticide residues, other organic compounds such as polychlorinated biphenyls, together with a range of pathogenic organisms. Actual sludge composition varies with time, and between treatment plants, depending on the types of sewage and wastewater received and the nature of the treatment process. Many urban sewage treatment works receive a mixture of domestic and industrial sewage as well as urban run-off from roads and other sealed surfaces. This makes prediction of sewage sludge composition extremely difficult. Composition of sludge from rural areas is likely to be less variable and less contaminated with sundry contaminants than urban sludges.

There are many reports of the beneficial effects on plant growth of applying organic wastes to land. Brechin and McDonald [5] demonstrated that pig slurry was as effective in increasing barley yield as conventional inorganic fertiliser. The major benefit appeared to be the increase in N concentration of the barley plant.

Land application of sewage effluent has also been reported to cause significant pasture yield increases. Quin and Woods [6] showed that pasture response to the applied nutrients (equivalent to 116kg N, 34kg P and 68kg S/ha-year) was greatest in summer and autumn.

Poultry manure has frequently been found to increase yields of pastures and crops and is a valued organic fertiliser used by horticulturalists and dairy farmers. However, when this has been applied at high rates (more than about 18t/ha), there are reports of damage to crops and pasture.

Long-term field trials in England have shown that N efficiency of cattle slurry applications declined once the application rate exceeded pasture plant requirements [7]. The cattle slurry was reported to be 90% as efficient as ammonium nitrate fertiliser when applied at 300kg N/ha compared with 63% efficient at 600kg N/ha and that the efficiency varied with season of application, being most effective in spring.

Although the residual effects of repeated applications of solid manure are known to persist for many years, the residual effects of slurry application are generally low [7,8]. Changes in plant species composition and plant density, however, can occur with continued waste application. Therefore, land application effects need to be carefully monitored to ensure that the plant-soil system is not damaged.

Cameron *et al.* [1] have reviewed the effects of land application of wastes on soils and discuss particularly organic compounds carbon, nitrogen, phosphorus, sodium, heavy metals, soil cation exchange properties, microbial populations, pathogens, earthworms, effect on plants, leaching of metals and organics. Of particular interest here are their reported values [9–13] for the concentrations of organic compounds in sewage sludges applied to land (Table 16.3).

Application of sludge on agricultural soils could result in these organics entering the food chain. As many of these compounds are toxic to humans or animals, their presence could be a constraint for the use of sludges as fertilisers. Some food companies have set soil limits above which crops grown on such contaminated soils are rejected [11]: Aldrin/Dieldrin, 0.1; DDT, 0.75; and Diuron, 0.3mg/kg soil.

Table 16.3 Typical concentrations (mg/kg) of selected organic found in sewage sludges

Compound	*Concn*	*Compound*	*Concn*
Pesticides		PCB[B]	<1
Aldrin	0.08	Halogenated aliphatics	
Dieldrin	0.1	Chloroform	1
DDT/DDE/DDD	0.3	Carbon tetrachloride	0.1
2,4–D	5	Tetrachloroethylene	0.4
Heptachlor	0.07	Trichloroethylene	1
Lindane	0.1	Vinyl chloride	6
Malathion	0.05		
Monocyclic aromatics		PCDD[C] and PCDF[D]	<0.01
Benzene	0.3	Phenols	
Toluene	0.03	Chlorophenol	3
Xylene	0.03	Pentachlorophenol	5
Ethylbenzene	6	Phenol	5
PAH[A]		Phthalate esters	
Pyrene	5	Diethylhexylphthalate	10^2
Fluoranthrene	9	Surfactants	
Benzo-[*a*]-pyrene	3	LAS[E]	10^4
Benzo-[*b*]-pyrene	8	Nonylphenol	10^3

Source: Own files

[A]Polycyclic aromatic hydrocarbons
[B]Polychlorinated biphenols
[C]Polychlorinated dibenzo-*p*-dioxins
[D]Polychlorinated dibenzofurans
[E]Linear alkylbenzenesulphonate

There is very limited information in the literature on the behaviour and fate of sludge-borne organics transferred to agricultural soils. Although some inferences may be drawn from studies involving the application of agrochemicals on their own, this should be done with caution because other sludge components may have a significant impact on the behaviour and fate of the organic present.

The fate of trace organic compounds applied to soil is controlled by several processes: volatilization, degradation, sorption, leaching and bio-accumulation. The significance of each process is affected by the physico-chemical properties of the organic compound, sludge and soil properties, and environmental conditions.

16.6 Factors affecting the fate of organic compounds applied to soil

Factors affecting the fate of organic applied to soil are discussed below.

16.6.1 Volatilization

Volatilization loss can be a significant dissipation pathway for organics applied to land. The rate of dissipation of organics is governed by the vapour pressure of the compound, and on soil and environmental conditions. Losses to the atmosphere may take place immediately if the organics are applied at the soil surface; if the organics are incorporated with the surface soil layer or injected below surface, the rate of volatilization loss is significantly reduced and is dependent on the rate of transport to the soil surface. As an example, 90% of Heptachlor applied on the soil surface may be lost in 2–7 days, in comparison to a 7% loss in 167 days when incorporated to 7.5cm [14].

Few studies have been carried out on the volatization rates of organics in land applied sludges. The presence of organic matter in the sludge may increase the sorption of organic compounds and thus reduce volatilization rates.

16.6.2 Sorption

Sorption refers to the processes by which an organic chemical in the soil sorbs onto soil solid surfaces or penetrates into the solid matrices. The predominant soil material contributing to sorption is the soil organic matter; therefore, the partition coefficient is usually normalised by the weight fraction of soil organic C, known as K_{OC} [15]. Most organochlorine compounds have very high sorption coefficients. The K_{OC} values are 5000 for Aldrin, 2×10^6 for DDT, 12 000 for Dieldrin, and 24 000 for Heptachlor [16].

Sludge organic matter applied to soil may increase the sorption capacity of the soil. However, the properties of the organic matter in sludges are different from the stabilised organic matter found in most soils, and may therefore have a different sorption capacity for organic chemicals. Sludges often contain solvents, and these solvents tend to increase the solubility of water-insoluble organic compounds, decreasing the partition coefficient. In addition, surfactants, which are often present at high concentrations in sludges, may increase the solubility of hydrophobic organic and decrease the sorption partition coefficient [17]. The surfactant effect on the apparent solubility of hydrophobic organic is particularly significant when the surfactant concentration is above the critical micelle concentration at which surfactants assemble to form ordered aggregates or micelles [18,19].

16.6.3 Degradation

Degradation is probably the most important process by which many organic chemicals dissipate in the soil. It refers to the breakdown of organic chemicals by biotic (biological or biochemical) or abiotic (e.g. hydrolysis or photolysis) processes. While degradation generally results in the detoxification of toxic organic chemicals in the soil, intermediate degradation products of some compounds may be just as toxic and persistent as the parent compound. Microbial degradation (biodegradation) is probably the most important degradation process for many chemicals in agricultural soils. The rate of degradation is usually described by first-order kinetics, and degradation half-lives are obtained by monitoring the decreases in concentration over time.

Degradation rates of most organic chemicals are site-specific and thus vary widely with soil properties (e.g. organic matter content, microbial activity, pH).

Data reported in the literature that have been obtained from laboratory experiments may not be appropriate for field conditions. Many of the degradation studies that have been carried out are for chemicals singularly applied to soils in the absence of sludges or effluents, and extrapolation of these results to sludge- or effluent-borne organic is questionable. The high organic matter and nutrient contents of sludges are likely to stimulate microbial activity and thus increase the degradation rate; on the other hand, organic matter contained in sludges may strongly sorb organic compounds and protect them from degradation.

Surfactants which are often found in waste sludges at high concentrations may increase the apparent solubility of hydrophobic compounds and have a significant impact on their degradation behaviour. Current reports of surfactant effects on organic chemical degradation are variable, with both enhanced and inhibited degradation reported, largely because the factors involved are many and the processes concerned are complex [20].

Wild *et al.* [21] reported a long-term study of the persistence of PAHs applied in sewage sludges to agricultural soils. The compounds with greater molecular weight were found to be generally more persistent. Following the land application of sludge, the less persistent compounds were rapidly degraded, leaving the recalcitrant residues which could persist in the soil for many years. The estimated half-life for the total PAHs was 19 years.

Estimated half-lives for chlorinated hydrocarbon pesticides vary widely [16,22]: Aldrin, 1–9; Dieldrin, 3–7; Chlordane, 1–8; Heptachlor, 1–4; and DDT, 3–10 years. Half-lives for PCBs range from one year to 16 years [23]. Other types of pesticides, e.g. organophosphates, triazines, carbamates and ureas, are generally less persistent [16, 24].

If the dissipation rate of a chemical is slower than the rate at which the chemical is applied in a soil, then the concentration of the chemical will increase, reaching a limit above which the soil may not be suitable for food production. A study by Jones *et al.* [25] at Rothamsted Experimental Station showed that the concentration of PAHs in the topsoil had been increasing steadily over the last century, with the total PAH increasing by about four-fold.

16.6.4 Leaching

The potential for an organic chemical to contaminate the groundwater is dependent on the mobility and persistence of the chemical in the unsaturated soil. Sorption of the organic chemical by the soil solid phase, particularly organic matter, reduces the velocity of chemical transport. Organic matter in sludges and effluent therefore, may reduce leaching through sorption. Surfactants and organic solvents present in sludges, however, may increase the mobility by increasing the apparent solubility of hydrophobic compounds. Degradation decreases the concentration of organics in the leaching solution and thus reduces their chances of leaching.

Groundwater contamination by agrochemicals from non-point sources has been well documented in a number of countries [26–28,30–32]. The pesticides that have been detected in regional council groundwater surveys include 2,4-D, Amitrole, Picloram, Simazine and Atrazine [20].

Muszkat *et al.* [28] studied the migration of organics in a soil which had received sewage effluent for about 20 years, and found that many of the organic compounds had migrated through 20m depth, e.g. aliphatic hydrocarbons, pesticides (e.g. Prometon, a triazine herbicide), solvents (toluene), organic acids and esters, and plasticisers (e.g, diisooctyl phthalate). The apparent mobility and deep penetration of the compounds were attributed to the enhancement of aqueous solubility of the organics by surface-active surfactants and dissolved humic and fulvic acids present in the effluents.

Health hazards to humans by sludge-borne organics result from contamination of food for human consumption. The main pathways include contamination of crops destined for human consumption, and contamination of animal feeds and subsequent transfer to animal food products.

Also, of course, impaction of organics from soil by rainwater into the aquifer and subsequent migration to rivers which might be an intake for potable water treatment plants is another possible source of health hazards to humans. In this connection there is considerable variation between, and uncertainty over, guidelines on permissible concentrations of pesticides in drinking and other water supplies. The EU drinking water directive which came into force in July 1985, sets a maximum permissible concentration for total pesticides and any individual pesticide of 0.5μg L^{-1} and 0.1μg L^{-1} respectively, while the WHO and US Environmental Protection Agency both make recommendations for some individual pesticide compounds, with differences between values for specific compounds, both allowing much higher concentrations than the EU directive for most (but not all) compounds.

Repeated applications of pesticides to crops and soil will result in a build-up in their concentration in the soil over a period of time, especially if they have a low solubility in rain water, more so if they are stable and of low volatility. If the pesticides have an appreciable solubility in rainwater, the build-up in their concentration in soil with repeated applications will be lower than is the case of relatively water insoluble pesticides. This is important from the point of view of transfer of insecticides from the soil to crops, grown on the land, or to watercourses, streams, rivers and perhaps eventually the oceans.

Two of the most commonly used types of pesticides in Britain are herbicides, of the carboxyacid and phenylurea groups, applied to cereal growing land (Table 16.4).

It is seen from Table 16.4, for example, that TCA a phenoxyacetic acid type of herbicide has a solubility exceeding 500 000μg L^{-1} while chlorotoluron, a phenyl urea herbicide has a water solubility of 70mg L^{-1}. Thus, due to rainfall over a period of time the concentration of TCA in soil will reduce considerably more rapidly than that of chlorotoluron.

Typical rates of pesticide application in Britain are currently in the range 1–10kg/ha/a active ingredient, depending upon compounds [33]. Generally no more than 10% of an application reaches the target area, normally the plant roots or soil insects [34]. Since evaporation losses are also generally likely to be small, due to relatively low air and soil temperatures in Britain at normal times of application (March–May and September–November), the major proportion of most pesticide applications remains for some time in the soil and can potentially be leached into underlying aquifers or, in less permeable soils via drainage networks to surface watercourses; the water solubility of most pesticide compounds being in excess of 10mg L^{-1} and in some cases of 1000mg L^{-1} (Table 16.4).

Table 16.4 Application of pesticides etc. to land, basic parameters

Compound name	*Chemical group*	*Current UK usage (tonnes/a)*	*Rate of application (kg/ha)*	*Water solubility*** *(mg/l)*	*Physical class*	*Pollution hazard class****	*Drink water guideline (µg/l)*
(A) Herbicides							
MCPP	carboxyacid	2000+	1–2	620	I–nV	4	7[b]
TCA	carboxyacid	2000+	10–30	>500 000	I–nV	5	20[b]
Chlorotoluron	phenylurea	1000–2000	1–2	70	nI–nV	3	8[b]
Isoproturon	phenylurea	1000–2000	1–20	50	nI–nV	3	0.4[b]
MCPA	carboxyacid	1000–2000	0.5–2	820	I–nV	4	10[b]
2,4-D	carboxyacid	*	0.5–2	620	I–nV	4	100[a]
Atrazine	basic triazine	*	1–5	30	I–nV	3	30[b]–150[c]
Simazine	basic triazine	*	2–20	<10	I–nV	3	30[b]–1500[c]
Paraquat	cationic	*	0.1–1	>500 000	I–nV	1	0.3[b]–60[c]
(B) Fungicides							
Metham-sodium	misc. ionic	1000–2000	?	<500 000	I–V	5	ns
Carbendazim	misc. ionic	500–000	?	30	I–nV	3	3[b]
(C) Insecticides							
Dimethoate	carbamate	20–50	0.1–1	25 000	nI–V	4	7[b]
Aldicarb	carbamate	*	1–5	6000	nI–V	4	ns

Source: Reproduced with permission from the British Geological Survey [33]

ns not specified; (n)I (non) ionic; (n)V (non) volatile; * believed to be relatively minor but included for comparison since mentioned in text
[a]WHO; [b]derived from WHO–Acceptable Daily Intake data; [c]US–EPA
**at 20°C
***based on partition coefficient (K_{OC}) data: running from 5 to 1 with decreasing hazard and corresponding to K_{OC} values <25, 25–100, 100–250, 250–1000, >1000

The fate of a given compound in the soil depends upon its partition between the soil particles themselves and the water, air and organisms it contains. Certain physicochemical properties of pesticides can be used to predict which are most likely to be leached. The most diagnostic properties are the sorption and degradation characteristics.

The first of these environmentally-important parameters can be expressed as a partition coefficient. In aqueous solution many, but not all pesticide compounds exhibit strong affinity for soil organic matter or concentrate in the lipid phase of soil organisms. Some, notably the cationic group, also exhibit marked affinity for clay or other mineral surfaces. An overall partition (or distribution) coefficient (k_D) can be defined:

$$k_D = \frac{\text{amount sorbed (µg) per unit mass of soil (g)}}{\text{amount In solution (µg) per unit mass of liquid (g)}}$$

Compounds with low k_D are not strongly sorbed and more liable to leaching from the soil; most are also relatively soluble in water. This property, however, gives only a general guideline of the vulnerability to leaching, and the amount leached will depend upon the soil type involved (especially its proportion of organic matter), pH and temperature. In relation to Table 16.4 it should be noted that for most compounds (other than the cationic compounds) k_D shows a direct relationship to the partition coefficients between water and soil organic matter (k_{OM}), soil organic carbon (k_{OC}) and octanol (k_{OW}) respectively, all of which are more readily determined and standardised upon [35].

An important anomaly in respect of subsurface pesticide mobility is the fact that some otherwise strongly sorbed compounds can be mobile when in the sorbed phase attached to colloidal minerals. This phenomena has been little investigated to date but could be the explanation for the presence of some otherwise highly immobile compounds in groundwater samples, especially from fissured aquifers.

The concentration of many pesticide compounds in soils is substantially reduced by degradation processes before they can be leached. Half-lives are normally quoted for each compound in a fertile clayey-loam soil and are normally less than one year and, nowadays, in many cases less than one month. Degradation may be by chemical hydrolysis in the case of some compounds and by bacteriological oxidation in the case of many others. However, certain compounds are either relatively resistant to such degradation or the derivatives of partial hydrolysis/oxidation may be equally toxic as the original compounds.

As a result of their sorption and degradation characteristics, pesticides tend to be retained and eliminated within the soil horizon. These processes are well illustrated by some British field data [36] relating to the fate of a winter application of two pesticides of contrasting properties, the nematocide Aldoxycarb and the herbicide Fluometruron, in relation to a

^{36}Cl tracer (Fig. 16.1) on a fallow plot of a structured clay loam soil (56% clay, 6% organic carbon).

A plot of time interval (in days) since application of the pesticide to soil versus percentage of original pesticide addition remaining in soil after various time intervals (31, 66 and 93 days) shows that, for example, after 30 days, in the case of the less water soluble Fluometretruron (solubility in water 200mg L $^{-1}$) some 30–40% remains in the soil, while in the case of the more water soluble Aldoxycarb (solubility in water 9000mg L $^{-1}$) some 13% remains in the soil. After 90 days exposure the corresponding figures are 18% and 1%.

It can be assumed that the rate of removal of these pesticides from soil by rain elution follows the usual exponential law, namely:

$I_n = \quad I_I (1 - P/100)^n$ or

$\log I_n/I_I = \quad n\log (1 - P/100)$

where

I_I (mg kg $^{-1}$) = concentration of pesticide in soil immediately following its application; assumed 1mg kg $^{-1}$ for Fluometruron, and Aldoxycarb.

I_n (mg kg $^{-1}$) = concentration of pesticide remaining in soil after a period of n months

p = percentage of original addition of pesticide eluted by rain during n months

Using the above data, namely

	Fluometruron	*Aldoxycarb*
I_n mg kg $^{-1}$	1	1
I_I mg kg $^{-1}$	0.4	0.13
n months	1	1

the following calculations ensue.

$$\log \frac{0.4}{1} = \log \left(1 - \frac{P}{100}\right) \qquad \log \frac{0.13}{1} - \log \left(1 - \frac{P}{100}\right)$$

$$\text{i.e.} -0.4 = \log \left(1 - \frac{P}{100}\right) \qquad -0.886 = \log \left(1 - \frac{P}{100}\right)$$

$$\therefore \left(1 - \frac{P}{100}\right) = 0.4 \qquad \left(1 - \frac{P}{100}\right) = 0.13$$

∴P = 60% of original addition per month eluted by rain

∴P = 87% of original addition per month eluted by rain

Thus of the original additions of Fluometruron and Aldoxycarb to soil (namely 1mg kg $^{-1}$) at the end of the first month some 60% (0.6mg kg $^{-1}$) of Fluometruron and 87% (0.87mg kg $^{-1}$) of the more water soluble Aldoxycarb are eluted from the soil by rainwater. At the end of the second

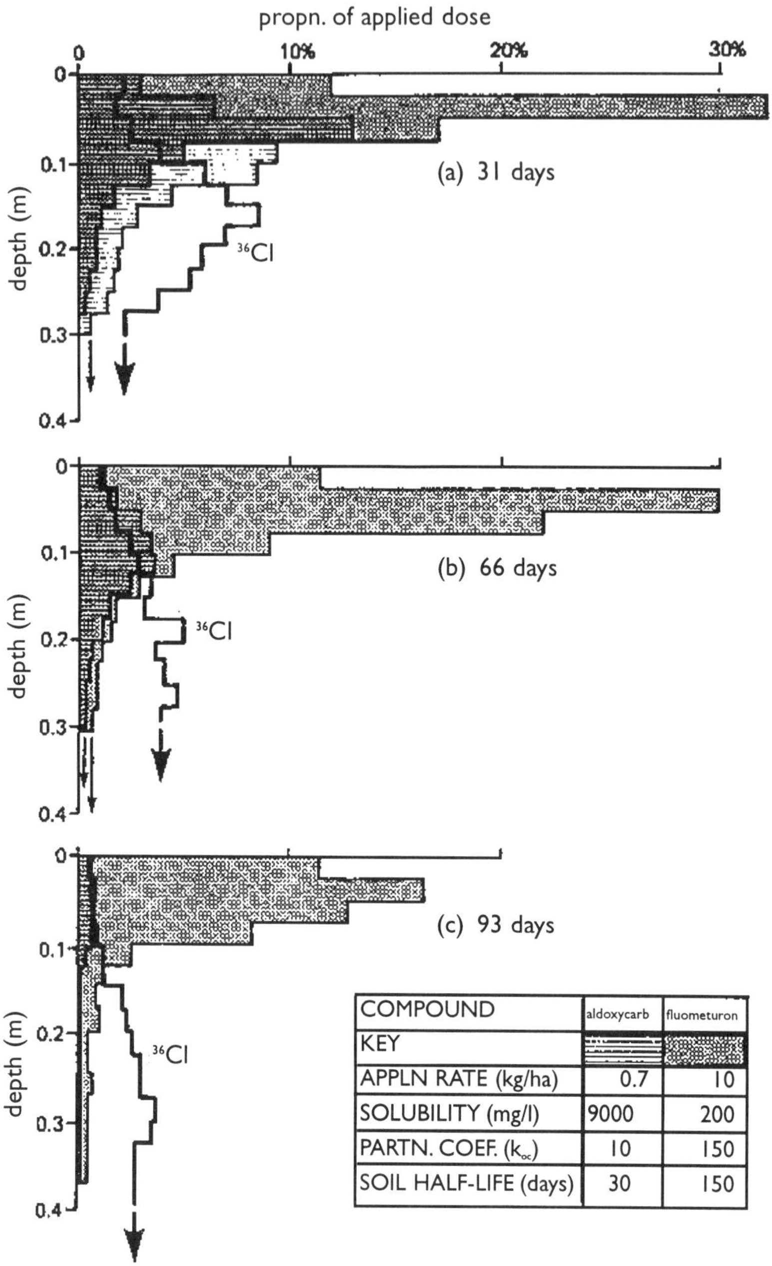

COMPOUND	aldoxycarb	fluometuron
KEY		
APPLN RATE (kg/ha)	0.7	10
SOLUBILITY (mg/l)	9000	200
PARTN. COEF. (k_{oc})	10	150
SOIL HALF-LIFE (days)	30	150

Fig. 16.1 Plot of percentage of applied dose of pesticide remaining in soil versus depth of soil at various time intervals after application. (a) 31 days, (b) 66 days, (c) 93 days after pesticide application
Source: Own files

month a further 60% of the remaining 0.4mg kg $^{-1}$ Flumetron (i.e. 0.24mg kg $^{-1}$) and a further 87% of the remaining 0.13mg kg $^{-1}$ Aldoxycarb (i.e 0.11mg kg $^{-1}$) are eluted from the soil i.e. total rainwater elutions in two months of Flumetron of 0.84mg kg $^{-1}$ leaving 0.16mg kg $^{-1}$ in the soil and of Aldoxycarb of 0.98mg kg $^{-1}$ leaving 0.02mg kg $^{-1}$ in the soil.

The sorption and degradation characteristics listed for most pesticide compounds in terms of partition coefficients and half-lives relate only to a (standard) fertile, organic clayey soil and must not to be taken as representative of the permeable sandy soils widely developed on aquifer outcrops. Thus leaching of 1% of original application rates, and perhaps significantly higher, could easily occur for certain compounds on permeable soils.

Pesticide compounds being leached from agricultural soils into the unsaturated zone of aquifers enter an environment which contains a much smaller proportion of clay minerals, is very low in organic carbon and has a much smaller population of indigenous bacteria. Thus the persistence and mobility of all compounds should be expected to be very many times greater in groundwater systems than in the standard agricultural soil, although no data are available on the corresponding parameters. Such parameters need to be determined for a short list of pesticides in common use in Britain in laboratory media representative of the principal British aquifers.

Moreover, the results of this laboratory work need to be corroborated by field sampling:

1 in the unsaturated zone beneath fields or plots on the main aquifer outcrops subjected to controlled and representative land use, weed and pest control practices;
2 at the phreatic surface in areas of shallow groundwater table and similarly representative land-use and agricultural practices.

In view of experience with the groundwater nitrate problem and the possibility of a large unsaturated time-lag in British aquifer systems (especially in the Chalk), it is necessary to address the following questions in the above investigation:

1 Are any pesticides being leached in significant quantities from permeable agricultural soils into the unsaturated zone of British aquifers in a similar manner to that observed for nitrate?
2 Are the concentrations of these pesticides being attenuated to below troublesome or detectable levels in passage through the unsaturated zone, and is this primarily due to sorption or degradation?

Preliminary investigations have been initiated by a collaborative BGS–ICI programme in the unsaturated zone of the Berkshire Chalk beneath thin permeable soils continuously cropped for winter cereals over an eight-year period under a minimal tillage regime using a wide range of pesticides, especially MCPP, MCPA and Paraquat. Chalk cores were taken from 0.6–3.4m depth, with 2 × 200ml pore-water samples being obtained by high-speed centrifugation. Any organic compounds were extracted by ether at different pHs and analysed by a linked GLC–MS system to a broad screening resolution of between 0.5–5.0μg L^{-1} depending upon the compound. No compounds were found to be present above this detection level at any depth, although it is accepted that such investigations need to be broadened to analyse samples of both the mobile (fissure and macropore) water and the aquifer matrix for any sorbed phase that may be present. Moreover, the procedure was not designed to determine the most water soluble and least volatile pesticide compounds, such as Paraquat and Carbendazim respectively, although a separate concentration and analytical procedure was employed for the former and it was still not detected.

Computer models have been developed and used to predict the groundwater contamination potential of organic chemicals (particularly pesticides) from non-point sources [36–39]. Di *et al.* [40] used a simple screening model with Monte Carlo simulation to assess the groundwater contamination potential of 29 organic pesticides commonly used on the sandy coastal plains of Western Australia, and found that Fenamiphos, Simazine, Metribuzin, Linuron, Fenarimol and Metalaxyl had high probabilities of 0.8–1.0 to contaminate the local groundwater at 300cm depth (Fig. 16.2). It must be remembered that these models have been designed for situations where the organic chemical is applied individually without the presence of organic sludges or effluents, and possible effects that the sludges may have on the fate of organic contaminants.

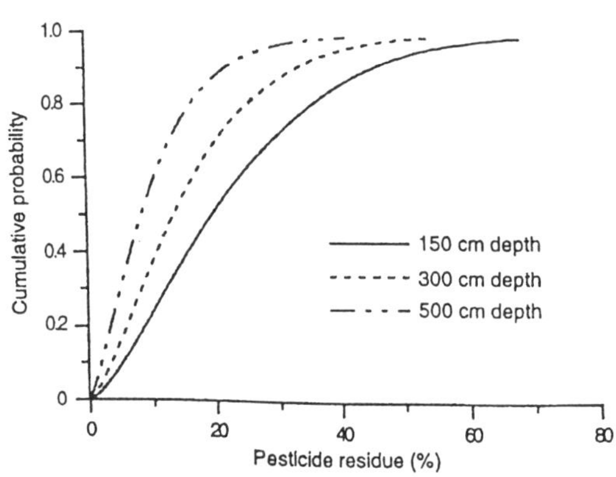

Fig. 16.2 Cumulative probability distributions of Metalaxyl residue at 150, 300 and 500cm soil depths, simulated on the basis of Western Australian coastal sandy soil conditions. The diagram shows that the cumulative probabilities are high for Metalaxyl to reach the three soil depths at significant residue fractions (% applied)
Source: Reproduced with permission from the Soil Society of America [40]

16.7 Consequences of repeated applications of sewage contaminated with organic compounds to land

Some 45% of the 1.24mt of dry sewage sludge produced annually in the UK and Wales is used as an agricultural fertilizer, the remainder being disposed of to rivers or dumped at sea.

This sludge contains organic substances (Table 16.3) also heavy metals. The possible deleterious effects to humans and animals of these applications to land are discussed above. Clearly, to avoid adverse effects on crops, animals and humans, control must be exerted on both the organic contaminant content of the applied sewage and its application rate and frequency.

Consider the case of 1ha of land (100m × 100m) of density dg cm^3 and thickness to which the sewage is applied tm. If the density is 1.5g cm^3 and the soil thickness is 0.2m then the top 0.2m of land weighs $10^4 \times 0.2 \times 1.5 = 3000$t. The application of 100t of sewage to the top 0.2m of soil represents a 3.3% addition. If the organic content of the sewage were 0.08mg kg^{-1} Aldrin (Table 16.3) then the Aldrin content of top 0.2 of sewage treated soil would increased by 3.3% of 0.08 = 0.00264mg kg^{-1}.

We discuss first the simple case in which it is assumed that once sewage has been applied to land, the organic contaminants present in the sewage remain in the land for prolonged periods of time, i.e. there is practically no loss with time due to rainwater leaching, volatilization or degradation. this is a worst case which is probably seldom if ever realized in practice. We then proceed in section 16.7.2 to the second case in which the organic contaminant content of the land due to sewage application gradually increases with repeated applications of sewage, but this effect is offset by a gradual decrease in the organic contaminant content of the land caused by leaching of the contaminant from the soil by rainwater or by volatilization of the contaminant or by oxidative or biological degradation of the contaminant to metabolites and eventually non-toxic breakdown products.

16.7.1 Simple case in which once sewage is applied to land no subsequent losses of organics occur

We consider the case of applying sewage containing Aldrin pesticide to soil.

If Aldrin content of a soil prior to addition of sewage =	M mg kg^{-1}
Aldrin content of sewage =	S mg kg^{-1}
Application rate of sewage to soil =	T tonne per hectare
Thickness of soil layer to which sewage added =	t, m
Density of soil to which sewage added =	d, g cm^3
Weight (w tonne) of one hectare of untreated soil =	$10^4 \times t \times d$
Weight (mg) of aldrin in $10^4 \times t \times d - T$ tonne untreated soil =	$1000\ M (10^4 \times t \times d - T)$ mg

Weight (mg) of aldrin in T tonne sewage = 1000ST mg

Weight (mg) of aldrin in $(10^4 \times t \times d - T) + T = 10^4 \times t \times d$ tonne of soil–sewage mixture immediately following addition of sewage to soil = $1000M(10^4 \times t \times d - T) + 1000ST$ mg aldrin per $10^4 \times t \times d$ tonne of mixture

∴ mg kg^{-1} of aldrin in soil–sewage mixture (I_I)

$$= \frac{M(10^4 \times t \times d - T) + ST}{10^4 \times t \times d} \quad (1)$$

To simplify this equation it is assumed that t = 0.2 m, d = 1.5g cm^3 and M = 0.01mg kg^{-1}.
Then Dieldrin content of sludge-treated soil (mg kg^{-1})

$$= \frac{30 + T(S - 0.01)}{3000} \quad (2)$$

Table 16.5 shows reported values of Aldrin content of treated soil that would be obtained by applying between 100 and 500t per hectare of sewage containing between 0.05 and 0.5mg kg^{-1} of Aldrin. It is seen that depending on the Aldrin content of the sewage and its application rate the sewage has imparted to the soil an additional 0.011–0.092mg kg Aldrin over the 0.01mg kg^{-1} assumed present in the untreated soil. These data apply to a single application of sewage. Higher levels would result if sewage applications were made yearly as is often the case. Clearly, there is a need for standards on what is an acceptable level of Aldrin in soil which would be applied to data of this kind in order for a decision to be made on the desirability of using sewage for land treatment.

Rearranging equation (1) it is possible to calculate the maximum permissible application rate of sewage of known Aldrin content to land (i.e. T_{max} tonne per hectare) that will permit the Aldrin content of the soil sewage mixture immediately after mixing to meet advised organic levels

$$T_{max} = \frac{(I_I - M) \times 10^4 \times t \times d}{S - M} \text{ tonne per hectare} \quad (3)$$

It is also possible by rearranging equation (1) to calculate the maximum allowed Aldrin content of sewage that is permitted for the Aldrin content of the treated soil not to exceed an advised limit

$$S_{max} \text{ mg kg}^{-1} = \frac{(10^4 \times t \times d) I_I - M(104^4 \times t \times d - T)}{T} \quad (4)$$

Thus in equation (3) if the advised maximum concentration of aldrin in treated soil was 0.05mg kg^{-1} (I_I) and the Aldrin content of the sewage (s) were 0.5mg kg^{-1} then the maximum permitted application rate of sewage to land would be 245t per ha.

Table 16.5 Aldrin contents of sewage treated soil = $\frac{30 + T\,(S - 0.01)\ \text{mg kg}^{-1}}{3000}$ (see equation (2))

Aldrin content of sewage (S), mg kg $^{-1}$	*0.05*			*0.1*			*0.5*		
Application rate of sewage, T tonnes per hectare	100	300	500	100	300	500	100	300	500
Aldrin content of treated soil mg kg	0.011	0.014	0.016	0.013	0.019	0.025	0.026	0.059	0.092
Aldrin content of untreated soil M 0.0 mg kg $^{-1}$, soil density of 1.5g cm ^{3}, soil thickness to which sewage applied, t, 0.2m									

Source: Own files

16.7.2 *Case in which losses of organic compounds by rainwater elution, volatilization and degradation run in parallel with gains in organic contaminated levels in soil caused by sewage addition*

Cause of losses of organics from soil

Rainwater elution

When a foreign substance is introduced into soil it may to varying degrees become bounded to soil particles or may not become so bonded to soil particles. The factors governing the rate of loss of organic contaminants from soil by rain elution are complex. Depending on the degree of bonding the foreign substance will to varying degrees be eluted from the soil by rain and will drain off to water courses, i.e. the concentration of this substance in the soil will decrease as a function of time and the amount of rain. Of course, some substances form such tight bonds with soil particles that they are never eluted by rain while other substances elute from the soil very quickly.

Volatilization

More volatile substances, e.g. hydrocarbons and chlorinated hydrocarbons can be lost from the soil by volatilization.

Degradation

Oxidative or bacterially induced reactions will reduce the concentration in soil of many types of organic contaminants to degradation products or metabolites, thereby reducing their concentration in the soil. This needs to be taken into account when assessing the toxicity of a soil as many of the breakdown products are themselves toxic. As shown in equation (1) the concentration of an organic compound I_1mg kg $^{-1}$ in a sewage treated soil immediately following application of sewage is given by:

$$I_1\text{mg kg}^{-1} = \frac{M(10^4 \times t \times d - T) + ST}{10^4 \times t \times d} \qquad (1)$$

where Mmg kg $^{-1}$ = concentration of organic in soil prior to sewage application
Smg kg $^{-1}$ = organic content of sewage
T tonne per hectare = application rate of sewage
t, m = thickness of soil layer to which sewage applied
d, g cm^3 = density of soil

If P% represents the annual loss of organic compounds caused by rainwater leaching etc. then

$$\frac{I_n}{I_I} = \left[\left(1 - \frac{P}{100}\right)^n\right] \tag{5}$$

where $I_n$mg kg $^{-1}$ = concentration of organic compound in sewage treated soil at end of year n

$I_I$mg kg $^{-1}$ = concentration of organic compound in treated soil immediately following application of sewage

Then combining equations (1) and (5)

$$I_n = I_I\left(1 - \frac{P}{100}\right)^n = \left[\frac{M(10^4 \times t \times d - T) + ST}{10^4 \times t \times d}\right]\left[\left(1 - \frac{P}{100}\right)^n\right] \tag{6}$$

I_n gives the net concentration of organic compound left in the soil at the end of time n years, corrected for losses by leaching etc.

When $I_n = 0.5\ I_I$
Then $I_n / I_I = 0.5$ and $\log I_n/I_I = \log 0.5 = n/2 \log (1 - P/100)$
Therefore

$$\frac{n}{2} = \text{half life} = \frac{\log 0.5}{\log(1 - P/100)} = \frac{-0.301}{\log(1 - P/100)} \text{ years} \tag{7}$$

The results in Table 16.6 show that if, immediately following addition of sewage contaminated with Aldrin to soil, the Aldrin content of the soil is 0.05mg kg $^{-1}$ and if the annual percentage loss is for example 30%, then at the end of three years the aldrin content of the treated soil reduces by 66% to 0.017mg kg $^{-1}$ i.e. the half life of the Aldrin is 1.94 years, i.e. the initial Aldrin content of 0.05mg kg $^{-1}$ reduces to 0.025mg kg $^{-1}$ in 1.94 years. Correspondingly, under similar circumstances an initial Aldrin addition of 0.5mg kg $^{-1}$ reduces to 0.17mg kg $^{-1}$ in three years, assuming 30% of the initial addition is lost per annum.

In actual fact the regime of adding sewage to land is usually an annual event, i.e. a certain weight of sewage is incorporated per hectare of the soil annually. So in parallel with the steady decrease in Aldrin content of the soil with time due to losses by elution, etc. with each further addition of sewage there occurs a further increase in Aldrin content. Estimates of the net Aldrin content of soil at any given time after the applications of sewage as a consequence of these two opposing mechanisms are possible by the following treatment.

Table 16.6 Concentrations of Aldrin in sewage treated soil as a function of time

Residual Aldrin content retained in sewage treated soil at end of stipulated year, n = 1, 2, 3 years old, $I_n = I_1 (1 - P/100)^n$. See equation (4)

P, annual % of aldrin lost from sewage treated soil by rain elution, volatilization, etc.	*Initial, i.e. aldrin content of sewage treated soil immediately after addition of sewage I_1mg kg^{-1}*	*Aldrin content of sewage treated soil at end of year 2 $= I_n$ mg kg^{-1} $= I_1 (1 - P/100)$*	*Aldrin content of sewage treated soil at end of year 2 $= I_n$ (mg kg^{-1}) $= I_1 (1 - P/100)^2$*	*Aldrin content of sewage treated soil at end of year 3 $= I_n$ (mg kg^{-1}) $= I_1 (1 - P/100)^3$*	*Half-life ($n_{½}$) of aldrin in years $n_{½} = \frac{-0.301}{\log (1 - P/100)}$ (see equation (5))*
0	0.050	0.050	0.050	0.050	–
10	0.050	0.045	0.040	0.036	6.54
20	0.050	0.040	0.032	0.025	3.10
30	0.050	0.035	0.024	0.017	1.94
50	0.050	0.025	0.0125	0.0062	1.01
70	0.050	0.015	0.0045	0.0013	0.57
90	0.050	0.005	0.0005	0.00005	0.30
0	0.500	0.500	0.500	0.500	–
10	0.500	0.450	0.400	0.360	6.54
20	0.500	0.400	0.320	0.250	3.10
30	0.500	0.350	0.245	0.170	1.94
50	0.500	0.250	0.125	0.062	1.01
70	0.500	0.150	0.045	0.013	0.57
90	0.500	0.050	0.005	0.0005	0.30

Source: Own files

If the initial concentration of Aldrin in the soil at the start of year 1 is $I_I$mg kg $^{-1}$ then by equation (6) the concentration of Aldrin (I_n mg kg $^{-1}$) at the end of years 1, 2, 3 are respectively given by I_I (1 – P/100) mg kg $^{-1}$ at the end of year one; I_I $(1 - P/100)^2$ mg kg $^{-1}$ at the end of year 2; and I_I $(1 - P/100)^3$ mg kg $^{-1}$ at the end of year 3. If at the end of years 1, 2 and 3, a further addition of sewage to soil adds a further I_I mg kg $^{-1}$ of Aldrin, then the net concentrations in the treated soil will be:

$$I_I + I_I\left(1 - \frac{P}{100}\right) = I_I\left[\left(1 + \left(1 - \frac{P}{100}\right)\right)\right] \text{ mg kg}^{-1} \text{ at the end of year 1}$$

$$I_I + I_I\left(1 - \frac{P}{100}\right)^2 + I_I\left(1 - \frac{P}{100}\right) = I_I\left[\left(1 - \frac{P}{100}\right)^2 + \left(1 - \frac{P}{100}\right) + 1\right] \text{ mg kg}^{-1}$$

at the end of year 2 and

$$I_I + I_I\left(1 - \frac{P}{100}\right)^3 + I_I\left(1 - \frac{P}{100}\right)^2 + I_I\left(1 - \frac{P}{100}\right) + 1$$

$$I_I\left[\left(1 - \frac{P}{100}\right)^3 + \left(1 - \frac{P}{100}\right)^2 + \left(1 - \frac{P}{100}\right) + 1\right] \text{ mg kg}^{-1}$$

at the end of year 3

If at the start of year 1 an amount of sewage is added to the soil which imparts 0.05mg kg $^{-1}$ (= I_I) Aldrin to the mixture and if at the end of that year and subsequent years a further amount of sewage is added which imparts an additional 0.05mg kg $^{-1}$ of Aldrin then the levels of Aldrin to be expected in the soil assuming various P values (i.e. percentage annual loss of Aldrin due to rain leaching, etc.) are illustrated in Table 16.7.

It is seen in Table 16.7 that if only 10% (P) of the Aldrin is lost annually by rain leaching, etc. then after three years of sewage treatment the Aldrin content of the soil has doubled. If, however, 70–90% (P) is lost annually then the Aldrin content of the treated soil remains practically constant despite annual sludge additions.

Calculations of the kind discussed in this section while not providing precise values are very useful in that they provide a feel for what happens to toxic organic substances incorporated into soil by sewage addition and effects of rainfall, etc. on these values over a period of time.

It must be understood that although a particular organic compound might be very soluble in rain and would be expected to leach quickly from the soil, if the seasonal rainfall is low then losses by leaching will be low leading to increased levels in crops. Also, when heavy rainfall does occur a sudden surge of toxicant leaching will occur to the water course leading possibly to fish kills in adjacent rivers.

Table 16.7 Net effect of sewage addition and leaching at losses on aldrin content of sewage treated soil

P, annual % of Aldrin lost from sewage treated soil by rain elution, volatilization, etc.	*Start of year 1* $I_1 = 0.05 mg\ kg$	*Aldrin content of treated soil, I_n mg kg^{-1}* End of year 1 $I_1 \left[1 - (1 - \frac{P}{100})\right]$	End of year 2 $I_1 \left[(1 - \frac{P}{100})^2 + (1 - \frac{P}{100}) + 1\right]$	End of year 3 $\left[I_1 \ (1 - \frac{P}{100})^3 + (1 - \frac{P}{100})^2 + (1 - \frac{P}{100}) + 1\right]$
10		0.095	0.135	0.171
30		0.085	0.109	0.126
70		0.065	0.069	0.071
90		0.055	0.0555	0.05555

Source: Own files

16.8 Uptake of toxicants from soil to crops

O'Connor *et al.* [13] provided an extensive review of plant uptake of organics from the soil and found that this was not an important mechanism of plant contamination. In general, the bioaccumulation factors (BCF, the ratio of organic chemical concentration in plant biomass to that in the soil) for most hydrophobic organic found in sludges, including PAHs, PCBs and chlorinated pesticides and hydrocarbons, are low, often <0.01 (dry weight). Compounds such as phthalate esters may be taken up by plants, but are metabolised in the plants to polar metabolites. High concentrations of aromatic surfactants, e.g. linear aklylbenzene sulphonates and nonylphenol, are often present in sludges, and high BCFs have been reported for these compounds (Litz *et al.* [41]), the significance of these compounds to plant contamination is not well understood. Volatile organic (e.g. Henry's constant $>10^{-4}$), such as toluene and benzene, may contaminate plant tops by volatilization rather than by plant uptake from the soil, but the significance of this mechanism is not clear.

An important route of plant contamination comes from direct contact during surface application of sludge. A study by Chaney and Lloyd [42] showed that sludge constituted 30% of forage dry matter immediately following spray application of liquid sludge on tall fescue (*Festuca arundinacea Schreb*). The sludge concentration decreased to 10% after 30 days.

Another entry point for accumulation of organic in animal tissue and animal food products is direct ingestion of contaminated soil by grazing animals [44]. The compounds of main concern are the halogenated aromatics, including PCBs, organochlorine pesticides, PCDDs and PCDFs, which are resistant to metabolization and tend to accumulate in animal fat. The bioaccumulation factor (the ratio of the concentration of animal tissue or produce to the concentration in the diet) can be as high as 5–6 [43]. Compounds such as PAHs and phthalate esters are readily metabolised and excreted by the animals and thus do not accumulate in animal tissue or products.

The optimal amount and frequency of sludge application on land will depend the concentration of the chemicals, particularly for the halogenated aromatics, persistence of these compounds in the soil environment and management practices. Incorporation of sludges into the plough layer (15cm) will reduce residues in animal tissue by 85% compared with surface application [23]. A ten day delay in grazing following sludge application can decrease animal intake by 80%. The chemical concentration in animal milk or fat tissue can be further reduced by supplemental feeding.

It should be expected that there is a relationship between the organic content of soil and the organic content of crops grown on the soil. Little discussion of such a relationship has been found in the literature.

Such a relationship does seem to exist in the case of heavy metals (Crompton [44]) and this has been applied to spinach, cornflour, wheat flour, potatoes and rice flour.

$$\log C_C = \log C_s - 2.1 \pm 0.8$$
$$\text{i.e. } \log C_c/\log C_s = -2.1$$

where
C_c = concentration mg kg $^{-1}$ in crops
C_s = concentration mg kg $^{-1}$ in soil

This is in fair agreement with the ratio for organics proposed by O'Connor [13] namely

$$C_c/C_s = < 0.01 \text{ i.e. } \log C_c/C_s = -2$$

Thus, the concentration for crops is about 1/100 that in the soil.

References

1 Cameron, K.C., Di, H.J. and McLaren, R.G. (1997) *Australian Journal of Soil Research*, **35**, 995.
2 UNEP (1993) *Organic Contaminants in the Environment, Environmental Pathways and Effects*, (ed. K.C. Jones) Elsevier, London, pp.275–289.
3 Carnus, J.M. and Mason, I. (1994) *Land Treatment of Tannery Wastes, Land Treat Collective Review*, **10**, 31–39.
4 Sommers, L.E. (1977) *Journal of Environmental Quality*, **6**, 225.
5 Brechin, J.-McDonald (1994) *Australian Journal of Exploratory Agriculture*, **34**, 505.
6 Quin, B.F.C. and Woods, P.H. (1978) *New Zealand Journal of Agricultural Research*, **21**, 419.
7 Smith, K.A., Unwin, R.J. and Williams, J.H. (1985) Experiments on the Fertilizer Value of Animal Waste Slurries, in *Long Term Effects of Sewage Sludge and Farm Slurries Applications* (eds. J.H. Williams, G. Gindi and P. D'Hermite) Elsevier, London, pp.124–135.
8 Cameron, K.G., Rate, A.W., Noonan, M.J. *et al.* (1996) *Agricultural Ecosystems and Environment*, **58**, 187.
9 Wild, S.R. and Jones, K.C. (1992) *Science of the Total Environment*, **119**. 85.
10 US Environmental Protection Agency (1985) *Summary of Environmental Profiles and Hazard Indices for Constituents of Municipal Sludge. Methods and Results.* US Environmental Protection Agency, Office of Water Regulations and Standards, Washington, DC.
11 Jacobs, L.W., O'Connor, G.A., Overcash, M.A. and Zabik, M.J. (1987) Effects of trace organic in sewage sludge on soil-plant systems and assessing their risk to humans, in *Land Application of Sludges* (eds. A.C. Page, T.J. Logan and J.A. Ryan) Lewis Publishers, Chelsea, MI, pp.101–143.
12 Webber, M.D. and Lesage, S. (1989) *Waste Management Research*, **7**, 63.
13 O'Connor, G.A., Chaney, R.L. and Ryan, J.A. (1991) *Reviews of Environmental Contamination and Toxicology*, **121**, 129.
14 Glotfelty, D.E., Taylor, A.W., Turner, B.C. and Zoller, W.H. (1984) *Journal of Agriculture and Food Chemistry*, **32**, 638.

15 Hamaker, J.W. and Thompson, J.M. (1972) in *Organic Chemicals in the Soil Environment*. Vol. 1. (eds. C.A.I. Goring and J.W. Hamaker) Marcel Dekker, New York, pp.49–143.

16 Augustijn-Beckers, P.W.M., Hornsby, A.G. and Wauchampe, A.D. (1994) *Reviews of Environmental Contamination and Toxicology*, **137**, 1.

17 Vigon, B.W. and Rubin, A.J. (1989) *Journal of Water Pollution Control Federation*, **61**, 1233.

18 Edwards, D.A., Luthy, R.G. and Liu, Z. (1991) *Environmental Science and Technology*, **25**, 127.

19 Smith, J.A., Tuck, D.A., Jaffe, P.R. and Mueller, R.T. (1991). Effect of Surfactants on the mobility of nonpolar organic contaminants in porous media, in *Organic Substances and Sediments in Water*. Vol. 1 (ed. R. Baker), Lewis Publishers, Chelsea, MI, pp.201–203.

20 Rouse, J.D., Sabatini, D.A., Suflita, J.M. and Harwell, J.H. (1994) *Critical Review of Environmental Science and Technology*, **24**, 325.

21 Wild, S.R., Waterhouse, K.S., McGrath, S.P. and Jones, K.C. (1990) *Environmental Science and Technology*, **24**, 1706.

22 Nash, R.G. and Woolson, E.A. (1967) *Science*, **157**, 924.

23 Fries, G.F. (1982) *Journal of Environmental Quality*, **11**, 14.

24 Wauchamp, R.D., Buttler, T.M., Hornsby, A.G. *et al.* (1992) *Reviews of Environmental Contamination and Toxicology*, 1**23**, 1.

25 Jones, K.C., Stradford, J.A., Waterhouse, K. *et al.* (1989) *Environmental Science and Technology*, **23**, 95.

26 Reinhard, M., Goodman, N.L. and Barker, J.F. (1984) *Environmental Science and Technology*, **18**, 953.

27 Lesage, S. and Jackson, R.E. (1992) *Groundwater Contamination and Analysis of Hazardous Waste Sites*. Michael Dekker, New York.

28 Muszkat, L., Rancher, D., Magaritz, M. *et al.* (1993) *Groundwater*, **31**, 556.

29 Rugge, K., Bjerg, P., and Christensen, T.H. (1995) *Environmental Science and Technology*, **24**, 325.

30 Leistra, M. and Boesten, J.J.T.I. (1989) *Environmental Economics*, **26**, 369.

31 Gustafson, D.I. (1993) *Pesticides in Drinking Water*, Van Nostrand & Reinhold, New York.

32 Brusseau, M.L. and Koakana, R.S. (1996) Transport and Fate of Organic Contaminants in the Subsurface, in *Contaminants and the Soil Environment in the Australasia–Pacific Region*. (eds. R. Naidu, R.S. Kookana, D.P. Oliver, S.Rogers and M.J. McLaughlin) Kluwer, Dordrecht, pp.95–125.

33 Lawrence, A.R., Foster, S.S.D. (1987) *The Pollution Threat from Agricultural Pesticides and Industrial Solvents. Hydrogeological Report No. 87/2*, British Geological Survey, Wallingford, Oxfordshire.

34 Riley, D. (1976) Physical Causes and Redistribution of Pesticides in the Liquid Phase. *British Crop Protection Monograph No. 17*, pp.109–115.

35 Briggs, G.G. (1981) *Journal of Agriculture and Food Chemistry*, **29**, 1050.

35 Nicholls, P.H., Bromilow, R.H. and Addiscott, T.M. (1982) *Pesticide Science*, **13**, 475.

36 Nicholls, P.H., Walker, A. and Baker, R.J. (1982) *Pesticide Science*, **12**, 484.

37 Jury, W.A., Spencer, W.F. and Farmer, W.J. (1983) *Journal of Environmental Quality*, **12**, 558.

38 Hutson, J.L. and Wagner, R.J. (1992) *Leaching Estimation and Chemistry Model. A Process Based Model of Water and Solute Movement, transformation, Plant Uptake and Chemical Reactions in the Unsaturated Zone*. Version 3. Dept. of Soil, Crop and Atmospheric Sciences, Series No. 92–3, Cornell University, Ithica, New York.

39 Di, H.J., Kookana, R.S. and Aylmore, C.A.G. (1995) *Australian Journal Soil Research*, **33**, 1031.
40 Di, H.J. and Aylmore, L.A.G. (1997) *Soil Science Society of America Journal*, **61**, 17.
41 Litz, N., Doering, H.W., Thiele, M. and Blume, H.P. (1987) *Ecotoxicology and Environmental Safety*, **14**, 103.
42 Chaney, R.L. and Lloyd, C.A. (1979) *Journal of Environmental Quality*, **8**, 407.
43 Fries, G.F. (1995) *Reviews of Environmental Contamination and Toxicology*, **141**, 71.
44 Crompton, T.R. Unpublished work.
45 Smith, G. (1995) *Water Wastes, New Zealand*, **86**, 50.

Appendix I

Instrument suppliers

(Cross-referenced with Chapter 1)

I.1.1.1 Visible ultraviolet and near infrared spectrometers

UV, visible and near infrared
PU 8620 basic instrument
Optional PU 8700 scanner for colour graphics
PU 8800 research applications

Pye Unican Ltd
York Street
Cambridge CB1 2PX
UK

Phillips Electronic Instruments
85 McKee Drive
Mahway NJ 07430
USA

Phillips Nederland BV
HSD Analysetechnicken VB3
Postbus 90050
5600 PB Eindhoven
Netherlands

Cecil Instruments Ltd
CE 2343 Visible range spectrophotometer
CE 243D Visible range spectrophotometer
CE 2393 Digital visible grating spectrophotometer
CE 2292 Digital ultraviolet spectrophotometer
CE 2303 Grating spectrophotometer
CE 2202 Ultraviolet spectrophotometer
CE 2373 Linear read out grating spectrophotometer
CE 2272 Linear read-out ultraviolet spectrophotometer
CE 594 Ultraviolet and visible double-beam spectrophotometer
CE 6000 Ultraviolet visible double-beam spectrophotometer with
CE 6606 Superscan graphic plotter

Cecil Instruments Ltd
Milton Industrial Estate
Cambridge CB4 4AZ
UK

Kontron Instruments
Unikon 860 Ultraviolet visible double-beam spectrophotometer
Unikon 930 Ultraviolet – visible graphics

Kontron Instruments AG
Bernerstrasse, SUD 169
8010 Zurich
Switzerland

Perkin–Elmer Ltd
Lambda 2 Ultraviolet-visible double-beam spectrophotometer
Lambda 3 Ultraviolet-visible double-beam spectrophotometer
Lambda 5 and Lambda 7 Ultraviolet visible spectrophotometers
Lambda 9 Ultraviolet visible – near infrared spectrophotometer
Lambda Array 3430 Spectrophotometer

Perkin–Elmer Ltd
Post Office Lane
Beaconsfield
Buckinghamshire HP9 1QA
UK

Perkin–Elmer Corporation
Analytical Instruments Division
761 Main Avenue
Norwalk, CT 06859
USA

Bodenseewerk Perkin–Elmer P 60 GmbH
Postfach 1120
D–770
Uberlingen
Germany

1.1.1.2 Luminescence instruments and spectrofluorimeters

Luminescence instrument LS–3B; luminescence instrument LS–5B; Accessories: low flow cell, cell holders, bioluminescence spectroscopy, fluorescence spectroscopy, recorder/printers, low-temperature luminescence, fluorescence plate reader, polarization accessory, microfilm fluorimeter LS–2B

Perkin–Elmer Ltd
Analytical Instruments Division
761 Main Avenue
Norwalk
Connecticut 068-59-0012
USA

Perkin–Elmer Ltd
Post Office Lane
Beaconsfield
Buckinghamshire HP9 1QA
UK

SFM–25 spectrofluorimeter
Kontron Instruments
Kontron AG
Bernerstrasse Sud 169
8010 Zurich
Switzerland

Chemi and bioluminescence; lumicon, luminescence instrument
Hamilton Co
PO Box 10030
Reno
Nevada 89520
USA

Hamilton Bonaduz AG
PO Box 26
CH–7402 Bonaduz
Switzerland

I.I.I.4 Fourier transform infrared spectroscopy

FTIR

Spectratech Europe Ltd
Genesis Centre
Science Park South
Birchwood
Warrington
Cheshire WA3 7BH
UK

Mattson Instruments Inc
1001 Fourier Court
Madison
Wisconsin W183717
USA

Mattson Instruments Ltd
Linford Forum
Rockingham Drive
Linford Wood
Milton Keynes
Buckinghamshire MK14 6LY
UK

FTIR Model 1800, 1700 series

Perkin–Elmer Corporation
Analytical Instruments Division
761 Main Avenue (MS–12)
Norwalk
Connecticut 06859
USA

Perkin–Elmer Ltd
Post Office Lane
Beaconsfield
Buckinghamshire
HP9 1QA
UK

Digilab TT57

Bio-Rad
53/56 Greenhill Crescent
Watford Business Park
Watford
Hertfordshire WD1 8QS
UK

PV 9800

Phillips Electronic Instruments
85 McKee Drive
Mohway
NJ 07430
USA

Phillips Nederland BV
Asd Analysetechnickin VB 3
Postbus 90050
5600 PB Eindhoven
Netherlands

I.I.2 Flow injection analysis

FIA star 5020
FIA star 5032
FIA star 5025 ion selective electrode meter
FIA star 5010

The Aquatec System

UK Supplier
EDT Analytical Ltd
14 Trading Estate Road
London NW10 7LN
UK

Tecator AB
Box 70
S 26301 Hanagas
Sweden

FIA System LGCI
Advanced Medical Supplies Ltd
Caker Stream Road
Mill Lane Industrial Estate
Alton, Hampshire GU34 2PL
UK

Chemlab FIA system
Chemlab Instruments Ltd
Hornminster House
129 Upminster Road
Hornchurch, Essex RM11 3XJ
UK

FIA
Skalpar BV
Spinvaid
LL 4815 HS Breda
PO Box 3237
NL 4800 DE Breda
Netherlands

Fialite 600
Fiatrode 400
Fiatrode 410
Fiatron Laboratory Systems
5105 South Wortington Street
Oconomowoc WI 53066
USA

Fiazyme 500 Series Automated Carbohydrate Analysis Models

I.1.3.1 Flame and graphite furnace atomic absorption spectrometry

IL and Video Series:
Thermoelectron Ltd
830 Birchwood Boulevard
Birchwood
Warrington
Cheshire WA3 7QT
UK

Thermoelectron Ltd
590 Lincoln Street
Waltham
MA 02254
USA

Perkin–Elmer 2280; 2380, 1100 and 2100
Perkin–Elmer Ltd
Post Office Lane
Beaconsfield
Bucks
HP9 1QA
UK

Perkin–Elmer Corporation
Analytical Instruments Division
761 Main Avenue
Norwalk
Connecticut 06856
USA

Varian Associates Spectr AA 30/40 and Spectr AA 10/20
Varian Associates Ltd
29 Manor Road
Walton on Thames
Surrey KT12 2QF
UK

Varian Techtron Pry Ltd
679 Springvale Road
Mulgrove
Victoria
Australia 3170

Varian Instruments Division
611 Hansen Way
Palo Alto
California 94303
USA

Varian AG
Steinlausertrasse CH–6300
Zug
Switzerland

GBC 903 and 902
GBC Scientific Equipment Pty Ltd
22 Brooklyn Avenue
Dandenong
Victoria
Austraila 3175

UK Agent
Techmation Ltd
58 Edgware Road
Edgware, Middlesex HA9 8JP
UK

Shimadzu AA 670 and AA 670G

Shimadzu Corporation
International Marketing Division
Shinjuki Mitsui Building, 1–1
Nishe–Shinjuku 2 Chrome
Shinsuku–ku
Tokyo 163
Japan

UK Agent
VA Howe Co Ltd
12–14 St Ann's Crescent
London SWI8 2LS
UK

Autosamplers

20.020 20 Position autosampler
20.080 80 Position autosampler and fraction collector atomic absorption model comprising:

PSA 20.080 SS Stainless steel probe
PSA 20.080 PP Polypropylene probe
PSA 20.080 CC Complete automatic control from computers
PSA 20.080 OA On-line dilution probe
PSA 20.080 FC Triple-probe fraction collector
PSA 20.080 VP Vial piercing option
PSA 20.080 PH PH Electrode assembly
PSA 20.080 TM Turrax mixer assembly and interface requirements:
TTL logic
RS 232 Random access
RS 232 Other requirements

PS Analytical Ltd
Arthur House
Cray Avenue
Orpington
Kent BR5 3TR
UK

Gilson 300 position programmable autosampler

Gilson International
Box 27
300 W Beltine
Middleton
Wisconsin 53562
USA

Gilson Medical Electronics (France) SA
BP 45 F–95400
Villiers-le-Bel
France

Zeeman atomic absorption spectrometry

Perkin–Elmer Zeeman 3030 and Zeeman 5000

Perkin–Elmer Ltd
Post Office Lane
Beaconsfield
Bucks
HP9 1QA
UK

Perkin–Elmer Corporation
Analytical Instruments Division
761 Main Drive
Norwalk
Connecticut 06856
USA

Varian Associates Spectr AA30/40 and Spectr AA 300/400

Varian Associates Ltd
28 Manor Road
Walton on Thames
Surrey KT12 2QF
UK

Varian Techtron Pty Ltd
679 Springvale Road
Mulgrove
Victoria
Australia 3170

Varian Instrument Division
611 Hansen Way
Palo Alto, CA 94303
USA

Varian AG
Steinhauserstrasse CH–6300
Zug
Switzerland

Hydride generator assemblies

PS Analytical Ltd Model PSA 10.002

Dr P B Stockwell
Varian Associates Ltd
Arthur House, Far North Building
Cray Avenue, Orpington
Kent BR5 2TR
UK

Varian AG VGA–76

PS Analytical Ltd
28 Manor Road
Walton on Thames
Surrey KT12 2QF
UK

Varian Techtron Pty Ltd
679 Springvale Road
Mulgrove
Victoria
Australia 3170

Varian Instrument Division
611 Hanson Way
Palo Alto
CA 94303
USA

Varian AG
Steinhauserstrasse CH–6300
Zug
Switzerland

I.1.3.2 Inductively coupled plasma optical emission spectrometers

Spectroflame

Spectro Analytical UK Ltd
Fountain House
Great Cornbew
Halesowen
West Midlands B63 3BL
UK

Spectro Inc
160 Authority Drive
Fitchbury 01420
Massachusetts
USA

PU 7450 and PV8050

Phillips Analytical Agents
Pye Unicam Ltd
York Street
Cambridge CB1 2PX
UK

Phillips Electronic Instruments
85 McKee Drive
Mahwah
New Jersey 07430
USA

Inductively coupled plasma mass spectrometers

Plasmaquad
VG Isotopes Ltd
Ion Path, Road Three
Winsford
Cheshire
UK

Elan 500
Perkin–Elmer Ltd
Post Office Lane
Beaconsfield
Buckinghamshire HP9 1QA
UK

Auto samplers
Intelligent autosampler 84100
R 5232 C serial
Communications interface
A B Labtam Ltd
43 Malcomb Road
Braeside
Victoria
Australia 3195

1.1.5.1 High-performance liquid chromatography

2000 series, 2500 series, 5000 series, 5500 series, 9060 diode array detector, 9060 LC autosampler

Varian
220 Humboldt Court
Sunnyvale
California 94069
USA

Varian Associates Ltd
28 Manor Road
Walton on Thames
Surrey KT12 2QF
UK

Series 10 chromatography LC–95 variable wavelength UV/visible detector, LC–90 variable wavelength UV detector, LC–135 and L–235 diode array detectors, LC 1–100 computing integrator, 1SS–100 intelligent sampling system, Series 410 LC pump

Perkin–Elmer Corporation
Analytical Instruments Division
761 Main Avenue
Norwalk CT 06859
USA

Perkin–Elmer Ltd
Post Office Lane
Beaconsfield
Buckinghamshire HP9 1QA
UK

System 400, comprising 420 and 414 pumps, 460 autosampler, 430 and 432 detectors, 450 data system, 480 column oven, 425 gradient former, Anacomp 220 data management, 306 autosampler, MSI 66 autosampler, 740 LC variable wavelength detector, 720 LC digital variable wavelength detector, 735 LCC variable wavelength detector

Kontron Instruments
Blackmore Lane
Croxley Centre
Watford, Hertfordshire WD1 8XQ
UK

Series 4500i

Dionex Corporation
PO Box 3063
Sunnyvale
California
USA

Dionex (UK) Ltd
Albany Park
Camberley
Surrey GU15 2PL
UK

2350 pump, 2360 gradient programmer, 2351 gradient controller V4, UA5 and 228 wavelength detectors, FL–2 fluorescence detector, Chem Research data management/gradient control system, 015A recorder, autoinjector, Foxy, Retriever II and Cygnet fraction collectors, Peak collection instrument

Isco
4700 Superior
Lincoln NE 67504
USA

2150 pump, 2249 gradient pump, 2152 controller, 2157 autosampler, 2154 injector, 2510 UV detector, 2151 variable wavelength detector, 2140 rapid spectral detector, 2142 refractive index detector, 2143 electrochemical detector, 2221 integrator, 2145 data system, 2210 and 2240 recorders, 2134, 2133, 2135, 2134, 2131 column

Pharmacia LKB
Bjorkgaten 30
75182 Uppsala
Sweden

Pharmacia Ltd
Pharmacia LKB Technology Division
Midsummer Boulevard
Central Milton Keynes
Buckinghamshire MK9 3HP
UK

LC 6A system comprising SCL–6A controller, SPD–6A and SPD–6AV spectrophotometric detectors, CTO–6A column oven, LC–6A pump, SIL–6A autoinjector

LC–8A preparative HPLC comprising LC–8A pump, FCV–110AL reservoir switching valve, FCV–130 AL valve/pump box, FCV–120AL recycle valve, SCL–8A system controller, SPD–6A UV detector, SPD, 6AV UV-visible detector, SIL–8A autoinjector 7125 manual injector, FCV 100B fraction collector, C–R4A data processor, PC–11L 3-pump interface, PC–30L pump interface, PC–24L interface for reservoir switching valve FCV 110 AL and recycle valve FCV 120 AL, PC–16N interface for fraction collector FCV 100B, PC–14N interface for data processor C–R4A, LC–7A bicompatible system comprising LC–7A pump, LC–7A gradient system, 7125/T sample injector, SPD–7A UV detector, SPD 7AV UV/visible detector

Shimadzu Corporation
International Marketing Division
Shinjuku Mitsui Buildings
1–1 Nishi Shinjuku 2 Chrome
Shinjuku ku, Tokyo 163
Japan

Dyson Instruments Ltd
Hetton Lyons Industrial Estate
Hetton
Houghton le Spring
Tyne and Wear DH5 3RH
UK

HP 1050 Series comprising programmable variable-wavelength detector, multiple-wavelength detector, pumping system, autosampler

Hewlett Packard
PO Box 10301
Palo Alto
California 94303 0890
USA

HPLC columns only
HPLC Technology
Wellington House
Waterloo St West
Macclesfield
Cheshire SK11 6PJ
UK

Spectroflow 400 system
HPLC pump only
Kratos Analytical Instruments
170 Williams Drive
Ramsey
New Jersey NJ 07446
USA

Chromo-A-Scope comprising UV visible detector and data processing system only

Barspec Ltd
PO Box 560
Rehovot 76103
Israel

Barspec Ltd
PO Box 430
Mansfield
MA 02048
USA

Roth Scientific Co Ltd
Alpha House
Alexandra Road
Farnborough
Hampshire GU14 6BU
UK

Hichrome Ltd
6 Chiltern Enterprise Centre
Station Road
Theale
Reading
Berkshire RG7 4AA
UK

Series 100 comprising CE 1100 pump, CE 1200 variable wavelength monitor, CE 1300 gradient programmer, CE 1400 refractive index detector, CE 1500 electrochemical detector, CE 1700 computing integrator, CE 1710 recorder and 1720 recorder, CE 1800 sample injector, 1200, 2000 column monitoring panel and sample valve.

Cecil Instruments Ltd
Milton Technical Centre
Cambridge CB4 4AZ
UK

Model 5100 A Coulochem electrochemical detector only

ESA Inc
45 Wiggins Avenue
Bedford MA 01730
USA

Severn Analytical
30 Brunswick Road
Gloucestershire GL1 1JJ
UK

Model RR/066 351 and 352 pumps: models 750/16 variable-wavelength UV monitor detector 750/11 variable filter UV detector, MPD 880S multiwave plasma detector, 750/14 mass detector 750/350/06 electrochemical detector refractive index detector; HPLC columns: column heaters, autosamplers, pre-columns derivatization systems, solvent degassers, preparative HPLC systems

Applied Chromatography Systems Ltd
The Arsenal
Heapy Street
Macclesfield
Cheshire SK11 7JB
UK

LCA 15 system; LCA 16 system
EDT Ltd
EDT Research
14 Trading Estate Road
London NW10 7LU
UK

Aspec Automatic sample preparation system; Asted automated sequence trace diazylate enricher; 231/401 HPLC autosampling injector: 232/401 automatic sample processor and injector:

Gibson Medical Electronics (France)
72 Rue Gambetta
BP45 F–95400
Villiers le Bel
France

SA Gilson Medical Electronics Inc.
Box 27, 300 010 Beltine Highway
Middleton WC 53562
USA

Isoflo HPLC radioactivity monitor, iso mix interface between HPLC column and Isoflo HPLC radioactivity monitor:

Nuclear Enterprises Ltd
Bath Road
Beenham
Reading
Berkshire RG7 5PR
UK

Advanced automated sample processor:

Varian AG
Steinhauserstrasse CH 6300
Zug
Switzerland

Varian Associates
28 Manor Road
Walton on Thames
Surrey KT12 2QF
UK

High-performance liquid chromatography–mass spectrometry

HP 5988 A Mass selective detector; HP 4987 A Mass selective detector:
Hewlett Packard
PO Box 10301
Palo Alto
California 94303–0890
USA

I.1.5.4 Supercritical fluid chromatography

501 SFC, 602 SFC, 622 SFC/GC

Lee Scientific
4426 SO Century Drive
Salt Lake City
Utah 84123–2513
USA

Dionex (UK) Ltd
Selmoor Road
Farnborough
Hampshire GU14 7QN
UK

Supercritical fluid chromatography pressure gradient solvent delivery system:
Pierce Chemicals
Life Science Laboratories Ltd
Sedgewick Road
Luton
Bedfordshire LU4 9DT
UK

CCS 5000
Severn Analytical Ltd
Unit 2B St Frances' Way
Shefford Industrial Estate
Shefford
Bedfordshire SG17 5DZ
UK

CW 14 Detector
Valco Instruments Co Ltd
PO Box 55603
Houston
Texas 77255
USA

Valco Europe
Untertannberg 7 CH–6214
Schenkon
Switzerland

1.1.5.5 and 1.1.5.7 Gas chromatography

There are numerous suppliers of gas chromatography equipment, a selection of which are given below
Carlo Erba Instruments
Strada Rivoltana
20090 Rodano
Milan
Italy

Models 8100, 8200, 8400, 8500 and 8700 sigma 2000 range
Perkin–Elmer Corporation
Analytical Instruments
761 Main Avenue
Norwalk
CT 06859–0012
USA

Perkin–Elmer Ltd
Post Office Lane
Beaconsfield
Buckinghamshire HP9 1QA
UK

GC 14A, GC 15A, 16A, GC 8A
Shimadzu Corporation
International Marketing Division
Shinjuku Mitsui Building
1–1, Nishi Shinjuku 2 Chrome
Shinjuku ku, Tokyo 163
Japan

Dyson Instruments Ltd
Hetton Lyons Industrial Estate
Hetton
Houghton le Spring
Tyne and Wear DH5 3RH
UK

Micromat HRGC 412
Nordion Instruments Co Ltd
PO Box 1
SF 003171 Helsinki
Finland

Silchromat 1–4 and Silchromat 2–8

Siemens AG
Instrumentation and Control Division
E687
Postfach 211262
D–7500 Karlsruhe 21
Germany

Siemens Ltd
VE6, Siemens House
Eaton Bank
Congleton
Cheshire CW12 1PH
UK

I.1.5.5 Gas chromatography–mass spectrometry

SSQ 70 Series Single stage gas chromatography quadruple mass spectrometer TSQ–70 series gas chromatograph triple-stage quadrupole mass spectrometer; MAT–90 gas chromatography high-resolution mass spectrometer; H SQ–30 hybrid mass spectrometer – mass spectrometer; series 700 ion-trap detector; Incas–50 gas-chromatography quadruple mass spectrometer; ChemMaster workstation, 1020 quadruple mass spectrometer, OWA–20/30B organics in water gas chromatograph–mass spectrometer (see above)

Finnigan MAT
355 River Oaks Parkway
San Jose
California CA 9513–1991
USA

Finnigan MAT Ltd
Paradise
Hemel Hempstead
Hertfordshire HP2 4TG
UK

Ion-trap detector

Perkin–Elmer Corporation
Analytical Instruments Division
761 Main Avenue
Norwalk, Connecticut 06859–0012
USA

Perkin–Elmer Ltd
Post Office Lane
Beaconsfield
Buckinghamshire HP9 1QA
UK

GCMS–QP 2000 gas chromatograph mass spectrometer, GCMS–Q 1000 and GC–QP 1000A gas chromatograph mass spectrometer; Mispack 200 GCMS QP series MS data system

Shimadzu Corporation
International Marketing Division
Shinjuku, Mitsui Building
1–1 Nishi Shinjuku-2-Chrome
Shinjuku-ku
Tokyo 163
Japan

Dyson Instruments Ltd
Hetton Lyons Industrial Estate
Hetton
Houghton le Spring
Tyne & Wear DH5 3RH
UK

I.1.5.6 Headspace samplers

HS 101, HS 100 and HS 6

Perkin–Elmer Corporation
Analytical Instruments Division
Main Avenue MS–12
Norwalk CT 06856
USA

HSS – 2A

Shimadzu Corporation
International Marketing Division
Shinjuku Mitsui Building
1–1 Nishi Shinjuku 2 Chrome
Shinjuku ku
Tokyo 163
Japan

Dyson Instruments Ltd
Hetton Lyons Industrial Estate
Hetton
Houghton le Spring
Tyne and Wear DH5 3RH
UK

Headspace 6

Siemens, AG
Instrumentation and Control Division
E687
Postfach 21 1262
D–7500 Karlsruhe 21
Germany

Siemens Ltd
VE6 Siemens House
Eaton Bank
Congleton
Cheshire CW12 1PH

Purge and trap concentrators

4460A

OIC Corporation
Graham Road at Wellborn Road
PO Box 2980
College Station
Texas 77841 2980
USA

Eden Scientific
1 Beechrow
Ham Common
Richmond
Surrey TW10 5HE
UK

LSC 2000 (concentrator) ALS 2016, 2032, (discrete automatic samplers) automatic sample heater ALS 2050 (vial sampling system)

Tekmar
10 Knollcrest Drive
PO Box 371556
Cincinnati
Ohio 45222–1856
USA

1.1.6 Total elements

Total halide (DX 80B), also total organic halogen (DX 20A), total sulphur and chlorine (MCIS 130/120) total nitrogen, (DN–100), total organic carbon DC–80, DC–180, DC–90, DC–85A, DC–88 and DC–54

Dohrmann Instruments
Rosemount Analytical Division
3240 Scott Boulevard
Santa Clara 90062, CA
USA

Sartec Ltd
Bourne Industrial Estate
Wrotham Road
Borough Green
Sevenoaks
Kent TN15 8DG
UK

Total sulphur, TS 02 and TN 02 and 702X
Mitsubishi Chemical Industries Ltd
Instruments Dept
Mitsubishi Building
5–2 Marunouchi-2-Chrome
Chyoda-ku
Tokyo 100
Japan

Total sulphur chlorine (TOX–10), total organic halogen (TSX–10), total nitrogen (TN–05, DN–10, DN–100 and N–100)
EDT Analytical Ltd
14 Trading Estate Road
London NW10 7LU
UK

Total nitrogen
Kjeltex System I with kjeltec system 1026 distilling unit also 1030 analyser also system 6120 kjeldahl digestor also system 12/40 kjeldahl digestor
Tecator AB
Box 70, S263–01 Honagas
Sweden

System for chemical oxygen demand
Tecator Ltd
Cooper Road
Thornbury
Bristol BS12 2UP
UK

Carbon, hydrogen, nitrogen 2400 CHN (combustion gas chromatography)
Perkin–Elmer Ltd
Post Office Lane
Beaconsfield
Buckinghamshire HP9 1QA
UK

Perkin–Elmer Corporation
Analytical Instruments Division
761 Main Avenue
Norwalk CT 06859–0012
USA

Nitrogen, carbon and sulphur NA–1500
Carlo Erba Instruments
Strada Rivoltana
20090 Rodano
Milan
Italy

Fisons Instrument
Sussex Manor Park
Gatwick Road, Crawley
West Sussex RH10 2QQ
UK

Total organic carbon – TOC–500
Shimadzu Corporation
International Marketing Division
Shinjuku Mitsui Building
1,1, Nishi Shinjuku 2 Chrome
Shinjuku ku
Tokyo 163
Japan

Dyson Instruments Ltd
Hetton Lyons Industrial Estate
Hetton
Houghton le Spring
Tyne and Wear DH5 ORH
UK

Total organic carbon
Model 700

0 1 Corporation
Graham Road at Wellborn Road
PO Box 2980
College Station
Texas 77841 2980
USA

Centronic Sales Ltd
Centronic House
King Henry's Drive
New Addington
Croydon, Surrey CR9 0B9
UK

Acid digestion systems for wet digestions
System 6/20
System 12/40

Tecator AB
Box 70
S-203-01 Hanagas
Sweden

Tecator Ltd
Cooper Road
Thornbury
Bristol BS12 2UP
UK

System 500
Skalar BV
PO Box 3237
4800 DE Breda
Netherlands

1.1.8 NMR Spectroscopy

Gemini Superconducting Fourier transform
NMR systems, VXR series 5

Varian Instruments
Sugar Lane
Texas
USA

NMR imaging spectrometer systems

Vis
1120 Auburn Road
Fremont
California 945 38
USA

1.1.9 Partially automated immunoassay systems

Rotary platers for petri dishes: RP 453, RP 454; Autospreader for petri dishes: A 450; Multipoint spreader for petri dishes: A400; Single-reagent dispenser for microplates: Weflfill 3; Multi-reagent dispenser for microplates: Wellfill 4; Single-reagent plate-stacking dispenser for microplates: Wellfill 5; Microplate innoculation: Wellrepp 2 automatic replicator; Shaker for microplates: Wellmix X1, Wellmix X2, Wellmix X3, Wellmix X4

Denley Instruments Ltd
Notts Lane
Billingshurst
West Sussex RH14 9EY
UK

Microlab AT automated pipetter diluter/distributor for microplates
Hamilton Bonaduz AG
PO Box 26
CH–7402 Bonaduz
Switzerland

Fluorimeters: Microfilm fluorimeter LS–2B

Perkin–Elmer Corporation
Analytical Instruments Division
761 Main Avenue
Norwalk, Connecticut 06859–0012
USA

Perkin–Elmer Ltd
Post Office Lane
Beaconsfield
Buckinghamshire HP9 1QA
UK

Unicam chemi- and bioluminescence luminometer

Hamilton Co
PO Box 10030
Reno
Nevada 59520
USA

Hamilton Bonaduz AG
PO Box 26
CH–7042 Bonaduz
Switzerland

Lambda reader automated microplate reader

Perkin–Elmer Corporation
Analytical Instruments Division
761 Main Avenue, Norwalk
Connecticut 06859–0012
USA

Perkin–Elmer Ltd
Post Office Lane
Beaconsfield
Buckinghamshire HP9 1QA
UK

Ependorf Geratebau
Netherler and Hius GmbH
PO Box 650670
D–2000 Hamburg 65
Germany

MDA 312 Multidetector radio-immunoassay analyser
Kontron AG
Bernerstrasse Sud 169
8010 Zurich
Switzerland

Cobra-one Auto Gamma 5012 or 5013 crystal plus bench top radio-immunoassay system

Packard Instrument Co
2200 Warrenville Road
Downers Grove, Illinois 60515
USA

Canberra Packard International SA
Peuggesstrasse 3
CH–8038 Zurich
Switzerland

Fully automated immunoassay workstations

Zymark Ryobotic Laboratory Automation system for inununoassays

Zymark Corp
Zymark Centre
Hopkinton 1
MA 01748
USA

Zymark Ltd
The Genesis Centre
Science Park South
Birchwood, Warrington
Cheshire WA3 7BH
UK

Macrogroupamatic

Kontron Instruments SA
2 Avenue du Manet
781 80 Montigny-le-Britonneux
France

Kontron AG
Bernerstrasse Sud 169
8101 Zurich
Switzerland

Biomek 1000 automated laboratory workstation

Beckman Instruments Inc
Spinco Division
1050 Page Mill Road
Palo Alto
California 94304
USA

Beckman RIIC Ltd
Progress Road
Sands Industrial Estate
High Wycombe
Buckinghamshire HP12 4JL
UK

14.4.1 Laboratory homogenizers

Planetary micromill Pulverisette 7, vibration micro pulverizer, Pulverisette 0, rotary speed mill, Pulverisette 14

Fritsch GmbH
Laborgeraetebau
Industriesstrasse 8
D–6580 Idar Oberstein
Germany

Laboratory comminuters: vibrating cup mill, Pulverisette 9, laboratory disk mill, Pulverisette 13, mortal-grinders, Pulverisette 2, centrifugal mill, Pulverisette 6, planetary mill, Pulverisette 5

Laboratory sieving devices: vibratory sieve shaken for micro-precision sieving, Analysette 3, rotary sieve shaker, Analysette 18

Christison Scientific Equipment Ltd
Albany Road
Gateshead
Tyne & Wear NH8 3AT
UK

Particle size distribution measurement:
(a) Sedimentation in gravitational field, Analysette 20;
(b) Laser diffraction, Analysette 22
(c) Sedimentation in centrifugal field, Analysette 21

Fritsch GmbH
Laborgeraetbau Industriestrasse 8
D 6580 Idar Oberstein
Germany

Christison Scientific Equipment Ltd
Albany Road
Gateshead
Tyne & Wear NH8 3AT
UK

Appendix 2

Methods of soil analysis published by the Ministry of Agriculture, Fisheries and Food, UK

The Analysis of Agricultural Materials, 2nd edn, R.B. 427. HMSO, London, (1979)

Method 2, p.6. Preparation of Samples of soil
Method 8, p.21. Boron, water soluble in soil
Method 57, p.134. Nitrogen in soil
Method 62, p.148. Organic matter in soil
Method 63, p.151. Particle size distribution in soil
Method 65, p.158. Bicarbonate extractable phosphorus extractable in soil
Method 76, p.183. Sulphur, sulphate in soil

Index